PERGAMON INTERNATIONAL LIBRARY
of Science, Technology, Engineering and Social Studies

The 1000-volume original paperback libra[...] industrial training and the enjoym[...]

Publisher: Robert Maxwell

BIOLOGICAL OCEANOGRAPHIC PROCESSES

BIOLOGICAL OCEANOGRAPHIC PROCESSES

THIRD EDITION

TIMOTHY R. PARSONS
Department of Oceanography, University of British Columbia

MASAYUKI TAKAHASHI
Institute of Biological Sciences, University of Tsukuba

and

BARRY HARGRAVE
Marine Ecology Laboratory, Bedford Institute

PERGAMON PRESS

OXFORD · NEW YORK · TORONTO · SYDNEY · PARIS · FRANKFURT

U.K.	Pergamon Press Ltd., Headington Hill Hall, Oxford OX3 0BW, England
U.S.A.	Pergamon Press Inc., Maxwell House, Fairview Park, Elmsford, New York 10523, U.S.A.
CANADA	Pergamon Press Canada Ltd., Suite 104, 150 Consumers Rd., Willowdale, Ontario M2J 1P9, Canada
AUSTRALIA	Pergamon Press (Aust.) Pty. Ltd., P.O. Box 544, Potts Point, N.S.W. 2011, Australia
FRANCE	Pergamon Press SARL, 24 rue des Ecoles, 75240 Paris, Cedex 05, France
FEDERAL REPUBLIC OF GERMANY	Pergamon Press GmbH, Hammerweg 6, D-6242 Kronberg-Taunus, Federal Republic of Germany

First edition 1973
Second edition 1977
Third edition 1984

Library of Congress Cataloging in Publication Data

Parsons, Timothy Richard, 1932–
Biological oceanographic processes.
(Pergamon international library of science,
technology, engineering, and social studies)
Bibliography: p.
Includes index.
1. Marine ecology. 2. Marine plankton. 3. Primary
productivity (Biology) I. Takahashi, Masayuki.
II. Hargrave, Barry. III. Title. IV. Series.
QH541.5.S3P37 1983 574.5′2636 83-8871

British Library Cataloguing in Publication Data

Parsons, Timothy R.
Biological oceanographic processes. —3rd ed.
(Pergamon international library)
1. Marine ecology
I. Title II. Takahashi, Masayuki
III. Hargrave, Barry
574.5′2636 QH541.5.S3

ISBN 0-08-030766-3 (Hardcover)
ISBN 0-08-030765-5 (Flexicover)

Printed in Great Britain by A. Wheaton & Co. Ltd., Exeter

ACKNOWLEDGEMENTS

ACKNOWLEDGEMENTS for copyright material reproduced in the 3rd Edition are given as follows:

Fig. 3 from *Sea Microbes*, Oxford University Press; Fig. 15 from Inter-Research; Fig. 20A and B and Fig. 21 from Plenum Press; Fig. 31 from Elsevier Scientific Publications; Fig. 40 from Pergamon Press; Figs. 54, 57, and 59 from the American Society of Limnology and Oceanography; Fig. 64 from Elsevier Biomedical Press; Fig. 66 from Academic Press; Fig. 74 from Duke University Press; Figs. 77, 110, 118, 119, and 120 from the *Canadian Journal of Fisheries and Aquatic Sciences;* Figs. 105 and 106 from Biologische Anstalt Helgoland; Fig. 108 from Biologische Anstalt Helgoland and British Crown Copyright; Fig. 121 from IRL Press Ltd.; Fig. 111 from the National Academy Press, Washington, D.C.; Fig. 117 from Macmillan Journals Ltd.

Acknowledgements for copyright material reproduced from the 2nd Edition are given as follows:

American Society of Limnology and Oceanography for Figs. 8, 13, 16, 22, 35, 45, and 51; Conseil International pour l'Exploration de la Mer for Figs. 7, 17, 43, and 46; National Academy of Sciences for Fig. 12; Fisheries Research Board of Canada for Figs. 24, 25, 65, and 103; International Association of Geochemistry and Cosmochemistry for Fig. 39; Microforms International Marketing Corp. for Figs. 9, 10, and 14A; Oceanographic Society of Japan for Fig. 4; Hokkaido University for Figs. 55 and 61; *Journal of Phycology* for Fig. 44; Springer-Verlag for Figs. 60 and 83; Scottish Marine Biological Association for Figs. 1 and 9; Oliver and Boyd, and Otto Koeltz Antiquariat for Fig. 71; Figs. 2 and 14B are copyright 1963 and 1969 respectively by the American Association for the Advancement of Science; Fig. 6 and parts of Fig. 3 were originally published by the University of California Press, reprinted by permission of the Regents of the University of California; Fig. 67 is with permission from the North-Holland Publishing Co.; Fig. 63 is with permission from Biologische Anstalt Helgoland; Figs. 84, 91, and part of Fig. 86 are with permission from Ophelia; parts of Fig. 85 are with permission from the English Universities Press, McGraw-Hill Book Co. and the Linnean Society; Fig. 89 is with permission from the Fresh-water Biological Association; Figs. 95 and 96 are with permission from Oikos; part of Fig. 104 is with permission of Harvard University Press; Fig. 90 is with permission from Blackwell Scientific Publications.

CONTENTS

INTRODUCTION

BIOLOGICAL oceanography may be defined as a study of the biology of the oceans, including the pelagic and benthic communities. Having adopted this broad definition we have found it necessary to limit our discussion of the subject in several different ways. These include largely omitting any reference to special marine environments, such as the littoral zone and coral reefs. The purpose of the book is to serve as an introduction to the field of quantitative biological oceanography. The title of the book has been chosen to reflect this fact and the book should not be regarded as a literature review. We have selected reference material illustrative of certain types of biological oceanographic processes which can generally be explained either in terms of an empirical equation or through the use of definite biological or chemical descriptions. By definition these empirical relationships are advanced as being among the most acceptable at the time of writing the text but it is to be expected that researchers will improve or disprove many of the processes discussed, in the light of further scientific advancement. Such is the nature of science. As an introduction to the subject, however, we feel that students, physical oceanographers, engineers, hydrologists, fisheries experts, and scientists in a number of other professions may require some quantitative expressions of biological oceanographic phenomena. In some cases we have drawn on examples from the freshwater environment; in practically all examples, however, we have referred to processes in the near-surface pelagic environment and the benthic community. The book does not cover descriptive oceanography or biological oceanographic models; rather it is intended to bridge the gap between these two subjects which are covered by other texts referred to at the end of this section.

In the first two chapters we have attempted to describe the plankton community in terms of its composition and distribution of organisms. We felt that these descriptions were necessary for the reader to gain an introduction to the processes described in Chapters 3 to 5. Chapters 1 and 2 give some indication, therefore, of the complexity of both plankton distributions and the chemistry of plankton. Added to this complexity, however, are the results obtained by different scientists using different techniques in widely separated areas of the world. Consequently a synthesis of results reported in the literature is difficult. Instead we have tended to present data which offer the reader examples of biological oceanographic variability rather than trying to convince people of the acceptability of any one author's results in an absolute sense.

Chapter 3 deals with the primary formation of particulate material which is the beginning of the food chain in the pelagic environment. Feeding processes and the kinetics of food exchange in the pelagic food web are discussed in Chapter 4. In Chapter 5 an attempt is made to relate various processes into cycles which emphasize the interdependence of all processes in the sea. Chapter 6 is intended to serve as an introduction to biological processes in the benthic community.

In the final chapter we have given a number of examples of problems in the marine environment which we feel require the particular attention of the biological oceanographer. These have been chosen as representing areas in which there are also interdisciplinary interests between biological oceanographers and other professionals working in the marine habitat, including physical oceanographers and fisheries experts. The chapter is not intended to solve any problems but to suggest, by example, where there is a basis for the solution of such problems, using some of the information contained in the first five chapters.

Symbols used in equations and figures have presented us with a problem since aquatic biologists have tended to use the same symbol for several different entities. Thus R is commonly used to represent 'ration' and 'respiration'; P is used to represent the element phosphorus, photosynthesis and production. Where possible we have kept the common usage of the symbols and defined each one in the immediate context of its use. Where two processes are defined by the same letter in the same equation we have differentiated between the symbols used. In referring back to the literature we felt that the preservation of the popular symbols would be less confusing than introducing a large number of new symbols. Biological oceanographers have also tended to use different units for measuring the same parameters. Added to these differences is the fact that there is a lack of uniformity in the use of abbreviations for the same units. For example, light has been reported both in energy units and units of illumination, while the distance of one micron is sometimes abbreviated as 1 μ or 1 μm. It is not our purpose to endorse any uniform use of units or abbreviations but it is our purpose to clarify the literature by showing how units can be converted and where any similarity in abbreviations may exist.

For more detailed coverage of specific subjects in biological oceanography, the reader will find a number of recent reviews in multi-author texts or in journals; also several books can be particularly recommended for coverage of the literature up to the date of their publication. The latter include *The Chemistry and Fertility of Sea Waters* by H. W. Harvey (Cambridge University Press, 1957); *Measuring the Production of Marine Phytoplankton* by J. D. H. Strickland (Queens Printer, Ottawa, 1960); *Plankton and Productivity in the Oceans* by J. E. G. Raymont (Pergamon Press, Oxford, 1963); *The Structure of Marine Ecosystems* by J. H. Steele (Harvard University Press, Cambridge, Mass., 1974); *Organic Materials in Aquatic Ecosystems* by H. Seki (CRC Press, 1982) and *Fundamentals of Aquatic Ecosystems* edited by R. S. K. Barnes and K. H. Mann (Blackwell Scientific Publications, 1980).

CHAPTER 1

DISTRIBUTIONS OF PLANKTON AND NUTRIENTS

1.1. TAXONOMIC, ENVIRONMENTAL AND SIZE SPECIFIC GROUPS OF PLANKTON

Organisms which are unable to maintain their distribution against the movement of water masses are referred to as 'plankton'. Included in this group are bacterioplankton (bacteria), phytoplankton (plants) and zooplankton (animals). Generally all plankton are very small and, in many cases, microscopic. However, relatively large animals, such as the jellyfish, are also included in the definition of plankton. Some plankters, including both plants and animals, are motile but their motility is weak in comparison with the prevailing movement of the water. Animals, such as fishes, which can maintain their position and move against local currents are known as 'nekton'. However, the division between plankton and nekton is not precise and some small fish, especially fish larvae, may be a part of the plankton community, while some large zooplankton, such as euphausiids, might also be thought of as 'micronekton'.

The biomass or weight of plankton or nekton per unit volume or area of water is referred to as the 'standing stock'; typical units used for standing-stock measurements are $\mu g/l$, mg/m^3, g/m^2, kg/hectare, etc., where the weight should be specified as referring to wet weight, dry weight, or carbon. The productivity of organisms is defined in terms of 'primary productivity', 'secondary productivity' and 'tertiary productivity'; units are the same as in standing-stock measurements when expressed per unit time (e.g. per hour, day or year). Ideally, primary productivity represents the autotrophic fixation of carbon dioxide by photosynthesis; secondary productivity represents the production of herbivorous animals and tertiary productivity represents the production of carnivorous animals feeding off the herbivore population. However, these definitions are not precise since some plants may utilize growth factors, such as vitamins (auxotrophic growth), and others are capable of taking up organic substrates as a source of energy (heterotrophic growth). Thus the particulate material grazed by secondary producers may be derived from a variety of processes and include phytoplankton and bacterioplankton. Similarly many filter-feeding zooplankton which might be nominally classed as herbivores may at times feed upon other small animals, such as Protozoa; thus the boundaries between components in the aquatic biosphere are difficult to define with the same precision as is used in chemistry or physics. Biological associations are better considered *in toto* as an ecosystem in which various components react with each other to a greater or lesser degree. Components of an ecosystem can be defined in terms of their taxonomy or chemistry; interactions between components can then be expressed quantitatively by empirical equations. Thus a phytoplankton standing stock may be described as consisting of 10^6 cells per litre of a species, *Skeletonema costatum,* or as being represented by a cholorophyll *a* concentration of 1 mg/m^3, or in a trophic sense as a ration for zooplankton, such as is represented in Chapter 4. Attempts to synthesize these various components of biological production in the sea have given rise to a variety of mathematical models. Walsh (1976) has reviewed the many types of models involved; he points out, however, that no model is a perfect representation of the real world. It is in fact impossible to know all the states of biological variables in an ecosystem.

However, where the ecosystem is sufficiently restricted in time and space it is possible to create an acceptable simulation of major events. For example, the upwelling ecosystem is relatively simple in the sense that it is possible both to formulate the principal biological interactions and at the same time to test the ecosystem model using fast, well-equipped research vessels and aerial coverage. More detailed discussions of biological modelling are given by Steele (1974) and by Platt *et al.* (1981).

The lowest forms of life found in the ocean are unicellular organisms; either prokaryotes (lacking a nuclear membrane) or eukaryotes (with a distinct nucleus). For detailed definitions of these two groups the reader is referred to Broda (1978). The marine prokaryotes include the bacteria and the blue-green algae while all other phytoplankton are eukaryotes. Thus in function both eukaryotes and prokaryotes can be either autotrophic or heterotrophic. In practice, however, the role of unicellular eukaryotes in the sea includes most of the autotrophic production by phytoplankton, some heterotrophic (and often phagocytic) production, particularly by certain flagellates, and production by unicellular animals (protozoa) — collectively this whole group may be referred to as the kingdom, Protistae (i.e. Protists). In contrast the ecological function of the unicellular prokaryotes is most apparent in the many activities of the heterotrophic and chemoautotrophic bacterioplankton, with the blue-green algae generally playing a minor role (in the total ecology of the sea) as autotrophic photosynthetic organisms. However, the latter have one notable function in oligotrophic waters which is their ability to fix nitrogen.

The numbers of bacterioplankton in the sea generally are maximal at the sea surface where they are associated with a high concentration of organic material making up part of the neuston (Sieburth, 1971; Tsyban, 1971). Values of 10^8 bacteria/ml may be encountered in the sea surface film while below the sea surface values of 10^5–10^6/ml would be more common in the euphotic zone. Bacteria in the ocean depths decrease by several orders of magnitude ($< 10^4$/ml) except in immediate association with sediments rich in organic material, or near hydrothermal vents. Generally, those bacteria suspended in the water column will be aerobes but microhabitats inside detrital particles, including fecal pellets, allow for the presence of anaerobic bacteria in otherwise oxygenated water columns. In the presence of hydrogen sulphide layers (e.g. the Black Sea or at hydrothermal vents), chemotrophic bacteria are present in large numbers (for recent reviews see Sieburth, 1979; Jannasch and Wirsen, 1979).

According to Wood (1965) the principal genera of bacteria represented in the oceans are the *Micrococcus, Sarcina, Vibrio, Bacillus, Bacterium, Pseudomonas, Corynebacterium, Spirillum, Mycoplana, Norcardia and Streptomyces.* Among these genera are various morphological differences including coccoid, rod and spiral forms. A large number of the most common bacteria are motile and gram negative. The bacterioplankton do not usually contribute significantly to the total biomass of particulate organic matter but in association with detritus (see Section 2.4) they may form an appreciable organic reserve during times of low phytoplankton density. Their role in the oceans is more important in recycling elements and organic material back into the food chain (see Chapter 5). MacLeod (1965) has discussed the specific identity of marine bacteria as compared with bacteria from a terrestrial origin. His findings showed that marine bacteria have special requirements for inorganic ions which include a highly specific need for Na^+ and a partial need for halide ions which could be satisfied by either bromine or chlorine ions. Mg^{2+} and Ca^{2+} were also required, usually at concentrations higher than are normally need for terrestrial bacteria. Oliver (1982) has devised a flow chart for the identification of marine bacteria. Based in part on Bergey's *Manual of Determinative Bacteriology,* the scheme involves a relatively short series of tests that can be employed to identify major taxa of motile and non-motile gram negative rods.

The taxonomy of phytoplankton has been in a state of revision for some time but it is believed that Table 1 represents the most recent revisions in the various class names. Of the thirteen classes represented in Table 1, four are the most important with respect to the total standing stock of

TABLE 1. REPRESENTATION OF ALGAL CLASSES IN MARINE PHYTOPLANKTON (Prepared by F. J. R. Taylor, Department of Oceanography, University of British Columbia)

Taxonomic class	Common name	Area(s) of predominance	Notes
CYANOPHYCEAE	Cyanobacteria/ blue-green algae	tropical (filamentous) cosmopolitan (coccoid)	Chiefly *Trichodesmium* (= *Oscillatoria*); N_2 fixer. Chiefly *Synechocystis,* minute (2 μm or less).
RHODOPHYCEAE	Red algae	v. rare, coastal	*Rhodosorus*; abundant benthic forms.
BACILLARIOPHYCEAE	Diatoms	all waters, esp. coastal	Major microplanktonic primary producers.
CRYPTOPHYCEAE	Cryptomonads*	cosmopolitan, mainly coastal	Much neglected, but often important, nanoplankters.
DINOPHYCEAE	Dinoflagellates*	all waters, esp. tropics	Autotrophs or heterotrophs; common red tide producers.
CHRYSOPHYCEAE	Chrysomonads* Silicoflagellates*	rare, coastal occasionally abundant	Important in fresh water, except for silico flagellates.
HAPTOPHYCEAE = PRYMNESIOPHYCEAE	Coccolithophorids* & prymnesiomonads*	oceanic (coccolith.) coastal (prymnesio)	Some (coccolithophorids) with $CaCO_3$ scales, others (e.g. *Chrysochromulina, Prymnesium*) without.
RAPHIDIOPHYCEAE	Chloromonads*	rare, but occasionally abundant, brackish	Some fish-killers (*Chattonella*).
XANTHOPHYCEAE	Yellow-green algae/ heterochlorids*	v. rare	A few coccoids; others are mostly benthic or f.w.
EUSTIGMATOPHYCEAE	-	v. rare	A few coccoids; others are mostly benthic f.w.
EUGLENOPHYCEAE	Euglenoids*	coastal	Occasionally common, e.g. *Eutreptiella.*
PRASINOPHYCEAE	Prasinomonads*	all waters	Flagellates are coastal, often tide pools (e.g. *Pyramimomas) 'phycoma' (cyst) phase pelagic (Halosphaera, Pterospema). Micromonas* is important nanoplankter.
CHLOROPHYCEAE	Green algae, volvocaleans*	v. rare, coastal	Mostly f.w. or benthic.

*Also classified as phytoflagellates of the protozoa.

phytoplankton in the ocean. These are the Bacillariophyceae, Dinophyceae, Haptophyceae and Cryptophyceae. Of these, diatoms and dinoflagellates are found extensively throughout the world oceans both in coastal and oceanic waters; the coccolithophorids are found more abundantly in oceanic waters while the cryptomonads are often numerous in coastal waters. Rarer classes of algae, including such organisms as the silicoflagellates, the prasinomonads, euglenoids and chloromonads have sometimes been found to be abundant in coastal waters. The blue-green algae may be abundant at times in tropical water (e.g. *Trichodesmium* blooms in the Red Sea) while Jeffrey and Hallegraeff (1980) have shown that in warm core ocean waters a prasinophyte, *Micromonas,* probably accounts for small persistent occurrences of chlorophyll *b* in photosynthetic pigment extracts. Similarly Johnson and Sieburth (1979 and references cited) have reported on the occurrence of small (*ca.* 1 μm diameter) cyanobacteria (blue-green algae) in open ocean waters at concentrations of 10^3–10^4/ml. Thus while the largest part of the primary productivity in the oceans is confined to a few classes of algae, it is probable that most other classes of algae in Table 1 are present as minor components which may occasionally form blooms under local conditions. In addition some of the colourless phytoplankton flagellates (e.g. dinoflagellates such as the genus *Noctiluca*) or microflagellates (e.g. see Haas and Webb, 1979 and references cited) may represent secondary productions due to their ability to consume either phytoplankton or bacteria. In addition, a number of classes of algae occur as symbionts in animals; these are known collectively as the 'zooxanthellae'.

The zooplankton include members of the animal kingdom (Metozoans) as well as of the kingdom Protistae (Protists). The classification of both kingdoms is a subject of continual revision. For example, the former animal phylum Protozoa, may now be regarded as a sub-kingdom of the Protists;

or, among higher animals, the phylum Ctenophora may be included by some authors as part of the phylum Coelenterata. For purposes of this text, the phylla represented have been taken from a more detailed summary by Newell and Newell (1963).

Phylum	*Some representatives among the zooplankton*
Protozoa (also regarded as a sub-kingdom)	Oligotrich and Tintinnid ciliates Radiolaria and Foraminifera
Coelenterata	Hydrozoa, Scyphozoa (Jellyfish)
Ctenophora	(Ctenophores)
Chaetognatha	(Arrow worms)
Annelida	(Polychaete worms)
Arthropoda (class) Crustacea	(Copepods, cladocerans, mysids, euphausiids, ostracods, cumaceans, amphipods, isopods)
(Sub-phylum) Urochordata	(Salps and appendicularians)
Mollusca	(Heteropods, Pteropods)

In addition to the above, there are many large invertebrates having larval planktonic stages (e.g. polychaetes, crustaceans, gastropods, lamellibranchs and echinoderms). Among vertebrates (Phylum: Chordata, Sub-Phylum: Vertebrata) fish eggs and larvae both occur as members of the plankton. Important commercial species which have a planktonic state include the herring, anchovy, tuna, and bottom feeders, such as cod and plaice. From among the types of zooplankton listed above, by far the most abundant group are the Crustacea, and of these, the copepods are the most predominant.

Biogeographical distributions of plankton have been based on the very early recognition that specific environmental factors, such as light, temperature, salinity, and nutrient requirements, to some extent determined the occurrence and succession of species. Smayda (1958 and 1963) has reviewed a number of the terms used to describe plankton from similar environments. Plankton with a tolerance to a wide range of temperatures is described as 'eurythermal', while a narrow range of temperature tolerance is described as 'stenothermal'; similarly salinity (euryhaline and stenohaline), pH (euryionic and stenoionic), and light (euryphotic and stenophotic). A further classification of light response has been used to obtain a vertical separation of plankton communities into those inhabiting the 'euphotic' zone (where the net rate of photosynthesis is positive) as opposed to the 'disphotic' and 'aphotic' zones, where there is enough light for biological detection and where no further light can be detected, respectively (see Chapter 3). Unfortunately, these words usually lack precise quantitative description but some knowledge of their meaning may be useful in reading other publications.

The terms 'oceanic' and 'neritic' have been used quite extensively in describing plankton associated with the oceans and with coastal waters, respectively. The classification may be particularly useful in reporting taxonomic data collected from commercial vessels, such as with a Hardy recorder as illustrated in Fig. 1. From this figure it is easy to see that certain 'indicator species' belong to each region. Over large areas of ocean there may be several oceanic groups. Bary (1959 and 1963) defined such plankton distributions in terms of their temperature and salinity tolerances (called T–S–P diagrams). He emphasized that the importance of such diagrams was in showing the distribution of plankton in certain water bodies rather than the exact geographical location of the samples. Fager and McGowan (1963) used an 'affinity index' to show relationships between groups of species. Their index was defined as

$$\frac{J}{\sqrt{N_A N_B}} \qquad \frac{1}{2\sqrt{N_B}}$$

where J was the number of joint occurrences of species; N_A and N_B were the total number of occurrences of species A and B respectively, where $N_A < N_B$. Pairs of species for which the index was arbitrarily $> 0{\cdot}5$ were considered to show affinity. North Pacific plankton were classified into six groups and showed interrelationships between groups as illustrated in Fig. 2. Perhaps the most important aspect of all such groupings is that the

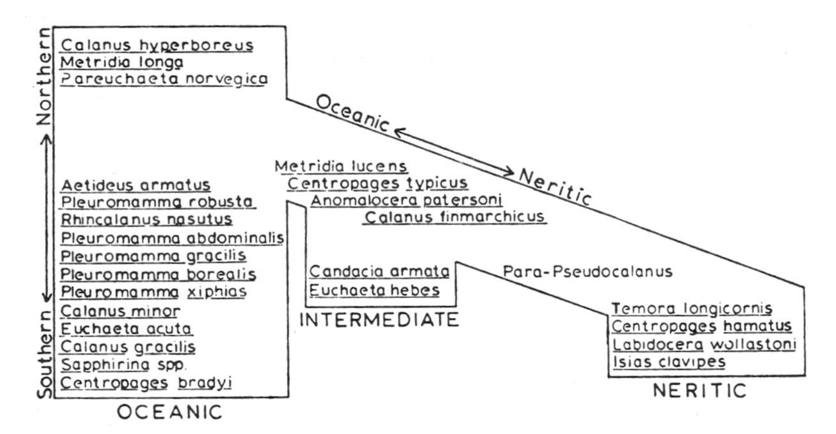

FIG. 1. Distribution series of copepods in the North Sea and the north-eastern Atlantic arranged in such a way that the distribution of each organism is most similar to those of the neighbouring organisms in the list (redrawn from Colebrook *et al.,* 1961).

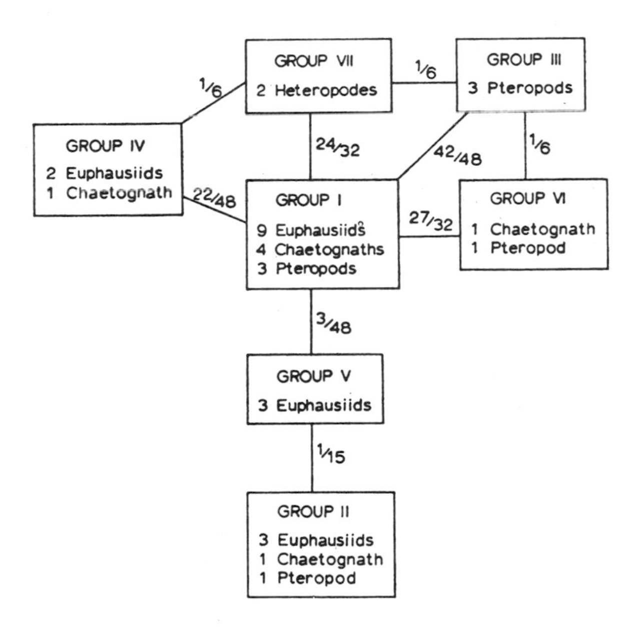

FIG. 2 Composition of zooplankton groups in the north Pacific. Fractions are the ratios of the number of observed species-pair connections between groups to the maximum number of possible connections; for example, there are six possible intergroup species pairs between group IV and VII but only one of these showed affinity at the 'significance' level used (redrawn from Fager and McGowan, 1963).

occurrence of species in another group (or changes in the between groups associations) may indicate changes in the ocean and coastal environments, such as might occur from a shift in the direction of a current. However, Fager and McGowan (1963) found that many of the usually measured properties of water (e.g. temperature, thermocline depth, etc.) were not closely correlated with differences in zooplankton abundance. From this it was concluded that the organisms reacted to a more complex interaction of known properties or to some environmental factors yet to be elucidated.

Phytoplankton species which produce resting spores or have a sedentary phase are known as 'meroplankton' as opposed to 'holoplankton' (Smayda, 1958). However, these terms were originally used to refer to zooplankton and in this sense 'meroplankton' refers to organisms which are only temporarily members of the plankton community (e.g. some bivalve larvae) while 'holoplankton' refers to a permanent member of the plankton community (e.g. most calanoid copepods). A general classification of plankton abundance based on availability of nutrients is used in describing waters as 'eutrophic', 'mesotrophic', and 'oligotrophic', in decreasing order of plankton abundance (see Hutchinson, 1969, for a further discussion of these terms). Plankton may be grouped by the depth zone in which they are found in the 'pelagic' or open-sea environment. These zones have received a number of different classifications but the simplest approximate definitions appear to be 'epipelagic' (0 to 150 m), 'mesopelagic' (150 to 1000 m), 'bathypelagic' (1000 to 4000 m), and 'abyssopelagic' (4000 to 6000 m) (see Hedgpeth, 1957, for further definitions). Plankton (or other particulate matter) produced within a designated ecosystem is referred to as 'autochthonous' while 'allochthonous' material is imported into the ecosystem. A large number of other groupings have been employed by systematists in describing plankton communities. In some cases these terms lack universal usage because of the specificity of their original definition; others have acquired common scientific usage while lacking a precise definition (e.g. see Smayda, 1958). Geographically it has been popular to refer to species as coming from warm or cold regions of the hydrosphere. These regions are not defined by latitude since warm or cold currents may cross such imaginary lines (e.g. the Gulf Stream). McGowan (1971) defines up to twelve subregions of hydrosphere in describing the distribution of many zooplankton. In general, however, the principal regions may be considered as follows:

Tropical	(> 25°C water all the year)
Subtropical	(*ca.* 15–30°C)
Subpolar	(*ca.* 5–15°C)
Polar	(*ca.* 0–5°C)

The combination of subpolar and subtropical waters also encompasses a region in which authors refer to 'temperate' species.

From the point of view of food-chain studies one of the most useful groupings for plankton and larger organisms is to consider all particulate material on a single size scale. The most recent effort aimed at providing a universal size scale for plankton and nekton is based on an earlier grade scale by Sheldon and Parsons (1967). The scale which has now been proposed by Sieburth *et al.* (1978) is shown in Fig. 3. While definitions of the

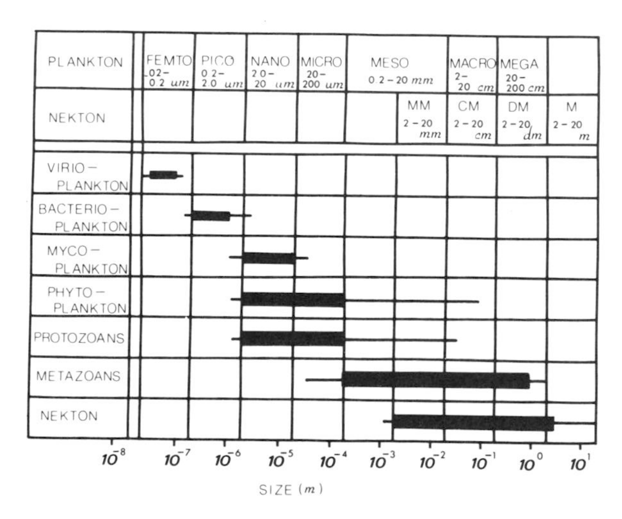

FIG. 3. A proposed size classification scheme for components of the pelagic ecosystem modified from Sieburth *et al.* (1978).

smaller size groupings given in Fig. 3 were originally based on microscopic measurements, or passage through filters of different pore size, it should be noted that neither of these earlier techniques are precise, although they may have considerable practical application, particularly under field conditions. A more precise approach to the representation of linear measurements of irregular-shaped bodies (i.e. plankton) is to consider the volume displaced by that body and to represent its dimension as the radius of a sphere having an equivalent volume. This is illustrated in Fig. 4 where the equivalent volume (V_1, V_2, V_3, V_4, etc.) of irregular particles have diameters (d_1, d_2, d_3, d_4, etc.) which follow a grade scale based on $2\frac{1}{3}$. The biomass in each size category is then ($n_i \times v_i$) and peaks in the size spectrum are seen when a particular plankton bloom occurs. Such measurements can be made with electronic counts (e.g. the Coulter Counter®). The total biomass of material measured with the Coulter Counter® is statistically related to such parameters as the weight of particles, chlorophyll *a* concentration and particulate carbon (e.g. Zeitzschel, 1970). The advantages of using a continuous size spectrum for particle distributions in the sea are (i) the size group of plankters contributing most to the total standing stock of plankton can be readily identified as peaks in the spectrum, (ii) biomass diversity indices (Wilhm, 1968) can be calculated from the spectrum, and (iii) the growth increment, or grazing loss, of different size categories can be determined independently of the total biomass of phytoplankton, zooplankton, and other particles (Parsons, 1969; Parsons and LeBrasseur, 1970). However, particle size spectra *per se* do not relate to taxonomic groups and microscopic identification of the principal components in a plankton crop is recommended when using this technique. Also the results include all detrital particulate material, and special methods are sometimes necessary to differentiate between detritus and growing cellular material (Cushing and Nicholson, 1966).

Example illustrations of common phytoplankton and zooplankton are shown in Figs. 5 A, B and C.

1.2 DIVERSITY

The diversity of a plankton community may be expressed using data on the number of species present, the distribution of biomass, the pigment

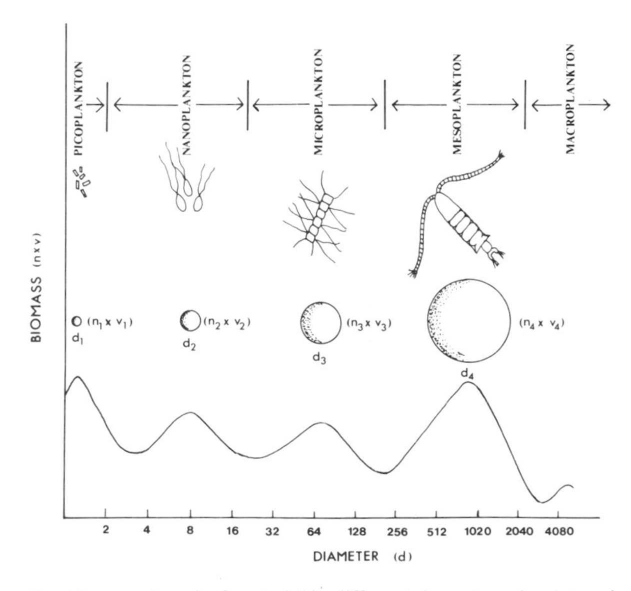

FIG. 4. Particle spectrum representing biomass ($n \times v$) of material in different size categories determined by the diameter (d) of a sphere equivalent in volume (v) to the original particle, times the number of particles (n).

composition, or a number of other parameters which are easily measured properties of plankton. From a heuristic approach, an index of diversity may be used in the same way as other environmental parameters, such as temperature and salinity, to characterize the environment. There are both theoretical and empirical bases for the use of specific diversity indices, but as Lloyd *et al.* (1968) have stated, "which one is 'best' depends upon which one proves in practice to give the most reliable, surprising ecological predictions and the greatest insight".

The simplest expression of diversity is to determine the percentage composition of species in a sample; the more species making up the total, the greater is the diversity of the organisms. However, this value is almost wholly dependent on the total number of individuals (N) and is therefore unsatisfactory as a diversity index. From some of the earliest quantitative studies in ecology it was recognized, however, that a relationship existed between the number of species in a population (S) and the logarithm of the total number of individuals (N), so that the simplest diversity index (d) can be expressed as:

$$d = \frac{S}{\log_{10} N}. \qquad (1)$$

This value will be very small under conditions of a plankton bloom and generally high in tropical plankton communities. A better expression which reduces to 0 when all the individuals are from the same population was given by Margalef (1951):

$$d = \frac{S-1}{\ln N}. \qquad (2)$$

Margalef (1957) introduced the idea that the 'information content' could be used as a measure of diversity in a plankton sample. Thus the diversity of a collection containing a total of N individuals and $n_1, n_2, \ldots n_i$ individuals of each species can be written as:

$$\frac{N!}{n_1! n_2! \ldots n_i!},$$

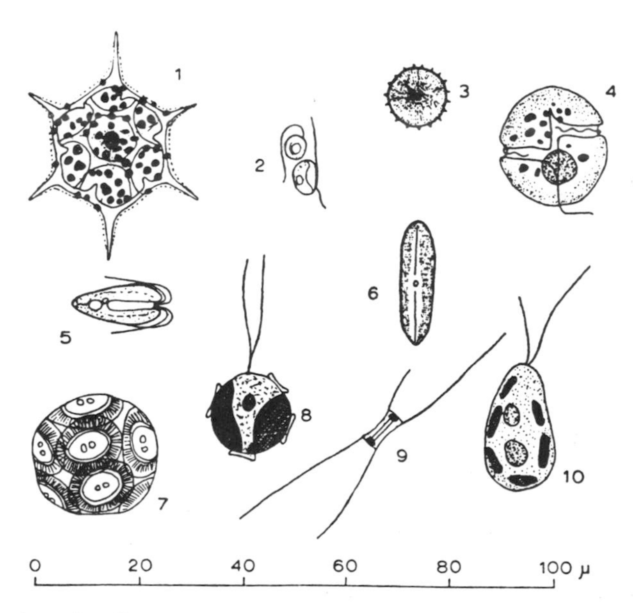

FIG. 5A. Examples of nanoplankton flagellates [*Distephanus* (1), *Thalassomonas* (2), *Gymnodinium* (4), *Tetraselmis* (5), *Coccolithus* (7), *Pontosphaera* (8), *Cryptochrysis* (10)], diatoms [centrate (3), pennate (6), *Chaetoceros* (9)] (redrawn from Wailes, 1939, Cupp, 1943, Fritsch, 1956 and Newell and Newell, 1963).

and information (H in 'bits')* per individual as:

$$H = \frac{1}{N} \log_2 \frac{N!}{n_1! n_2! \ldots n_i!} . \quad (3)$$

The information content as expressed above, eqn. (3), can be interpreted as the degree of uncertainty involved in predicting the species identity of a randomly selected individual. If N is large and none of the n_i fraction are too small, information content per individual (H' in bits) can be approximated from the expression:

$$H' = -\Sigma\, p_i \log_2 p_i \quad (4)$$

where $p_i = n_i/N$ and is the proportion of the collection belonging to the ith species (Shannon and Weaver, 1963). Under some circumstances eqn. (4) will be more easily determined than eqn. (3) but Lloyd *et al.* (1968) have provided examples and tables for the solution of both equations for values of N from 1 to 1050. Margalef (1961) made a statistical comparison between the diversity of plankton samples calculated from the diversity index [eqn. (2)] and the theoretical diversity [eqn. (3)]. There was a highly significant correlation between the two although there was a difference in the regression line depending on whether diversity was determined for a diatom or dinoflagellate population. Using eqn. (4) Lloyd and Ghelardi (1964) related diversity to the maximum possible value for a given number of species if they were all equally abundant. This term was called the 'equitability (ϵ) and was expressed as the ratio of H' to a theoretical maximum (M) for the same number of species, where $n_1 + n_2 + \ldots n_i$. *A table of M* values for 1 to 1000 species is given by the authors. The value of ϵ in describing a collection may be more useful if units of biomass rather than number are employed to determine diversity (e.g. Wilhm, 1968).

'Equitability', as defined by Lloyd and Ghelardi (1964), is the opposite of 'dominance' which expresses the most abundant species in a population. Hulburt *et al.* (1960) expressed the dominance of a plankton community as the ratio of

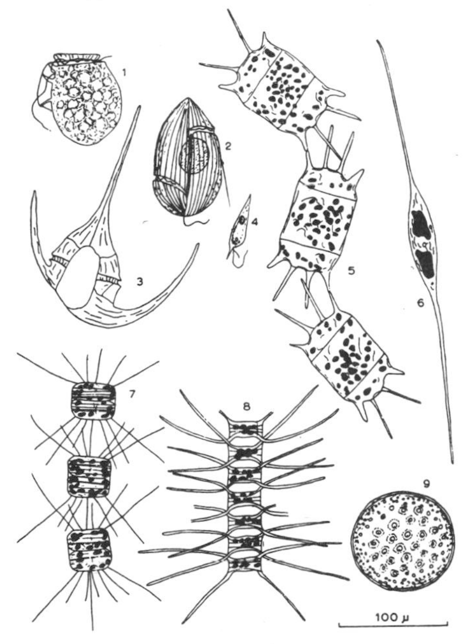

FIG. 5B. Examples of microphytoplankton: dinoflagellates [*Dinophysis* (1), *Gyrodinium* (2), *Ceratium* (3), *Prorocentrum* (4)], diatoms [*Biddulphia* (5), *Nitzschia* (6), *Thalassiosira* (7), *Chaetoceros* (8), *Coscinodiscus* (9)] (redrawn from Wailes, 1939, and Cupp, 1943).

*Information is commonly expressed in 'bits' when using $\log_2$, or 'nats' when using $\log_e$. The following conversion may be useful: $\log_{10} = \log_e \times 2{\cdot}303$ and $\log_2 = \log_e \times 1{\cdot}443$.

the concentration of the most abundant species to the total cell concentration. Also if the presence of one species in a population is nearly always accompanied by another species, the amount of information gained is small. This can be expressed as the 'redundancy' (R) in terms of the equation (Patten, 1962a):

$$R = \frac{H_{max} - H}{H_{max} - H_{min}}, \tag{5}$$

where H_{max} is the diversity when the species are equally distributed and H_{min} is the diversity when all the individuals belong to one species. The value R varies between O and 1 and is also partially an index of 'dominance'.

The general use of diversity of type shown in eqns. (3) and (4) above has been discussed by Pielou (1966) who points out that the diversity of a sample should not be regarded as the diversity of a larger population from which it was obtained; the sample itself may, however, be treated as a population and defined. Patten (1959) discusses the absolute diversity of an aquatic community in terms of its information content. As an approximation he calculated that in a Florida lake, the community could be described in terms of 3×10^{24} bits/cm²/year. If the average information content of a printed page is 10^4 bits, it is apparent that the amount of information required annually to describe the Florida lake community is many orders of magnitude larger than the information contained in the largest libraries! Thus diversity, as discussed in this section, is a property of the entity from which

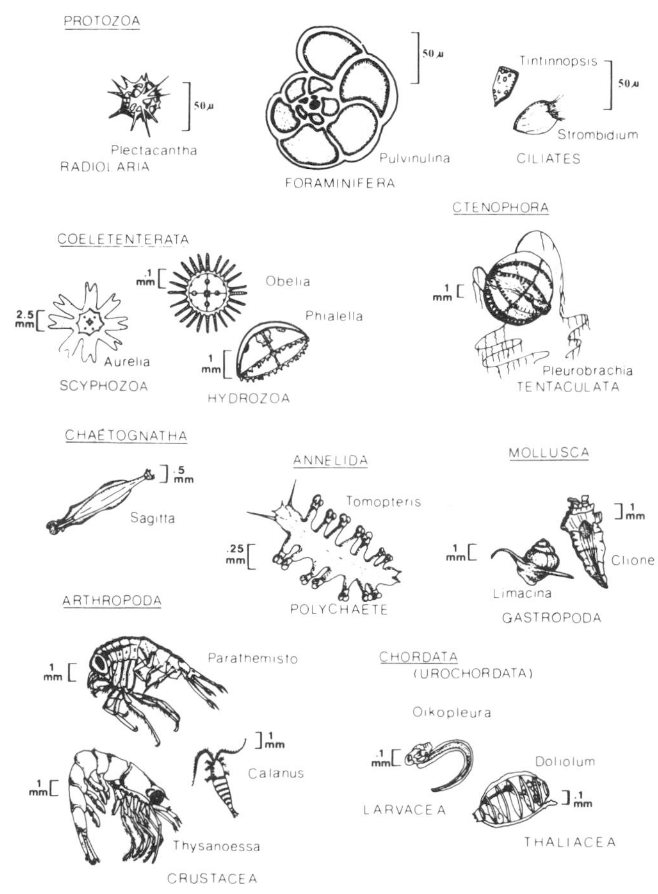

FIG. 5C. Illustrations of the major phyla of zooplankton (redrawn from Lebrasseur and Fulton, 1967; Wailes, 1937 and 1943; Cushman 1931).

the data are collected and not of the whole environment.

The species diversity indices discussed above [e.g. eqn. (2)] are dependent to some extent on the size of the sample, especially for small numbers ($N < 100$). In a method used by Sanders (1968), however, samples of benthic fauna from the same environment, ranging in size from 35 to 2514 individuals, showed no tendency for smaller samples to be less diverse. The technique is described as the 'rarefaction' method and it depends on determining the shape of the species abundance curve rather than obtaining an expression for the absolute number of species per sample.

Changes in the diversity index of samples from a plankton community are shown in Fig. 6. From these data Margalef (1958) recognized three stages of succession. Stage 1 was typical of turbulent waters in which a few species survived and in which there was an occasional bloom of diatoms; stage 3 was characteristic of highly stratified waters in which there was a mature phytoplankton crop and a high diversity following nutrient depletion. Stage 2 was characteristic of inflowing waters which may have transported allochthonous species into the area of study, thus increasing the diversity of organisms present in any sample. Hulburt *et al.* (1960) studied changes in species diversity in the Sargasso Sea and recognized a succession of three species groups. These consisted of a sparse population with a normal distribution of abundant and rare species, a winter period in which a single species was dominant over all other species, and a period of thermal stratification in which dominance was shared by several species. The most extensive field tests of various diversity indices have been carried out by Travers (1971) in the Mediterranean. From this study the author concluded that the degree of maturation and of organization of an ecosystem can be appreciated by means of several diversity indices. From the use of different indices he concluded that a diversity index based on plankton pigments (Margalef, 1965) was a poor method, especially where plankton levels were low; diversity indices based on information theory were considered the best measure of structure although less laborious calculations of diversity can often be used [e.g. eqn. (2)]. Heip and Engels (1974) have compared diversity indices from the point of view of the statistical significance of observed differences or similarities. They conclude by recommending use of the Shannon–Weaver function [eqn. (4)] together with a new index of 'evenness' or 'equitability'.

1.3 SPATIAL DISTRIBUTIONS

1.3.1 Statistical Considerations

In carrying out a series of replicate analyses on a single, well-mixed sample of sea water, small

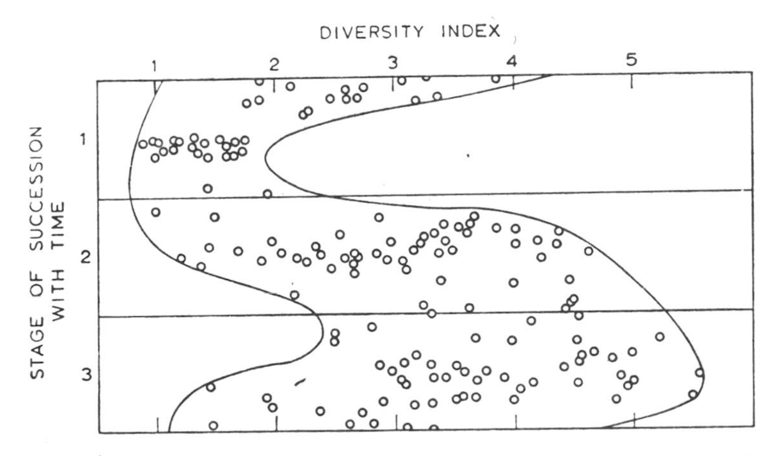

FIG. 6. Stage of succession dependence of diversity index [eqn. (2)] in a number of samples from the surface water at the bay of Vigo (redrawn from Margalef, 1958).

differences in the values obtained may be attributed to analytical technique. Such errors are caused by a lack of instrument reliability, sub-sampling, and slight variations in the way an individual analyst repeats each analysis. It may be assumed that these errors are randomly distributed and that if a large number of replicate analyses are made on one sample, the mean and standard deviation (s) of the analysis can be determined. The precision of the method can then be expressed with a 95% confidence limit for n determinations, as

$$m \pm 2s/\sqrt{n}\,, \tag{6}$$

where m is the mean of the replicate samples. The principal exception to the use of these statistics for analytical techniques is in the counting of plankton from a settled volume of sea water. In this case the distribution of plankton cells may not be random and special methods may have to be used in order to determine the degree of contagion (Holmes and Widrig, 1956).

The collection of samples of sea water, or plankton from the ocean, introduces much larger differences between replicate samples than can be ascribed to analytical errors alone. Thus Cushing (1962) has summarized a number of reports and showed that in calm weather the % variability (expressed as the coefficient of variation $s/m \times 100$) of the number of three species of plankton in individual hauls varied from 15 to 70%; under conditions of rough weather this variability was increased up to 300%. Cassie (1963) estimated that the coefficient of variation for large samples is most often in the range of 22 to 44%, with obvious exceptions being made for rough weather, or highly stratified environments. Wiebe and Holland (1968) have summarized data on the 95% confidence limits for *single* observations of zooplankton abundance; the range for most data was between *ca.* 40 and 250%.

While the use of sampling gear itself may contribute a small amount of variability to ocean sampling (Cushing, 1962, assigns *ca.* 5% variability to gear operation), the principal cause of variability in replicate samples is due to the non-random or patchy distribution of plankton, and other non-conservative properties, such as the concentration of nutrients. The mechanisms leading to these differences in spatial abundance are many and diverse. They include the physical accumulation of particles by the vertical and horizontal movement of water masses (e.g. divergence and convergence), differences in growth rates of individual plankters, and nutrient uptake and predation patterns of the food chain. These processes are discussed at other points in the text (see Section 1.3.5) and the following discussion (primarily from Cassie, 1962a) deals only with the extent and not the cause of distributions.

In random distributions, two or more samples of sea water of a given volume are equally likely to contain the same organism. The expected distribution of samples with n_1, n_2, n_3, etc., individuals is given by successive terms of the binomial expansion

$$(q+p)^k,$$

where k is the maximum number of individuals a sample could contain, p is the probability of an organism's occurrence and $q = 1 - p$. The population mean (μ) and variance (σ^2) of a binomial distribution are

$$\mu = kp \text{ and } \sigma^2 = kpq$$

from which

$$\sigma^2 = \mu - u^2/k. \tag{7}$$

Since for plankton in a sea water sample $k \rightarrow \infty$, the variance become equal to the mean,

$$\sigma^2 = \mu \tag{8}$$

The above relationship expresses a special case of the binomial distribution in which the probability of an organism occurring ($p = \mu/k$) is small; this is known as the Poisson distribution and the experimental value of the variance (s^2) and mean (m) for replicate plankton collections can be expressed as s^2/m and used to determine if the plankton are randomly or 'over-dispersed'. Theoretically if the value σ^2/μ is greater than 1, the population will be over-dispersed. However, in practice if s^2/m is calculated for a series of samples their distribution can be represented as $\chi^2/N-1$; thus with twenty samples (19 degrees of freedom), s^2/m should be less than $30 \cdot 14/19 = 1 \cdot 6$ for a 95% probability that the organisms are distributed randomly (Holmes and Widrig, 1956). However, in most cases involving the collection of plankton over an area, the value s^2/m will be significantly greater than 1; thus the ratio can be used as a dispersion coefficient (Ricker, 1937) which along with other parameters (e.g. diversity indices) may be useful in characterizing a body of water.

When populations are over-dispersed the presence of one organism in a sample increases the probability of additional organisms of the same species occurring in the same sample. This is the opposite of a binomial distribution and it can be expressed theoretically as the negative binomial distribution which is given by an expansion of the expression

$$(q-p)^{-k}$$

where $q = 1 + p$. The variance (σ^2) is

$$\sigma^2 = \mu + \mu^2/k. \tag{9}$$

As Cassie (1962b) has pointed out, since p and k are negative, they cannot have the same meaning as they did in the binomial distribution. In particular, however, k appears as a useful parameter for expressing the degree of patchiness, or contagion, in a population. Cassie (1962a) has given an estimate of $1/k$ as $\hat{c}$, where

$$\hat{c} = \frac{s^2 - m}{m^2}, \tag{10}$$

s^2 and m being the sample variance and mean, respectively. The expression [eqn. (10)] was used by Cassie (1959) as a coefficient of dispersion and he concluded that $\hat{c}$ was better than s^2/m, since $\hat{c}$ was not strongly correlated with the mean. This allows for a comparison of dispersion to be made between samples with different means.

In practice it may be found easier to establish an empirical relationship than to fit raw data to a theoretical distribution. Barnes (1952) and Cassie (1962a) have discussed transformations which may be suitable for marine biological data. The most convenient transformation is to convert the raw data to logarithms and the transformed frequencies may then have a log-normal distribution; an example of transformed data is shown in Fig. 7. The mean (m') of the transformed data is the geometric mean and after taking antilogarithms, the value m' will be less than the arithmetic mean, m. Variability about the mean can be expressed as the 'logarithmic coefficient of variation' (Winsor and Clarke, 1940):

$$V' \text{ per cent} + 100(10^{s'} - 1), \tag{11}$$

where s' is the standard deviation calculated from the logarithms of the raw data and V' is the

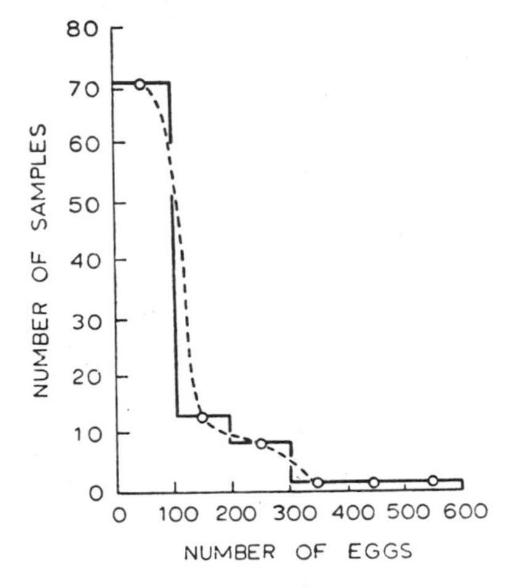

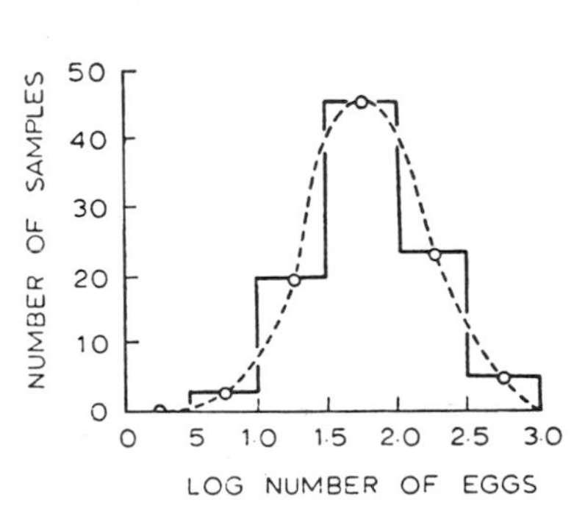

FIG. 7. Frequency distribution of pilchard egg counts in arithmetic and logarithmic forms (redrawn from Barnes, 1952).

logarithmic coefficient of variation. The advantage of using V' has been illustrated by Cassie (1968) who considered two samples with means of 100 and coefficients of variation V (normal distribution) and V' (log-normal distribution) both of 110%. The mean with one standard deviation for the normal distribution was,

$$100 \pm (1{\cdot}10 \times 100) = -10 \text{ to } 210$$

and for log-normal

$$100 \overset{\times}{\div} 2{\cdot}1 = 48 \text{ to } 210.$$

The negative lower limit in the arithmetic range above is meaningless in the context of a normal distribution.

In a detailed study of spatial heterogeneity, Platt *et al.* (1970) separated the sources of variation in a single analysis as follows:

$$\sigma_T^2 = \sigma_0^2 + \sigma_1^2 + \sigma_2^2\,, \qquad (12)$$

where σ_T^2 was the total variance, σ_2^2 was the variance due to real differences between stations, σ_1^2 was the variance between repeated samples taken at the same station and σ_0^2 included sub-sampling and other analytical errors. The authors found that σ_0^2 and σ_1^2 were about the same size and accounted for *ca.* 10% of the variance each. Real differences between stations were generally much larger, both in time and space. Thus the log-coefficient of variation for a single chlorophyll *a* observation in a near-shore community increased rapidly from 14% at 0·625 sq. mile, to 70% at 1 sq. mile, thereafter remained relatively constant out to 4 sq. miles. During a study of temporal variations over a period of 5 weeks, the log-coefficient of variation varied from 11 to 111% (mean 42%) at nine stations covering 13 sq. miles in the same location. During this period phosphate varied from 10 to 60%. Rapid temporal variations in V' sometimes occurred over a few days and ranged from 21 to 45% in the case of chlorophyll and from 32 to 64% for phosphate. Since this study was conducted under relatively ideal weather conditions, the values quoted may be considered to be representative of maximum σ_2^2 values, but minimal σ_0^2 and σ_1^2 values. Thus Cassie (1962b) showed that patchiness was greatest in calm seas and least at times of turbulent mixing, such as during storms; conversely the operation of gear over the side at a single point, as well as laboratory analyses on board vessels, becomes much more difficult during rough weather and consequently the terms σ_0^2 and σ_1^2 may be expected to increase under such conditions.

Most of the discussion above pertains to overall descriptions of the patchiness of plankton communities in terms of plankton numbers, chlorophyll *a* or nutrients. Quite a different distributional problem arises when a description is required of the total number of species that might define the fauna of a particular area. The question is then a matter of how large a sample must be filtered in order to include all the species. In experiments conducted while following a current drogue, McGowan (1971) showed that for three groups of plankters (fish larvae, molluscs and euphausiids) the amount of water to be filtered varied with the group. In the case of euphausiids there was no increase in the number of species with the volume of water filtered over the range 128 to 1510 m^3, but for fish larvae and molluscs the amount of water to be filtered in order to include all species was in excess of 10,000 m^3. However, the relationship between the number of species and the volume filtered was logarithmic; for example, *ca.* 80% of the fish larvae species were found in *ca.* 1000 m^3 of water. The amount of water filtered by a plankton net can be increased either by increasing the size of the net or the length of the tow. Wiebe (1971) studied the precision of replicate tows with different nets and towing distances and his conclusion was that the length of the towing distance was considerably more important in determining precision than the size of the net.

While the concept of obtaining statistically reliable results from field observations follows a classical pattern common to many biological observations, it is important to note that purely statistical evaluations may often hide the true nature of a biological event. The simplest example of this would be the statistical fitting of a linear relationship to a non-linear function. The introduction of time series data collections can lead to further statistical misrepresentation if a biasing of data occurs (i.e. when the sampling frequency is greater than the actual frequency of an event: e.g. a 36-hr sampling strategy of a 24-hr event generates a 72-hr frequency in the data). However, the most important event which is masked by purely distributional statistics is the patchiness of distributions and the temporal spatial scales on which it varies. Platt (1972) was the first person to consider this problem in respect to plankton communities and the technique of spectral analysis which he employed has become widely used; a theoretical discussion of the technique has been given by Fasham (1977) and some practical results of the analyses are discussed below.

1.3.2 Areal Distributions

A large amount of data has been collected on the areal distribution of plankton and nutrients. These data are generally the result of observations carried out at points along the cruise track of a research vessel; consequently while the data are useful for surveys of very large areas (e.g. seas, oceans and the hydrosphere) they are of very little use in trophic studies. The latter subject is discussed later in the text, but for the present it must be apparent that plankton distributions mapped from samples collected miles apart are probably not representative of the food supply for a larval or juvenile fish, which may travel less than 100 m in a day. Thus small-scale plankton distributions are particularly important in assessing the food supply and hence, in part, the survival of very young fish.

It has been recognized from the time of the earliest explorers of the hydrosphere that plankton may sometimes occur in dense swarms or blooms (see Bainbridge, 1957, for historical references). Scientific observations (e.g. Barnes, 1949; Barnes and Marshall, 1951) showed that in general planktonic organisms were more often clumped or aggregated than randomly distributed. For example, Cassie (1959) showed in a study on the occurrence of plankton over a distance of 1 m that the distribution of the diatom *(Coscinodiscus gigas)* was non-random. The problem of collecting detailed samples over appreciable distances was solved in 1936 for larger plankters with the invention of apparatus which could be towed behind ships and continuously collect plankton or a slowly moving fine-mesh belt (Hardy, 1936). An adaptation of this apparatus (Longhurst *et al.*, 1966) for studying micro-distributions of plankton has been particularly important for trophodynamic studies. Using this apparatus Wiebe (1970) was able to show areal patchiness in the distribution of zooplankton species over distances of less than 20 m. Some of the results obtained by Wiebe are shown in Fig. 8. From these data it is apparent that there is both a small-scale and a larger-scale patchiness in the distribution of species reported. Due to the mechanical ability of the apparatus, the minimum distance over which the plankton patches could be detected with a Longhurst–Hardy recorder was *ca.* 14 m.

The most extensive descriptions of large-scale plankton species distributions are contained in reports from the Oceanography Laboratory, Edinburgh (published in *Bulletins of Marine Ecology*). Zooplankton data are obtained from samples collected with Hardy plankton recorders towed behind commercial vessels (Hardy, 1936); an example of the descriptive data is given in Fig. 9. The numbers of zooplankton are reported as averages by rectangular sub-divisions for the North Sea and the Atlantic approaches to the British Isles. Similar data are collected for the larger phytoplankton species which are reported as a percentage incidence for each species.

Some data on regional differences in the plankton on an oceanic scale have been reported on for all of the world's oceans. For example, Omori (1965) has defined three oceanic regions in the north Pacific based on the distribution of three species-groups of copepods. These are (1) a cold off-shore water region characterized by *Calanus plumchrus–C. cristatus,* (2) a warm offshore region associated with *Calanus pacificus* and, (3) a neritic water mass region represented by *Pseudocalanus minutus–Acartia longiremis.* The latter region is oceanographically very complex and large differences in plankton concentrations are encountered on oceanic approaches to neritic environments. Additional information on the oceanic distribution of certain plankton species can be obtained from a study of sediments. For example, in the case of coccolithophores, the calcium carbonate coccoliths are often preserved both in the surface sediments, and in fossil remains. McIntyre and Bé (1967) have used this technique to describe species-specific zones of coccolithophore production in the Atlantic Ocean and similar maps have been drawn to show the distribution of diatoms and planktonic foraminifera.

Various attempts have been made to summarize productivity data on a global scale. Koblentz-Mishke *et al.* (1970) have reported primary productivity data for the world's oceans, based on a review of a large number of reports, and a modified version of their original figure has been redrawn in Fig. 10. From these results it is apparent that over large areas of the Pacific and Atlantic Oceans, primary production is relatively low but that higher primary productivities are generally found in the proximity of land masses. There are some exceptions to this, such as where the South Equatorial current in the Pacific Ocean causes a band of relatively high primary productivity to occur along the equator. Platt and Subba Rao (1975) have provided a detailed summary of primary productivity estimates in different oceans and seas of the world. Their summary indicates that the total primary productivity of the world's oceans is *ca.* 31×10^9 tons carbon per year. Although the Pacific Ocean accounted for approximately one-third of this value because of its area, the Atlantic Ocean was considered to be more productive per unit area, while continental shelf areas were two to three times more productive than the open ocean, as generally indicated by Fig. 10.

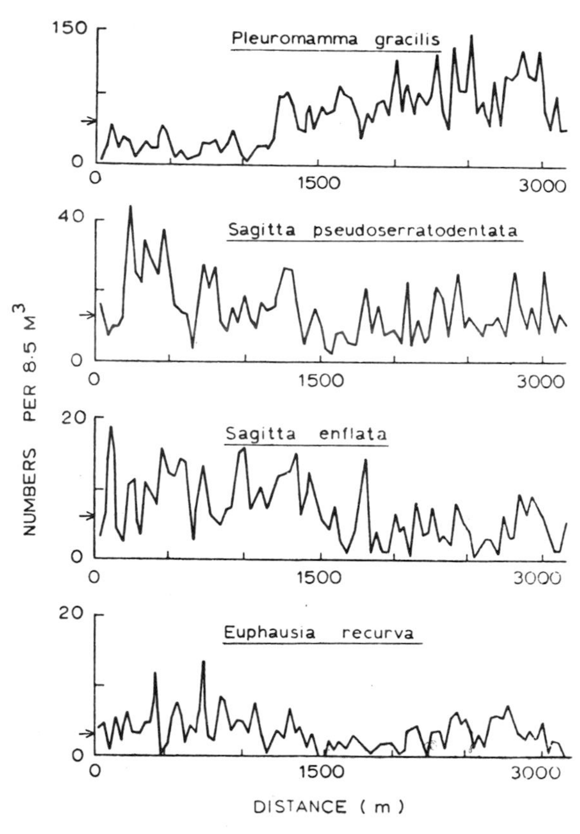

FIG. 8. Plots of abundance versus distance illustrating the presence of large-scale patchiness on which is superimposed smaller-scale patchiness (redrawn from Wiebe, 1970).

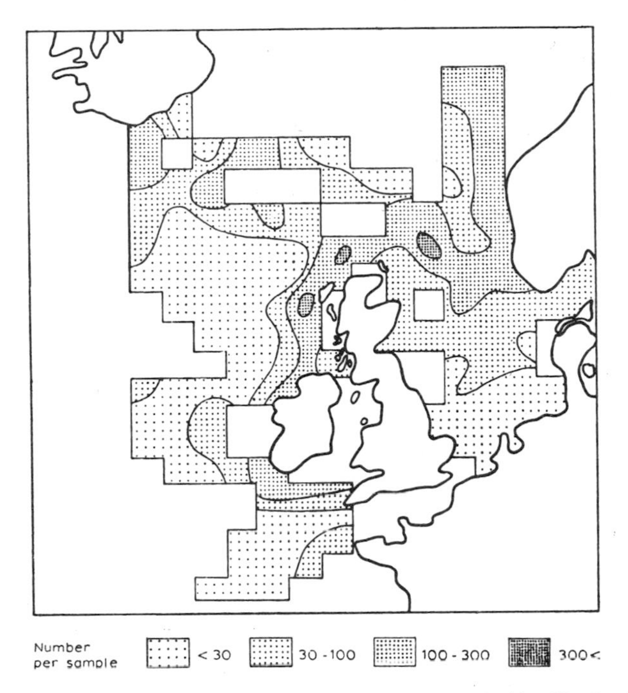

FIG. 9. The distribution of *Calanus finmarchicus*, stages V and VI, from data obtained with a Hardy plankton recorder. Data show the number of animals per sample; blank rectangles indicate insufficient data (redrawn from Colebrook *et al.*, 1961).

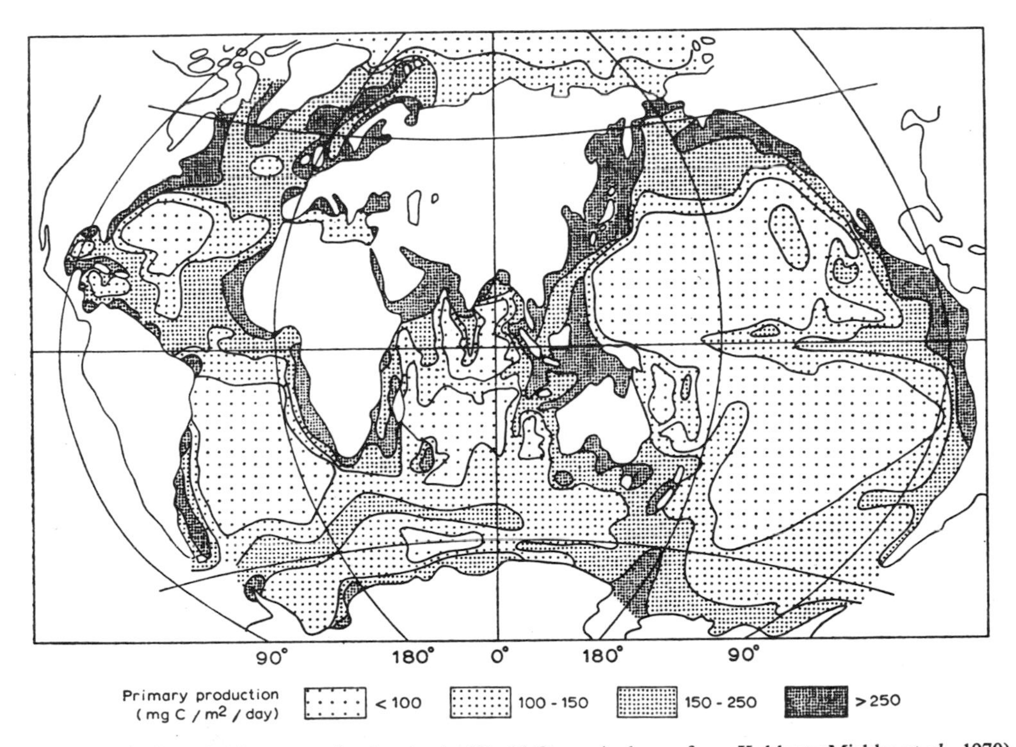

FIG. 10. Distribution of primary production in the World Ocean (redrawn from Koblentz-Mishke *et al.*, 1970).

Differences in biological production associated with near-shore processes will generally give rise to greater patchiness of plankton in neritic compared with oceanic environments. This is true both for large-scale difference in productivity, such as the influence of an estuary on biological production (e.g. Ketchum, 1967) and for small-scale patchiness (e.g. Venrick, 1972). In the latter reference it is shown, for example, that over a 10-mile distance, a lack of aggregation was observed in some species of oceanic phytoplankton; this is interpreted to mean that the randomizing of phytoplankton by turbulent processes is generally greater, relative to biological factors causing differences in productivity, in oceanic compared with neritic environments. Differences in coastal productivity may be associated with several factors — including tides (e.g. Kamykowski, 1973), local morphogeography (e.g. LaFond and LaFond, 1971), and offshore physical processes (e.g. Platt *et al.*, 1972). A special case of higher biological production associated with islands is sometimes referred to as the 'island mass effect'. Gilmartin and Revelante (1974 and references cited therein) generally attribute increased oceanic production in the vicinity of islands to nutrient enrichment caused by local turbulent upwelling in the island passages as well as to the effects of possible nutrient additions from local runoff.

The development of automated analysers has greatly assisted in the description of nutrient distributions in the ocean. An illustration of nitrate distribution obtained with an Autoanalyser® during eight transects of an approximate 10-sq.-mile area is shown in Fig. 11. The illustration has been chosen to show changes in nutrient concentration in an area where nutrient depletion was general but in which there was some upwelling. Over larger areas of ocean, trends in nutrient concentration can be seen which are generally larger than the small-scale differences shown in Fig. 11. This is illustrated in Fig. 12 for nutrient observations carried out with an Autoanalyser® on two cruise tracks across the Pacific Ocean. The greater variability in the results in the western Pacific compared with the Gulf of Alaska reflects

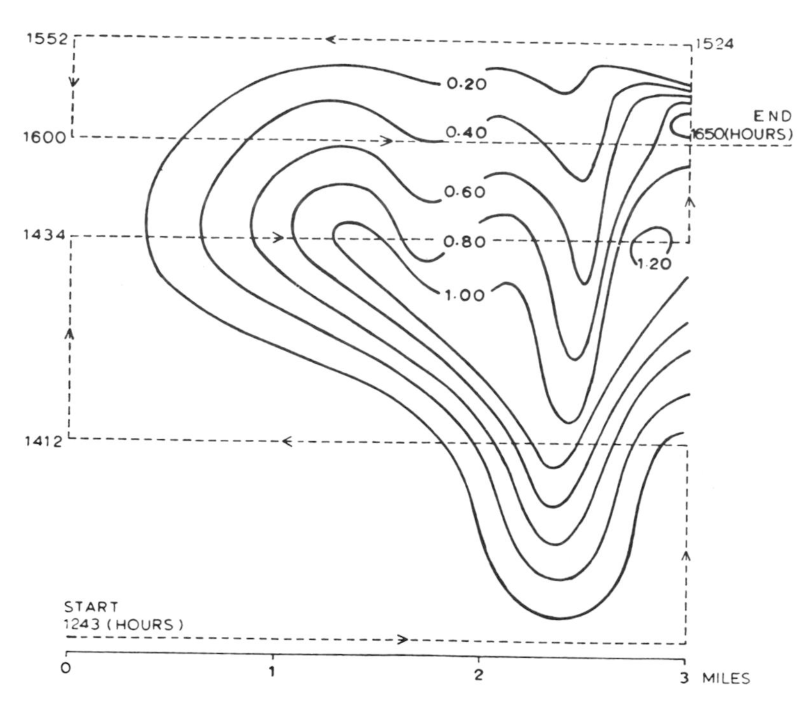

FIG. 11. Nitrate (μg at/l) at the surface off Punta Colnett, Baja California (29° 57′N, 116° 20′W), 9 July 1965. ——— nitrate concentration; — — — cruise track (redrawn from Armstrong *et al.*, 1967).

FIG. 12. Silicate and nitrate at 3 m as measured along an eastward and westward cruise track in the subarctic Pacific Ocean during April 1969 (——— west-bound, — — — east-bound, Victoria, B.C., to Tokyo (redrawn from Stephens, 1970).

the mixture of the two very different water masses (the Oyashio and the Kuroshio Currents) off the coast of Asia.

1.3.3 Vertical Distributions

Until quite recently, most vertical profiles of biological parameters were made either with water bottles, which collected samples from discrete depths, or with plankton nets designed to open over some depth interval. The use of automated sampling gear, as well as recent advances in echo-sounding equipment, have greatly improved the data which are now being collected on vertical distributions. Strickland (1968) made direct comparisons between nutrients and chlorophyll *a* as measured in samples pumped from 0 to 75 m and the same data represented by a standard hydrographic cast. Differences between profiles integrated from standard casts and continuously recorded data were particularly marked in the case of chlorophyll *a*; for example the chlorophyll *a* peak at *ca*. 20 m in Fig. 13 measured 2·9 mg/m^3 when detected in pumped samples using a fluorometer but was only 1·3 mg/m^3 according to an integrated curve based on bottle casts at standard depths. Thus the total amount of chlorophyll *a* per m^2 integrated from a bottle cast or a continuous profile also tends to be different; to some extent, however, these errors are smoothed out and variations in chlorophyll *a* per m^2 were found by Strickland (1968) to be less than 25%. Nutrient analyses carried out at the same time showed less variability than the chlorophyll *a* data; nutrient data for integrated bottle and pump samples (i.e. per m^2) generally differed by less than 10%.

The vertical distribution of chlorophyll in the sea generally shows a maximum which may sometimes be found near or at the surface and at other times, at or below the apparent euphotic depth (Steele and Yentsch, 1960). A deep chlorophyll maximum appears to be a seasonal feature of summer vertical profiles as far north as 45° to 50° in both the Atlantic and Pacific Oceans. Anderson (1969) found the chlorophyll maximum off the Oregon coast at *ca*. 60 m was formed by photosynthetically active cells which were apparently adapted to very low light intensity. South of 40°N Venrick *et al.* (1973) have described a deep chlorophyll maximum at 100–150 m. This appears to be a more or less permanent feature of oceanic latitudes as far south as the region of tropical upwelling at 10°, north and south of the equator. In the southern hemisphere the deep chlorophyll maximum starts again south of 10°S and is sometimes found below 200 m.

Large differences in the concentration of zooplankton at specific depths have been encoun-

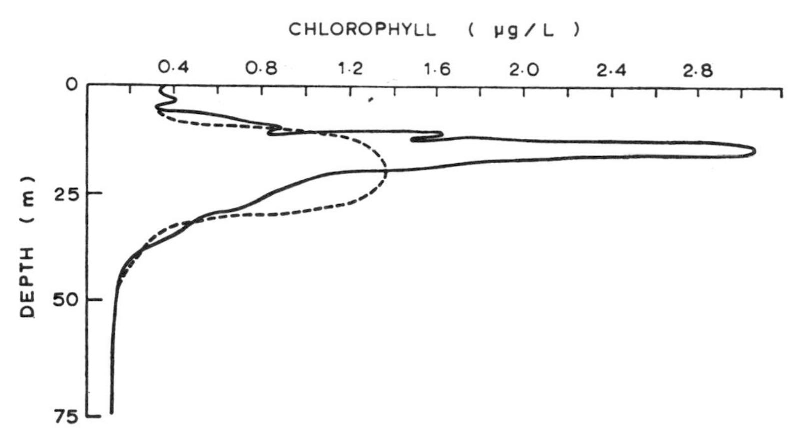

FIG. 13. Chlorophyll profiles integrated from standard bottle casts (— — —) and continuously recorded (———) data (redrawn from Strickland, 1968).

tered using the Longhurst–Hardy recorder (Longhurst *et al.*, 1966), which collects samples continually by integrating catches over very short intervals of *ca.* 10 m. An example of the zooplankton concentrations down to 400 m as measured with this apparatus is shown in Fig. 14A. The approximate 6-fold increase in zooplankton biomass at 300 m is a unique feature of this profile which would have been difficult to observe using data collected with a conventional plankton net. A similar result can be obtained with high-frequency echo sounders as is illustrated in Fig. 14B from Barraclough *et al.* (1969). In this example the presence of echo-sounding material is indicated as a discrete band on the echogram. When the layer was sampled with conventional nets, however, the discrete accumulations of zooplankton appear more as a smoother maximum in concentration. This is due to the generally unavoidable collection of zooplankton at intermediate depths while nets are being lowered and brought up from specific sampling depths. One solution to this problem is the use of specially designed nets (e.g. the Clarke–Bumpus sampler) which can be made to open and close at the beginning and end of a specific sampling period. However, these samplers still represent integrated concentrations for the distance over which the net is towed when open. Alternatively discrete *in situ* location of zooplankton now appears possible using a towed electronic particle counter (Boyd, 1973). Results obtained with this apparatus showed that aggregations of zooplankton were associated with the thermal microstructure in a 250-m vertical profile. However, not all zooplankton aggregations are associated with temperature structure (e.g. Fasham *et al.*, 1974) and other causes are discussed at the end of this chapter.

A number of detailed studies have been carried out on sound scattering layers (for review, see Hersey and Backus, 1962); in particular there are many reports on deep scattering layers (DSL) which are usually found between 100 and 500 m in the oceans. Animals which have been found in scattering layers include squids, euphausiids, fish and certain siphonophores. Although the types of animals which occur at discrete depths in the ocean may vary, it is probable that sound scattering with low-frequency sounders (12 kHz) is only caused by certain specific animals, including large fish and particularly animals containing gas bubbles. Data from echograms, Longhurst–Hardy recorders, and closing nets all indicate that species of zooplankton and nekton generally occur over quite limited depth ranges and the general classifications of organisms by depth zone (e.g. epipelagic, mesopelagic, etc., Section 1.1) may be employed in referring to the vertical distribution of animals. In some cases, however, animals may migrate vertically over distances of up to 1000 m in one day; in these cases specific depth location has little meaning.

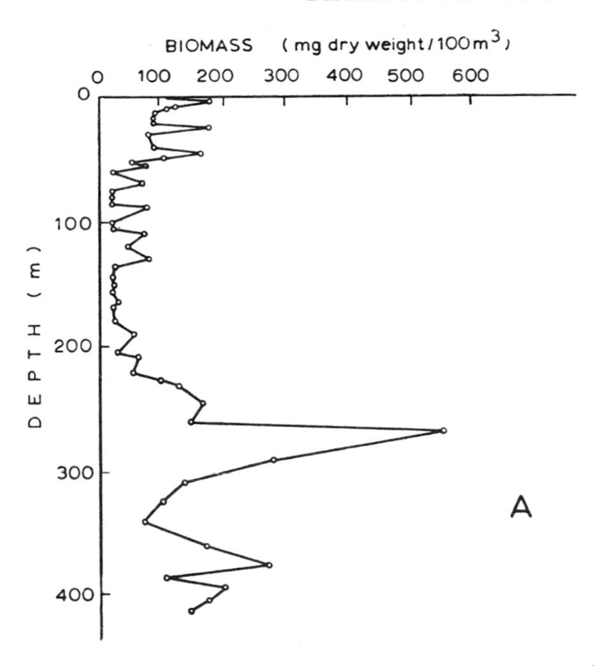

FIG. 14A. Vertical distribution of zooplankton biomass in the eastern Pacific Ocean; data obtained using a Longhurst–Hardy recorder (redrawn from Longhurst *et al.*, 1966).

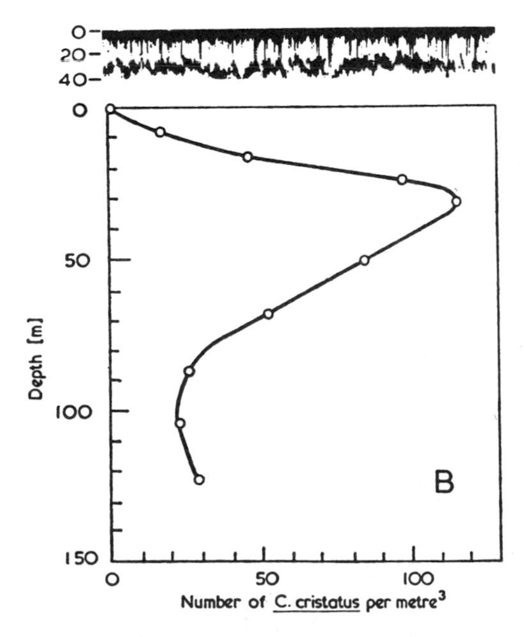

FIG. 14B. Depth profile of *Calanus cristatus* showing the actual echogram obtained with a 200-kHz recorder (top) and depth samples (o) using Miller nets (bottom) (redrawn from Barraclough *et al.*, 1969).

The vertical distribution of biological properties other than phyto- and zooplankton has been studied extensively. In general the standing stock of particulate organic carbon (= POC, which is composed of phytoplankton and detritus) shows a maximum in the near surface layers (*ca.* 50–200 μg C/l) and decreases exponentially below the maximum. This has been described by Nakajima and Nishizawa (1972) as

$$C = C_o e^{-k(z_0 - z)} \qquad (13)$$

where C is the concentration of carbon at depth z and C_0 is the maximum concentration of POC at depth z_0 (usually between 0 and 30 m). The exponent k reflects the rate of decomposition of POC with depth. A similar exponential decay in the standing stock of particulate organic nitrogen (PON) is encountered except that below depths of *ca.* 1000 m, PON appears to disappear faster than POC (Gordon, 1977) which gives rise to an increase in the C:N ratio of particulate organic matter below *ca.* 1000 m.

Holm-Hansen *et al.* (1966) give a good illustration of relative changes in a wide variety of biological parameters with depth down to 1300 m. Inorganic nutrients (NO_3^-, $PO_4^{\equiv}$) which were very low at the surface increased below 50 m; vitamin B_{12}, on the other hand, showed a maximum in approximate association with the bottom of the pycnocline (*ca.* 200 m); dissolved organic carbon (DOC) was high in the surface layers (*ca.* 800 μg C/l) but decreased rapidly to < 500 μg C/l below 50 m. These example results appear to be very similar to other reports in the literature if one allows for differences in the depth of the pycnocline, euphotic zone and for seasonality in other parts of the oceans.

Szekielda (1967) has drawn attention to the fact that while there is a general decrease of POC with depth, intermediate water masses in the gulf of Aden showed higher values for POC than were generally encountered immediately above or below the *ca.* 600-m POC maximum. Similar results were found by Menzel (1964) in the western Indian Ocean although the same author (Menzel, 1967) found constant POC (15 ± 5 μg/l) and DOC levels (0.6 ± 0.1 mg/l) in the tropical Atlantic, with slightly lower but constant values for the tropical Pacific Ocean, below 200 m.

While eqn. (13) reflects the change in standing stock of POC with depth, it has recently been shown that the vertical flux of organic carbon is much greater than the standing stock. This has been possible by the deployment of sediment traps at various depths in the ocean. For shallow water, Sasaki and Nishizawa (1981) showed that the POC flux down to *ca.* 200–300 m was several hundred milligrams per day (see Fig. 15). At 500 m this daily flux had decreased to 28 mg C/m²/day. In deeper water, values for the vertical flux at 1000 m range from *ca.* 0.6 to 12 mg C/m²/day while at 5000 m the flux is < *ca.* 3 mg C/m²/day, depending on the ocean and the results of different investigators, as summarized by Handa and Tanoe (1980).

The vertical distribution of the bacterioplankton has been studied by a number of authors (e.g. ZoBell, 1946; Kriss *et al.*, 1960; Sorokin, 1964b). Methods used for determining the presence of bacteria have led to very different assessments of the

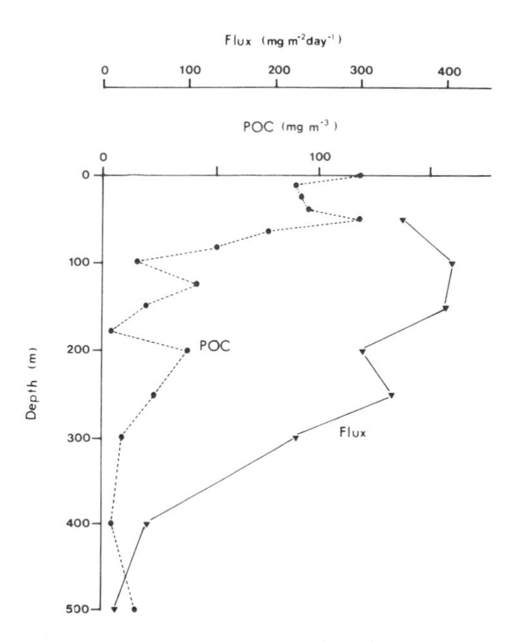

FIG. 15. Vertical variation of particulate organic carbon fluxes and vertical profile of POC concentrations sampled off the coast of Japan (from Sasaki and Nishizawa, 1981).

total biomass of bacterioplankton in the water column. Thus Kriss *et al.* (1960) used submerged glass slides which were liable to collect bacteria from a number of depths as they were lowered and raised through the water column. The results of this work may have some relative significance, however, and in particular it is claimed that discrete water types at depths down to 4000 m could be identified by their bacterial content. Sorokin (1964b) used direct counts of bacteria collected in sterile water samplers as a method for determining the total biomass of bacteria in the water column. From studies in the central Pacific, Sorokin (1964b) found that the biomass of bacteria in the euphotic zone was between approximately 10 and 50 mg/m^3 which represents only a few per cent of the total particulate carbon. However, Sorokin *et al.* (1970) have shown that concentrations of bacterioplankton at least an order of magnitude higher (i.e. 300 to 700 mg/m^3) may exist in discrete layers, often below or at the bottom of the euphotic zone and associated with thermocline. Below the euphotic zone the number of bacteria decreased sharply to 2000 or 3000 per ml, while below 1000 m the number of bacteria was generally less than 1000 per ml or a total biomass of $< 0.2\ mg/m^3$. The general form of these results agrees with biomass estimates of total living material made by Holm-Hansen (1969a) using ATP analyses as an indication of the amount of living organic carbon; however, the total amount of living carbon determined from ATP analyses was considerably greater than that found from direct bacterial counts alone. This indicates that bacterioplankton, both in the euphotic zone and in deeper water, are generally a small fraction of the total biomass of living material and only a few per cent of the total particulate organic carbon in the water column. Relative changes in the biomass of bacterioplankton are generally more important, therefore, as an indication of recycling processes than as a direct source of food. However, the methodology of bacterial counting still leaves a lot to be desired and it is probably correct to assume that many of the detrital particles in the ocean have bacteria accumulated on their surfaces and that this material is not included in direct counts of bacterial cells and clumps.

The presence of bacteria in the deep oceans (below 3000 m) has been reviewed by ZoBell (1968) who concluded that bacteria were widely but unevenly distributed at all depths. Numbers ranged from nil to 10^6/ml, the most dense populations having been found in materials from the sea floor. In nearshore environments, sedimentary material may be very rich in bacteria and it is believed that bacteria on the sedimented particles are the chief source of food for some benthic animals (Newell, 1965: Seki *et al.,* 1968).

Pelagic populations or organisms in the uppermost surface of the sea are referred to under the general term of neuston. Zaitsev (1961) first drew attention to the importance of neuston in the marine environment and a review of the subject has recently been given by Hempel and Weikert (1972). These authors subdivide the depth distribution of the neuston community into the 'euneuston' — organisms with maximum abundance in the surface where they stay night and day; 'facultative neuston' — organisms which concentrate at the surface only during certain hours, mostly during darkness; 'pseudoneuston' — the maximum concentrations of these organisms do not lie at the surface but at deeper layers; however, the range of their vertical distribution reaches the surface layer at least during certain hours. The largest change in the population of neuston organisms occurs during the evening and at night when many organisms among the facultative neuston and pseudoneuston join the comparatively few euneuston species. Hempel and Weikert emphasize that while the population density of neuston organisms may be very high, the total biomass of neuston in the water column is small since only a very thin layer is occupied by this community. Organisms which are permanently fixed to the sea surface by their own buoyancy and subject to wind drift are referred to as 'pleuston'. This group would include such organisms as the seaweed *Sargassum natans* in the Sargasso Sea and the coelenterates *Physalia* and *Velella*. Some other organisms, such as the blue-green alga *Trichodesmium* which forms blooms at depth in tropical seas (e.g. Marumo and Asaoka, 1974), may float to the surface and form dense mats of senescent organisms. As such they can be temporarily described as pleuston. Banse (1975) has drawn attention to some inconsistencies in the use of the words pleuston and neuston. In order to present a clear definition of the difference between these two words Banse (1975) has proposed that pleuston be defined as organisms specialized to live on or below, but close to the surface and neuston as organisms actually attached to, above (epi-) or below (hypo-) the surface film. In this sense pleuston becomes the general word for a surface-living organism (i.e. living in the surface zone of 'pleustal') and neuston becomes a subordinate zone connected only with the surface film.

1.3.4 Temporal Changes in Plankton Communities

The most rapid temporal changes in a plankton community can be observed by continuous monitoring at a fixed point; these changes are not due to changes within the plankton community *per se* but are caused by internal waves. Armstrong and LaFond (1966) studied changes in nutrients, transparency, and temperature at a fixed point in a highly stratified near-shore environment; continuous 3-hr records showed correlated fluctuations as illustrated in Fig. 16. The changes shown in Fig. 16 correspond to internal waves up to 5 m high with periods of around 10 min. Since the water intake used to collect these data was located at 9 m near the principal thermocline, the data reflect maximum changes due to an abrupt vertical gradient (e.g. from 0 to 12 μg at NO_3^-/1 between 6 and 12 m). Superimposed on the short time scale changes in Fig. 16 are other oscillations caused by tide and alternating wind speed and direction.

Temporal changes within a plankton community itself are largely determined by the growth, mortality, sinking, and migration rates of the individual plankters and their predators. The most rapid growth or reproduction rates for phytoplankton are of the order of several hours but whole populations generally require at least a day or more to double in size. Bacterioplankton may generate within a matter of hours, depending on temperature and substrate concentration. Zooplankton growth rates vary enormously from less than a week for some protozoa to 2 years for Antarctic euphausiids.

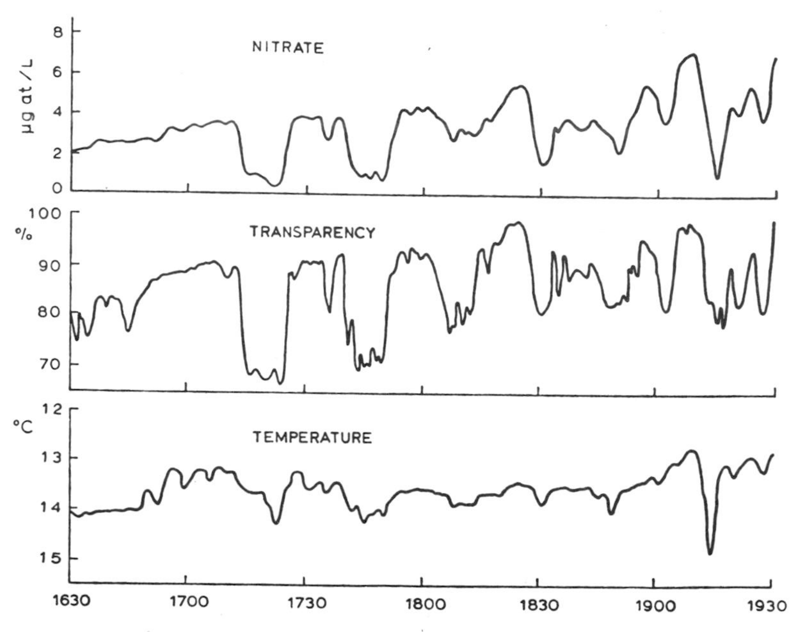

FIG. 16. Continuous 3-hr records of nitrate, transparency, and temperature at a fixed point (9·2 m from the sea floor) in a highly stratified environment off S. California (redrawn from Armstrong and LaFond, 1966).

Large populations of temperate copepods having one generation per year normally grow from egg to adult in 2 or 3 months, depending on food supply and temperature. Average mortality rates are generally less than growth rates or species would become readily extinct; however, over short periods, predation by planktivorous animals may cause mortality to exceed growth in a population. Sinking rates are discussed in Section 1.3.5; these, together with the rate of animal migrations, may range from less than a meter to several thousand meters per day. Consequently observed temporal changes in a plankton community may vary depending on the frequency of an investigator's observations. Changes which occur in a plankton community regularly every 24 hr are referred to as 'diel' changes; daily and nightly occurrences are called 'diurnal' and 'nocturnal', respectively. Included among diel changes are animal migrations, changes in photosynthetic potential, and, inshore, changes in plankton communities associated with the tidal cycle.

A detailed account of changes in an animal community caused by diel migrations is given by Bary (1967) as an example of a number of studies in this field. Using a 12-kHz sounder, Bary (1967) has described four stages in the ascent of organisms from a deep scattering layer (DSL) in a coastal environment. These stages started with a gradual vertical spreading of the DSL, 1 or 2 hr before sunset, followed by a period of slow ascent which started just before and ended after sunset. A period of rapid ascent followed about 1 hr after sunset; during this period the migrating organisms came up at speeds ranging from *ca.* 1 to 8 m/min. The final stage in the ascent migration was characterized by a reduced rate of ascent as the animals approached the surface, followed by a period when the animals gradually dispersed themselves in the surface layers. The descent followed the reverse process and commerced 1 or 2 hr before sunrise.

Diel variations in the rate of photosynthesis are quite apparent in that photosynthetic organisms require light for autotrophic growth; however, less obvious diurnal changes occur in the physiological response of phytoplankters to light. Shimada (1958) showed, for example, that photosynthesis of a plankton community reached a maximum during the early morning and declined during the rest of the day. This was apparently caused by a decrease in the amount of chlorophyll *a* during the latter part of the day. Steemann Nielsen and Jørgensen (1962) attributed this daily chlorophyll *a* rhythm to the fact that under laboratory conditions, no chlorophyll *a* was synthesized during the latter part of the day and consequently if herbivore grazing remained constant, a net decrease in chlorophyll *a* would be observed in nature. However, Sournia (1967) and others have shown that the photosynthetic index (mg C assimilated per mg Chl *a*) of phytoplankton is higher before noon than after noon. More recent data reported by Malone (1971) have indicated that changes in the photosynthetic index are complex and related to at least three factors including the time of day, the size of cells, and the availability of nutrients. Thus the photosynthetic index of microphytoplankton in tropical waters was highest after noon while the same index for nanoplankton was highest before noon. However, in eutrophic waters nanoplankton showed their highest photosynthetic index after noon. These differences were not attributable to changes in cellular chlorophyll *a*. Thus apart from grazing effects, there is an obvious physiological change in photosynthesis which may effectively slow down any potential increase in phytoplankton standing stock during different parts of the day (see also Chapter 3).

Tidal changes in near-shore communities cause very marked fluctuations in the relative abundance of plankton and nekton. This is particularly apparent in estuarine communities where there is a large change in the type of water at a fixed point over a single tidal cycle. Welch and Isaac (1967) showed that chlorophyll *a* values could vary up to 800% in an estuary during 24 hr. Since the tidal cycle has a fortnightly component (spring/neap tides) as well as a diel component, it is to be expected that chlorophyll *a* and productivity cycles based on a *ca.* 14-day cycle might also exist. Such cycles have been reported by Sinclair (1978) for Puget Sound (Washington) and by Webb and D'Elia (1980) for Chesapeake Bay — a review of this subject is given by Legendre (1981).

Large-scale temporal variations are associated

with seasonal cycles in oceanic and neritic environments. A summary of seasonal cycles in plankton communities has been prepared by Heinrich (1962) and is represented in Fig. 17 with some modifications as suggested by the work of Sournia (1969). Four different seasonal cycles may generally be recognized in oceanic environments based on changes in standing stock of phytoplankton and zooplankton. The first of these (Fig. 17, 1) is characteristic of Arctic or Antarctic waters where the amount of light is only sufficient for a single plankton bloom during the summer. The second seasonal cycle is characteristic of North Atlantic temperate waters (Fig. 17, 2) where breeding of zooplankton cannot start until an increase in primary productivity has occurred in the spring. Thus as a result of the increased primary production, there is a temporary increase in the standing stock of phytoplankton followed by a decrease, as the grazing pressure from an increased standing stock of zooplankton becomes effective. In these areas there are generally two maxima in phytoplankton and zooplankton standing stocks, a large one occurring in the spring and a smaller one in the autumn. Theoretical discussions of factors causing an increase in plankton during the spring are given in Chapter 3; the occurrence of a second plankton maximum in the autumn is associated with similar factors including the amount of nutrient mixed into the water column after summer stabilization, the presence of sufficient light to cause photosynthesis and the occurrence of zooplankton species capable of taking advantage of an autumn increase in the standing stock of phytoplankton.

A third type of seasonal increase in plankton (Fig. 17, 3) is found in the north Pacific Ocean. In this environment neither the beginning of the zooplankton breeding nor the size of the zooplankton standing stock is dependent on the presence or abundance of phytoplankton in the early spring. Nauplii of the zooplankton species, *Calanus plumchrus* and *Calanus cristatus*, which predominate in this area are hatched from adults which have wintered in deep water (> 200 m) and laid their eggs without feeding in the spring. Consequently the young stages of these animals can take immediate advantage of any increase in primary productivity. In such environments it is difficult to observe any change in the phytoplankton standing stock (Fig. 17, 3) except possibly during the autumn when a relaxation in zooplankton grazing allows a small increase in phytoplankton standing stock before winter.

A fourth type of seasonal cycle (Fig. 17, 4) is

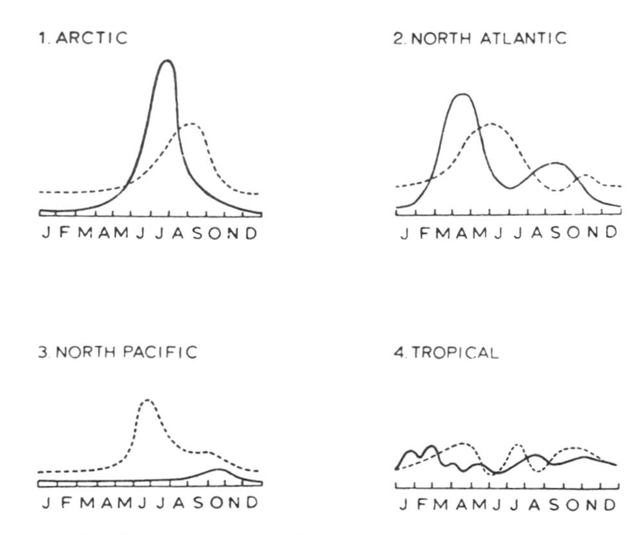

FIG. 17. Summary of seasonal cycles in plankton communities (——— changes in phytoplankton biomass; ------ changes in zooplankton biomass (modified from Heinrich, 1962)).

found in tropical oceans. According to Sournia (1969) and Blackburn *et al.* (1970) there is very little evidence for predominant maxima and minima associated with seasonal events. A succession of small increases and decreases in phytoplankton and zooplankton standing stocks may occur throughout the year and these are largely determined by local weather conditions and the movement of water masses. Thus where a condition of upwelling occurs, such as in the Pacific approximately at the equator, a relatively high standing stock of plankton can be observed which is largely determined by the exact location of the equatorial counter-current (Blackburn *et al.,* 1970). However, Owen and Zeitzschel (1970) observed a statistically significant seasonal change in primary production in the eastern tropical Pacific; apparently this change was not reflected in the plankton standing stock as discussed above. While the seasonal cycles in Fig. 17 indicate clear differences between different oceanic environments, it is also apparent that there are large zones of transition in the oceans. In sub-tropical waters at 32°N, for example, Menzel and Ryther (1960) observed a definite seasonal cycle in primary production in spite of an almost constant euphotic zone depth of 100 m.

While the illustrations in Fig. 17 indicate a smooth transition from periods of low to periods of high biomass, actual data often show considerable scatter and it would be more correct to describe the onset of seasonal conditions as occurring in a series of pulses. In addition the biomass changes shown do not indicate any seasonal change in species composition. For phytoplankton such a seasonal progression might involve a change in predominant species from diatoms, to dinoflagellates, to blue-green algae during the period spring to autumn (e.g. Bodungen *et al.,* 1975).

Annual differences in the quantity of plankton measured at a fixed location may also be due to the strength of currents. This has been illustrated by Wickett (1967), who showed that annual variations in the concentration of zooplankton in the surface layers off California varied directly with the southerly transport of water during the previous year. Quantitative changes in zooplankton biomass and growth over a 25-year period are illustrated in Fig. 18 from Glover *et al.* (1974). These results show that both the total number of copepods and the zooplankton biomass have significantly decreased. These decreases have either been paralleled by, or may in part be tied to, a delay in the advent of the spring phytoplankton bloom from March to April and a consequent shortening of the zooplankton growing season by about 1 month. The results in Fig. 18 are for the North Sea but similar results were obtained for the northeast Atlantic. The longest time series changes in a plankton community have been documented by the Plymouth laboratory (Russel *et al.,* 1971; Southward, 1974). Their data from the English Channel show an approximate 40-year cycle of events. Starting in the 1920s, the biological community was characterized by a high-plankton abundance, a herring fishery and the indicator organism, *Sagitta elegans;* these all virtually disappeared during a 40-year period through to the late 1960s when they returned. During the 40-year period they were replaced by low plankton abundance, a pilchard fishery and the indicator organism, *Sagitta setosa.* While a number of theories have been put forward for these changes, it is generally believed that they are related to climatic changes. Warming of the Arctic would affect circulation in the North Sea by allowing Atlantic water to extend farther north; cooling would have the reverse effect. At the same time, population centres of more northerly and southerly plankton populations and fisheries would shift their position (Russel *et al.,* 1971).

Conditions governing seasonal changes in neritic environments are much more complex than in oceanic environments due to the added effects of local geography, river discharge, and tides. Barnes (1956) described the seasonal occurrence of phytoplankton blooms and development of mero-planktonic barnacle larvae in the Firth of Clyde. From a study extending over 10 years the author concluded that the normal spring increase in phytoplankton could be heavily suppressed in some years by strong winds which resulted in a sequential catastrophic reduction in the development of barnacle larvae. In the vicinity of large rivers the neritic oceanographic climate may be generally modified to provide nutrient entrainment and a

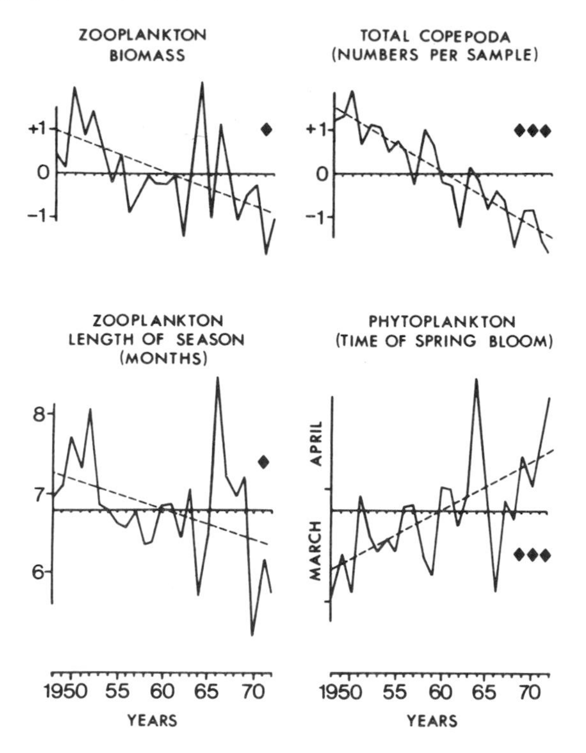

FIG. 18. Fluctuations in the plankton abundance in the North Sea showing (top) mean abundance plus or minus one standard deviation and (bottom) monthly variation from 25-year mean value. Calculated trend lines drawn and the significance of the fit shown by one, two or three ◆ corresponding to $P = < 5\%$, $<1\%$, and $<0{\cdot}1\%$ respectively (redrawn from Glover *et al.,* 1974).

more stable water column resulting in a higher standing stock of plankton. This is apparent, for example, in data presented by Anderson (1964) on seasonal changes in chlorophyll *a* off the Washington and Oregon coasts. In this study it is shown that chlorophyll *a* was higher in an area influenced by fresh water from the Columbia River (the Columbia River plume) than in either adjacent oceanic or neritic areas. Ice cover, and the effect of melting ice on the stability of the water column, may also cause earlier seasonal changes in plankton abundance in polar regions than in temperate oceanic waters located at lower latitudes (Marshall, 1958; Bunt and Lee, 1970).

Oscillatory variations in permanently stratified seas removed from the immediate influence of land have been reported and are more difficult to explain in terms of known physical processes. Steven and Glombitza (1972) showed that *Trichodesmium* blooms occurred in the tropical Atlantic with a frequency of about 120 days, measured over a 3-year period. This they suggest could neither be explained by some unknown variation in deep-water circulation, or more simply, by assuming that 120 days is the minimum time for biological factors controlling productivity to produce a bloom of *Trichodesmium.*

Studies on the temporal variation in bacterial numbers and species have been reviewed by Sieburth (1968) using specific examples from Narragansett Bay, Rhode Island. The author showed that there were apparent effects of phytoplankton species, solar radiation and temperature on the number and type of bacteria present. For example, two different colonies of flavo-bacteria were present during the year, a yellow-pigmented colony being present during periods of low radiation and an orange-pigmented colony being present in larger numbers during periods of high solar radiation. Also inverse relationships between genera were found to occur independently of season. Thus pseudomonads appeared to be dependent on phytoplankton blooms and a percentage increase in the isolates of pseudomonads was accompanied by a percentage decrease in the isolates of arthrobacters. While these relationships appear as real temporal variations, it should be emphasized that their cause may be more complex than indicated by seasonal factors or by the presence of other organisms.

1.3.5 Some Processes Governing Spatial Patchiness of Plankton Distributions

Causes of patchiness among marine organisms can be broadly divided into physical and biological effects. The physical effects have their greatest influence on the plankton while the biological effects are most pronounced among the nekton. Thus advection and turbulence will affect the distribution of plankton while the social behaviour of some species of fish causes them to school in large aggregates. However, neither physical nor biological effects are mutually exclusive in their impacts on the marine biota. For example, the boundary between salmon and tuna populations is a physical boundary in which salmon are found in water of $< ca.$ 14°C and tuna are found in water of $> ca.$ 14°C. Conversely the growth rate of plankton, a biological property, will vary with the physical environment and itself give rise to patchy distributions of phytoplankton. Furthermore, as noted by Platt *et al.* (1975), the patchiness of plankton is always difficult to completely dissociate from the sampling method used.

Physical effects which will be described below occur on various scales from small-scale turbulence to large-scale advection. The effects are not mutually exclusive of one another except in the sense of the scales involved. In general they may be summarized as consisting of the formation of gyres, the occurrence of tidal fronts, Langmuir circulation, convergent and divergent current systems and finally, on the smallest scale, random turbulence.

Two basic physical processes which govern the lateral dispersion of plankton are advection and turbulence. By advection it is generally understood that large-scale movements of water masses are involved in which plankton are imbedded. Two extreme differences in water masses are to be found in convergent and divergent systems; such systems may cover areas of several thousand kilometers. Thus the plankton ecology of a convergent system, such as the Sargasso Sea, is very different from a divergent system, such as areas of upwelling in the Peru or Bengala currents. Smaller-scale differences caused by advective processes are to be found in cyclonic and anticyclonic rings and in coastal frontal zones. The former are characteristic of open ocean while the latter are characteristic of continental shelf and slope waters. In both cases the physical processes involved may extend over several hundred kilometers and there has been considerable recent work on the biology of these systems.

A description of cyclonic and anticyclonic rings occurring in the Gulf Stream is given by Richardson (1976) and some aspects of their biology are given by the Ring Group (1981). An idealized cross-section of a cold core ring has been drawn in Fig. 19 on the basis of information in the latter reference. Essentially the cold core ring is a cyclonic gyre which has an upwelling component as seen from the temperature structure in Fig. 19. The advection of deep water towards the surface brings with it a supply of nutrients which is the basis for the increased primary productivity within the ring as seen in Fig. 19 in terms of an increased standing stock of chlorophyll *a*. Thus patchiness on this scale of *ca.* 200 km is a result of a relatively small body of water being spun off a major current system (i.e. the Gulf Stream).

An excellent description of coastal zone fronts has been given by Dr. R. D. Pingree (e.g. Pingree, 1978 and references cited). In his explanation of large-scale plankton patchiness in coastal waters, Pingree (1978) considered that an area of low production due to high vertical stability could be changed into a region of high primary production if the stable water flowed over a region where a shallowness of the bottom both increased the water velocity and caused a frictional response with the bottom, resulting in vertical turbulence and transfer

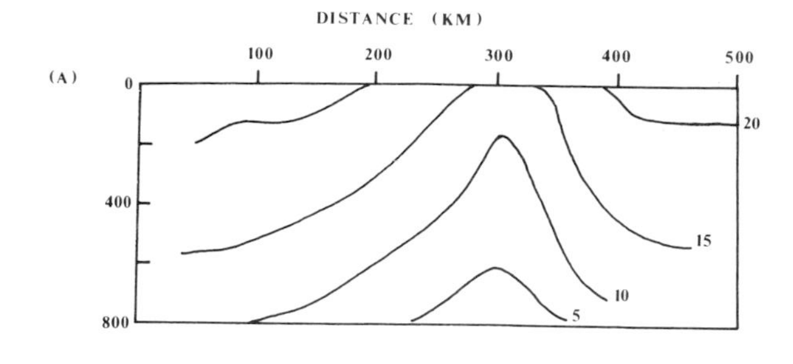

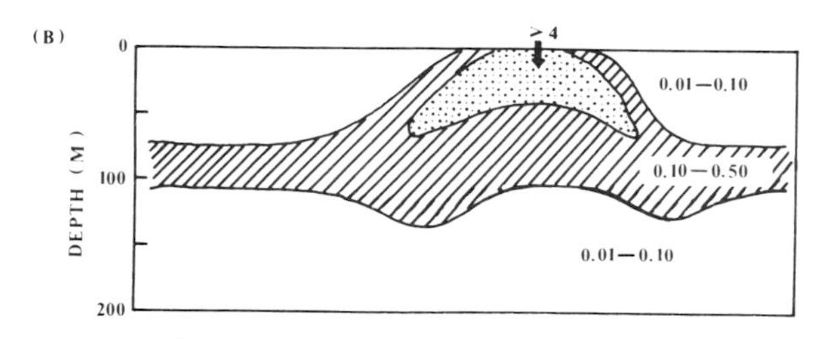

FIG. 19. Approximate vertical section through a cyclonic ring in a subtropical ocean. (a) Temperature (°C). (b) Chlorophyll *a* (μg/l).

of nutrients into the euphotic zone of the water column. This process can be expressed by considering the ratio (R) of the production of potential energy (PE) in maintaining well mixed conditions to the rate of tidal energy dissipation (TED) in a water column of unit cross sectional area:

$$R = \frac{PE}{TED}, \tag{14}$$

The two forms of energy in the above equation can be formulated in terms of a number of parameters, most of which are constants for a restricted area. However, two important terms which are not constant are the average water velocity, $|\bar{U}|$, and the water depth, h. From the physical formulation, a stratification parameter (S) or index can then be extracted to give:

$$S = \log_{10} \frac{h}{C_D|\bar{U}|^3} \text{ (c.g.s. units)}, \tag{15}$$

where C_D is the drag coefficient which can usually be approximated, and in Pingree's evaluation it was 0·0025. The stratification parameter as given in eqn. (15) is then easily calculated for a coastal region and in general it is seen to have values ranging from −2 to +2. From eqn. (15) low values of S will reflect turbulent mixing and high values, stratification. Pingree (1978) found that there was an intermediate value of $S = ca.$ 1·5 where the degree of stratification and turbulence were just sufficient, on the one hand, to stabilize the water (i.e. $D_{cr} > D_m$) and on the other hand to supply sufficient new nutrient to the mixed layer for enhanced primary productivity. An illustration of Pingree's results is shown in Fig. 20 where a predicted $S = 1{\cdot}5$ isoline (A) is seen to correspond approximately with a surface chlorophyll *a* maximum in the Celtic Sea (B). The vertical structure of a frontal zone showing the near-surface chlorophyll maximum and its extension into the deep chlorophyll maximum of more stable water is shown in Fig. 21. Departure from the $S = 1{\cdot}5$ prediction can be accounted for in terms of differences in extinction coefficients for adjacent water masses, the role of freshwater in stabilizing production and in particularly shallow or deep shelf areas where production may occur at values of $< 1{\cdot}5$ or $> 1{\cdot}5$, respectively. Examples of phytoplankton patchiness using Pingree's approach and taking into consideration the above departures from the predicted S parameter of 1·5 are found in shallow water off Flamborough Head (U.K.), in deep water between the Orkney and Shetland Islands in the North Sea (Pingree *et al.*, 1978) and in the Strait of Georgia off the coast of British Columbia (Parsons *et al.*, 1981) where the influence of freshwater and different extinction coefficients caused a shift in the expected frontal zone areas as predicted from the $S = 1{\cdot}5$ value. In all these cases, however, the frontal zone analysis as suggested by Pingree (1978) was an effective way to describe the process involved.

The distribution of plankton under conditions of Langmuir circulation is another important mechanism which accounts for the type of long thin patch first described by Bainbridge (1957) and discussed by Steele (1976). In this case the response of the plankton to the physical effect is largely passive and results in patches many kilometers long but only one meter or less in width. As observed by Stommel (1949), in any boundary region where there is a

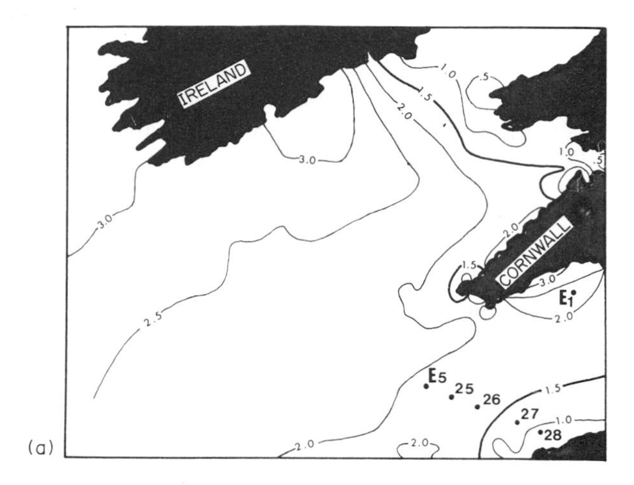

FIG. 20A. Stratification parameter (S) contained for the Celtic Sea.

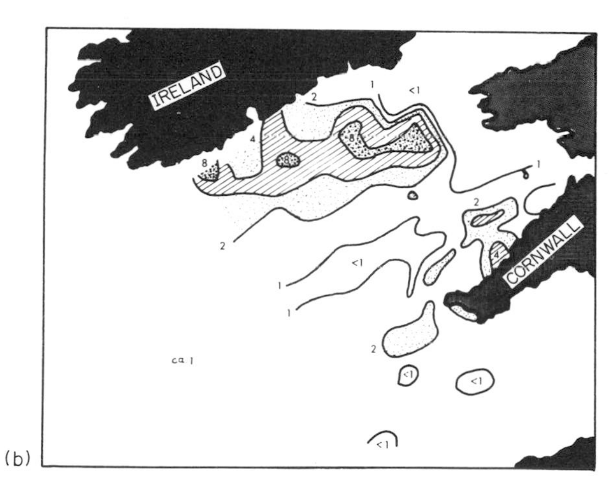

FIG. 20B. Surface distribution of chlorophyll, 9–12 April 1975, for the Celtic Sea (from Pingree, 1978).

convergence, particles which tend to float will accumulate along a line of downwelling, while those that sink will be buoyed up in the upwelling region. In this connection, the small-scale association of the effects of wind on plankton distributions was first observed by Nees (1949), who discovered that plankton samples collected parallel with the wind direction were more variable than samples collected at right angles to the wind. The reason for this is that under conditions of light wind, plankton accumulate in rows parallel to the wind (a property which can be observed visually in the Sargasso Sea, where *Sargassum* weed floats on the surface forming familiar wind rows). Thus if plankton are collected across the direction of the wind, the samples result in an integrated collection but if the samples are collected parallel to the wind, the resulting samples may come from a column of water in which plankton are concentrated, or from the relatively barren water in between the wind rows. The exact explanation for plankton patchiness under conditions of light wind may include several factors but in general it appears to be related to Langmuir circulation (Sutcliffe *et al.*, 1963; Stavn, 1971). The principal property of this circulation is that the water tends to move in vortices which result in 'micro-zones' of upwelling and downwelling water (Fig. 22 and Faller, 1971, for a discussion of the physics of Langmuir circulation). Conditions for particle accumulations A, B and C are based

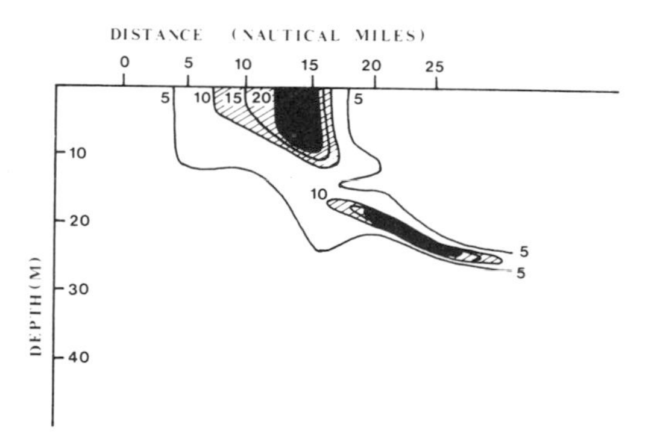

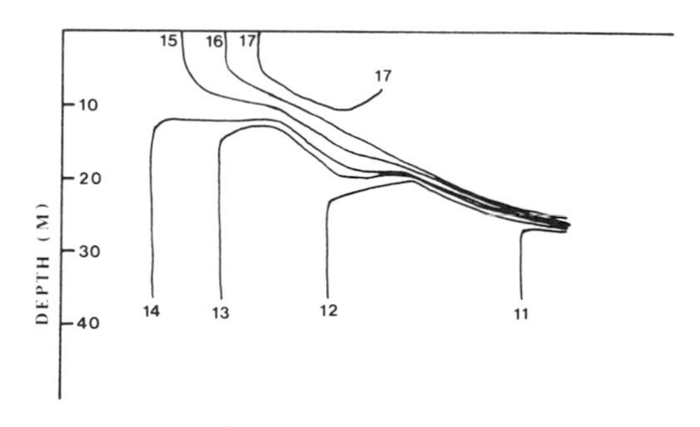

FIG. 21. (top) Chlorophyll a through a frontal region. (bottom) Corresponding temperature through a frontal region (from Pingree, 1978).

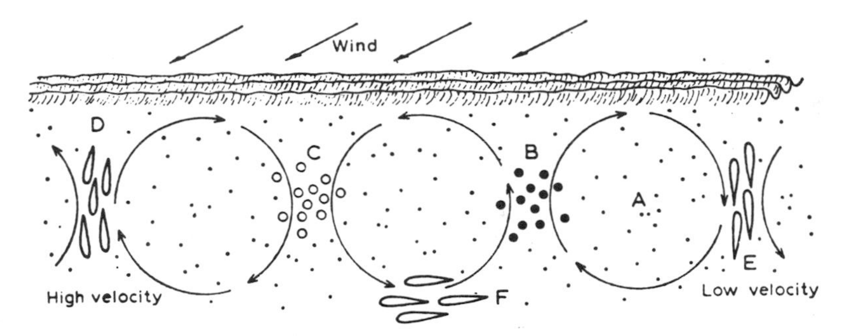

FIG. 22. Langmuir vortices and plankton distributions redrawn from Stavn (1971). A — neutrally buoyant particles randomly distributed. B — particles tending to sink, aggregated in upwellings. C — particles tending to float, aggregated in downwellings. D — organisms aggregated in high velocity upwelling, swimming down. E — organisms aggregated in low-velocity downwelling, swimming up. F — organisms aggregated between downwellings and upwellings where there is less relative current velocity than within the vortices.

only on the buoyancy of particles as suggested by Stommel (1949); these conditons may apply particularly to phytoplankton and detrital accumulations or to all planktonic organisms in the absence of light. Under conditions D, E, and F, however, experimental results obtained by Stavn (1971) are included; these take into account the relative current velocities of the water and the swimming speed of the animals together with their light reaction which (in the case of *Daphnia*) may be used to sense the currents. Thus during the day planktonic animals may swim into a current (low velocity) since this is the most stable swimming position. In this case the animals will be aggregated in downwelling water and swimming up. At intermediate current velocities the plankton will be clumped in between the spirals and swimming horizontally against the current. At high current velocities animals may be trapped in a high-velocity upwelling where they are attempting to swim down in order to avoid surface light which is generally too bright for many zooplankters (negative phototaxis).

Large-scale patchiness, such as is depicted in Fig. 10 for phytoplankton production in the hydrosphere, can be mostly accounted for in terms of ocean circulation. LaFond and LaFond (1971) have discussed the patterns of water movement which lead to areas of high and low plankton productivity. The general mechanism involves some process which carries nutrient-rich deep water to the surface where there is adequate light for photosynthesis. Of particular importance in this respect is the influence of a land mass on circulation. LaFond and LaFond (1971) recognized three types of land influences on water movement which result in an upwelling of deep water; islands which create large eddy currents on the leeward side of the principal current flow, land promontories which cause similar eddies, and changes in underwater topography which cause turbulence in the flow of near-surface currents.

In addition to these physical land-mass effects, boundary conditions between currents moving in opposite directions may lead to large-scale upwelling. This is particularly apparent along the equator in the Pacific Ocean where the equatorial current in the northern and southern hemispheres tends to move in a northerly and southerly direction, respectively (due to the Coriolis force, see Pickard, 1964, Fig. 27). This results in a divergence near the equator and an upwelling of deep water. Similarly a commonly occurring seasonal upwelling may be produced by winds blowing parallel to a coastline which can result in water being displaced offshore (an upwelling, or divergence) or onshore (a downwelling, or convergence). These observations are in accordance with physical oceanographic theory and are illustrated schematically in Fig. 23. According to Wooster and Reid (1963) the principal areas in the hydrosphere where an exchange occurs between near-surface and deeper waters are (1) in high latitudes, (2) along the equator, and (3) in

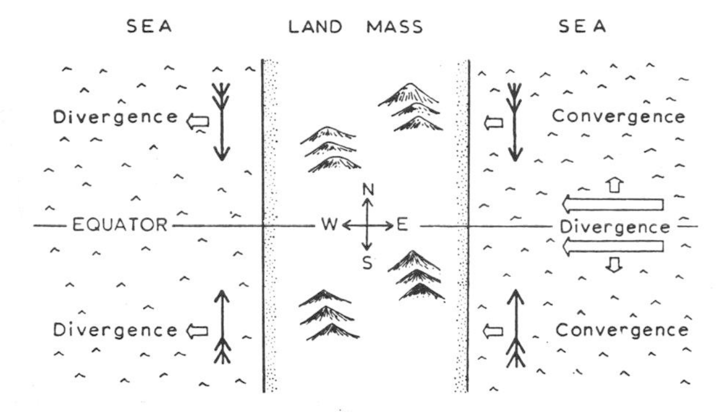

FIG. 23. Schematic effect of wind (⋙⟶) and water (⟹) movement in producing areas of upwelling (divergence) and downwelling (convergence). Wind direction shown from the south toward the equator in the southern hemisphere and from the north toward the equator in the northern hemisphere. Water movement caused by this wind pattern results in a divergence off the west coast and a convergence off the east coast. Water movement is reversed with a reversal in wind direction. Currents along the equator in the northern and southern hemispheres similarly tend to move water towards or away from their boundary area, depending on the principal direction of water flow.

coastal regions, particularly on the eastern sides of oceans. This is apparent in the higher biological productivity of waters in such areas as the California current, the Peru current, and the Benguala current.

The smallest scales of plankton patchiness are influenced (if not wholly decided) by turbulence. Platt (1972) found that the variance of phytoplankton abundance followed a frequency similar to the turbulence recorded in the dissipation of tidal energy, over length scales of 10 m to 1 km (sometimes referred to as the $-5/3$ power relationship). This implies that on small length scales, plankton (mostly phytoplankton) will be distributed according to local turbulent conditions and that the growth rate of the organisms will not influence that distribution. However, exceptions to this concept have been clearly defined in the case of Langmuir circulation and are also suggested by Lekan and Wilson (1978). Although the interpretation of data in the latter reference is questionable, the possibility is raised of the chlorophyll *a*/frequency distribution being a -1 slope; this would imply that the growth rate of the phytoplankton had time to react to differences in turbulent fields on space scales of less than 5 km. In order for this to occur, the plankton would have to have division rates of a few hours. Although this is not impossible, the *in situ* division rates of phytoplankton are generally closer to a day than to hours. Reviews of the role of stability in determining phytoplankton production are given by Margalef (1978) and Legendre (1981).

The absolute size of a phytoplankton patch is determined in its upper dimensions by the total area of high or low productivity (e.g. the Peruvian upwelling or the Sargasso Sea, respectively). The lower size scale is determined by the growth rate of plankton and the degree of turbulence. This problem has been considered on theoretical grounds by Kierstead and Slobodkin (1953), Steele (1976) and Orubo (1977 and references cited therein). The simplest consideration in determining the smallest size scale of a plankton patch is to consider the growth rate of the phytoplankton versus the rate of diffusion of plankton away from the patch. Obviously if diffusion exceeds the growth rate, the patch is destroyed. From theoretical considerations, Okuda (1977) has summarized various estimates of patch size and concludes that for a phytoplankton growth rate of 1 division per day, phytoplankton could maintain a patch size of *ca.* 1–2 km but that for growth rates of 0·1 division per day the patch size has to be 15–50 km in relatively calm waters. These studies are in keeping with observations made on both phytoplankton and zooplankton patches by Bainbridge (1975), Steele (1974), and Cushing (1955). In the last reference Cushing followed phytoplankton and zooplankton patches which maintained themselves for a period of at least 3 months. Obviously in this situation the zooplankton predator patch has a modifying influence on the phytoplankton patch; in addition zooplankton may show some social behaviour which may increase the patch density as discussed below. In all these observations, however, it should be noted that smaller or larger scales of plankton patchiness which can be observed in the ocean may also contain the temporary effect of physical turbulence. In some cases, the transitory appearance of these patches may be important on the biological scale of plankton and fish feeding.

Patchiness of zooplankton populations appears to involve more biological factors than growth rate which was the only biological factor considered above for phytoplankton. In considering both motile species of phytoplankton (e.g. dinoflagellates) and, more especially, all zooplankton, a variety of swimming techniques in response to different stimuli may maintain patch integrity which would otherwise be destroyed by turbulence. This has been demonstrated in a general sense by Mackas and Boyd (1979) who compared the spectral analysis of zooplankton, chlorophyll, and temperature. Their results clearly show that the power spectrum of the zooplankton was less steeply sloped than that of chlorophyll or temperature. This was interpreted to show that zooplankton form a more finely grained pattern which suggests that they have some ability to resist turbulent diffusion. Such a general observation has been carried much further by Omon and Hamner (in press) who give specific examples of relatively small (< 100 m), highly aggregated (up to 1000 times average density) populations of zooplankton. The latter observations were only possible through different forms of visual contact by researchers and could not be obtained with plankton nets. The conventional plankton net must in fact be regarded as highly destructive both of many fragile organisms as well as of the true small-scale distribution of zooplankton in the sea.

The vertical component of plankton patchiness is determined by a variety of factors including light intensity, density gradients and the availability of nutrients at the surface.

The vertical migration of zooplankton and their aggregation at specific depths has been reviewed by Banse (1964). In general vertical movement appears to be related either to the life history of a species or to its feeding habits. The former generally cause seasonal differences in the vertical distribution of plankton while the latter cause diel changes in vertical abundance. Vinogradov (1955) postulated in the latter case food material was transported to great depths by a series of overlapping animal migrations. Thus it was considered that some bathypelagic animals obtained their food by migrating into the mesopelagic zone while some mesopelagic animals migrated into the epipelagic zone; the exact depth ranges being dependent on the species.

Several theories have been offered as explanations for the vertical migration of zooplankton populations. One suggestion (McLaren, 1963) is that an animal can conserve energy by reducing its metabolism during part of the day (or life cycle) if it migrates into cold water below the seasonal thermocline. Several theories on the optimization of phytoplankton resources have been advanced (e.g. Petipa and Makarova, 1969; McAllister, 1970; Kerfoot, 1970). These theories, while not precisely the same, have in common a separation of phytoplankton from zooplankton during maximum light thus allowing the phytoplankton a period in which they can increase exponentially in abundance instead of being continually grazed by zooplankton. In another theory (McLaren, 1974) it is suggested that for populations in which size and fecundity are negative functions of temperature, it is advantageous to migrate vertically in thermally stratified waters.

Seasonal changes in the vertical distribution of phytoplankton in temperate waters are discussed in Chapter 3 with respect to different light and nutrient regimes. The persistent occurrence of the deep chlorophyll maximum, particularly in tropical and sub-tropical seas, has been a subject of special interest which requires some explanation.

The vertical distribution of chlorophyll in the sea often shows a single maximum and occasionally two or more maxima (see Fig. 13). The concentration of chlorophyll typically increases by a factor of 2 to 20 in the maximum, although much larger increases have been reported. The existence of chlorophyll maxima has been known for some time (cf. Riley *et al.*, 1948) but the truly ubiquitous nature of chlorophyll maxima became evident only after the introduction and widespread application of the *in-vivo* fluorescence technique (Lorenzen, 1966). In temperate latitudes a chlorophyll maximum is usually found only in stratified water columns during the summer. In tropical and subtropical latitudes a chlorophyll maximum is a consistent feature throughout most or all of the year, except in the region of tropical upwelling between *ca.* 10°N and 10°S. The chlorophyll maximum at midlatitudes is probably a continuous layer across both the Pacific and Atlantic Oceans.

The depth of the chlorophyll maximum is quite variable. It is usually found at the greatest depth (100–250 m) in the oligotrophic central gyres and shallower (20–100 m) near-shore, near the region of equatorial upwelling and seasonally during the summer months at higher latitudes (⩾40°). The depth of the chlorophyll maximum can also change dramatically over much shorter periods in response to local disturbances such as internal waves (see Fig. 16).

Although the depth of the chlorophyll maximum varies both spatially and temporally, it generally occurs near the bottom of the euphotic zone, associated with the nitracline (i.e. the region of maximum change in nitrate concentration). The chlorophyll maximum is usually found at depths where the light intensity is less than 10% and frequently less than 1% of the light intensity at the sea surface. Although most studies have not obtained nutrient measurements with a resolution comparable to that of the *in-vivo* or *in-situ* fluorescence measurements, it is likely that chlorophyll maxima are usually located in the nitracline, and at or near the primary nitrite maximum. Primary productivity is generally positive in the chlorophyll maximum, accounting for 10–40% of the total carbon fixation integrated over the euphotic zone.

A variety of explanations have been suggested to account for the formation and maintenance of chlorophyll maximum layers. These include a reduction in the sinking rate of phytoplankton due to increases in nutrient concentration, density changes of the water column, or vertical mixing; differential grazing of phytoplankton above the chlorophyll maximum by herbivores; active growth of phytoplankton at or near the depth of the chlorophyll maximum; an increase in the amount of chlorophyll per cell as a result of adaptation to low light intensity or photodegradation of chlorophyll in cells above the chlorophyll maximum; horizontal layering of water masses; and active aggregation of motile phytoplankton in response to light and/or nutrient concentrations. In most cases the formation and maintenance of chlorophyll maxima probably results from a combination of these processes, although a single process may clearly dominate at any given time or location. The trophic significance of chlorophyll maxima is still equivocable, although it clearly can be an important factor in the reproductive capacity of copepods (cf. Checkley, 1978) and survival of some larval fish (cf. Lasker, 1975). Further progress on this question requires better measurements of phytoplankton biomass, herbivore distributions, and grazing rates.

The distribution of POC in the ocean appears to be derived from a balance between production on one hand and sinking and decomposition on the other. The sinking and decomposition processes together give rise to a second phenomenon which is the oxygen minimum, found under stable conditions mostly in subtropical and tropical oceans. The depth of the oxygen minimum is usually between 200 and 500 m; it coincides with a maximum in the apparent oxygen utilization (AOU) and the source of organic material is believed to be largely the POC flux as shown in Fig. 15 (Kroopnick, 1974; Sasaki and Nishizawa, 1981). The POC flux material has been shown to be different in chemical composition to the standing stock of POC (Tanoue and Handa, 1980). The mechanism by which the flux is decomposed in the depth range 200–500 m appears to be related to the rapid sinking of fecal pellets and their reconsumption (coprophagy) by bathypelagic zooplankton (Paffenhofer and Knowls, 1979; Sasaki and Nishizawa, 1981). Evidence for the rapid sinking rate of fecal pellets is presented by Turner (1977) and Beinfang (1980). The former gives an average sinking rate of 66 m/day for copepod fecal pellets while the latter reports average sinking rates of fecal pellets formed from flagellates as 88 m/day and those from diatoms as 123 m/day. By contrast the sinking rate of phytoplankton is at least an order of magnitude less (*ca* 0·3 to 1·7 m/day — Beinfang, 1981). Thus in order to satisfy a daily flux of several hundred mg C/m^2/day (Fig. 15) below the euphotic zone (i.e. > 100 m) it is necessary to suggest that the flux is largely composed of rapidly sinking particles. Since such large particles could not be decomposed in a short time span by bacteria alone, it is necessary also to suggest that the fecal pellets are reconsumed by a bathypelagic zooplankton community. In contrast, the standing stock of POC appears to be a residual, highly refractory component of the food chain. The distribution of this material, which sinks very slowly, follows a vertical gradient which is more in keeping with largely bacterial decomposition processes, the bacterial activity being greatest at the POC maximum (0–30 m) and diminishing where only the refractory organic compounds remain at greater depths. The distribution of bacterioplankton in the water column as discussed earlier would tend to support their role in determining the exponential decline in the standing stock of POC with depth.

Some modifications to the general concepts described above for the distribution of zooplankton, phytoplankton and detritus in the vertical gradient are apparent in the literature. For example, different species of zooplankton may respond with greater or less sensitivity to salinity changes from 1‰ to 10‰ (Lance, 1962; Harder, 1968). Further, the sensitivity of zooplankton to light (e.g. McNaught and Hasler, 1964) may include colour differentiation and the detection of currents. Smith and Baylor (1953) showed that the freshwater cladoceran, *Daphnia,* was sensitive to different coloured light and that light of over 500 nm caused upward swimming while light of less than 500 nm caused a reversed behaviour. Stavn (1971) showed that there was a minimum light intensity of 70 ergs/cm^2/s below which *Daphnia* did not respond to water movement; the author concluded that orientation in moving water was affected through visual detection of currents.

Mechanisms which tend to modify the sinking rate of phytoplankton will also affect their vertical distribution in the water column. Such changes include cell buoyancy and, among flagellates, the ability to swim in response to stimuli (e.g. light). Since some small flagellates can be observed under the microscope to swim a distance of their body length in less than a second, maximal swimming speeds for some species may be around 100 to 1000 μm per second, or in the approximate range of 1 to 10 m per day. Loeblich (1966) measured the swimming speed of *Gonyaulax polyedra* and from his observations, as well as others which he quotes, it appears that large dinoflagellates may swim at speeds from *ca.* 2 to 20 m per day, in other words, 25 to 50 times the cell length per second (Throndsen, 1973).

Changes in the buoyancy of phytoplankton cells may be accomplished through a change in the cellular constituents. Thus an increase in lipids or a change in the ionic content of the cell will alter the cell density. Of these two mechanisms, Smayda (1970a) considers that the latter mechanism is the more important, especially in cells having large vacuoles. The author reports on the work of Beklemishev, Petzikova and Semina who analysed cell sap in the diatom, *Ethmodiscus rex,* and found that the total ion content of living cells was 23·9 mg/ml compared with 33·3 mg/ml in dead cells and 33·8 mg/ml in the sea-water medium. The selective exclusion of heavy ions from cell sap as a means of buoyancy was first proposed by Gross and Zeuthen (1948) but it is apparent from observations made by Eppley *et al.* (1967) and theoretical calculations made by Smayda (1970a) that the theory does not entirely account for the range of density changes observed among the phytoplankton. Other physiological mechanisms yet to be fully explored are changes in buoyancy due to nutrient enrichment (Steele and Yentch, 1960) and the effect of light (Smayda and Boleyn, 1966).

CHAPTER 2

CHEMICAL COMPOSITION

2.1. SEA WATER

The following section deals with a few important facts regarding the chemistry of sea water, particularly in relation to properties of biological importance. A good summary of the chemical properties of sea water is given by Riley and Chester (1971) while a more thorough treatment is contained in a series of volumes entitled *Chemical Oceanography* (Academic Press, edited by J. P. Riley and G. Skirrow). Texts on the analysis of chemical constituents have been prepared by Strickland and Parsons (1972) and Grasshoff (1976).

Sea water is generally characterized in terms of its salt content or salinity. Earlier definitions of salinity referred to the dry weight of salts per kilogram of sea water. However, since this proved impossible to measure in practice, later definitions related salinity to the chloride content and finally to conductivity. The most recent practical definition of salinity (Perkin and Lewis, 1980) is based on the conductivity of a sample of sea water in comparison with that of a standard solution of potassium chloride (KCl), in which the mass fraction of KCl is 32.4356×10^{-3}, the temperature is 15°C, and the pressure is one standard atmosphere. For most practical purposes this definition is satisfactory for biologists but, since it is no longer related to the chemical content of the water, it is possible at very low salinities for two samples to have the same salinities as determined by conductivity, but for different ions to be present in each sample (cf. discussion by Gieshes, 1982). Other properties of sea water which can generally be derived from a knowledge of the salinity (or the salinity and temperature combined) are the chloride content,

$$S‰ = 0.03 + 1.8050\ Cl‰^* \qquad (16)$$

where S‰ and Cl‰ are the salinity and chlorinity in parts per thousand, the density, the alkalinity, the refractive index, freezing point, and osmotic pressure. With the exception of the alkalinity, these properties are sometimes referred to as the 'conservative properties' of sea water. Since the salinity of the sea has been determined over geological time, the only change in salinity and other conservative properties is seen as being due to physical processes, such as advection or loss of fresh water. By contrast the 'non-conservative properties' are those which can be controlled over short time intervals by the biology of the sea or by some near-shore or sediment processes. These properties include the alkalinity of sea water, the nutrient content, the organic content and the extinction coefficient (transparency).

Oceanic waters are generally high in salt content and the composition of major ions in a sample of ocean water containing about 35 g of salt per kilogram of sea water is shown in Table 2. In addition to these major ions, sea water contains practically every element in the periodic table,

TABLE 2. CHEMICAL COMPOSITION OF THE MAJOR CATIONS AND ANIONS IN SEA WATER OF 35‰ SALINITY (Cl = 19·374‰) BASED ON 1967 ATOMIC WEIGHTS (AFTER MORCOS, 1973)

Cations	g/kg	Anions	g/kg
Na^+	10·76	Cl^-	19·35
Mg^{2+}	1·30	SO_4^{2-}	2·71
Ca^{2+}	0·41	Br^-	0·07
K^+	0·40	HCO_3^-	0·14
Sr^{2+}	0·01	H_3BO_3	0·03

**Note:* The intercept in this equation is based on riverine fresh water; for oceanic samples diluted with rain water, the intercept is not valid and the slope becomes 1·80655.

although some of these may be at concentrations which are difficult to analyse. More important for the biologist are the nutrients in sea water; these include such macronutrients as nitrate, phosphate, silicate and bicarbonate as well as micronutrients such as iron, manganese, cobalt, vitamins (e.g. B_{12}, thiamine, and biotin) and unspecified organic compounds. The concentration of macronutrients is generally expressed in μg at/l and typical ranges for nitrate are 0 to 30 μg at/l and phosphate <0.1 to 3 μg at/l. Micronutrients, such as the vitamins, may be present in $\mu\mu$g/ml* while soluble organic compounds (including a very wide spectrum of metabolizable and refractory materials) may range from *ca.* 0.2 to 2 mg C/l. In addition sea water contains much microscopic particulate matter; this includes the phytoplankton and organic detritus (sometimes expressed collectively as the particulate organic carbon, POC) and inorganic particles derived from the land or sediments, as well as the skeletons of microscopic organisms. The total weight of all particulate materials which can be collected on a filter of specified pore size (e.g. 0.45 μm) is referred to as 'seston', although in fact the size range of particles in the sea is probably continuous below the arbitrary pore size chosen.

Seasonality in temperate climates and geographical location largely determine the biological properties of sea water which lend themselves to chemical analyses. In many tropical and subtropical environments as well as in temperate waters following a phytoplankton bloom, the concentration of nitrates and phosphates may be below the limit of detection by conventional methods; in such cases it may be possible to demonstrate that the principal source of nitrogen is present in low concentrations (<5 μg at/l) of ammonia, urea, or some organic nitrogen compounds. Thus changes in the nutrients of sea water are usually reflected in a redistribution of biologically important elements; this is illustrated in Fig. 24, which shows changes in the concentration of three forms of phosphorus in a coastal environment over a period of 1 year. The occurrence of plankton blooms in themselves cause

*1 $\mu\mu$g = 1 picog = 10^{-3} nanog = 10^{-6} microg = 10^{-9} mg = 10^{-12} g.

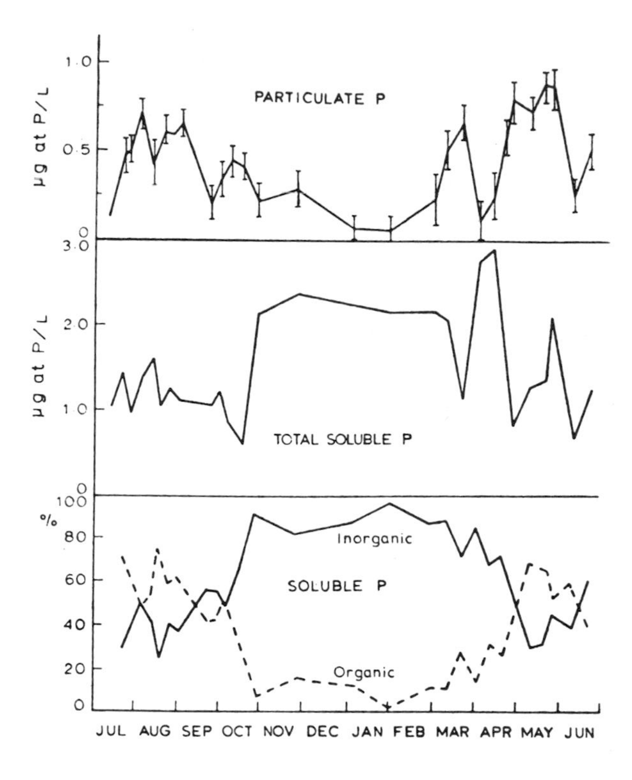

FIG. 24. Annual variations in the presence of three forms of phosphorus in a coastal environment, Departure Bay, British Columbia (redrawn from Strickland and Austin, 1960).

subsequent changes in the chemistry of sea water, not only through the redistribution of macronutrients, but also by modifying the dissolved organic carbon (DOC) chemistry including the release of organic compounds by phytoplankton (e.g. oxalic acid) and the production of vitamins by bacteria.

The two gases dissolved in sea water which are changed by biological activity are carbon dioxide and oxygen. Carbon dioxide is present in sea water, mostly as bicarbonate in concentrations of *ca.* 25 mg C/l; this level is a large excess over the amount generally required for plant growth. The pH of sea water, which is largely determined by the bicarbonate/borate concentration, is usually in the range 7·5 to 8·5. Changes in pH within this approximate range are due to photosynthesis and respiration of marine organisms and low pH values are reflected in high concentrations of total carbon dioxide. Since carbon dioxide and oxygen are the products of biologically opposite processes, it is also usual to find high concentrations of oxygen (*ca.* 4 to 8 ml/l) and higher pH values in areas of high photosynthesis, and lower oxygen concentrations ($<$4 ml/l) where respiratory processes predominate.

While temperature and salinity are important properties in determining the physical history of different water masses (cf. Pickard, 1964) other examples (e.g. Park, 1967) can be given of the use of non-conservative properties to follow different water masses in the sea. The amount of oxygen which will dissolve in sea water at the sea surface is a function of sea water temperature and salinity. Any difference between the measured oxygen content of sea water and that expected from the known solubility at the temperature and salinity of the sample is called the 'apparent oxygen utilization' or AOU (Redfield, 1942). AOU may be negative (due to photosynthesis) or positive (due to mineralization). For example, if surface water is caused to sink at a convergence, oxidation or organic material will proceed at aphotic depths and the oxygen content of the water will be lowered (positive AOU). Assuming an elemental atomic ratio in plankton of O:C:N:P = 276:106:16:1, the AOU can be expressed in terms of the oxidized state of these elements in a sea-water sample. Thus the *measured* nitrate and phosphate ($P_{meas.}$ and $N_{meas.}$) in a sea-water sample can be expressed as

$$P_{meas.} = P_P + P_{ox} \quad (17)$$

and

$$N_{meas.} = N_P + N_{ox} \quad (18)$$

where P_P and N_P are the levels of preformed phosphate and nitrate, and P_{ox} and N_{ox} are the levels of phosphate and nitrate derived from the apparent oxygen utilization (AOU), respectively. Equations (17) and (18) can therefore be rewritten in terms of AOU, as

$$P_{meas.} = P_P + 0{\cdot}0036\text{AOU} \quad (19)$$

and

$$N_{meas.} = N_P + 0{\cdot}058\text{AOU} \quad (20)$$

where all units are in numbers of atoms (e.g. μg at/l). Sugiura (1965) has suggested that the preformed nutrients are a conservative property of sea water; as such, they can be used to characterize water masses by solving eqns. (19) and (20). The use of AOU values is not without difficulties, however, since it is assumed that no oxygen is lost from the surface to the atmosphere (Stefánsson and Richards, 1964). Loss of oxygen occurs when the surface concentration is greater than saturation. Since AOU is calculated from the difference between saturation and measured concentration, difficulties arise when the water leaves the surface with an oxygen content different from the equilibrium solubility.

The nature of chemical changes caused by the admixture of river water to some extent depends upon the specific chemistry of local river waters. In general, however, both inorganic and organic constituents of sea water will be altered in the immediate vicinity of a river. Silicate is usually higher in river water than sea water and certain trace elements, such as thorium and cerium, may be several orders of magnitude more concentrated in river water (Goldberg, 1971). The natural heavy metal concentration of river waters has been difficult to determine in view of the widespread industrial use of metals such as copper, lead, and mercury. A sharp gradient in heavy metal concentration appears to exist in coastal waters, particularly in areas where there is some accountable source of heavy metal pollution (Abdullah *et al.*, 1972). The total dissolved organic constituents of large rivers entering the sea are generally higher than the surrounding sea water. While the types of organic compounds contributed by rivers are diverse and poorly investigated, particular biological interest has been placed on the vitamin content (e.g. Vitamin B_{12}; Burkholder and Burkholder, 1956) and the presence of chelating compounds. The latter may be closely associated with the humic acid content of river water; aside from their chelating properties, Prakash (1971) has indicated that humic acids may also enhance plant growth through certain specific physiological and biochemical reactions.

The presence of river water in the surrounding sea water can be detected by a change in salinity. The area in which this salinity change is most marked is referred to as the river 'plume'. The most obvious visual difference which is generally apparent in the plume waters is the coloration of the sea due to the sediment load. This may vary in extent depending on the volume of river flow and the river sediment load. The largest river flow in the world is the Amazon (mean flow: 175,000 m^3s^{-1}) while the rivers carrying the most sediment are the Hwang-Ho, entering into the Yellow Sea, and the Ganges/Brahmaputra entering into the Bay of Bengal. Both rivers carry over 10^9 tons of sediment per year. From these extremes to smaller, local rivers the effect of flow, sediment, and dissolved substances on the surrounding marine environment generally results in an increased biological production and a decrease in the diversity of organisms when compared with waters outside the plume. A useful summary of river inputs to ocean systems has been prepared by UNEP/UNESCO (1980).

In order to determine the fraction (F) of freshwater in any sample of plume water, one can use a simple expression such that

$$F = \left(1 - \frac{S}{S_0}\right) \tag{21}$$

where S is the salinity of the sample and S_0 is the salinity of the source water (i.e. deep water off-shore from the estuary). The expression can be used for a preliminary examination of departures from expected chemical constituents to be found in a sea water sample of salinity, S, in order to examine for abnormal (pollution) changes in estuarine chemistry. Thus if total phosphorus is treated as a conservative property, the sample S should have a total phosphorus content P_s; a measured value greater than P_s would indicate a riverine source of phosphorus.

2.2 PHYTOPLANKTON

Table 3 shows the percentages of protein, carbohydrate, and fat found in different species of phytoplankton harvested during the exponential phase of growth. Identification of major metabolites for the species reported was 100 ± 10% of the total organic matter. Considering the diversity of size and taxonomic groups involved, the cellular composition of these species appears to be remarkably similar and shows, in contrast to terrestrial plants, a general predominance of protein over other constituents. However, among the two representatives of the Dinophyceae and the one Myxophyceae, the ratio of protein to carbohydrate, or lipid, is lower than among the diatoms. Haug and Myklestrad (1973) attributed the low protein to carbohydrate ratio of dinoflagellates to carbohydrate material associated with cell walls. In their results the lipid fraction of both diatoms and dinoflagellates was less than 10% and the protein to carbohydrate ratio was suggested as a measure of the physiological state of the diatoms, as indicated by data in Table 10. Myklestrad (1974) has also shown that while different species of phytoplankton have similar composition during the logarithmic phase of growth as shown in Table 3, the composition changes appreciably between species during the stationary growth phase (i.e. after exhaustion of nutrients). Thus the ability of cells to make glucan was shown to range from 25 to 60% of the dry weight among six different species of diatom in the stationary growth phase.

The relationship between cell size and total organic content of phytoplankton has been studied extensively by Strathmann (1967 and references cited therein), who gives two regression equations for the carbon content of different cell volumes. The first of these for phytoplankton, other than diatoms, is given as:

$$\log C = 0{\cdot}866 \log V - 0{\cdot}460. \tag{22}$$

Since diatoms contain a vacuole, the quantity of carbon per unit volume is less than for other phytoplankton species and is given as:

$$\log C = 0{\cdot}758 \log V - 0{\cdot}422, \tag{23}$$

where C is the carbon per cell in picograms and V is the cell volume in cubic microns.

For energy budgets it is necessary to convert the carbon content of phytoplankton to their calorific value. Platt and Irwin (1973) determined the calorific value for mixed blooms of phytoplankton having different proportions of the major organic components and found a simple regression equation ($r^2 = 0{\cdot}91$) such that

TABLE 3. PHYTOPLANKTON COMPOSITION OF MAJOR METABOLITES (FROM PARSONS *et al.*, 1961)

Species	Approx. cell volume (μ^{3c})	Percentage composition, ash free dry wt. Protein[a]	Carbo-hydrate[b]	Fat[c]
Prasinophyceae				
Tetraselmis maculata	310	68	20	4
Chlorophycae				
Dunaliella salina	400	58	32	7
Bacillariophyceae				
Skeletonema costatum	1390	58	33	7
Chrysophycae				
Monochrysis lutheri	28	53	34	13
Dinophyceae				
Amphidinium carteri	740	36	39	23
Exuviaella sp.	780	35	42	17
Myxophyceae				
Agmenellum quadruplicatum	1.5	44	38	16

[a]Nitrogen × 6·25. [b]Anthrone reaction. [c]Saponifiable fraction only.

$$\text{cal/mg dry wt.} = 0{\cdot}632 + 0{\cdot}086\,(\%\text{C}). \quad (24)$$

From acid hydrolysates of different phytoplankton species. Cowey and Corner (1966) showed that most of the organic nitrogen could be accounted for in terms of amino acid nitrogen and that in confirmation of earlier work, the amino acid spectrum of different algae was very similar, as illustrated in Table 4. By comparison with protein of known nutritional value (casein), algal proteins show some differences in the proportion of amino acids but in general there is a well-balanced distribution which includes all the essential amino acids. Feeding experiments with rats, using the dinoflagellate, *Gonyaulax polyedra,* have further confirmed the dietary adequacy of phytoplankton protein (Patton *et al.*, 1967). In a very extensive examination of the amino acid spectra in thirty-one species of marine phytoplankton, Chau *et al.* (1967) and Chuecas and Riley (1969) agreed with the general distribution of amino acids found by earlier workers but noted also a number of additional characteristics. These included the occurrence of amino acid derivatives of butyric and adipic acids, the frequent occurrence of serine as the principal amino acid in diatoms, and differences between diatoms and other algae in the relative proportions of a number of other amino acids.

The principal lipids found in three species of marine diatoms are shown in Table 5 (Lee *et al.*, 1971). The particular characteristics of these data are the high concentration of phospholipid and the absence of wax esters. The latter are often the principal lipids in copepods which may feed upon diatoms. Fatty acids are included in the triglyceride, free fatty acid, and phospholipid fractions in Table 5 and their distribution in eight classes of phytoplankton is shown in Fig. 25 (from Ackman *et al.*, 1968). Early data on the fatty acid composition of plankton are difficult to interpret since most of the long-chain fatty acids were missed in analyses. However, in general marine phytoplankton fatty acids contain between 10 and 30% palmitic acid (C_{16}) as well as appreciable amounts of C_{18}, C_{20}, and C_{22} polyunsaturated acids.* Among the Bacillariophyceae the major saturated acids are C_{14} and C_{16} and the unsaturated acids are 16:1, 16:3 and 20:5. As

*C_{18} indicates a chain length of 18 C atoms; 16:1 indicates a chain length of 16 C atoms and 1 double bond; 20:5 ∞ 3 indicates 3 carbon atoms between the terminal methyl group and the middle of the double bond nearest the terminal methyl group.

TABLE 4. AMINO ACID COMPOSITION OF UNICELLULAR ALGAE: WHOLE CELL HYDROLYSATES EXPRESSED AS g AMINO ACID N/100 g TOTAL N (FROM COWEY AND CORNER, 1966)

	Organism					
Amino acid	*Skeletonema costatum*	*Phaeo-dactylum tricornutum*	*Monochrysis lutheri*	*Cricosphaera elongata*	*Chlorella ellipsoida*	Whole[a] casein
Aspartic acid	7·92	7·05	6·13	6·52	5·7	7·1
Threonine	4·22	3·71	3·59	3·23	3·6	4·9
Serine	5·63	4·55	4·29	3·23	4·4	6·3
Glutamic acid	7·15	7·99	6·32	5·75	7·9	22·4
Proline	4·05	4·71	3·18	3·23	3·3	10·6
Glycine	8·89	6·69	6·72	5·61	7·9	2·0
Alanine	7·03	7·08	8·25	5·74	6·2	3·2
Valine	5·25	5·75	4·96	3·99	3·8	7·2
Methionine	0·71	1·29	1·59	2·27	0·8	2·8
Iso-leucine	4·14	3·33	2·91	2·04	2·4	6·1
Leucine	5·94	5·70	6·67	5·47	5·1	9·2
Tyrosine	1·51	1·80	2·19	2·55	2·1	6·3
Phenylalanine	3·22	3·03	2·79	2·95	3·0	5·0
Lysine	7·93	8·32	8·16	5·75	7·1	8·2
Histidine	3·08	3·28	3·55	3·29	3·2	3·1
Arginine	10·67	10·79	11·34	10·62	14·4	4·1
Tryptophan	—	—	1·74	0·98	2·1	1·7
Cystine/2	—	1·12	1·07	—	0·8	0·3
Total	87·34	86·19	91·94	78·53	89·9	110·5

[a]From Gordon and Whittier. 1966: (—) not reported.

TABLE 5. LIPID COMPOSITION OF DIATOMS FROM LEE *et al.*, (1971)

	Lauderia borealis	*Skeletonema costatum* (% recovery by weight)	*Chaetoceros curvisetus*
Hydrocarbon	11	16	2
Wax ester	0	0	0
Triglyceride	16	14	12
Sterol	17	15	10
Free fatty acid	5	12	16
Phospholipid	51	57	50
Total % dry wt.	13·2	8·6	9·1

an exception to other classes of algae, C_{18} acids appear to be very minor constituents of the Bacillariophyceae. The Chrysophyceae differ from the Bacillariophyceae in containing more polyunsaturated acids, particularly C_{18} and C_{22}. The Dinophyceae appear to be characterized by 16:0 and polyunsaturated C_{20} and C_{22} acids. However, Chuecas and Riley (1969) showed that two species of Dinophyceae contained large amounts of 16:0 fatty acid, while in a third the principal fatty acid was 16:1. Four members of the Cryptophyceae examined by Chuecas and Riley (1969) were similar and the principal acids were 16:0, 16:1, 18:3, 18:4, 20:1 and 20:5. These results differ from the results found by Ackman *et al.* (1968), who reported a virtual absence of 16:1 and 20:1 acids while Chuecas and Riley (1969) described a relatively high concentration of the latter as a characteristic of the

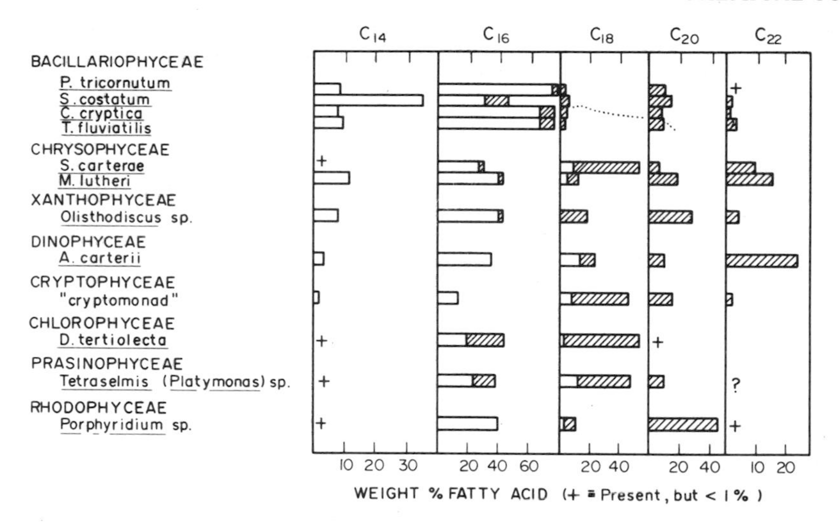

FIG. 25. Fatty acid spectrum of twelve marine phytoplankters, [open bar] satured or one double bond; [hatched bar] polyunsaturated acids (redrawn from Ackman *et al.*, 1968).

Cryptophyceae. The representatives of the Chlorophyceae and Prasinophyceae shown in Fig. 22 have very similar compositions; the principal components were 16:0 and polyunsaturated C_{16} and C_{18} acids. In *Porphyridium* sp. the principal fatty acids were palmitic acid and C_{20} polyunsaturated acids including the unusual occurrence of over 20% of 20:4. The taxonomic position of the Xanthophyceae, *Olisthodiscus* sp. is uncertain, but its fatty acid composition was similar to the two Chrysophyceae. Parker *et al.* (1967) investigated the fatty acid composition of eleven representatives of the Cyanophyceae and found a predominance of 16:0, 16:1, 18:1 and 18:2 acids; however, there were obvious differences between species including one species in which half of the fatty acids were C_{10}. Branch-chained fatty acids were absent from the Cyanophyceae but were a major component in marine bacteria.

Both diatoms and coccolithophores contain large amounts of the sterol 24-methylcholesta-5,22-dienol (diatomsterol). Other sterols, including cholesterol which is abundant in higher organisms, may also be present in small amounts in some phytoplankton. The identity of these minor constituents appears to vary both with species and growth conditions. For example, dinoflagellates appear to contain a unique 'dinosterol' while the quantity of 'diatomsterol' appears to be related to the growth activity of diatoms (e.g. Orcutt and Patterson, 1975; Alam *et al.*, 1967; Volkman *et al.*, 1981).

The carbohydrates of marine phytoplankton occur either as storage products of cellular metabolism or as constituents of cell wall material; small quantities of other carbohydrates may be associated with cellular metabolic processes. Using *Skeletonema costatum,* Handa (1969) demonstrated that approximately half of the algal carbohydrate was water soluble (Table 6) and that this consisted primarily of a glucose polymer, β-1,3-glucan; the water-insoluble residue consisted of mannose with lesser amounts of rhamnose, fucose, and xylose. The carbohydrate storage product of diatoms is very similar to laminarin, the reserve polysaccharide of the brown seaweeds. The diatom polysaccharide is known as chrysolaminarin which differs from laminarin in that although it is a glucose polymer it contains no mannitol which is a constituent of laminarin. Members of the Chrysophyceae may also contain chrysolaminarin (Beattie *et al.*, 1961, and references cited therein). The dinoflagellate, *Thecadinium inclinatum,* has been reported to contain an α-1,4-glucan (Vogel

TABLE 6. CARBOHYDRATE FRACTIONS OBTAINED FROM *Skeletonema costatum* (FROM HANDA, 1969)

Carbohydrate fraction	Yield (%)	Principal sugar
Water–extractable	45·6	91% glucose
Soluble in ethanol-acetone	(10·8)	Glucose and oligosaccharides
Insoluble in ethanol-acetone	(34·8)	β-1,3-glucan
Residue	54·4	55% mannose

and Meeuse, 1968); water-soluble extracts of the Rhodophyceae may also contain an α-glucan, floridean starch.

The carbohydrate constituents of cell wall material are complex but are usually known to include homopolysaccharides, such as cellulose, mannan, and xylan, or some heteropolysaccharides, such as glucuromannan found in the Bacillariophyceae (Ford and Percival, 1965). From these observations Handa and Tominaga (1969) concluded that the cell wall carbohydrate, represented by the residue in Table 6, was not a single compound but that the diverse monosaccharide constituents found in acid hydrolysates represented sugars from several polysaccharides.

Monosaccharides obtained from acid hydrolysates of whole cells are shown in Table 7 for algal species harvested during the exponential phase of growth. The amount of 'crude fibre' found in different species may be interpreted as indicating the proportion of the total carbohydrate employed as cell wall material. Glucose, galactose, and ribose were found in all species analysed; glucose was always the predominant sugar which agrees with

TABLE 7. MONOSACCHARIDE COMPOSITION OF WHOLE HYDROLYSATES OF UNICELLULAR ALGAE (FROM PARSONS *et al.*, 1961)

	Crude fibre (percentage of total carbo-hydrate)	Principal sugars (percentage dry weight of cells)										
		Glucose	Galactose	Mannose	Ribose	Xylose	Arabinose	Rhamnose	Fucose	Fructose	Hexosamine	Hexuronic acids
CHLOROPHYCEAE												
Dunaliella saline	9·8	17·2	11·8	–	1·7	–	–	–	–	–	–	+
PRASINOPHYCEAE												
Tetraselmis maculata	12·6	11·9	2·3	–	0·95	–	–	–	–	–	–	+
CHRYSOPHYCEAE												
Monochrysis lutheri	3·6	22·1	4·4	–	1·3	3·5	–	–	–	–	–	+
HAPTOPHYCEAE												
Syracosphaera carterae	1·7	9·2	7·1	–	1·5	0·8	1·9	–	–	–	–	+
BACILLARIOPHYCEAE												
Chaetoceros sp.	22·8	3·3	1·5	0·79	0·71	0·4	–	2·8	+	–	–	+
Skeletonema costatum	9·6	16·4	1·8	0·87	1·2	–	–	1·0	0·9	–	–	+
Coscinodiscus wailsii	29·0	2·1	0·4	0·41	+	–	–	0·7	0·5	–	–	+
Phaeodactylum tricornutum	2·5	10·7	2·7	3·7	0·72	0·7	–	1·5	–	–	–	+
DINOPHYCEAE												
Amphidinium carteri	2·0	19·0	8·4	–	0·9	–	–	+	–	–	–	–
Exuviaella sp.	37·0	26·8	8·3	–	+	+	+	+	–	–	–	–
MYXOPHYCEAE												
Agmenellum quadruplicatum	17·4	17·4	3·2	–	1·5	–	–	–	–	3·5	0·3	+

+Sugars detected but not estimated. –Sugars not detected.

more recent results for three species of Bacillariophyceae analysed by Handa and Yanagi (1969). However, these authors found approximately 30% of the monosaccharides to be mannose which is a considerably larger fraction than is shown for the Bacillariophyceae in Table 7. Further differences between the results in Table 7 and other authors may also occur. For example, Lewin *et al.* (1958) reported the presence of fucose in *Phaeodactylum tricornutum*. A note of caution is required in accepting data presented in Table 7 since these analyses were performed using paper chromatography and better methods are now available. Haug *et al.* (1973) used gas/liquid chromatography to examine the monosaccharides in extracts from natural phytoplankton. They found that an acid extractable fraction contained β-1,3-glucan but that an alkali extractable fraction contained a complex group of monosaccharides, similar to data given in Table 7.

A soluble polysaccharide excreted by the diatom, *Chaetaceros affinis*, has been shown by Myklestad *et al.* (1972) to consist of a sulphate ester; monosaccharides tentatively identified in an acid hydrolysate consisted primarily of rhamnose and fucose with lesser amounts of arabinose and galactose. Another extracellular polysaccharide from the diatom, *Nitzschia frustulum*, was also shown to be a sulphate ester but to contain different sugars, including large amounts of rhamnose and mannose with lesser amounts of six other sugars (Allan *et al.*, 1972).

Major metabolites and some minor cellular constituents in five species of unicellular algae have been reported by Ricketts (1966a). Among minor cellular constituents, nucleic acids ranged from 1 to 7%; DNA 0·31 to 0·86%; RNA 0·7 to 6·65%; phospholipids 1·0 to 4·0%; acid-soluble phosphorus 0·13 to 0·69%; phosphoprotein phosphorus 0·02 to 0·14%; total phosphorus 0·49 to 1·19% of the dry weight. Total phosphorus has been analysed by a number of authors and may range from about 0·5 to 3% of the dry weight depending to some extent on the supply of phosphate in the surrounding water. Holm-Hansen (1969b) measured the DNA content of ten species of marine phytoplankton and found that the amount ranged from 0·01 to 200 pg per cell. This range was strongly correlated with the amount of carbon per cell and the ratio of DNA to cell carbon was 1:100.

The ash content of some phytoplankton species may be very high due mainly to cell wall constituents, such as calcium carbonate in coccolithophores and silica in diatoms. Silicon is widely distributed in small amounts throughout all plants and is generally present in phytoplankton at levels between 0·1 and 1% dry weight; among diatoms the silica frustule may represent up to *ca.* 40% of the dry weight of the cells but this can vary by a factor of at least 5 depending on the availability of silicon in the surrounding water (Lewin, 1957; Vinogradov, 1953; Parsons *et al.*, 1961). In a series of papers Coombs *et al.* (1967a, b, c) have shown that silicon uptake and deposition in the cell wall of the diatom *Navicula pelliculosa* required ATP and that metabolic changes accompanied silicon starvation. These included a decrease in the net synthesis of proteins, carbohydrates, chlorophyll, and fucoxanthin, and an increase in the net synthesis of diadinoxanthin and lipid. Among coccolithophores, the calcium carbonate coccoliths originate from within the cell (Paasche, 1962) and may form a dense cover over the cell wall or at other times they may be totally absent (e.g. Braarud, 1963). The distribution of eighteen trace metals in fifteen species of phytoplankton has been studied by Riley and Roth (1971) and compared with earlier work by Vinogradova and Koval'skiy (1962); some of the former analyses are reported in Table 8. The principal conclusion reached from analysing species from the same class was that trace element distribution was not correlated with taxonomy and further that the concentration of individual elements may vary considerably between species. From further experiments Riley and Roth (1971) were able to show that in general the trace metal content of algae could be increased by increasing the concentration of metals in the medium in which the organisms were grown.

The photosynthetic pigments of unicellular algae are probably their most exhaustively analysed components. Early work has been summarized by Strain (1951) and Goodwin (1955); later contributions include reports by Goodwin (1957), Strain

TABLE 8. DISTRIBUTION OF TRACE ELEMENTS IN PHYTOPLANKTON (FROM RILEY AND ROTH, 1971)

	Chlorella salina[a]	*Asterionella japonica*[a]	*Phaeo-dactylum tricornutum*[a]	Sea plankton[b]
% ash, dry wt.	9·5	24·1	7·9	45·8
Element (ppm)				
Mn	48	54	73	118
Zn	301	115	325	282
Cu	25	105	110	36
Ag	4·6	10	6·6	3·3
Pb	<10	<20	46·3	900
Sn	<9	35	101	34
Ni	3·1	<12	6·2	48
Be	2·7	<4	3·8	<6
V	3·7	<5	<2	3·1
Al	118	1750	490	<5000
Ti	13·5	85	16	940
Ba	70·5	75	95	248
Cr	<3	5·5	4·4	7·5
Sr	31·4	<7	7·6	70

[a]Grown on synthetic medium. [b]Sample from Irish Sea.

(1958), Haxo and Fork (1959), OhEocha and Raftery (1959), Allen *et al.* (1960 and 1964), Dales (1960), Jeffrey (1961), Parsons (1961), Jeffrey and Allen (1964), Rickets (1967 a and b, 1970), Riley and Wilson (1967), and Riley and Segar (1969). Some of the more recent reports have included extensive quantitative data. However, apart from the use of quantitative data on chlorophyll *a* and total carotenoids in ecological studies, the chief importance of pigment data lies in the obvious differences in pigment composition of different taxonomic groups. A summary of these data is presented in Table 9, which shows that all marine algae contain chlorophyll *a* and some accessory pigments, including chlorophylls and carotenoids. Secondary chlorophylls appear to be absent in the Myxophyceae and possibly in the Xanthophyceae. The tentative presence of chlorophyll *c* in the latter is based on its presence in *Olisthodiscus,* which may have been wrongly placed in the Xanthophyceae (Riley and Wilson, 1967). Chlorophyll *c* occurs in the Bacillariophyceae, Dinophyceae, Chrysophyceae, Cryptophyceae, and Haptophyceae; chlorophyll *b* occurs in the Chlorophyceae and the Prasinophyceae. Beta carotene is the predominant carotenoid in all the unicellular algae except in the Crytophyceae where α-carotene predominates. The Prasinophyceae appear to be unique in containing several carotenes and possibly lycopene (Ricketts, 1970). Fucoxanthin is usually the predominant xanthophyll in the Bacillariophyceae, Chrysophyceae, Xanthophyceae, and Haptophyceae while peridinin characterizes the Dinophyceae. Other xanthophylls are more or less characteristic of the taxonomic classes. However, the Prasinophyceae contain a large number of xanthophylls and many of these have not been properly identified (e.g. see Ricketts, 1970). In addition, inconsistencies occur; for example, Riley and Wilson (1967) have reported on the occurrence of fucoxanthin as the principal xanthophyll in *Gymnodinium veneficum*—peridinin was absent from this species; astaxanthin, which is usually a characteristic pigment in some marine animals, was found by Jeffrey (1961) to occur in a Chlorophyceae; antheraxanthin was found by Parsons (1961) to be the principal xanthophyll in a Myxophyceae in which both myxoxanthin and myxoxanthophyll were apparently absent. The

TABLE 9. PIGMENTS OF MARINE PHYTOPLANKTON

Pigment		Bacillario-phyceae	Dino-phyceae	Chryso-phyceae	Chloro-phyceae	Myxo-phyceae	Xantho-phyceae	Crypto-phyceae	Prasino-phyceae	Hapto-phyceae
Chlorophyll	*a*	+++	+++	+++	+++	+++	+++	+++	+++	+++
	b				++				++	
	c	++	++	+			(+)	++		++
Carotene	α							+++	+	
	β	+++	+++	+++	+++	+++	+++		+++	+++
	γ								+	
Xanthophylls										
Fucoxanthin		+++	(+)	+++			+++			+++
Neofucoxanthin		++		++						++
Diadinoxanthin		++	++	++			++			++
Diatoxanthin		+					+			+
Dinoxanthin			+							
Peridinin			+++							
Neoperidinin			+							
Lutein					+++				++	
Zeaxanthin					+				+	
Flavoxanthin					+					
Violaxanthin				+	+				++	
Neoxanthin					+					
Alloxanthin (1 + 2)								+++		
Monodoxanthin								+		
Crocoxanthin								++		
Myxoxanthin						++				
Myxoxanthophyll						++				
Anthraxanthin						(+)				
Siphonaxanthin					+[a]				+	
Number of unidentified pigments			2		1		1	2	8	1
Phycobilins						++		++		

( ) Presence uncertain, +++ principal pigment, ++ generally reported as present, + sometimes reported as present.
[a]Only in the siphonous (non-planktonic) members.

reporting of chlorophyll *c* as a single compound in Table 9 may be open to question since Jeffrey (1969) has identified two spectrally different components in chlorophyll *c* preparations. In general, however, it remains the conviction of pigment chemists (e.g. Strain, 1966) that there is a relationship between the occurrence of particular pigment systems and the taxonomic classification of an organism. In this respect it has been suggested (Ricketts, 1966b) that the very diverse spectrum of pigments found in the Prasinophyceae may reflect their possible position at a branching point in protistan evolution.

From a consideration of photosynthetic mechanisms the diversity of pigments discussed above can be divided into three major pigment systems. These are known as

1. the chlorophyll a and *b* system,
2. the chlorophyll *a, c* and carotenoid system,
3. the chlorophyll *a* and phycobilin system.

Photosynthetic processes in the Chlorophyceae, Bacillariophyceae, and Myxophyceae are characteristic of the three systems, respectively. Jeffrey (1980) has provided a summary of algal pigments based on the possible phyletic relationship between different algal classes, including the separation of chlorophyll *c* into c_1 and c_2 components.

The chemical composition of unicellular algae

can be greatly affected by changes in environmental conditions; one of the earliest illustrations of this was with *Chlorella* (Spoehr and Milner, 1949). Subsequent investigations have been carried out by a number of authors and some differences between the results of different analysts may be attributed to a lack of clear definitions of the environmental conditions under which the algae were grown. Environmental conditions which may affect composition include nutrient levels, light, temperature, and salinity. An illustration of changes in the composition of a phytoplankton bloom is given in Table 10. The results are taken from Antia *et al.* (1963). In the experiments reported, a large volume of sea water was enclosed *in situ* within a plastic sphere and changes in the sea water chemistry and phytoplankton were followed over a period of 30 days. The principal phytoplankton species which grew during the experiment were a mixture of diatoms (*Thalassiosira rotula, T. aestivalis, Skeletonema costatum, Stephanopyxis turris, Chaetoceros pelagicus, Navicula* sp.) and a dinoflagellate (*Gyrodinium spirate*). From the results of the experiment it may be seen that the nitrogen content of the phytoplankton decreased when nitrate became exhausted; carbohydrate and lipid increased as organic nitrogen decreased. A summary of the overall changes in cell components, together with a summary of values obtained from the literature (Strickland, 1960) are presented in the bottom half of Table 10. Similar studies have been carried out on the time course change in chemical composition of two species of algae, the chrysophycean, *Monochrysis lutheri* (*Pavlova lutheri*), and the diatom, *Skeletonema costatum* (Sakshang and Holm-Hansen, 1977).

The ratios reported in Table 10 are useful in ecological studies involving trophic relationships where it is often impossible to make separate measurements of plant and detrital particulate material *in situ.* More recently, Banse (1974 a and b, 1977) has re-examined some aspects of the literature on the ratios of the principal elements and compounds in phytoplankton. From his studies it is clear that some ecological data have led to incorrect interpretations and one of these has been corrected in Table 10. However, a major point made by Banse (1977) is that all particulate material ratios are subject to the time frame of sample collection. Thus,

TABLE 10. CHANGES IN THE CHEMISTRY OF SEA WATER AND PHYTOPLANKTON DURING AN ALGAL BLOOM[a] (FROM ANTIA *et al.*, 1963)

				Phytoplankton ratios							
Day	Nitrate (μg at/l)	Phosphate (μg at/l)	Silicate (μg at/l)	Carbohydrate / Carbon	Lipid / Carbon	C[b] / N	C / P	Si / C	C / Chl *a*	Carotenoids / Chl *a*	N / P (atoms)
12	17	1·5	48	0·2	0·1	3	17	0·5	37	1·0	13
14	9	0·9	42	0·3	0·2	3	22	0·6	23	0·9	16
16	0	0·2	15	0·6	0·3	15	27	1·0	25	0·9	18
18	0·5	0·2	10	1·1	0·3	15	31	0·9	49	1·0	19
20	0·5	0·3	2	1·0	0·4	15	37	0·9	49	1·0	19
24	1·0	0·4	1	1·2	0·3	15	33	—	52	1·1	16
27	0·5	0·3	2	1·1	0·4	15	32	—	66	1·2	15
30	0	0·5	4	0·9	0·3	15	35	—	79	1·2	17
Vigorously growing phytoplankton with excess nitrate in the water				0·2	0·15	3	20	0·6	25	0·95	12-15
Unhealthy phytoplankton in nitrate-depleted water				1·1	0·35	15	33	1·0	60	1·15	15-18
Values suggested by Strickland (1960)				—	—	6 ±2	40 ±15	0·8	30	0·9	14·5

[a]Predominantly diatom. [b]Ratios suggested by Banse (1974a).

for example, an actively growing phytoplankton population in recently upwelled water may have a carbon:chlorophyll ratio of 30 but as the water ages the increase in detrital and microzooplankton particulate carbon may lead to an observed increase in the carbon:chlorophyll ratio before phytoplankton themselves have departed from a ratio of *ca.* 30, due to nutrient exhaustion. From these observations it is apparent that while the cellular composition of phytoplankton is heavily influenced by environmental conditions, the accurate collection of data on such changes may be difficult under field conditions.

Changes in the ratio of major elements in Table 10 have been discussed by Goldman *et al.* (1979). The authors point out that under oceanic conditions there is a remarkable constancy in the C:N:P ratio (i.e. the Redfield ratio of *ca.* 106:16:1 by atoms) even under conditions of apparent nutrient depletion (i.e. in stable tropical waters). Since this ratio is largely maintained under maximal growth rates it implies that in stable, nutrient impoverished waters there is a steady state condition of maximal growth rate, coupled to maximal grazing and nutrient regeneration. Furthermore, variations in the N:P ratio (× 1·5) which do occur in nature are observed to be less than those involving carbon, particularly the C:N ratio (× 5) compared to the C:P ratio (× 2). This observation further implies that the N:P ratio of phytoplankton is determined by the relative constancy of the N:P ratio of nutrients in the surrounding oceanic waters but that the C:N ratio is determined by physiological rather than aqueous chemistry. The lower variability in the C:P ratio compared to the C:N ratio was also taken to indicate that phosphorus is less likely to be limiting in oceanic environments (due to more rapid regeneration) as compared with nitrogen.

Studies carried out by Handa (1969) on chemical changes in a pure culture of *Skeletonema costatum* showed that the transfer of healthy cells from light to dark resulted in a 57% decrease in carbohydrate followed by a 28% decrease in protein and a 44% decrease in lipid. The principal glucose constituent utilized was the water-soluble fraction (β-1,3-glucan) with little change occurring in the quantity of cell wall constituents. Ackman *et al.* (1968) studied time-dependent changes in the fatty acid content of three species of phytoplankton. Indefinite results were obtained with one species but *Dunaliella tertiolecta* showed a decrease in C_{16} unsaturated acids and an increase in C_{18} unsaturated acids over a 2-week period. The effect of temperature on fatty acid synthesis in *Monochrysis lutheri* was also investigated; total polyunsaturated acids were found in twice the quantity at 10°C compared with 20°C cultures, and 18:1 ω 7 was reduced at lower temperatures from 2·5 to 0·8%. There is an apparent parallel in these results and results which have been obtained with terrestrial plants (Ackman *et al.*, 1968). Degens (1970) studied changes in protein and amino acids of three species of unicellular algae grown at different temperatures in nutrient-enriched water and following respiration in the dark for up to 19 days. The author concluded that there was a gain in protein content with increasing water temperature but that the total protein content decreased with the length of respiration. The conclusion reached by Degens (1970) that in the initial stages of respiration, protein is lost more rapidly than carbohydrate, is not borne out by Handa's results with respect to the rapid utilization of the water-soluble carbohydrate fraction during respiration.

The effect of light quality on the chemical composition of phytoplankton has been studied extensively by Wallen and Geen, both in the laboratory (1971 a and b) and under natural conditions (1971c). From laboratory studies it was apparent that the colour of light influenced the pathway of $^{14}CO_2$ metabolism; for example, blue or green light favoured protein synthesis while white light favoured carbohydrate synthesis. Minor constituents, such as DNA, RNA, and pigment concentrations, were also influenced by the colour of light. These results were largely borne out by field studies. For example, the relative activity of ^{14}C in the phytoplankton carbohydrate fraction decreased with depth while the ^{14}C incorporated into the protein fraction increased with depth. From these results the authors concluded that depth differences in light quality were independent of changes in light intensity in determining the chemical composition of phytoplankton *in situ*.

It has been suggested (Blumer *et al.*, 1971, and references cited therein) that differences in the hydrocarbon content of different species of marine phytoplankton are sufficiently great to be used as a means of identifying algal classes and possibly species. This suggestion may be particularly useful in determining the composition of mixed phytoplankton populations. In addition, since hydrocarbons are not readily destroyed in passing up the food chain, it appears that a useful technique may be available for diagnosing food preferences of zooplankton and higher organisms, *in situ.* The predominant hydrocarbon found in the Bacillariophyceae, Dinophyceae, Cryptophyceae, Haptophyceae, and Euglenophyceae was the unsaturated n-21:6 heneicosahexaene (abbr. HEH). This compound was absent in the Cyanophyceae, Rhodophyceae, Xanthophyceae, and Chlorophyceae, which were generally characterized by n-14 to n-17 hydrocarbons and in particular by either n-pentadecane or n-heptadecane. The hydrocarbons of phytoplankton appear to be very different from the hydrocarbons of zooplankton (mostly C_{19} and C_{20} isoprenoid alkanes and alkenes) and of mineral oils, in which olefins are absent but which contain branched chain, alicyclic, and aromatic compounds in relatively high proportions.

2.3 ZOOPLANKTON

Major chemical constituents of the zooplankton from three areas are shown in Table 11. While some disagreement between similar groups may be due to differences in analytical technique (especially drying), it is apparent that the data show general similarities. Thus the dry to wet weight ratio of copepods and euphausiids appears to be in the range 10–20% with possibly higher values found among amphipods; high dry weights in pteropods may be associated with inorganic shell material, which is generally reflected in an ash content of greater than 20%. On a dry weight basis, the ash content of 70% encountered among the ctenophores and tunicates is due to inorganic salts

TABLE 11. MAJOR CHEMICAL CONSTITUENTS OF ZOOPLANKTON

Organism or group	Dry wt. as % wet wt.	Carbon	Nitrogen	Hydrogen	Phosphorus	Ash	Comment
		(all constituents expressed as a % of dry wt.)					
Copepods[a]	11·6–16·3	35–48	8·2–11·2		0·7–0·8		Sargasso Sea plankton from Beers (1966)—range for each group
Euphausiids & mysids[a]	14·5–18·0	35–43	9·4–10·5		1·4–1·6		
Chaetognaths[a]	6·0–7·4	22–34	6·3–9·4		0·5–0·7		
Fish/fish larvae[a]	11·9–16·0	33–42	8·3–10·7		0·9–1·8		
Polychaetes[a]	5·7–27·0	16–44	4·4–11·2		0·4–1·8		
Siphonophores[a]	0·3–6·1	3–16	1·0–4·4		<0·1–0·2		
Hydromedusae[a]	0·3–10·1	5–10	1·4–6·2		0·1–0·4		
Pteropods[a]	22–32	21–25	2·7–4·2		0·2–0·4		
Copepods[b]	10·2–15·8	32–42	4·7–7·1		0·4–0·8	18–23	Continental shelf off New York, from Curl (1962)—range for each group
Euphausiids[b]	19·0–20·0	33–37	5·2–7·1		0·9–1·2	19–1·2	
Ctenophores[b]	4·7–5·0	(6·4)	0·2–1·1		0·1–0·2	70–75	
Pteropods[b]	3·5–19	26–28	2·2–5·0		0·3–0·6	24–64	
Tunicates[b]	4·0–4·1	7–11	0·3–1·5		0·1–0·3	71–77	
Copepods[a]	9·2–33·9	39–66	5·1–13·1	6·7–10·3		2–6	North Pacific from Omori (1969)—range for each group
Amphipods[a]	18·4–36·6	26–48	4·4–8·2	4·4–7·6		10–37	
Euphausiids[a]	20·2–21·3	39–47	10·0–10·7	6·7–7·6		8–9	
Chaetognaths[a]	11·6–14·1	44–48	10·7–11·1	7·2–7·6		4–5	
Pteropods[a]	25·0–36·4	17–29	1·5–6·0	1·1–3·8		29–43	

[a]Dried at 60°C. [b]Dried at 105°C.

contained in the gelatinous bodies of these animals. The carbon content of a wide variety of zooplankton ranges from *ca.* 30 to 40% of the dry weight, except when a large amount of ash is present; similarly nitrogen and phosphorus values generally lie in the range 5–10 and 0·5–1·0%, respectively. The C/N ratio for more than 80% of the samples analysed by Omori (1969) was in the range 3–8.

Platt *et al.* (1969) found a high degree of correlation ($r = 0·94$) between the carbon content and caloric equivalent of marine zooplankton. Using their own data and data from other reports, the best fitting equation was given as

$$y = -3370 + 136x - 0{\cdot}514x^2, \qquad (25)$$

where y is in cal/g dry wt. and x is the % organic matter in the dry plankton. If data on ash content were included a higher correlation ($r = 0·98$) was obtained. For the authors' area of study this was given as

$$\text{cal/g dry wt.} = 1351 + 106(\%\ \text{carbon}) - 21{\cdot}2\ (\%\ \text{ash}). \qquad (26)$$

Raymont *et al.* (1969a and references cited therein) have analysed a large number of zooplankton species (decapods, mysids, and euphausiids) for their biochemical constituents in terms of total protein, carbohydrate, and lipid. The authors concluded that in all their zooplankton analyses, protein was high (53–64% dry weight), carbohydrate was extremely low (1–3%), and lipid content was variable. Some areal and vertical differences were encountered; neritic mysids had a high protein (70–72%) and low lipid (13-14%) content, while some deep-sea and offshore species had a lipid content of greater than 20%. Ash content was highest in deep-sea decapods (15–24%) and lowest in neritic mysides (7–8%). The chitin content of all species of crustacea ranged from 3 to 8%. Differences reported above may in part be due to seasonal changes in composition for a single species. Thus Raymont *et al.* (1969 a, b) showed that for the euphausiid, *Meganyctiphanes norvegica*, seasonal changes in lipid content varied from 10 to 30%, and in protein content from 50 to 60%; the seasonal relationship between protein and ash content was the reciprocal of the lipid content. With *Euphausia superba* it was shown that the lipid content of animals increased seasonally from spring to summer and with body weight. An accompanying decrease in the body protein gave large animals a protein to lipid ratio of nearly 1 (Ferguson and Raymont, 1974).

The amino acid constituents of two species of euphausiid (from the south and north Pacific) and one species of copepod from the English Channel are compared with the amino acids from casein in Table 12. According to Cowey and Corner (1963) the proportions of amino acids found in *Calanus helgolandicus* are very similar to those found in phytoplankton and particulate matter on which the animals feed. There are some differences between the proportion of amino acids in the two euphausiids compared with *Calanus helgolandicus*, particularly with reference to the amounts of glycine and alanine and the sulfur amino acids, cystine and methionine. However, in general the amino acid spectrum of the three crustaceans appears well balanced and comparable to a nutritionally reliable animal protein, casein. Raymont *et al.* (1973) analysed the amino acid composition of carnivorous zooplankton including *Sagitta setosa* and *Pleurobrachia pileus*. They found that except for minor differences in the predominance of one or two amino acids, the spectrum from protein hydrolysates of whole animals was remarkably similar to other planktonic organisms assayed.

Jeffries (1969) has studied the free amino acid pool in a temperate zooplankton community. From these studies it was shown that free amino acids were generally higher in summer than in winter and that major changes occurred during periods of community stress, such as those accompanying environmental changes during spring and early winter. Differences were reflected in quantitative changes in the concentration of free amino acids per unit weight of tissue, while relative amino acid composition was the same both in summer and winter zooplankton. The most abundant free amino acids were taurine, proline, glycine, alanine, and arginine. The changes reported are consistent with metabolic changes in other animals where the free

TABLE 12. AMINO ACID COMPOSITION OF ZOOPLANKTON

Amino acid	Species (g amino acid/100 g protein)			
	Euphausia pacifica[a]	*Euphausia superba*[a]	*Calanus helgolandicus*[b]	Whole casein[c]
Alanine	5·61	5·46	8·1	3·2
Glycine	5·35	4·67	8·9	2·0
Valine	5·19	5·90	6·6	7·2
Leucine	7·83	7·70	7·8	9·2
Isoleucine	5·16	5·10	4·8	6·1
Proline	3·47	4·21	4·1	10·6
Phenylalanine	6·50	6·47	4·1	5·0
Tyrosine	4·15	4·06	1·5	6·3
Tryptophan	1·57	1·50	—	1·7
Serine	4·82	4·95	4·1	6·3
Threonine	4·83	4·70	4·0	4·9
Cystine/2	1·35	1·45	0·7	0·3
Methionine	3·25	3·03	1·2	2·8
Arginine	5·95	6·22	7·8	4·1
Histidine	2·22	2·30	1·8	3·1
Lysine	7·84	8·58	8·1	8·2
Aspartic acid	13·7	12·2	9·4	7·1
Glutamic acid	14·7	14·6	11·8	22·4
Glucosamine	2·04	3·45	—	—
Amine N	1·40	1·37	—	—
Taurine	—	—	2·1	—

[a]From Suyama *et al.* (1965). [b]From Cowey and Corner (1963), and converted from % amino N to % amino acid. [c]From Gordon and Whittier (1966).

amino acid pattern is reported to represent a picture of an organism's metabolic activities; the author suggests that these changes could be used to monitor subtle environmental changes.

The major lipid fractions in two copepods are shown in Table 13. The most characteristic feature of this table is the presence of large amounts of wax esters, which are not generally found among the phytoplankton. Lee *et al.* (1971) found that the lipid content, and spectrum of fatty acids and alcohols, was largely dependent on the copepods' diet. Thus there was a linear correlation between the amount of food fed and total lipid content of the zooplankton; further, the composition of wax esters was changed with increased diet. The most obvious feature of this change was an increase in the amount of C_{30} ester from 12 to 37% in well-fed animals to *ca.* 50% or greater in animals on a minimal diet. The chain length and degree of saturation of fatty acids and alcohols of the wax esters are shown in Table 14, together with the fatty acid components of the phospholipids. The authors found that the triglyceride and free fatty acids generally resembled the dietary fatty acids but that the long-chain alcohols did not correspond either in chain length or degree of saturation with the dietary fatty acids. From this it was concluded that the alcohols were derived from dietary fats by a variety of different

TABLE 13. COPEPOD LIPIDS (FROM LEE *et al.*, 1970)

Fraction	*Calanus helgolandicus*[a] (% recovery by weight)	*Gaussia princeps*[a]
Hydrocarbons	3	Trace
Wax esters	30	73
Triglycerides	4	9
Polar lipids[b]	17	17 (Polar lipids + Phospholipids)
Phospholipids[c]	45	
Total lipid as % dry weight	15	28·9

[a]Collected off La Jolla, California.
[b]Free acids, cholesterol, mono- and diglycerides.
[c]Lecithin and phosphatidyl ethanolamine.

TABLE 14. *Calanus helgolandicus* LONG-CHAIN ALCOHOL AND FATTY ACIDS (FROM LEE *et al.*, 1971)—% WEIGHT AS METHYL ESTERS

	Wax esters		
Component	Alcohols	Fatty acids	Phospholipid
14:0	1·5	11·8	1·5
16:0	13·0	29·0	39·8
16:1	1·0	13·2	tr
16:2	—	—	0·2
16:2	—	tr	tr
17:1	1·2	0·6	—
18:0	0·8	4·3	4·1
18:1	3·8	12·1	2·7
18:2	0·2	3·6	0·4
20:1	18·4	7·3	0·2
20:2	tr	5·1	0·3
20:3	—	0·6	—
20:4	40·9	7·6	0·3
20:5	—	2·0	13·3
22:3	8·4	—	—
22:6	8·1	2·1	36·5
24:3 to 7	1·7	—	—

metabolic pathways. The fatty acids of the phospholipid fraction (Table 14) were dissimilar to the fatty acids of the wax esters as indicated primarily by the large amounts of 20:5 and 22:6 fatty acids found in the former. These fatty acids were not present in large quantities in the diet fed to the copepods and it is assumed that they must have been synthesized by the animals. The observation from feeding experiments that the phospholipid fatty acids were not affected by changes in the amount or type of food ingested is compatible with their structural function in animal metabolism; in contrast the wax esters appear to be entirely storage products of copepod metabolism. While these results have dealt almost exclusively with one species of copepod found off California, similar results regarding the high wax ester content of copepods has been found by Yamada and Ota (1970) with respect to *Calanus plumchrus* in the western subarctic Pacific Ocean. Two North Atlantic tunicate filter feeders, *Pyrosoma* and *Salpa cylindrica,* have been shown (Culkin and Morris, 1970) to contain appreciable amounts of 14:0 (myristic), 16:0 (palmitic), 16:1 (palmitoleic), and 18:1 (oleic), as well as the polyunsaturated 20:5 and 22:6 acids which were also found in the phospholipid fraction of *Calanus helgolandicus.* Lewis (1969) has reported that minor fatty acids in the amphipod, *Apherusa glacialis,* included branched-chain acids and positional isomers; small amounts of branched-chain fatty acids were also found in the two tunicates by Culkin and Morris (1970) who suggested that they may be synthesized within the zooplankton. In a study of fatty acids in phytoplankton, zooplankton, and fish, Williams (1965) showed that the proportional transfer of fatty acids from phytoplankton to zooplankton, discussed above, was also maintained between zooplankton and fish. Thus throughout the three levels of production in the pelagic food chain there is a ubiquitous predominance of palmitic acid with generally large proportions of myristic, palmitoleic, and oleic acids in many organisms.

The presence of wax esters in marine crustaceans has been found to increase in animals from deep water and in near-surface animals taken from subarctic and arctic environments (Lee *et al.,* 1971; Morris, 1972; Gatten and Sargent, 1973). The reason for the elaboration of large amounts of wax esters is believed to be associated with efficient energy storage among animals living in nutritionally sparse environments (e.g. deep water or the arctic). In these areas it is suggested that a brief input of food has to serve for long periods of food shortage. However, in this sense it is questionable why animals should store waxes instead of fats. According to Benson and Lee (1975) waxes and fats are similar in their physical properties, both having similar densities, caloric value per unit volume and compressibility. They appear to differ, however, in their coefficient of thermal expansion. Thus wax expands more than fat when warmed and this may have some effect in increasing buoyancy during diel vertical migrations.

Jeffries (1970) has discussed the effect of seasonal changes in the environment on the fatty acid composition of zooplankton in Narragansett Bay, Rhode Island. The ratio of palmitoleic to palmitic declined during thermal warming from 2·0 to *ca.* 0·3, which reflected a dietary change from diatoms to dinoflagellates. The two acids were found to vary reciprocally throughout the year, as were oleic and steraidonic acids. The author suggests that such variations in chemical patterns could be useful in studying biological organization at the community level.

Mayzaud and Martin (1975) have measured the mineral composition of marine zooplankton and compared their results with the mineral content of phytoplankton. They found that zooplankton had less of an ability to concentrate metals than phytoplankton and that metal concentrations in zooplankton varied with species and metal. Iron, manganese, copper, and zinc were all concentrated in zooplankton compared with sea water but the concentration factor was approximately 10 times less for zooplankton compared with phytoplankton. Since concentrations of these elements are dependent on their local concentrations in sea water, Mayzaud and Martin (1975) suggested that the basis for a comparison of results between species as measured by different authors should be made in terms of a 'discrimination factor'. This term was defined as the ratio of the concentration factor of one element to the concentration factor of another element. From a comparison of discrimination factors it was shown that the ratios were different for each category of plankton covering three trophic levels from phytoplankton, herbivore (*Calanus*) to carnivore (*Sagitta*). The authors believed that these differences indicated that the elements were being chemically adsorbed in various amounts rather than being incorporated into tissue.

2.4 PARTICULATE AND DISSOLVED ORGANIC MATERIALS

In this section particulate and dissolved organic carbon (POC and DOC, respectively) will be considered as being part of the non-living components of sea water. The actual division between what is alive and what is dead among the microscopic particles in sea water is difficult to make, either by microscopic or chemical analyses. Obviously part of the POC in sea water is composed of phytoplankton and bacteria; however, from microscopic examination the presence of small organisms (especially bacteria) adhering to fragments of organic or inorganic 'detritus' is difficult to detect. A chemical differentiation is sometimes attempted based either on the chlorophyll *a* concentration or the amount of adenosine triphosphate (ATP). The former, times a factor which may range from *ca.* 25 to 250 depending on environmental conditions (e.g. see Table 10), is taken to represent the amount of phytoplankton carbon; the latter times a factor (*ca.* 285, Holm-Hansen, 1970) is taken to represent the amount of living organic carbon, since ATP is destroyed rapidly in dead organisms.

The non-living particulate debris of the sea is sometimes referred to as detritus and has been the subject of a number of reviews (e.g. Nishizawa, 1969; Riley, 1970). Microscopic observations of particulate material from sea water collected on a membrane filter, or in a settling cylinder, always show the presence of a large number of particles of irregular shapes and sizes. These particles are collectively referred to as detritus although Odum and de la Cruz (1963) have pointed out that scientists use a wide variety of terms to describe particulate material in aquatic environments. Thus the terms 'organic debris', 'suspended matter', 'particulate organic and inorganic material', 'leptopel', and (in lakes) 'tripton' may be assumed to refer to detritus unless specifically defined in another sense (e.g. the word leptopel in geology is sometimes used to describe a fine mud or clay). The term 'seston' should be used in referring to all particulate material including living organisms (plankton) and detritus. Detritus may be further qualified as being inorganic detritus and organic detritus, or bio-detritus; the latter terms imply that the detritus has originated from dead organisms. However, detritus has micro-organisms associated with it and might be more appropriately thought of as microcosm consisting of a particulate substrate in which bacteria and other micro-organisms may be embedded. From this concept the term 'organic aggregate' has often been used as a better description of detritus seen under the microscope. Kane (1967) differentiated between two types of aggregates found in the Ligurian Sea; a 'typical' aggregate was composed of a substrate to which various recognizable particles, such as bacteria and phytoplankton, adhered. These aggregates were usually brownish-yellow and were generally 30 and 50 μm in their longest dimension. 'Granular' aggregates were composed of small inorganic grey-black granules and were considered to be a possible early stage in the development of 'typical' aggregates. Nemoto and Ishikawa (1969) used various stains to identify the nature of detrital aggregates in the East China Sea. Acid fuchsin, which is a general stain for cytoplasm, was found to stain many particles at all depths; however, the ratio of particles stained with Millon's reagent (protein stain) to particles stained with acid fuchsin increased with depth. A few particles stained with α-naphthol (carbohydrates) and with Sudan black (fats). From similar studies carried out in the north Atlantic, Gordon (1970a) differentiated between detrital particles which appeared as aggregates, flakes and fragments; judging from reactions to histochemical stains he concluded that the aggregates were chiefly carbohydrate, the flakes were chiefly protein and the fragments were entirely carbohydrate.

There is a lack of agreement among analysts on the ratio of carbon to nitrogen in detrital material from below the euphotic zone. The C:N ratio for compounds such as urea is less than 1 while for most proteins it is about 4:5; the latter value may also characterize some bacteria (Porter, 1946). In phytoplankton the ratio of C:N is 3 to 6 (see Table 10) while sediments generally have a C:N ratio of 10 or more (e.g. Seki *et al.,* 1968; Degens, 1970). It would be logical to expect, therefore, that deep-water detritus would have a C:N ratio between that of phytoplankton and sediments. Some authors (e.g. Holm-Hansen *et al.,* 1966; Handa, 1968; Gordon, 1970b) have obtained C:N ratios of deep-water detritus of greater than 10; others (e.g. Parsons and Strickland, 1962a; Menzel and Ryther, 1974; Dal Pont and Newell, 1963) have obtained values of less than 5. Differences also exist in the C:N ratio of soluble organic material. For example, Duursma (1960) found values in deep water of *ca.* 3 while Holm-Hansen *et al.* (1966) have reported values of *ca.* 10. While these values may be real, it must also be considered that differences in analytical procedures may have led to different results. For example, if the C:N ratio of deep-water soluble organic material is assumed to be low (*ca.* 3) and if this fraction is partly included with the particulate material (e.g. through adsorption on $MgCO_3$-coated membrane filters, such as were used by Parsons and Strickland, 1962a) it would tend to decrease the apparent C:N ratio of the particulate material. Alternatively if bacteria form an appreciable fraction of the detrital biomass they will tend to cause a lower C:N ratio than if the detritus is derived entirely from phytoplankton. In experimental studies on the decomposition of phytoplankton in the absence of zooplankton, Otsuki and Hanya (1968) showed that the C:N ratio of dissolved material released by phytoplankton gradually decreased over a period of 200 days from *ca.* 10 to <3. Decomposition of organic nitrogen in phytodetritus was rapid in the first 30 days; release of carbon and nitrogen during this period amounted to 65–70% of the total. Approximately 10% was converted to dissolved substances and the rest remineralized as CO_2 and NH_4^+. The rapid decomposition of particulate protein has also been observed *in situ* by Garfield *et al.* (1979). From observations taken in the Peru upwelling system, the authors found particulate protein (PP) concentrations of >100 μg/l at or near the surface which decreased rapidly to <20 μg/l below 500 m. The rate of decay could be described by an exponential function similar to that for POC [eqn. (13)] such that

$$PP_z = PP_0 e^{kz} \qquad (26)$$

where z was the depth below the surface and k ranged from −0·0032 to −0·0046 m^{-1} at different stations.

Williams and Gordon (1970) studied the $^{13}C/^{12}C$ ratios (expressed relative to a standard) in dissolved and particulate organic material down to 4000 m in the Gulf of Mexico and off the coast of southern California. For deep water they found similar ratios (−21·2 to −24·4) regardless of location or season; further, the ratio for dissolved organic material was very similar to the ratio for particulate organic material (−22·0 to −24·3). It was also shown that these ratios corresponded most closely to the cellulose and 'lignin' fraction (residue after all other extractions) of phytoplankton (−22·4 and −23·1, respectively). These ratios were also similar to the ratio for organic material from sediments (−20·8 to −22·3) but were quite different from organic material derived from the Amazon River (−28·5 to −29·5). From these detailed studies the authors concluded that deep-water organic detritus is derived primarily from marine plankton and that the soluble and particulate fractions are similar in chemical composition. In another report Williams *et al.* (1969) determined the age of deep-water soluble organic materials being 3400 years old; this value agrees within an order of magnitude with values derived by Skopintsev (1966) from a theoretical approximation based on input and the rate of decomposition.

Various methods have been used to determine how much of the detrital particulate matter is biologically utilizable, particularly in the euphotic zone where it is associated with large populations of filter-feeding animals. The principal difficulty in this respect has been to separate or differentiate between the detrital particulate material and the living particulate material (mostly phytoplankton). In a study on the biological oxidation of organic detritus in which a correction was made for the amount of phytoplankton, Menzel and Goering (1966) found that in samples from 1 m in the north Atlantic, between 16 and 52% of the detritus was biodegradable; applying the same technique to samples taken from below the euphotic zone (200 to 1000 m) the authors could not detect any oxidation of organic material and concluded that the material was essentially all refractory. From feeding experiments, Paffenhöfer and Strickland (1970) found that *Calanus helgolandicus* would not feed directly off detritus. This is similar to the result obtained by Seki *et al.* (1968), who showed that *Artemia* would not feed directly off sedimented plant material but that they could feed off bacterial aggregates which were grown using the detritus as a substrate.

Gordon (1970b), using proteases, showed that approximately 20 to 25% of deep organic detritus was hydrolysable. Holm-Hansen and Booth (1966) determined the amount of ATP in particulate material at various depths off the coast of California. In deep-water samples, 500 to 1000 m, the authors concluded that 3% or less of the particulate material was alive; above 100 m, 14 to 79% was living while at intermediate depths approximately 6% was living with one exception of 27%. Thus it appears that the deep-water hydrolysable material found by Gordon (1970b) represented dead biodegradable detritus; this is in contrast with the results found by Menzel and Goering (1966) using a different technique.

Direct chemical analyses were made on particulate material taken from 400 m in the north Pacific (Parsons and Strickland, 1962a). From the discussion presented above, it may be concluded that most of the organic material in this fraction was dead. The analyses showed that the material was composed of protein and carbohydrate; the principal amino acids were glycine and alanine with glutamic acid, aspartic acid, lysine, arginine, serine, and proline also being detected. Degens (1970) has reported in greater detail on the amino acid composition of deep-water detritus in the Atlantic. Below 200 m there was an apparent increase in the proportions of serine, glycine, lysine, and arginine, and a decrease in alanine with depth down to 2500 m.

The carbohydrate fraction of deep-water detritus analysed by Parsons and Strickland (1962a) was 70% insoluble to treatment with weak acid and alkali (in contrast with 'crude fibre' values for healthy phytoplankton in Table 7). The principal sugars following total acid hydrolysis were glucose, galactose, mannose, arabinose, and xylose. The quantity of fat present was less than 1%. Glucosamine and hexuronic acids were not detected, indicating the lack of appreciable amounts of chitin from crustaceans or hexuronides from marine plants, respectively. However, Wheeler (1967) has reported on the presence of copepod carcasses between 2000 and 4000 m in the north Atlantic; as chitinous material these would contribute to the total organic detritus, but on the basis of their concentration per m^3, they would only account for between *ca.* 0·5 and 5% of the total organic carbon in deep water. Handa and Tominaga (1969) and Handa and Yanagi (1969) have carried out detailed carbohydrate analyses of particulate materials down to 700 m in the northwest Pacific. Their results show that the water-soluble carbohydrate fraction of phytoplankton disappeared between 50 and 300 m and that between 300 and 1000 m, only water-insoluble carbohydrates remained. Detrital material from these depths contained 50% less glucose than phytoplankton, but correspondingly higher proportions of galactose, mannose, xylose and, in contrast to Parsons and Strickland (1962a), appreciable quantities of glucuronic acid.

The inorganic fraction of particulate detritus may be a variable fraction of the total dry weight but generally it amounts to at least 70% (Wangersky, 1965). The exact chemical nature of the inorganic material has not been defined but calcium carbonate particles are known to occur in the open ocean (Wangersky and Gordon, 1965) and these may be intimately associated with the organic material in sea water (Chave, 1965 and 1970). In coastal areas inorganic particles are associated with clays and other minerals derived from the land; these may include appreciable amounts of silicon, iron, aluminium, and calcium (Armstrong and Atkins, 1950). More detailed data on the inorganic fraction of detritus can be derived from literature on marine sediments (e.g. Griffin *et al.*, 1968).

The quantity of chlorophyll *a* in particulate material decreases with depth so that below the euphotic zone most of the chlorophyll *a* has either disappeared or been converted to phaeophytin or phaeophorbide (Lorenzen, 1965). Saijo (1969) reported on chlorophyll pigments down to 4000 m in the northwest Pacific. The total concentration of chlorophyll *a* in waters below 400 m was from $<0·001$ to 0·003 μg/l; phaeo-pigments were often present, however, at 10 times the concentration of chlorophyll *a*. Currie (1962) showed earlier that digestion of phytoplankton by zooplankton resulted in the almost total conversion of chlorophyll *a* to phaeo-pigments; this has been used by some scientists as a measure of zooplankton grazing (e.g. Levi and Wyatt, 1971; Malone, 1971b). From studies in the western Indian Ocean, Yentsch (1965) showed that the ratio of chlorophyll *a* to phaeo-pigments decreased with light intensity. This was concluded to be a reversible reaction since the chlorophyll to phaeophytin ratio of dark-adapted phytoplankton could be increased when they were restored to the light. Jeffrey (1980) has shown changes in the absorption spectra of phytoplankton pigments down to 100 m in the ocean. Changes in the maximum absorption of chlorophyll *a* at 430 nm to phaeophytin/phaeophorbide at 410 nm were taken as indicative of detrital material. This change was most pronounced below 50 m. Carotenoids appear to be much more resistant to decomposition both within the water column (Yentsch and Ryther, 1959) or as a result of the digestive processes of animals. In the latter case, Fox *et al.* (1944) found an average content of 24 mg of carotenoid per 100 g dry weight of feces from mussels, but negligible quantities of chlorophyll. However, these authors did not measure chlorophyll *c* and while chlorophyll *a* is biologically chemically unstable, chlorophyll *c* has been found (Jeffrey, 1974) to persist in detrital material and senescent cells.

Considerable biological importance has recently been attached to the surface film of organic material in the sea (Garrett, 1965; Harvey, 1966). Using a specially constructed surface skimmer, Harvey

(1966) showed that the surface film contained large amounts of living nanoplankton, structural components of disintegrated organisms, surface active substances, chlorophyll, and carotenoid pigments. Garrett (1964, 1967) showed that major chemical components of this layer were fatty acids, fatty acid esters, alcohols and hydrocarbons. Taguchi and Nakajima (1971) measured the ratio of particulate carbon in the surface layer (150 μm thick) compared with subsurface (10–15 cm deep) samples. The ratio of surface to subsurface particulate carbon was generally 2 to 5 with a maximum concentration factor of 17·6 in coastal environments. Similar high ratios were found by Nishizawa (1971) in the open ocean waters of the equatorial Pacific. Both Sieburth (1971), and Taguchi and Nakajima (1971) have drawn attention to the concentrations of living organisms (neuston) associated with the detrital surface film. Bacterial counts were generally one or two orders of magnitude higher in the surface film than just below the surface; the species of phytoplankton and zooplankton were also quantitatively greater and to some extent qualitatively different, in the surface layers compared with subsurface samples. These observations have led to the conclusion that a neuston community exists at the sea surface which constitutes a micro-environment rich in a specific microflora and fauna. This is particularly illustrated by Sieburth *et al.* (1976) chemical analyses of material in the upper 150 μm layer of the sea. They found very high concentrations of organic matter (mean 2·9 g/l) with a relatively high (28%) fraction of carbohydrate. Most of the organisms found at the surface by these authors were bacteria and phagotrophic protists.

The presence of this neuston community may in part be due to an accumulation of material floating on the sea surface. However, it may also be formed by the peculiar properties of the air–water interface. Sutcliffe *et al.* (1963) showed that when air is bubbled through filtered sea water, organic particles are formed. It was further shown that these particles could support the growth of brine shrimp (Baylor and Sutcliffe, 1963). Carlucci and Williams (1965) showed that the action of bubbling sea water tended to concentrate bacteria at the surface. However, Menzel (1966) and Barber (1966) have questioned some of the experimental evidence for these findings. In the latter reference it is reported that neither bacteria nor bubbling alone cause a significant increase in the amount of particulate material in sea water containing organic materials with a molecular weight of less than 100,000.

While most observations on detritus have been made on samples collected in water bottles it is apparent that the form of detritus as observed under the microscope may bear little relation to its appearance *in situ.* From some of the earliest *in situ* observations made under the sea it was noted that much of the detritus was aggregated into clumps or long streaks which were easily disintegrated by the action of sampling bottles. Suzuki and Kato (1953) described its appearance *in situ* as ‘marine snow’. Aggregation of particulate materials has been studied by a number of authors (e.g. Sheldon *et al.*, 1967; Parsons and Seki, 1970) and it appears that bacteria are probably involved in this process, either through certain species which tend to form clumps or through the release of organic polymers which tend to attach to other particles. The effect of aggregation is to increase the size and therefore the sinking rate of the particles. However, as Riley (1970) and Kajihara (1971) have observed, the sinking of detrital aggregates does not obey Stoke’s law since there is a decrease in the specific gravity of particles with increased size due to aggregation. Direct measurements on the sinking rate of ‘marine snow’ carried out by Shanks and Trent (1980) indicate that these aggregates contribute to the organic flux (as discussed in Section 1.3.3) as much as do fecal pellets. The average sinking rate for ‘marine snow’ particles was 68 m/day which is much greater than for phytoplankton and comparable to fecal pellets. However, the authors estimated that a relatively small portion of the total daily flux (<5% POC, <22% PON) was accounted for as ‘marine snow’. Chemical analysis of ‘marine snow’ has been carried out by Alldredge (1979) who reported a significantly higher ash content and a significantly higher C:N ratio than in total particulate organic matter (POM). The average C:N ratio for marine snow was 12·5 while for total POM it was 7·7.

The division between particulate (POC) and dissolved (DOC) organic carbon is usually made on the arbitrary basis of filter pore size. Filters most frequently used to separate particulate from dissolved substances in sea water have a pore size of 1 μm, plus or minus 0·5 μm depending on individual choice. However, Sheldon and Sutcliffe (1967) showed that there was a considerable difference in the size of particles retained by commercially available filters compared with the advertised pore size. Sharp (1973) has shown that the arbitrary division of particulate and soluble organic matter in sea water is actually unjustified. The author showed that a continuous size distribution exists of microscopic, submicroscopic and colloidal particles over the size range 10^{-3} to 10^3 μm. In spite of this, it is still convenient to refer to POC and DOC as a functional division of organic matter in sea water. This is because from the point of view of microscopic examination of sea water, as well as for many filter feeding organisms, there is a definite distinction between what can and cannot either be seen or filtered. For analytical purposes this distinction is generally made by considering everything larger than *ca.* 1 μm diameter to be particulate.

From a summary of values in the literature, Williams (1975) concludes that near-surface (<100 m) dissolved organic carbon values range from 0·6 to 2·0 mg C/l while deep-water values range from 0·4 to 1·5 mg C/l depending to some extent both on geographic location and the method of analysis used. Seasonal changes in the amount of DOC have shown that the maximum generally occurs about a month after the phytoplankton bloom in temperate waters and that at least a 3-fold seasonal difference in total DOC may be observed.

The composition of the dissolved fraction in sea water can be divided into two principal categories of compounds on the basis of their biological activity. The first group of compounds are generally present at extremely low, or 'threshold', concentrations because at higher concentrations they are rapidly metabolized by heterotrophic activity. Compounds that belong to this group include monosaccharides, such as glucose, amino acids, acetate, and other simple compounds such as those found in the tricarboxylic acid cycle. Jannasch (1970) found that experimentally determined threshold concentrations of lactate, glycerol, and glucose for single species of bacteria were very high (*ca.* 1 mg/l). In the sea the total carbohydrate is generally found to be less than 1 mg/l (e.g. Handa, 1970) so that individual sugars must be present at much lower concentrations. This discrepancy is explained by Jannasch (1970) by assuming that in the sea, the many species of bacteria present with different uptake efficiencies will selectively reduce the concentration of any one readily metabolizable substrate to very low concentrations (e.g. Iturriaga and Zsolnay, 1981). Such concentrations of specific compounds have to be analysed by enzymatic methods; for glucose the concentration in sea water is generally found to be less than 10 μg glucose C/l with a few higher near-surface values of *ca.* 50 μg glucose C/l (e.g. Vaccaro *et al.,* 1968; Andrews and Williams, 1971). The natural concentration of hydrocarbons in sea water as measured from twenty-three water samples taken between Nova Scotia and the Gulf Stream was 4·9± 0·92 μg/l. Variation was associated with the chlorophyll concentration indicating that the hydrocarbons move from the biota and not from petroleum products (Zsolnay, 1977).

Oruga (1972a) has discussed the rate of decomposition of organic material from dead phytoplankton. He found that microbial decomposition rate of dead *Scenedesmus* was 0·02 to 0·05 day^{-1} during the first 15 days and 0·0004 day^{-1} for the next 100 days. Comparable decay rates for recently dead phytoplankton are found for dissolved organic carbon in surface waters. Decay rates for deep organic carbon are more comparable to the second value above (i.e. for phytoplankton after an initial brief period of rapid decomposition).

The second group of compounds found in the dissolved fraction are refractory materials which probably make up the bulk of the total organic carbon dissolved in sea water (e.g. Ogura, 1972b). From the ageing of DOC in deep water (*ca.* 3400 years, see Section 5.2) it is apparent that these compounds are not readily metabolized. While their exact nature has not been thoroughly investigated it is probable that these compounds include high molecular weight polymers which may include such compounds as lignins, humic acids, and proteins (e.g. see review by Williams, 1975).

Skopintsev (1971) has attempted to give an organic carbon budget for the oceans based on the average concentration of organic matter in the water column and the average rate of primary production in the hydrosphere. Assuming the average phytoplankton production to be 120 g $C/m^2/yr$ (neglecting products of exudation), then the total production was calculated as $3{\cdot}84 \times 10^{16}$ g C (assuming a surface area of $3{\cdot}2 \times 10^{14}$ m^2). From experimental data and oxidation rates in the sea it was assumed that 92% of this production was respired back by higher trophic levels, 5% settled out on the bottom and 3% was contributed to the more refractory pool of organic carbon in the ocean. The latter figure amounts to $11{\cdot}52 \times 10^{14}$ g C to which Skopintsev added $1{\cdot}8 \times 10^{14}$ g C from terrigenous origin. The total input of organic carbon was then $13{\cdot}3 \times 10^{14}$ g which represents approximately 0·1% of the total organic carbon in sea water (i.e. assuming a concentration of 1 g C/m^3 and a total ocean volume of $1{\cdot}3 \times 10^{18}$ m^3). Hence the residence time of refractory organic matter in the ocean is in the order of 10^3 years, as calculated earlier by Skopintsev (1966).

Considerable biological importance has recently been attached to the surface film of organic material in the sea (Garrett, 1965; Harvey, 1966). Using a specially constructed surface skimmer, Harvey (1966) showed that the surface film contained large amounts of living nanoplankton, structural components of disintegrated organisms, surface active substances, chlorophyll, and carotenoid pigments. Garrett (1964, 1967) showed that major chemical components of this layer were fatty acid esters, alcohols and hydrocarbons. Taguchi and Nakajima (1971) measured the ratio of particulate carbon in the surface layer (150 μm thick) compared with subsurface (10–15 cm deep) samples. The ratio of surface to subsurface particulate carbon was generally 2 to 5 with a maximum concentration factor of 17·6 in coastal environments. Similar high ratios were found by Nishizawa (1971) in the open ocean waters of the equatorial Pacific. Both Sieburth (1971), and Taguchi and Nakajima (1971) have drawn attention to the concentrations of living organisms (neuston) associated with the detrital surface film. Bacterial counts were generally one or two orders of magnitude higher in the surface film than just below the surface; the species of phytoplankton and zooplankton were also quantitatively greater and to some extent qualitatively different, in the surface layers compared with subsurface samples. These observations have led to the conclusion that a neuston community exists at the sea surface which constitutes a micro-environment rich in a specific microflora and fauna. This is particularly illustrated by Sieburth *et al.* (1976) chemical analyses of material in the upper 150 μm layer of the sea. They found very high concentrations of organic matter (mean 2·9 g/l) with a relatively high (28%) fraction of carbohydrate. Most of the organisms found at the surface by these authors were bacteria and phagotrophic protists.

The presence of this neuston community may in part be due to an accumulation of material floating on the sea surface. However, it may also be formed by the peculiar properties of the air–water interface. Sutcliffe *et al.* (1963) showed that when air is bubbled through filtered sea water, organic particles are formed. It was further shown that these particles could support the growth of brine shrimp (Baylor and Sutcliffe, 1963). Carlucci and Williams (1965) showed that the action of bubbling sea water tended to concentrate bacteria at the surface. However, Menzel (1966) and Barber (1966) have questioned some of the experimental evidence for these findings. In the latter reference it is reported that neither bacteria nor bubbling alone cause a significant increase in the amount of particulate material in sea water containing organic materials with a molecular weight of less than 100,000.

CHAPTER 3

THE PRIMARY FORMATION OF PARTICULATE MATERIALS

3.1 AUTOTROPHIC PROCESSES

In the ocean there are algae and some bacteria which can synthesize high-energy organic compounds from low-energy inorganic compounds such as water and carbon dioxide. The source of energy for these organisms is either light, or chemical energy derived from the oxidation of inorganic compounds; such organisms do not require organic materials as a source of energy. This life style is called 'autotrophy' and the organisms are called 'autotrophs'. When one considers the cycling of organic material in the oceans, autotrophic organisms are referred to as 'primary producers' because it is these organisms which are the only producers of original autochthonous organic material in the sea. The organic material produced by the primary producers is referred to as 'primary production' and primary production per unit time in a unit volume of water (or under a unit of area) is called 'primary productivity'. On the basis of differences in energy source for organic matter synthesis, autotrophy is divided into two different categories known as 'photosynthesis' (light energy) and 'chemosynthesis' (chemical energy).

3.1.1 Basic Photosynthetic Reactions

The fundamental relationship governing the photosynthetic process can be summarized in the following equation:

$$nCO_2 + 2nH_2A \xrightarrow{\text{light}} n(CH_2O) + 2nA + nH_2O \qquad (27)$$

where reduced compounds, such as H_2O, H_2, H_2S, $H_2S_2O_3$, and some organic compounds may be used as the H-donor in H_2A but only light is used as the energy source.

The whole photosynthetic process is not a single reaction expressed by eqn. (27). For example, the photosynthetic process can further be described by three different steps: (1) capturing light energy and transferring the energy into chemical forms, (2) further changing the chemical forms into another suitable chemical form for biochemical reactions (ATP and NADPH, see p. 85 for their complete names), and (3) fixing CO_2 using ATP and NADPH produced by the former steps. The first two steps are distinctive only for the photosynthetic organisms, but the third step is observed widely in all autotrophic organisms including chemolithotrophs (see Section 3.1.7).

Photo-autotrophs in the ocean include representatives of the algae as well as photosynthetic bacteria; both types of these organisms are usually widely distributed in the ocean. However, quantitatively the algae are the most important photo-autotrophs in the ocean, with a few exceptions to this generalization to be found in neritic regions.

Photosynthetic algae require H_2O as the H-donor, and eqn. (27) can be modified for algal photosynthesis as follows:

$$nCO_2 + 2nH_2O \xrightarrow{\text{light}} n(CH_2O) + nO_2 + nH_2O. \qquad (28)$$

This process requires energy of *ca.* 112 kcal per mole of carbohydrate formed. The energy is derived through the absorption of light by photosynthetic pigments, which absorb the light mainly in the visible region from 300 to 720 nm. Each photosynthetic pigment has a distinctive light (precisely, photon or quanta) absorption characteristic depending on their molecular structure. Each group of organisms contain chlorophyll *a* and several accessory* pigments in the thylakoid membranes in the chloroplasts in the cells (in procaryotic algae, such as the blue-green, there are no obvious intracellular organelle and thylakoid membranes are suspended directly in the cells). Thus the light-absorption patterns are different in each algal group depending on their pigment systems (see Table 9). Fig. 26 shows light-absorption spectra of some

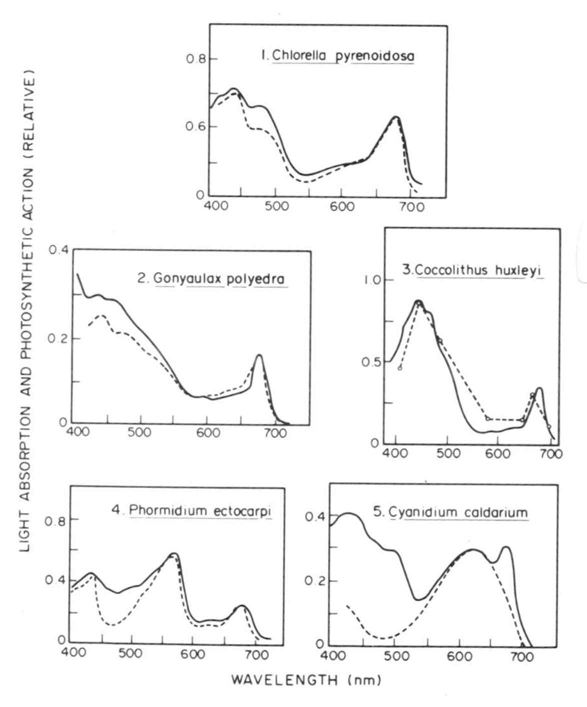

FIG. 26. Light absorption of intact cells (solid line) and photosynthetic action spectra (dotted line). 1, 2, 4, and 5 redrawn from Haxo (1960); 3, after Paasche (1966).

*'Accessory' may not be a suitable word because these pigments are the main light energy acceptors for the photochemical system II in the basic photosynthetic light reaction.

intact marine algal cells. Light of wavelengths shorter than 600 nm is mainly absorbed by chlorophyll *a* and accessory pigments. Above 600 nm, light for photosynthetic processes is only absorbed by chlorophyll. The latter absorption peak is normally observed at 680 nm; however, by improved techniques, such as derivative spectrophotometry and spectrofluorometry at very low temperatures, the existence of two other peaks, one at 670 and the other at 695 nm, were recognized (Kok and Hoch, 1961). These multiple peaks of chlorophyll *a* are distinctive for intact cells and are not observed in chlorophyll *a* extracted in organic solvents. From these findings it became clear that chlorophyll *a* has a complex stereo-structure within the thylakoid membranes. The function of these absorption peaks is not fully understood but two separate photochemical reactions are now recognized based on light absorption at the shorter wavelength (called chlorophyll *a* 670) and the longer wavelength (called chlorophyll *a* 680).

Energy absorbed at the longer wavelength (chlorophyll *a* 680) is used directly for photochemical reactions or emitted as fluorescence (fluorescence at 730 nm; Fl 730), but energy absorbed at the shorter wavelengths is transferred by the accessory pigments to chlorophyll *a* 670 before being used or emitted as fluorescence (fluorescence at 684 nm and 695 nm; Fl 684 and Fl 695, respectively). The energy accepted by both types of chlorophyll *a* is used for photochemical reactions in two photosynthetic systems, I and II (Fig. 27). These two photosynthetic systems are conjugated by a series of electron transfers, involving quinone and cytochrome. System I, which is primarily mediated through energy derived from chlorophyll *a* 680, is mainly involved in electron transfers. The photochemical reaction catalysed by pigment System II liberates oxygen from water and transfers electrons to plastoquinone (PQ in Fig. 27). Energy transferred from the two photochemical reactions is used for (1) the reduction of nicotinamide adenine di-nucleotide phosphate (NADP) and (2) photophosphorylation of adenosine diphosphate (ADP) into the high-energy compound, adenosine triphosphate (ATP). This series of reactions is carried out in the light and they

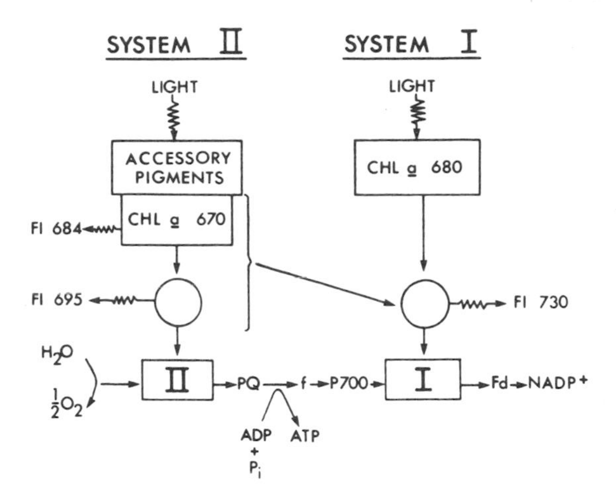

FIG. 27. Schematic presentation of photosynthetic system in algae, represented by two photochemical systems (I and II); PQ, plastiquinone; f, cytochrome; Fd, ferredoxin; NADP reductase; ADP and ATP, adenosine di- and tri-phosphates; P_i, inorganic phosphate (redrawn from Fujita, 1970).

are collectively referred to as the 'light reaction'. It is known that a herbicide, DCMU (3-(3,4-dichlorophenyl)-l,l dimcthyl urea), blocks electron transport just beyond the photosystem II trap so that the energy absorbed through the photosystem II is released as fluorescence and results in the increase of chlorophyll fluorescence. The ratio of the relative fluorescence yield before and after the addition of DCMU has been proposed as a measure of photosynthetic capacity of phytoplankton (Samuelsson and Öquist, 1977). The DCMU photosynthetic capacity gave a good indicator of growth rate of natural phytoplankton assemblage when an algal bloom was formed (Fukazawa *et al.*, 1980). The fluorescence yield of cells after DCMU addition has also been said to be 'a constant function of cellular chlorophyll *a*' (Slovacek and Hannan, 1977). However, Harris (1980) pointed out that great care must be exercised in the study of fluorescence because results obtained often depend on the methods and instruments used.

The reducing power of NADPH and the energy of ATP promote the reduction of CO_2 and produce carbohydrate as well as synthesizing proteins and fats. These reactions are carried out in the dark and are referred to collectively as the 'dark reaction'. Together the light and dark reactions are generally included in the term photosynthesis and this whole process takes place within the chloroplasts. The metabolic processes of the dark reaction were first demonstrated by Calvin and his colleagues, and thus it is also known as the 'Calvin–Benson cycle'. The details of the Calvin–Benson cycle can be found in the treatise of Calvin and Baasham (1962). The first product of CO_2 assimilation by the Calvin–Benson cycle is the three-carbon compound 3-phosphoglyceric acid (PGA) catalysed by the RuDP carboxylases. Recently another pathway was found for the dark reaction, the Hatch–Slack pathway (Hatch and Slack, 1970), in which the first products of CO_2 assimilation are not the three- but the four-carbon compound, oxaloacetic acid, catalysed by the PEP carboxylase. The carbon of the oxaloacetic acid is then decarboxylated. The CO_2 produced is fixed through by the Calvin–Benson cycle, and the three-carbon compound, phosphoglyceric acid, produced as a side product is then used again as the carbon carrier after phosphorylation. The plants which have the Hatch–Slack pathway are called the 'C4 plants' based on the carbon numbers of their first products of CO_2 assimilation. On the other hand, the plants which only have the Calvin–Benson cycle are called the 'C3 plants'. C4 plants are widely known in many species of terrestrial plants in which there is no obvious phylogenical relation recognized; even plants belonging to the same genus have different first products of photosynthesis. In algae, however, there is still only obscure information on which algal species belong to the C4 plants.

The basic photochemical reaction of photosynthesis is carried out by 'photons'; four photons (quantum energy) are required to produce one mole of ATP and NADPH. Since the quantum energy (E) is a function of wavelength,* the shorter wavelength photon has the greater energy. Considering a reaction requiring four photons (quantum energy), actual energy requirement will be much greater in the reaction with short-wavelength than with long-

*$E = h\nu = hC/\lambda$. h, Planck's constant, $6{\cdot}63 \times 10^{-34}$ joule/sec; ν, frequency; C, the speed of light in vacuum, 3×10^8 m sec^{-1}; λ, wavelength.

wavelength light. The efficiency of quantum energy transfer from pigments to photosynthetic systems is not always the same. The transfer efficiency can be estimated from the 'quantum yield, ϕ' which expresses how many moles of CO_2 are fixed (or O_2 are produced) by one photon of light absorbed by pigments ($\{CO_2\}/h\nu$). The energy equivalent of ϕ is called 'quantum efficiency, ϕ''' which has no dimension. Some examples of wavelength dependency of quantum yields are shown in Fig. 28 for different marine algae. The transfer efficiency of light to chlorophyll *a* is highest because the energy is transferred directly to the photosynthetic system; light is transferred via the accessory pigments with variable efficiencies. For example, the accessory pigments of diatoms, dinoflagellates, and coccolithophorids transfer light energy with an efficiency similar to that of chlorophyll *a*, but the accessory pigments of green and blue-green algae have relatively low transfer efficiencies. From a consideration of the photosynthetic efficiency of pigments, Fujita (1970) classified marine algal groups into (1) chlorophyll *a* and *b* type for green and euglenoid algae, (2) chlorophyll *a*, *c*, and carotenoid type for diatoms, dinoflagellates, and brown algae, and (3) chlorophyll *a* and phycobilin type for red and blue-green algae. The actual light utilization spectra of algae can be obtained by combining the light absorption of intact cells with the quantum yield; such a curve is called the 'action

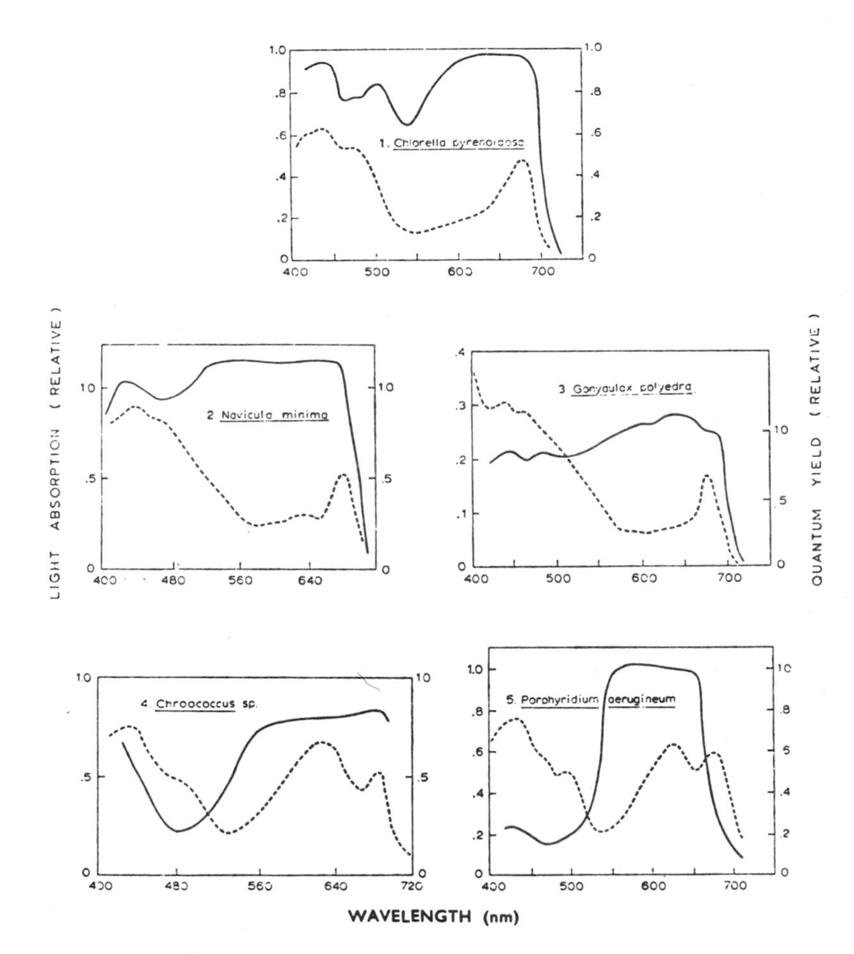

FIG. 28. Light absorption of intact algal cells (dotted line) and quantum yield (solid line). 1, 3, and 5 (light-absorption spectrum) redrawn from Haxo, 1960; 2 redrawn from Tanada, 1951 (quantum yield is absolute); 4 redrawn from Emerson and Lewis, 1942.

spectrum' (see Fig. 26). The action spectrum is considered to show the photosynthetic light utilization efficiency of a cell and it is one of the important characteristics of a species since it determines the ability of phytoplankton to adapt to different light regimes in the ocean. Halldal (1974, 1981) compared relative action spectra of algae in the three pigment types mentioned above (Fig. 29).

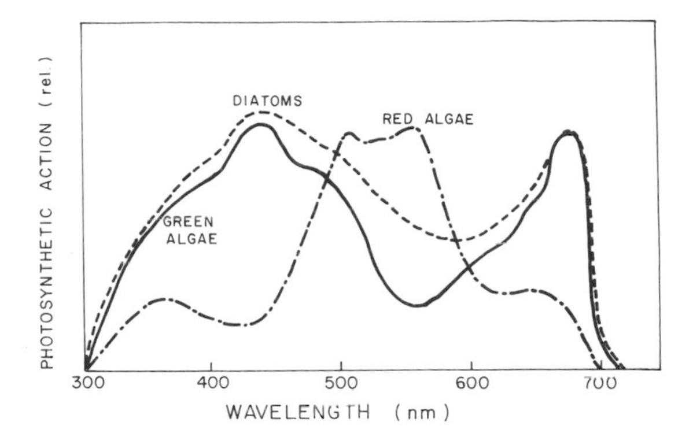

FIG. 29. Relative patterns of action spectra of three major algal groups in the sea. 'Diatom', 'red algae', and 'green algae, represent 'chlorophyll *a, c,* carotenoids', 'chlorophyll *a,* phycobilins', and 'chlorophyll *a, b*' types, respectively (from Halldal, 1980).

The chlorophyll *a, b* type shows active photosynthesis around 435 and 765 nm. On both sides of these peak wavelengths photosynthetic rate decreases, particularly sharp decline occurs above 675 nm. Low photosynthetic rate is noticed around 550 nm which is about one-third of the peak values. The chlorophyll *a, c* and carotenoid type has a similar action spectrum as the chlorophyll *a, b* type although the minimum in the green-yellow-orange light occurs around 580 nm and a lesser decrease to one-half of the peak value is noticed. The chlorophyll *a,* phycobilin type shows rather low photosynthetic rate in blue and red light regions and has three distinctive peaks between 500 and 560 nm due to phycobilins. The action spectrum is not stable, however, but sometimes changes depending on growth conditions such as differences in illumination, wavelength of light and nutrient concentration. Extreme examples of change are observed in the blue-green algae (Halldal, 1958; Fujita, 1970).

The final product of photosynthesis shown in eqns. (27) and (28) is carbohydrate, $n(CH_2O)$. This means that the photosynthetic quotient expressed as the ratio of evolved O_2 to absorbed CO_2 (PQ) is close to unity. The photosynthetic quotient becomes higher if other organic compounds are produced (e.g. *ca.* 1.25 for proteins and 1·43 for lipids). The photosynthetic quotient is not a stable property of a cell but it changes depending on the past history of the species and environmental conditions (e.g. Myers, 1953, observed changes from 1·04 to 2·50 in *Chlorella pyrenoidosa*).

Photosynthetic products are partly consumed by basic respiration occurring in mitochondria; this is generally the reverse reaction of photosynthesis. Respiration takes place both in the light and in the dark; however, it can usually only be detected experimentally as O_2 consumption or CO_2 production in the absence of light (i.e. with no disturbance either by photosynthetic O_2 production and CO_2 uptake). Measurement of phytoplankton respiration in the ocean, particularly open ocean, is practically impossible, firstly because of its very low activity and secondly because of the respiration of other micro-organisms (e.g. bacteria and zooplankton). Steemann-Nielsen and Hansen (1959b) proposed an indirect estimation approach of algal respiration using $^{14}CO_2$ uptake technique, which gives a reasonable estimate in coastal waters but a very high estimate for open ocean samples (Ryther, 1954).

According to many investigators (e.g. McAllister *et al.,* 1964; Humphrey, 1975), the basic dark respiration of algae obtained from many different species and growth conditions will be around 10% of maximum gross photosynthesis (P_{max}). When phytoflagellates are abundant in the water the ratio of respiration to P_{max} should be changed because high respiration rates of 35 to 60% P_{max} were observed in cultures of *Exuviaella cordata, Gymnodinium wulffi, Peridinium trochoideum,* and *Prorocentrum micans* (Moshkina, 1961). These high respiration rates were attributed to the motility of the flagellates.

Whether or not algal respiration is the same in the

light as in the dark has been a big question for a long time. By applying the non-radioactive mass spectrometric technique using $^{18}O_2$, it has become possible to have direct measurement of algal respiration in the light. From these experiments it has been found that respiration increased in the light due both to additional basic respiration (mitochondrial respiration) and photorespiration; the latter is defined as a light-dependent O_2 uptake and CO_2 release that occurs in photosynthetic cells. At the time of writing, available information on photorespiration has been mainly collected from cultured (freshwater) green and blue-green algae. Unfortunately none of these becomes dominant in the sea, except for a few special cases. However, it may be anticipated that photorespiration will occur commonly in many algae, including marine dominant groups (cf. Tolbert, 1974). The following description of the process is given even though the information was obtained entirely from experiments using green or blue-green algae. As shown schematically in Fig. 30, photorespiration is divided into two steps: (1) to produce glycolate (C_2 compound) from ribulose diphosphate (C_5 compound) which is one of the photosynthetic products, and (2) to oxidize glycolate into CO_2. The whole process is sometimes called the 'glycolate pathway': it occurs in chloroplasts, does not conserve energy as ATP, and does not utilize substrates involved in

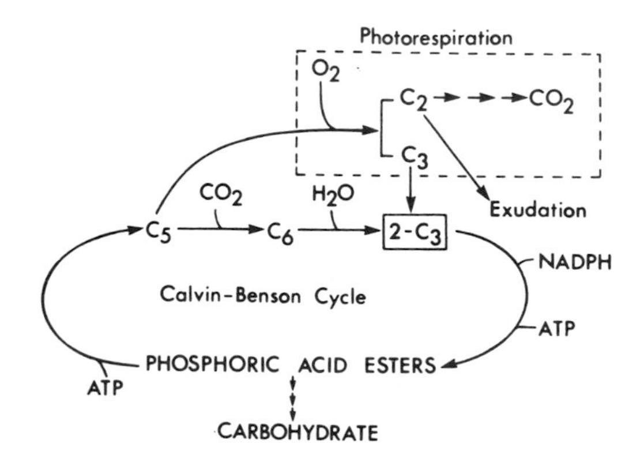

FIG. 30. Photosynthetic carbon fixation pathway of C_3 plant. The number of 'C' indicates the number of carbon atoms in the compound; C_2, phosphoglycolic acid; C_3, phosphoglyceric acid; C_5, ribulose diphosphate, etc (for details, see text).

the ordinary respiratory processes (i.e. tricarboxylic acid cycle). Some glycolate produced is also lost from cells by exudation. In most cases glycolate is the major photosynthetic product exudated by algae, but by no means the only compounds (Fogg, 1966, also see Section 3.2.1). Glycolate produced is also expected to be the major source for protein synthesis in many species of algae. The magnitude of photorespiration during photosynthesis in algae can vary, depending upon many parameters, from a few per cent to nearly 100% of total CO_2 fixation (Baasham and Kirk, 1962). High O_2, low CO_2, high light intensity, high temperature and high pH of the medium all favour increased photorespiration. Among these factors, all except for pH and light intensity (particularly near the surface) are rather unfavourable for photorespiration in the marine environment compared with the terrestrial environment. The light quality in the water column seems also unfavourable for photorespiration because photorespiration is sensitive to red and white light but insensitive to blue light which is the most predominant in the marine environment (cf. Lord *et al.,* 1970). Furthermore, algae do not lose much CO_2 during photorespiration because they refix the CO_2 by photosynthesis (Tolbert, 1974). However, all the factors mentioned above only reduce the possible magnitude of photorespiration among marine phytoplankton but it is not possible to eliminate its potential importance. It seems still too early to have quantitative verification of actual photorespiration of phytoplankton in natural aquatic environments.

3.1.2 Light Environment in the Sea

The basic aspects of light having biological importance are quantity and quality. Both of these characters of light in the sea fluctuate, sometimes in great magnitude, depending on time (i.e. daily, seasonally, and annually), space (different location on the earth, and depth), weather condition, angular distribution (including direction of maximum flux and degree of diffusion and polarization). The control of many of these aspects originates above or at the surface (such as the change in flux due to rising and setting of the sun), but for other aspects it originates within the water (such as changes in diffusion due to suspended matter and spectral changes due to selective absorption).

Light is qualitatively described by its spectral distribution depending on the difference in wavelength (units:* mμm, nm, and Ångstrom). Light of wavelength longer than 760 nm is, roughly speaking, considered to be infrared (IR), and that of wavelength shorter than about 300 nm is designated as ultraviolet (UV). The wavelength range between UV and IR is called visible (VS), which is the most important fraction for biological aspects such as photosynthesis and visual sense of organisms. Some organisms are known to be able to sense UV and IR light as well.

From the photosynthetic action spectra shown in Figs. 28 and 29, it is obvious that the light energy required by algal photosynthesis is restricted to the wavelengths between 300 and 720 nm. Total radiation at this wavelength is called 'photosynthetically available radiation' (PAR or PhAR). This definition does not prejudge the possible usefulness of this energy for phytoplankton because all photons, regardless of wavelength, within the defined spectral band must be counted. Considering the actual utilization of radiant energy through photosynthesis, Morel (1978) has proposed two additional definitions for radiation. One is the 'photosynthetically usable radiation' (PUR) as the fraction of radiant energy which is actually absorbed by the algae. PUR depends entirely on the pigment composition of the algal population as well as on the spectral composition of the submarine radiant energy. Only a fraction of PUR is really used in the photosynthetic process and the fraction is defined as the amount of radiant energy converted into and stored as chemical energy, in the form of organic matter; 'photosynthetically stored radiation' (PSR). There exists an obvious relation among these three quantities:

$$\text{PSR} < \text{PUR} < \text{PAR}.$$

*1 mμ = 1 nm = 10 Ångstrom (Å) = 10^{-9} m.

Quantitative assessment of light is done by intensity measurement, which is generally taken in three different ways (or in different units) depending on its use and also the instruments available; (1) illumination, (2) energy, and (3) quanta.

Illumination is the indicator of brightness which only includes the visible portion of the energy spectrum (400–750 nm) and is defined as flux per unit area. The basic unit for luminous intensity is the international candle (cd), which is the intensity of a standard light detected by a standard observer who has a distinctive sensitivity for different wavelength luminosity factors; highest sensitivity is at 555 nm and becomes less sensitive towards the longer and the shorter wavelengths, over the total spectrum, 400 to 750 nm. The luminous flux per unit area is defined for illumination as follows:

1 international candle on 1 m^2 = 1 lux (lx),
1 international candle on 1 ft^2 = 1 ft cd.
Then 1 ft cd $\simeq$ 10·76 lx.

Considering that most biological events occur in the visible light range, it is to be expected that the illumination unit is a convenient measure for biological problems in the sea. However, one should always remember that the basic luminous intensity (cd) is detected through a distinctive selective light filter (multiplied by a luminosity factor) as mentioned before, and therefore *precise* comparison of illumination units is only valid for the same type of light source used. If the spectral distributions in the visible range are known for the different light sources which are used (e.g. energy distribution actually measured or calculated based on the colour temperature of the light source), a suitable correction factor for the comparison of the illumination units can be estimated using the curve of the visual sensitivity characteristics.

Energy units for light-intensity measurement, on the other hand, apply for the whole energy spectrum including UV, VS, and IR light, and are expressed by the units such as watts·sec, erg, and gram-calories. All these energy units are interconvertible as follows:

$$1 \text{ g cal} = 4{\cdot}185 \times 10^7 \text{ ergs*} = 4{\cdot}185 \text{ watt·sec.}$$

The energy flux with the dimension of units time and unit area is the most suitable energy expression for the biological processes in the sea (e.g. g cal/cm^2/min or g cal/cm^2/day, etc.). The g cal/m^2 (i.e. without the time dimension) is defined as langley (ly):

$$1 \text{ g cal/cm}^2 = 1 \text{ ly.}$$

The measurement of light energy is entirely dependent on the quality of light (i.e. spectral differences). Thus when an energy unit is used for biological events, the light source (especially the energy spectrum distribution in the entire wavelength) should always be specified as well as additional information on experimental conditions, such as the thickness of water for incubation.

Only a rough conversion from an illumination unit to an energy unit (or vice versa) can be made; thus 1 lx = approx. 6×10^{-6} ly/min for the sunlight at the sea surface (Strickland, 1958), approx. 86×10^{-6} ly/min for a tungsten lamp (Hill and Whittingham, 1955), approx. 5×10^{-6} ly/min for white fluorescent lamps (Westlake, 1965).

The measurement of light in quanta measures the number of photons of a particular energy which is related to the wavelength:

$$\begin{aligned} 1 \text{ g cal} &= 2{\cdot}11 \times 10^{15} \times \text{Å quanta} \\ &= 3{\cdot}50 \times 10^{-9} \text{Å einsteins.} \end{aligned}$$

Then

$$1 \text{ einstein} = 6{\cdot}02 \times 10^{23} \text{ quanta} = \frac{2{\cdot}86 \times 10^8}{\text{Å}} \text{ g cal}$$

where Å is the wavelength in Ångstrom (10^{-10} m). This equation indicates that the quantum energy decreases with increasing wavelength. In the range of visible light (4000–7000 Å), the average energy of each photon will be estimated as follows (employing the average wavelength of visible light 5500 Å): For visible light:

$$1 \text{ einstein} = \frac{2{\cdot}86 \times 10^8}{5500} \text{ g cal} \simeq 52 \times 10^3 \text{ g cal.}$$

*10^7 ergs = 1 joule.

Considering that the basic photosynthetic reaction is carried out by photons, the quanta unit may be the most ideal measurement for photosynthetic events.

In the natural environment, solar radiation is the most important light source, but it can show order of magnitude changes temporally, spatially, and geographically. The subject has already been reviewed by a number of authors including Strickland (1958), Clarke (1965), and Jerlov (1968).

The intensity of solar radiation reaching the horizontal and perpendicular surface to the solar beams at the distance of $1{\cdot}5 \times 10^8$ km (average distance between the sun and the earth) lies mostly between 1·90 and 1·94 ly/min; on penetrating the atmosphere, energy is lost due to scattering and absorption, such as from water vapor, carbon dioxide, ozone, and dust. These substances are not equally distributed in the atmosphere (e.g. changes with cloud coverage). The angle of incoming solar light also affects the light penetrating into the sea surface. The lowering of the altitude of the sun makes the angle of incidence smaller and the path of light through the atmosphere longer with a corresponding reduction in intensity. Such changes in the sun's altitude result from differences in latitude as well as from changes in the seasons and in the time of day.

Theoretical solar radiation values, depending on latitude, season, and a clear sky, have been determined (e.g. Haltiner and Martin, 1957, Fig. 7–3); actual solar radiation data are collected by a number of countries using instruments known as pyranometers (formerly, pyrheliometers). In Canada radiation is measured at thirty-four different stations and published monthly by the Department of the Environment. A comparison of these values and theoretical values shows, for example, that at 50°N in June and December the maximum theoretical radiation is 769 and 131 ly/day, respectively, while actual measurements at Nanaimo, B.C. (*ca.* 50°N) during 1969 were maximum in July (622 ly/day) and minimum in December (44 ly/day).

The duration of the period of irradiation is also an important factor (latitudinal differences of irradiation period and incoming sun angle are tabulated in the Smithonian Meteorological Tables, List 1958). On the equator the day is always 12 hr long, but in the temperate region the day grows longer as spring progresses. This effect is accelerated at higher latitudes, and the day becomes 24 hr long during the summer in the polar region. Up to moderately high latitudes the increase in length of day during summer has more effect on the total amount of light received per day than the reduction in solar radiation due to the greater angle of irradiance. This indicates that a considerable amount of solar radiation energy reaches the sea surface even at high latitudes during a limited time of the year. This is quite important in understanding seasonal differences in biological parameters (see Section 1.3.4).

It is known that the energy spectrum of solar radiation will also change depending on cloud coverage in the sky (Fritz, 1957) and the height of the sun. On a cloudless day, approximately 50% of the total solar radiation reaching the sea surface is the photosynthetically active radiation (Strickland, 1958). Total energy in the wavelength shorter than 400 nm at the sea surface is less than 10% on a sunny day and disappeared quickly with depth in the water column. Consequently PAR in the aquatic environment should be practically concerned in the wavelength between 400 and 700 nm. Longer wavelength radiation energy such as >700 nm is also absorbed by water effectively, and the ratio of visible to total radiation increases with cloud and becomes to be almost 1 (100% of incoming radiation is PAR) at the overcast condition (Vollenweider, 1969).

Some solar energy is lost by true reflection, and by scattering from particles (including foam) at the sea surface. The actual value for surface varies considerably with conditions of the sea surface and sun angle; tables given by von Arx (1962) should be consulted for detailed values. On a fine day in summer, with a sun angle to the horizon of over 30°, the surface loss would be only a few per cent under conditions of complete calm; this value increases to 5–17% with light winds and to over 30% for moderate to strong winds. As the sun angle decreases to less than 10°, reflection increases rapidly to over 30%. For field work, a mean value of 15% for total surface losses may be used as an approximation for conditions under which it is usually possible to carry out photosynthetic measurements.

Light penetrating into the water is reduced by selective absorption and scattering due to the seawater itself, and dissolved and suspended matter in the water. The reduction of light in the water column can be expressed in terms of the (vertical) extinction coefficient* (k, also called attenuation coefficient, generally defined as m^{-1}):

$$I_d = I_0 \, e^{-kd} \tag{29}$$

where I_0 is the incoming light intensity, I_d the light intensity travelling a distance of (d). The value (k, m^{-1}) varies with the wavelength of light, being large for ultraviolet and infrared light; 0·033 at 425 nm, 0·018 at 475 nm, and 0·288 at 650 nm for the extinction coefficient of pure water (Jerlov, 1968, 1976). In clear oceanic waters blue-green light (maximum around 480 nm) can only penetrate to any appreciable depth (Fig. 31). However, under turbid conditions due to particulate material in the water, blue light is selectively scattered and the spectral peak of transmitted light is moved up towards the red (maximum at *ca.* 550 nm). Such a difference of attenuation characteristics of radia-

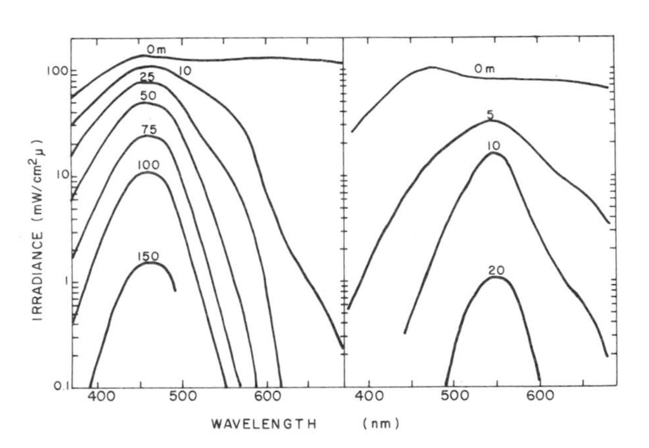

FIG. 31. Downward irradiance for solar elevation of 55 to 60° in Sargasso Sea (left) and Baltic Sea (right) (from Jerlov, 1976).

*Distinctive from the light-absorption coefficient.

tion may give some influence on the pigment composition of algal cells. A further discussion of these optical changes in ocean and coastal waters is given by Jerlov (1968, 1976).

The extinction coefficient in the water column (k) can be defined as follows:

$$k = k_w + k_p + k_s \tag{30}$$

where k_w, k_p, and k_s are possible diffusion and scattering of light energy due to water (w), suspended particles (p), and dissolved matter (s), respectively. The suspended particles include many different forms such as clay particles, organic detritus, and organisms varying in size from less than 1 μm to a few mm. Each of these extinction coefficients are highly dependent on wavelength as shown in Fig. 32. However, for the purpose of most biological events, the average extinction coefficient in the wavelength of PAR rather than the value at particular wavelengths is probably the most practical. The average extinction coefficient defined is distinguished as k' instead of k. According to Clarke and James (1939), the suspended particles reduce the light transmission by less than 7% (actual transmittance, m^{-1}) in the continental shelf water, but they reduce 30 to 70% in the coastal water. They also observed the same trend in the effects for the fraction of dissolved matter, although the actual reducing effects were much smaller than for the particulates. Then it can be expected that the optical properties of a given water column are significantly controlled by the quantity and quality of the total suspended particles in the sea. Details on this subject are also given in Section 3.1.5.

A summary of the conditions of light in the sea and its critical limiting effects is given in Fig. 33. Here the maximum depths are indicated for the growth of phytoplankton. Since the mean illumination necessary for the vision of aquatic animals is so very much smaller, the depth limitation is at a much greater level. Animals can respond to the difference between day and night at somewhat greater depths. Below this level no perceptible light from the surface penetrates and the water is completely dark

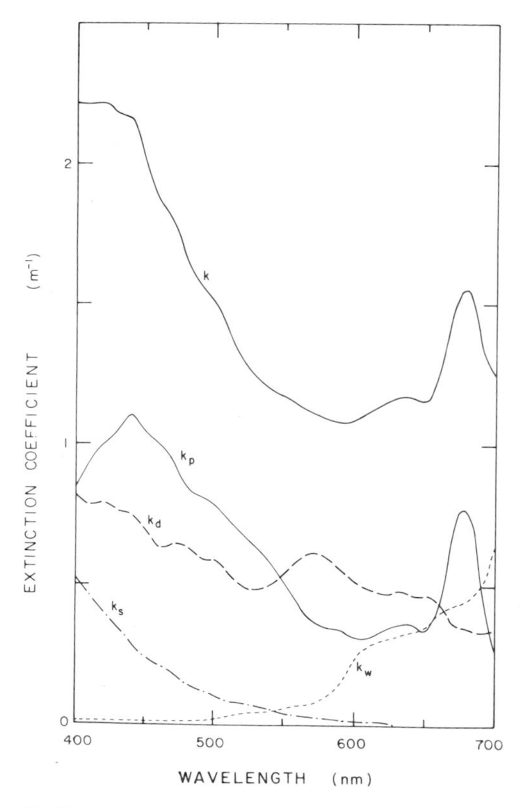

FIG. 32. Wavelength dependence of light extinction coefficients of phytoplankton (k_p), sea water (k_w), detritus (k_d), dissolved matter (k_s), and the total water (k) (from Kishino, 1980).

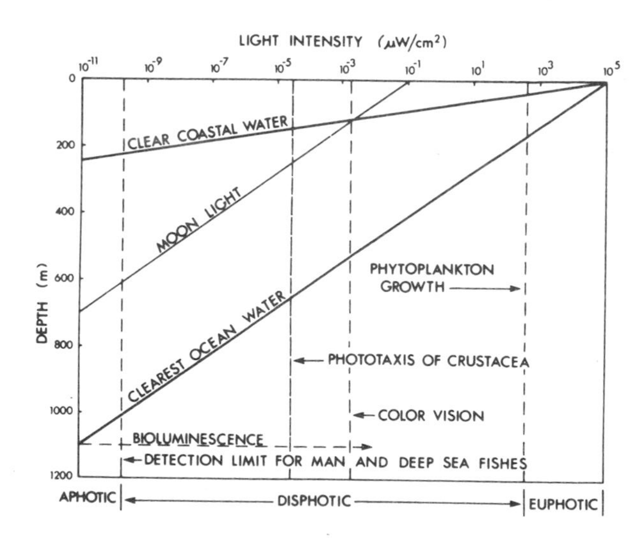

FIG. 33. Schematic diagram to show the penetration of sunlight into the clearest ocean water ($k = 0{\cdot}033$) and into clear coastal water ($k = 0{\cdot}15$) in relation to minimum intensity values for some biological light receptors (redrawn from Clarke and Denton, 1962).

except for light provided by luminescent organisms. Terms used in deep bodies of water for zones based on the light factor are as follows:

Euphotic zone: sufficient light for photosynthesis.

Disphotic zone: insufficient light for photosynthesis but sufficient light for animal responses.

Aphotic zone: no light of biological significance from the surface.

The actual depth limits of these zones differ widely according to transparency as indicated in Fig. 33, for the penetration of light in coastal waters.

For the measurement of solar radiation reaching the sea surface, one can use different types of sensors depending on what kind of radiation one wants to measure. For the total energy measurement, a pyranometer (almost no sensitivity difference at different wavelengths) is suitable. A photometer (detection range 400–700 nm with maximum sensitivity at 600 nm when a selenium cell is used) and quanta meter are suitable for illumination and quanta measurements, respectively. Recently several different types of instruments have become available in order to determine underwater radiation, both in energy and in quanta, at different wavelengths of PAR. One type has a series of interference filters and each of the filters allows radiation over a certain range of wavelength to penetrate. The other type has a grating mechanism which makes a measurement of radiation energy throughout the PAR range at a certain narrow wavelength range. Actual measurements can be done within a second to a few minutes which varies depending upon the type of system used. By using any one of these instruments, one can determine the radiation spectrum over a range of PAR and consider various efficiencies at each wavelength as well as for the total PAR energy.

There are two approaches (one direct and one indirect) to estimate the (average) extinction coefficient (k') in a given water column. The direct approach involves the actual measurement of light intensity at different depths using a suitable sensor (mentioned above, but embedded in a water-tight box); the extinction coefficient can then be calculated using eqn. (29). Since all energy meters are relatively insensitive, a photometer (responding to as low as 10^{-3} of sunlight using photovoltaic cells, and 10^{-12} of sunlight using photomultiplier tubes) or a quantum meter, is generally used for the measurement of underwater irradiance.

The indirect estimation can be done from Secchi disc reading (D_s, in meter), using an empirical relation which was originally proposed by Poole and Atkins (1929) as follows:

$$k' = \frac{1{\cdot}7}{D_s}. \qquad (31)$$

According to Idso and Gilbert (1974), the constant (1·7) always gave fairly close extinction coefficients when compared to optically measured coefficients in a wide range of water visibility (turbid to clear ocean water) covering D_s between 1·9 and 35 m.

Morel and Smith (1974) found that the ratio of (total quanta)/(total solar energy), Q:W ratio, in the range of 400 and 700 nm was fairly constant at the sea surface, $2{\cdot}77 \times 10^{18}$ quanta s^{-1} $watt^{-1}$ under various weather conditions with sun altitudes above 22°. By passing through the water column, the Q:W ratio decreased down to $2{\cdot}3 \times 10^{18}$ quanta s^{-1} $watt^{-1}$ depending upon the differences in the colour of water which were able to be determined by the absorption properties of water and of chlorophyll *a*. Blue water containing lesser amounts of chlorophyll *a*, $<0{\cdot}1$ mg m^{-3}, and the maximum attenuation of 440–485 nm had a smaller Q:W ratio, and green water containing more than 1 mg m^{-3} of chlorophyll *a* and 500–570 nm of the maximum attenuation showed a larger Q:W ratio. Blue-green water was in between those two waters. Even though the optical type varies in a wide range, the Q:W ratio varies by no more than ±10% and is $2{\cdot}5 \times 10^{18}$ quanta s^{-1} $watt^{-1}$. Once either the maximum attenuation wavelength or the average chlorophyll concentration is known in the photic zone, the Q:W ratio can be estimated within 5% variability. This result supports the fact that one can indirectly estimate total photosynthetic quanta (400–700 nm) from the measurement of total energy at those wavelengths or vice versa. This is quite convenient when one considers the possible experimental difficulty of building and accurately calibrating a quantum meter compared with an energy meter.

The solar beams penetrating the water column are expected to keep their incoming angle (refracted slightly at the air/sea water interface) at least near the surface. Since most organisms have three-dimensional shape, the actual light absorption by organisms might be closer to the absorption pattern given by 4π collectors (sphere, three dimensions) rather than 2π collectors (flat plate, two dimensions). For this reason 4π collectors as opposed to 2π collectors may be advisable for biological events in the aquatic environments.

3.1.3 Effects of Light on Photosynthesis

Light intensity strongly affects the rate of photosynthesis (usually expressed as mg C/mg Chl *a*/hr). Methods for the measurement of photosynthetic rate usually involve a measurement of either the carbon dioxide taken up or the oxygen produced per unit time. The ^{14}C-method first proposed by Steemann Nielsen (1952) is usually used for the measurement of CO_2 taken up since it is possible to detect very low photosynthetic rates with this method. This is particularly important in oceanic areas where photosynthetic rates are especially low and where an experimenter may only have a few hours in which to make measurements. Unfortunately some doubt may exist as to the interpretation of measurements made by this method as follows (cf. Steemann Nielsen and Willemoës, 1971): (1) $^{14}CO_2$ may be taken up into cells but not incorporated in organic compounds, (2) the rate of $^{14}CO_2$ assimilation may not be the same as that of $^{12}CO_2$, (3) some $^{14}CO_2$ fixed may be lost due to respiration which takes place simultaneously with photosynthesis, (4) some organic matter (^{14}C) may be lost by exudation during the experiment, and (5) some organic matter (^{14}C) may also be lost when the algae are filtered off from the experimental medium. Although the importance of none of these may be particularly pronounced, it is appropriate to introduce small corrections (cf. Steemann Nielsen, 1958). It is generally assumed, however, that in open waters, the ^{14}C-method measures the rate of increase in particulate carbon (e.g. Antia *et al.*, 1963). If a separate measurement is made of the amount of carbon lost through exudation, then the sum of the increase in particulate carbon, plus the loss of dissolved organic carbon, would be a measure of net photosynthesis. If the loss of dissolved organic carbon is small, then the ^{14}C-method will approximate net photosynthesis. The true result may be higher or lower according to the species of phytoplankton and environmental conditions. Production estimates based on oxygen evolution can be made using either an oxygen electrode or the Winkler titration technique, but both of these methods are usually an order of magnitude less sensitive than the ^{14}C-technique. Thus the oxygen technique is not particularly suitable for use in oceanic waters, but it may be used in some coastal areas, or in high-latitude oceanic waters having a high density of algae. Recently the Winkler technique has been improved by the use of colorimetric (Bryan *et al.*, 1976) or amperometric end points (Talling, 1973), which could increase the sensitivity a few to 10 times compared with the conventional Winkler titration. The merits and actual procedures for both methods are described in detail in the following two manuals: Strickland and Parsons (1972) and Vollenwieder (1969). In some nations, the use of radioisotope is highly restricted, particularly in natural environments because of its radioactive hazard. Hama *et al.* (1983) has established the use of non-radioactive isotope, ^{13}C, instead of ^{14}C, and has had great success to determine the photosynthetic rates in various trophic waters, including oligotrophic subtropical blue water. Since the natural abundance of ^{13}C is about 1·1%, which is not negligible compared with ^{14}C, the detection sensitivity of ^{13}C, using a mass-spectrometer requires a significant enrichment, reaching about 5% of the total inorganic carbon. One other advantage of using ^{13}C is that dual, or possible multiple, use of different isotopes can be simultaneously made in a given sample. Simultaneous uptake rates of ^{13}C and ^{15}N were determined by Slawyk *et al.* (1977).

The photosynthesis/light curve (or *P* vs. *I* curve)

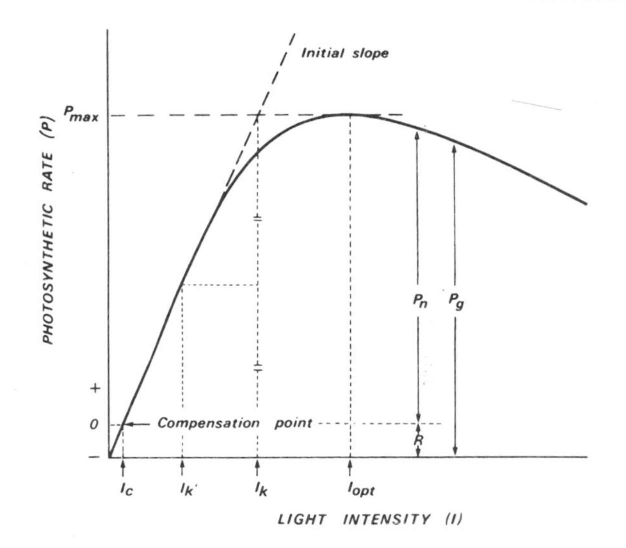

FIG. 34. Diagrammatic photosynthesis–light relationship. (P_{max}, photosynthetic maximum; I_c, light intensity at the compensation point; R, respiration; P_n, net photosynthesis; P_g, gross photosynthesis; I_{opt}, light intensity at P_{max}; I_k, see text).

shown in Fig. 34 is a convenient reflection of environmental effects on photosynthesis and can be used to diagnose certain properties of algal species, or natural samples of phytoplankton. Rabinowitch (1951), Steemann Nielsen and Jørgensen (1968), and Yentsch and Lee (1966) are recommended as references on the general interpretations of P vs. I curves. From Fig. 34 it is apparent that photosynthesis increases with increasing light intensity up to some asymptotic value, P_{max} where the system becomes light saturated. The two most important properties of the curve are the slope ($\Delta P/\Delta I$) and P_{max}, in which the latter is also called the 'assimilation number' (or index). The initial slope is a function of the light reaction (see Section 3.1.1) and is not usually affected by other factors. In plant physiology, the initial slope of the P vs. I curve has been defined as 'the quantum yield, ϕ' in which light intensity is expressed in the quantum unit. The quantum yield can then be expressed as the number of moles of oxygen evolved (or of carbon incorporated) per unit light intensity (in einsteins). It is now currently thought (e.g. Kok, 1960; Rabinowitch and Godvindjee, 1969) that the maximal value for the quantum yield is 0·125 mole oxygen evolved per einstein absorbed. In other words, the minimum quantum requirement, $1/\phi$, in order to produce 1 mole of oxygen is 8 einsteins. Considering 112 kcal per 1 mole of organic carbon and 570 kcal of 8 einsteins at 400 nm, the efficiency of photosynthesis can be calculated as 20%. Energy content of 8 einsteins at 700 nm is 326 kcal which makes an efficiency of 34%; 456 kcal of natural noon light (PAR) makes an efficiency of 25%.

The yield for carbon is likely to be slightly less than that for oxygen, as a result of nitrate reduction and the formation of some carbon compounds more reduced than carbohydrates [photosynthetic quotients (O_2/$-CO_2$) for natural populations suggested by Ryther (1965b) are 1·1 to 1·3]. The quantum yield mentioned is the approximate upper limit when the wavelength of illumination and the physiological state of the organisms are both optimized. In natural environments, several factors can be anticipated to lead to a lower value for the quantum yield. First, quantum yield and action spectra indicate that some phytoplankton accessory pigments sensitize photosynthesis less efficiently than does chlorophyll (Figs. 26 and 28); the average spectral yield is generally 10 to 20% less than that at wavelengths absorbed solely by chlorophyll. Secondly, the quantum yield changes depending upon the physiological state and growth stage of organisms, and their growth condition. Bannister (1974) has suggested a yield of 0·06 mol CO_2 $\cdot$einstein^{-1} absorbed for natural (healthy) populations. P_{max}, on the other hand, is a function of the dark reaction (see Section 3.1.1) provided no environmental factors are causing photosynthetic inhibition. If other environment factors are operative, such as low temperatures or nutrient limitation, P_{max} becomes a function of the environmental inhibitor. As a combination of the initial slope and P_{max}, Talling (1975b) proposed I_k which is the light intensity at the intersection of an extension from the initial slope and P_{max} (Fig. 34).

Recently several authors have tried to estimate the *in situ* quantum efficiency of natural phytoplankton populations based on the quantity of CO_2 fixation measured *in situ* and the quanta actually absorbed by the phytoplankton populations (photosynthetically usable radiation, PUR) (Tyler, 1975; Dubinsky and Berman, 1976; Morel, 1978). PUR is indirectly determined by multiplying PAR with the average light extinction due to chlorophyll, k'_p: PUR (d) = k'_p PAR (d), in which PUR (d) and PAR (d) are values at a given water depth, d. Chlorophyll-related light extinction, k'_p, can be expressed with the average extinction coefficient of a unit amount of chlorophyll *a*, k'_2 (mg Chl a/m^2)$^{-1}$, and the total concentration of chlorophyll *a*, $\bar{C}$ (mg Chl a m^{-2}), as follows: $k'_p = k'_2\,\bar{C}$. As shown in Table 20 and in Dubinsky and Berman (1979, Table 3), k'_2 varies almost 4 times. Morel (1978) has proposed to use a value of 0·0125 mg chla^{-1} m^2 for k'_2 tentatively. Since k'_2 gives a significant effect on the estimation of PUR as pointed out by Kishino (1981), a further evaluation of k'_2 is urgently required. Considering the above mentioned entire process to be treated in the quantum unit, the *in situ* quantum yield, ϕ, is then estimated as

$$\phi = P/PUR \text{ (mg C einstein}^{-1}\text{ absorbed).}$$

The efficiency obtained represents physiological photosynthetic responses of phytoplankton under natural conditions which include both internal and external effects mentioned earlier, and is also time averaged; this is because efficiency changes with light intensity which shows a marked daily variation. The efficiencies in Dubinsky and Berman's (1976) observations varied from 0·001 to 0·07 mole CO_2 fixed einstein^{-1} (= molecules CO_2 fixed quantum^{-1}) absorbed depending on the light intensity (Table 15), and showed an increase at greater depth where lesser amount of radiation penetrated. Extensive *in situ* quantum efficiencies have also shown by Morel (1978) in the subtropical waters.

Figure 34 also shows the difference between total photosynthesis (or gross photosynthesis, P_g) and net photosynthesis (P_n), which is the fraction of P_g minus respiration (R). An approximate P_n can be estimated directly by the oxygen method because oxygen is consumed by respiration at the same time as it is produced through photosynthesis in the light. However, care should be taken in interpreting the oxygen method because photorespiration (see Section 3.1.1) sometimes causes an over-estimation of respiration in the light. When P_g equals R, P_n is zero and the photosynthetic system is at the 'compensation point'. The light intensity at the compensation point is called the 'compensation light intensity', I_c, and photosynthetic micro-organisms held at the compensation light intensity should theoretically show no growth. In nature, phytoplankton are subjected to continually varied light conditions and, except during the summer in extreme latitudes, to no light at all during the night. Thus at noon the amount of light reaching a cell may still be above the compensation light intensity; however, over a period of 24 hr there may be a mean light level below which the algae will decrease in weight. The compensation point is therefore best expressed on a 24-hr basis and is usually determined as the 24-hr mean light intensity in ly/day or ly/min

TABLE 15. PHOTOSYNTHETICALLY AVAILABLE RADIATION (PAR), QUANTUM YIELD (ϕ), ENERGY EQUIVALENT QUANTUM YIELD (ϕ') AND LIGHT UTILIZATION EFFICENCY (ϵ) AT VARIOUS DEPTHS IN LAKE KINNERET (DUBINSKY AND BERMAN, 1976)

Depth (m)	PAR (cal m^{-2} h^{-1})	ϕ (moles C einstein^{-1} absorbed)	ϕ' (%)	ϵ (%)
0	$3{\cdot}63 \times 10^5$	0·00144	0·32	
1	$2{\cdot}3 \times 10^5$	0·00303	0·67	0·46
3	$8{\cdot}48 \times 10^4$	0·00581	1·28	0·5
5	$3{\cdot}22 \times 10^4$	0·0145	3·2	0·48
10	4×10^3	0·0317	6·97	0·42
16·5	$1{\cdot}2 \times 10^2$	0·07	15·34	0·54

Mean value of ϵ from 0 to 16·5 m = 0·48 (SD±0·046).

(Strickland, 1958). From studies on natural phytoplankton populations, not excessively contaminated with other microflora or fauna, an average 24-hr compensation light intensity appears to be in the range 0·002 to 0·009 ly/min or some 3 to 13 ly/day in temperate seas (e.g. Strickland, 1958; McAllister *et al.,* 1964; Hobson, 1966). The value is dependent on the rate of respiration which has been found to be approximately 10% of P_{max} (see Section 3.1.1). However, the basic assumption that respiration is constant in the light and dark is difficult to check and may not be valid for cells which spend a long time below the euphotic zone. For example, adaptation to photosynthesis at very low light intensities (*ca.* 0·00014 ly/min) has been reported for phytoplankton growing under ice (Wright, 1964). Since respiration is temperature dependent, the value for I_c may be expected to be larger at higher temperatures. In field work the compensation depth can be approximated from the depth of 1% of the surface radiation; this value can either be determined using a bathyphotometer or may be approximated as three times the depth of the Secchi disc visibility (Strickland, 1958; see also Section 3.1.4). Compensation depths may vary from a few meters in turbid coastal waters to over 150 m in some tropical seas.

Respiration rate of phytoplankton has sometimes been expressed in terms of biomass units (e.g. O_2/Chl *a*/time) but these values usually vary over a great range depending on species, past growth history, and physiological status of a cell (e.g. Ryther and Guillard, 1962). Another way for expressing algal respiration rate has been made on the simple assumption that respiration rate is proportional to the maximum photosynthetic rate, P_{max}, as follows:

$$R = rP_{max} \qquad (32)$$

where r is the proportionality constant (dimensionless), called the 'loss factor', which characterizes the respiration economy of species or populations. Since P_{max} depends on environmental conditions, growth history and physiological status of algae, no correction is generally needed in order to apply this relation, providing P_{max} is measured under the same conditions for estimating respiration. In most cases eqn. (32) is used with an assumption that respiration rate changes independently of light. According to literature (see Section 3.1.1), the loss rate (r) for phytoplankton is $\sim$ 0·1.

However, the assumption mentioned above may not be entirely true because it has been known that respiration rate increases in the light, in proportion to the amount of light. Tooming (1970) has extended eqn. (32) to account for additional respiration loss occurring under light by assuming that the additional carbon loss in the light was proportional to gross primary productivity which changes dependently of light intensity:

$$R = r_1 P_{max} + r_2 P_g \qquad (33)$$

where both r_1 and r_2 are (dimensionless) proportionality constants. Basic respiration constant, r_1, is the factor for respiration loss due to maintaining the photosynthetic and non-photosynthetic functions of algae, and it is the same as the 'loss factor' in eqn. (32). Light respiration constant, r_2, on the other hand, is the factor for respiration loss due to organic synthesis and growth which generally occurs under the light. Respiration loss due to 'photorespiration' will also be included in the factor of 'r_2'. Unfortunately at the time of writing, no datum is available for estimating 'r_2'.

If micro-organisms are exposed to a strong light above the point at which they are light saturated, the P vs. I curve may show a depression in photosynthetic rate. This phenomenon is named ('high) light inhibition' or 'photoinhibition'. Photoinhibition is not generally observed over short periods of time (e.g. 10 min) but may result from longer exposures and may also increase in magnitude with time (Takahashi *et al.,* 1971).

It appears from early work (e.g. Ryther, 1956) that there may exist a broad division between taxonomic groups with respect to their P vs. I curves. Ryther (1956) reported P vs. I curves for fifteen species of marine algae representing three taxonomic groups (green algae, diatoms, and dinoflagellates). Measurements were made under sunlight on a clear day at 20°C and the results showed a remarkable similarity in the photosynthetic behaviour of organisms within each taxonomic group, but a rather striking difference

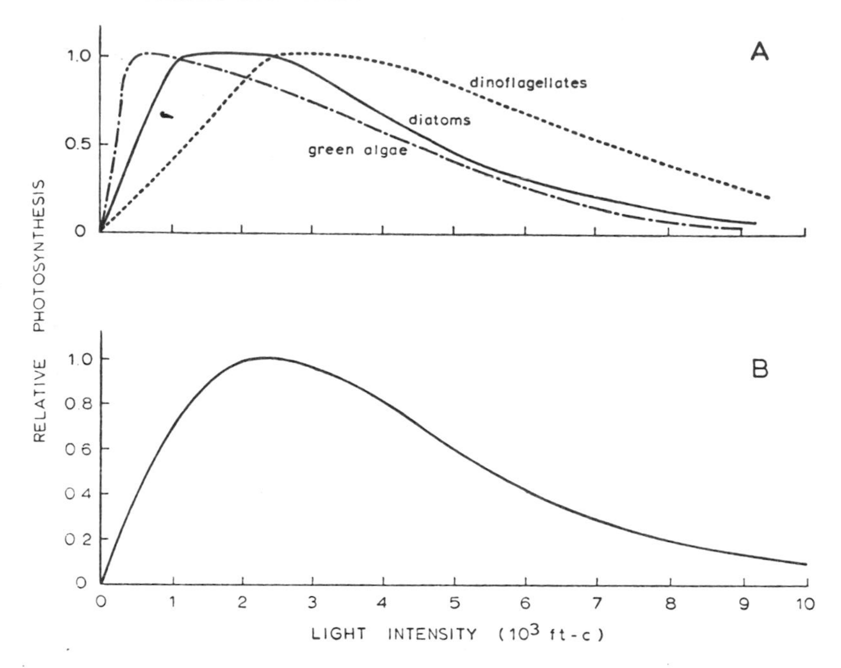

FIG. 35A. Relative photosynthesis–light curves in some marine phytoplankton. Green algae: *Dunaliella euchlora, Chamydomonas* sp., *Platymonas* sp., *Carteria* sp., *Mischococcus* sp., *Stichococcus* sp., and *Nannochloris* sp. Diatoms: *Skeletonema costatum, Nitzschia closterium, Navicula* sp. and *Coscinodiscus excentricus.* Dinoflagellates: *Gymnodinium splendens, Gyrodinium* sp., *Exuviaella* sp., and *Amphidinium klebsi* (redrawn from Ryther, 1956).
FIG 35B. Mean curve from FIG. 35A.

between those of different groups. A summary of Ryther's results is shown in Fig. 35; from these results it may be seen that the saturation light intensities for green algae are between 500 and 750 ft–c (or 3·3 to 4·9 × 10^{-2} ly/min), between 2500 and 3000 ft–c (or 16 to 20 × 10^{-2} ly/min) for the dinoflagellates, and at intermediate intensities for the diatoms. Photo-inhibition is apparent in all three algal groups within about 1000 ft–c (0·066 ly/min) of saturation. At intensities of 8000 to 10,000 ft–c (52 to 66 × 10^{-2} ly/min), which is comparable to full sunlight, the photosynthetic rate in green algae and diatoms is only 5 to 10% of that at saturation, while the photosynthetic rate for dinoflagellates is still 20 to 30% of P_{max}. The I_k of each curve in Fig. 35 is 400 ft–c (2·6 × 10^{-2} ly/min) for green algae, 1000 ft–c (6·6 × 10^{-2} ly/min) for diatoms and 2400 ft–c (16 × 10^{-2} ly/min) for dinoflagellates. In the sea, high I_k values are observed during the summer and in shallow algal communities, and low I_k values are observed during the winter and in deep-water communities (Steemann Nielsen and Hansen, 1959a and 1961; Ichimura *et al.,* 1962).

I_k gives a measure of the radiant energy or illumination at light saturation but it does not express the photosynthetic efficiency; consequently plants or phytoplankton communities may have the same I_k values but differ appreciably in the rate of photosynthesis at I_k. In terrestrial communities plants are divided into 'sun-' and 'shade-types' (Boysen Jensen, 1932) and a similar division is employed in algal communities. Thus sun-type algal communities are those who can utilize high light intensities with high efficiencies while photosynthesis in shade-type communities is generally depressed by high light intensities. However, the absolute photosynthetic rates of shade-type communities are usually higher than those of the sun-type communities at low light intensities. These

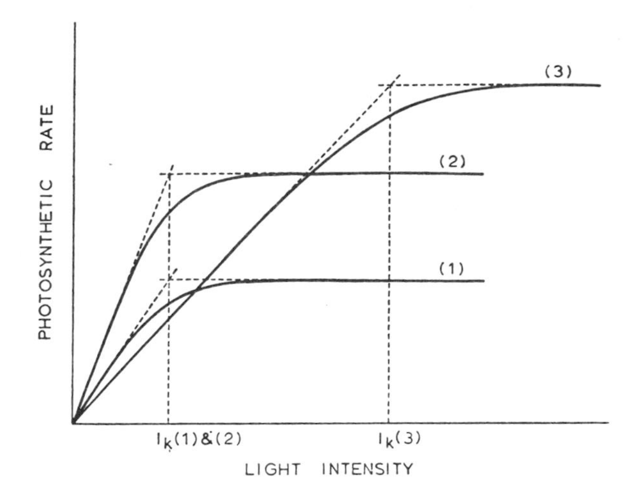

FIG. 36. Three types of P vs. I curves. (1) and (2) shade-type algae showing similar I_k values but with higher photosynthetic efficiency in (2) than (1). Sun-type community (3) showing lower photosynthetic efficiency than (1) or (2) at lower intensity.

differences in I_k values and photosynthetic efficiencies are illustrated in Fig. 36.

Table 16 shows some accumulated values for the initial slopes of P vs. I curves measured under incandescent and natural light. Algal cultures of different species show values between 0·33 and 2·2 (average 1·3) mg C/mg Chl *a*/ly. Those of natural samples range between 0·14 and 2·2 (average 1·1 mg C/mg Chl *a*/ly. The table shows that certain species or populations are adapted to low light intensities and others to high light intensities. Differences which are sometimes found between species, or within the same community over a period of time, may be due to inactivation of cells or changes in photosynthetic mechanisms (e.g. Ichimura *et al.,* 1968).

There are many reviews on the evaluation of P_{max} (Steemann Nielsen and Hansen, 1959a; Ichimura and Aruga, 1964; Yentsch and Lee, 1966). P_{max} is primarily influenced by environmental conditions (see Section 3.1.3) but under conditions of optimum temperature and sufficient nutrient P_{max} can be employed as the 'potential P_{max}' or 'potential photosynthetic rate'. The P_{max} of phytoplankton populations is usually measured at *in situ* temperature and nutrient concentrations; the results of some of these measurements, together with values for cultures, are shown in Table 17. The values range from between 1·1 to 6·2 mg C/mg Chl *a*/hr for cultures and from 0·1 to 6·0 mg C/mg Chl *a*/hr in natural samples. The data show some geographical differences in P_{max}; thus the values for polar waters are generally low, temperate waters are quite variable, and tropical waters have the highest values. Some exceptions to the general range of values have been reported; for example, *Skeletonema costatum* has sometimes been reported to give values during a bloom of 9·0 to 16·9 mg C/mg Chl *a*/hr (Hogetsu *et al.,* 1959).

From the photosynthetic action spectra shown in Fig. 26, it is to be expected that light energy available for algal photosynthesis is restricted to wavelengths between 400 and 720 nm. Within these wavelengths the light absorbed by the phytoplankton pigments can be divided into two parts: (i) the light of >600 nm which is mainly absorbed by chlorophyll *a* and (ii) the light of <600 nm which is mainly absorbed by accessory pigments. Except in some blue-green and red algae (e.g. curve 5 in Fig. 26), the action spectra at >600 nm of all algae are similar, while those at <600 nm are quite different and depend mainly on the light absorbed by accessory pigments. Since most of the light penetrating to depth is in the region 400 to 600 nm, it is quite apparent that the accessory pigments are most important as light absorbers in the ocean environment.

Another effect of light on photosynthesis is a diel rhythm. Although this subject, as recently reviewed by Sournia (1974), has not been fully understood yet, tentative generalizations only can be made. Diel oscillations are known for rates of photosynthesis (Doty and Oguri, 1957; Ichimura, 1960), of chlorophyll synthesis (Yentsch and Ryther, 1957), and rates of nutrient uptake (Goering *et al.,* 1964). The diel maximum in photosynthetic rate is recorded for early morning in the ocean near the equator (Doty and Oguri, 1957), for later in the day in lake (Lorenzen, 1963), and in inshore marine environments (Newhouse *et al.,* 1967). The amplitude of the diel oscillations generally decreases with increase in latitude (Doty, 1959) and is less pronounced on cloudy days and during the

TABLE 16. INITIAL SLOPES OF P vs. I CURVES OF ALGAL CULTURES AND NATURAL POPULATIONS***

Algal cultures

Species	Exp. temp. (°C)	Initial slope (mg C/mg Chl *a*/ly	Author
Chlorella vulgaris	20	1·7*	Steemann Nielsen (1961)
C. pyrenoidosa		1·6	Steemann Nielsen and Jørgensen (1968)
C. ellipsoidea	5–30	1·7**	Aruga (1965b)
Scenedesmus sp.	10–20	2·2**	Aruga (1965b)
Skeletonema costatum		1·7	Steemann Nielsen and Jørgensen (1968)
Skeletonema costatum	20	1·5**	Nakanishi and Monsi (1965)
Chaetoceros sp.	20	0·43**	Nakanishi and Monsi (1965)
Coccolithus huxleyi		0·43*	Jeffrey and Allen (1964)
Hymenomonas sp.		0·33*	Jeffrey and Allen (1964)

Natural populations

Situation	Dominant species	Exp. temp. (°C)	Initial slope (mg C/mg Chl *a*/ly	Author
Tokyo Bay	*Skeletonema costatum*	20	1·6**	Aruga (1965b)
Pond	*Synedra* sp.	10–30	1·8**	Aruga (1965b)
Pond	*Anabaena cylindrica*	10–30	0·69**	Aruga (1965b)
Lake	*Cryptomonas* sp.		*ca.* 2·2	Ichimura *et al (1968)*
Arctic		4·5–6	1·0	Steemann Nielsen and Hansen (1959a)
Tropical, temperate (summer, winter), Northern & Arctic (general)			1·0	Steemann Nielsen and Hansen (1959a)
Kuroshio (general)			0·14	Ichimura *et al.* (1962)
Oyashio (general)			0·10	Ichimura *et al.* (1962)
Mixed regions of Kuroshio & Oyashio (general)			0·47	Ichimura *et al.* (1962)
Antarctic			1·2	Burkholder and Mandelli (1965)

*mg C/mg Chl ($a + b$)/ly.
**mg C/mg Chl/ly.
***The initial slope and P_{max}, as calculated from chlorophyll and photosynthetic data in the early literature, might contain some over-estimations for chlorophyll(s) (about 25%: Banse and Anderson, 1967) and photosynthesis (up to 30%; Steemann Nielsen, 1965) because of the methods used. The former will increase the initial slope (or P_{max}) and the latter decreases it. Consequently, it is fortuitous that the results quoted here appear to be approximately correct. Light intensity originally determined in lux was converted to ly by conversion factors of 1 lux = 6×10^{-6} ly/min for the sunlight (natural population) and 1 lux = 5×10^{-6} ly/min for white fluorescent lamps (culture population).

winter (Saijo and Ichimura, 1962). The main question is how are these rhythms evolved? Two hypotheses may be considered for possible explanations of photosynthetic rhythms. The first is that the photosynthetic potential oscillates in reponse to intrinsic organization of the cell with a light–dark cycle. In the second hypothesis, photosynthetic rate oscillates in response to a forcing from the external concentration of limiting nutrients while photosynthetic potential of the cell

TABLE 17. P_{max} OF ALGAL CULTURES AND NATURAL POPULATIONS***

Algal cultures

Species	Culture conditions temp. (°C)	Culture conditions light int. ly/hr	Exp. temp. (°C)	Exp. light int. ly/hr	P_{max} (mg C/mg Chl *a*/hr)	Author
Chlorella ellipsoidea	20		30	6·0	5·5**	Aruga (1965b)
C. vulgaris		0·9		1·8	1·1*	Steemann Nielsen (1961)
		9·0		9·0	3·8	
Scenedesmus sp.	20		30	3·0	5·0**	Aruga (1965b)
Skeletonema costatum	20	2·4	20	6·0	6·2**	Nakanishi and Monsi (1965)
	20		20	3·9	3·6**	Aruga (1965a)
Synedra sp.	20		20	4·5	2·0**	Aruga (1965b)
Cyclotella meneghiniana		0·9			2·1–3·4	Jørgensen (1964)
		9·0			3·4–4·4	Jørgensen (1964)
Chaetoceros sp.	20	2·4	20	6·0	1·5**	Nakanishi and Monsi (1965)
Anabaena cylindrica	20		30	3·0	1·7**	Aruga (1965b)
Coccolithus huxleyi	14	11		>11	>2·2*	Jeffrey and Allen (1964)
Hymenomonas sp.	14	11		>11	>1·9*	Jeffrey and Allen (1964)

Natural populations

Situation	Depth (m)	Exp. temp. (°C)	P_{max} (mg C/mg Chl *a*/hr)	Author
Arctic (summer)	0		1·0–1·5	Steemann Nielsen and Hansen (1959a)
Antarctic (phytoplankton)			2·3	Burkholder and Mandelli (1965)
Antarctic (ice flora)		−1·6	2·6	Burkholder and Mandelli (1965)
Antarctic (ice flora)		−1·5	0·4	Bunt (1964b)
Northern (general) no vertical stability			2·9–3·4**	Steemann Nielsen and Hansen (1959a)
Western north Pacific (summer)	10	8–20	1·4–2·0	Takahashi *et al.* (1972)
Temperate (summer)	0		4·0–4·2**	Steemann Nielsen and Hansen (1959a)
Temperate (winter)	0		1·5**	Steemann Nielsen and Hansen (1959a)
Oyashio (general, summer)			3–6**	Ichimura and Aruga (1964)
Kuroshio (general, summer)			0·3–0·7**	Ichimura and Aruga (1964)
Bays and coastal waters near Japan (general)			2–6**	Ichimura and Aruga (1964)
Tokyo Bay (*Skeletonema* bloom, summer)		20–25	9·0–16·9**	Hogetsu *et al.* (1959)
Tropical (general)			8·0**	Steemann Nielsen and Hansen (1959a)
Tropical Pacific (autumn)		23–27	1·1–5·2	Takahashi *et al* (1972)
Tropical Pacific (spring, general) nitrogen-poor water			3·15	Thomas, (1970a)
Tropical Pacific (spring, general) nitrogen-rich water			4·95	Thomas (1970a)
Tropical Atlantic	10	20	3·0–4·0**	Yentsch and Lee (1966)
Lake (general)				Ichimura and Aruga (1964)
eutrophic			2–6**	
mesotrophic			1–2**	
oligotrophic			0·1–1**	

*mg C/mg Chl $(a + b)$/hr.

**mg C/mg Chl/hr.

***The initial slope and P_{max}, as calculated from chlorophyll and photosynthetic data in the early literature, might contain some over-estimations for chlorophyll(s) (about 25%; Banse and Anderson, 1967) and photosynthesis (up to 30%; Steemann Nielsen, 1965) because of the methods used. The former will increase the initial slope (or P_{max}) and the latter decreases it. Consequently, it is fortuitous that the results quoted here appear to be approximately correct. Light intensity originally determined in lux was converted to ly by a conversion factor of 1 lux = 5×10^{-6} ly/min for white fluorescent lamps.

remains constant. These two hypotheses were tested in natural populations using elegant models by Stross *et al.* (1973). They have concluded that photosynthetic rhythms could result from both an intrinsic and a nutrient (forcing) oscillation.

P_{max} may change with the physiological state of algae themselves (e.g. age) even under constant environmental conditions (Jørgensen, 1966). From experiments with synchronous cultures of *Skeletonema costatum,* it was shown that P_{max} increased in young cells (just after cell division) and reached a maximum value in full grown cells, just prior to cell division. Cell division started 6 hr after illumination and continued 4 hr after cells were placed in the dark.

From these observations it is apparent that adaptive changes in photosynthesis occur in algal cells in response to surrounding light conditions. This phenomenon is known as 'light adaptation', i.e. a physiological adjustment to surrounding conditions which has been observed to involve at least one of the following morphological or biochemical changes within the cell: (1) change in total photosynthetic pigment content, (2) change in the ratio of photosynthetic pigments, (3) change in the morphology of the chloroplast, (4) change in the arrangement of the chloroplasts, and (5) change in the availability of enzymes for the dark reaction. Specific examples of these changes have been demonstrated; for example, Fujita (1970) showed that in blue-green algae, red light induced phycocyanin synthesis and blue light increased the amount of phycoerythrin. In diatoms, the shrinking of chloroplasts and their aggregation under the influence of strong light has been observed as a reaction which is reversed under weak light (Brown and Richardson, 1968).

The time dependence of light adaptation of algae in the ocean is important in determining the day-to-day effects of light variation. According to Steemann Nielsen *et al.* (1962) *Chlorella* required about 40 hr to adapt to a change in light intensity from 0·9 to 9 ly/hr at 21°C. Algal populations taken from the surface of strong vertically mixed water masses off Friday Harbor, Washington, U.S.A., took 3 days to adapt to low light conditions of about 5% of the surface illumination (Steemann Nielsen and Park, 1964). If algae are kept in the dark there is a gradual loss in photosynthetic activity which in the case of *Nannochloris* amounted to 50% of the photosynthetic activity in about 40 hr at 20°C (Yentsch and Lee, 1966).

3.1.4 Nutrient and Temperature Effects on Photosynthesis and the Growth Requirements of Phytoplankton

Photosynthesis of algae is also controlled by factors other than light, such as nutrients and temperature, and to a lesser extent a variety of factors, such as pH and salinity. Liebig (1840) postulated a simple rule for the effect of various factors on yield (i.e. net photosynthesis, P_n). His statement was that "growth of a plant is dependent on the minimum amount of foodstuff presented" and this has come to be known as "Liebig's law of the minimum". Sixty years after Liebig's work, Blackman (1905) suggested that a generalized form of this law could be applied to photosynthesis in order to explain field and experimental observations. He took from Liebig the idea that the rate of a biological process (in this case, photosynthesis) is determined, under given conditions, by a single limiting factor. However, in addition to the supply of material ingredients (the only kind of factors with which Liebig was concerned), Blackman considered also light intensity and temperature as limiting factors. He suggested that the rate of photosynthesis increased with an increase in the value of any one of these factors (F_1), as long as the particular factor was rate limiting, and that it ceased to be dependent on F_1 when one of the other factors ($F_2, F_3 \ldots$) became limiting. In other words, the plot of photosynthesis, P_n, versus a variable F_1 (at constant values of all other kinetic factors) was postulated by Blackman to have the shape 1, 2, 3 in Fig. 37. In actual fact the curve approaches the maximum asymptotically, without a sudden break as indicated by the point 2. It should be noted also that an excessive amount of the factor, F_1, eventually causes a depression in P_n. Assuming F_1 to be light intensity, the P_n vs. F^1 relation is the same as the P vs. I curve shown before in Section 3.1.3. In order that other factors, $F_2, F_3 \ldots$ should conform to the same rule, it is apparent that for different values of a second parameter (F_2), photosynthesis (P_n) as a function of F_1 will be represented by a sequence of solid lines, such as 1.2.3, 1.4.5, 1.6.7, in Fig. 37. These coincide at low F_1 values (part 1.2 in the figure) but are distinguished at higher F_1 values by the position at which the ascending part of the curve becomes horizontal. This value is determined by the second factor (F_2), which causes an increase in P_n as represented by levels 3, 5, and 7. The horizontal plateau in Fig. 37 does not extend indefinitely, however, and P_n declines when F_1 causes an inhibition. The initial slope, which is dependent on F_1, tends to be similar for the sequence of curves but the decline may be more rapid with an increase in F_2. The maximum and minimum values F^{max} and F^{min}, in Fig. 37 are called the 'upper lethal limit' and the 'lower lethal limit' respectively, and these can be generally recognized in applying any factor to a growth process. This gives a numerical evaluation to descriptive terms which have been used to express a relative degree of tolerance. Thus the series of expressions which have come into general use in ecology (see Section 1.1) and which utilize the prefixes 'steno-' meaning narrow and 'eury-' meaning wide, with reference to the tolerance of a process, or an organism, can be given definite meaning in terms of Fig. 37. Furthermore, the ascending and descending slopes, together with the width and height of the plateau, are also important physiological characters to consider in the ecological adaptation of certain algal populations to a given aquatic environment.

An example of the relationship between the seasonal photosynthesis of mixed algal populations, and different phosphate concentrations and temperatures, is given by Ichimura (1967) for phytoplankton in Tokyo Bay (Fig. 38A). From these results it is apparent that phosphate regulation on P_{max} is governed by temperature; the relationship is quite similar to the Blackman-type limitation shown in Fig. 37, although the higher concentrations of phosphate were not sufficient to

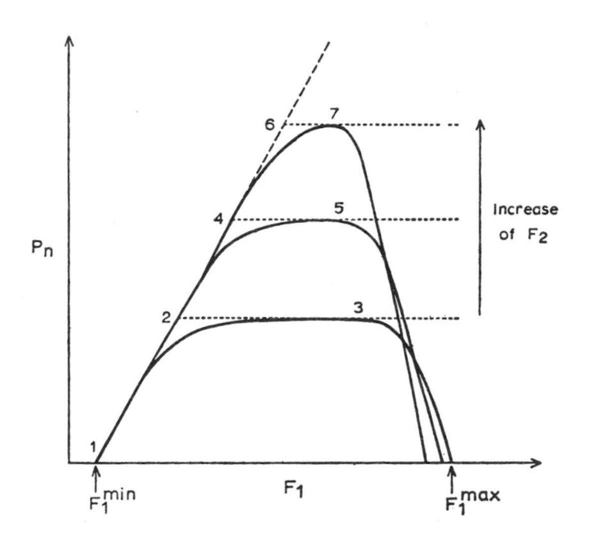

FIG. 37. Changes in photosynthetic rates, P_n, with a change in two environmental factors, F_1 and F_2, varied independently.

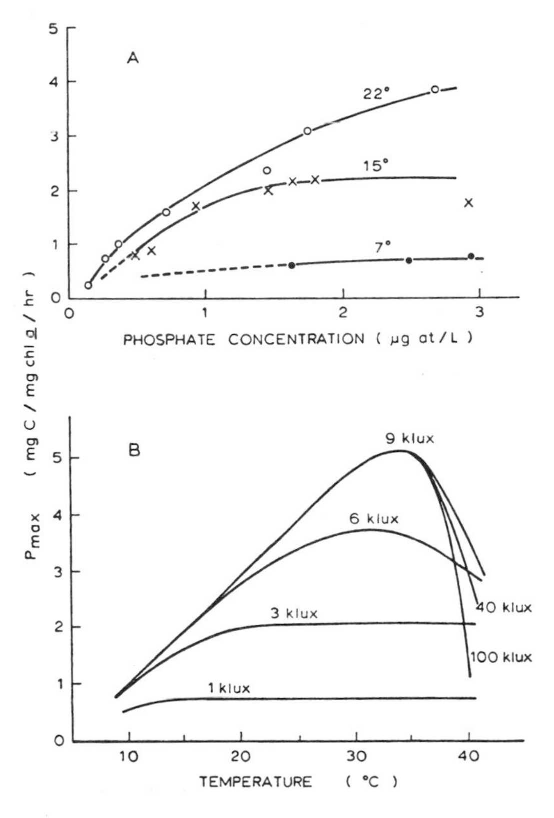

FIG. 38. Photosynthesis regulation by phosphate (A) and temperature (B). (A) Natural populations taken from Tokyo Bay (redrawn from Ichimura, 1967); (B) cultured *Scenedesmus* sp. (redrawn from Aruga, 1965a).

cause a break in the curve. Figure 38B from Aruga (1965b) serves as a second example of the general relationship in Fig. 37. In this case the influence of light and temperature on photosynthesis gives a more complete Blackman-type response including depression of P_{max} at high light intensities. While these results were obtained for local populations which had been reconditioned to a specific environment, it is also known that populations can adapt to some extent to different environments. Steemann Nielsen and Jørgensen (1968) showed, for example, that a phytoplankton population could become adapted to a new temperature regime with a few days. For some species, however, temperature adaptation has to be made in a series of small steps (e.g. 5°C changes) in order to exclude harmful effects caused by a sudden change in temperature. Experimental data on the effect of temperature on photosynthesis are generally scarce but some studies have been made in the Antarctic (Bunt, 1974 a, b) and in the Pacific (Aruga *et al.*, 1968; Ichimura *et al.*, 1962); the latter results indicated that temperate Pacific phytoplankton had their highest P_{max} at about 20°C in spite of *in situ* temperatures which varied between –0·9 and 17·9°C.

P_{max} measured at *in situ* temperatures is a good indicator of what kinds of factors (other than light) are limiting photosynthesis. As an example, seasonal variations in P_{max} from Tokyo Bay are shown in Fig. 39. From the results it is apparent that from October to June, temperature regulates photosynthesis; during July to September P_{max} is depressed by the lack of nutrients which increase again due to autumn mixing in October. On the other hand, the potential photosynthetic rate, which is measured under optimum temperature and nutrient conditions, is constant throughout the year except in February and August. Low potential photosynthetic rates in these months have been attributed to a changing phytoplankton flora during the early spring and late summer. In high latitudes where there are low temperatures and high nutrient concentrations, P_{max} will be largely regulated by temperature. Phytoplankton communities near river mouths may also be regulated by temperature throughout most of the year (Ichimura, 1967). On the other hand, P_{max} in tropical and sub-tropical communities is more likely to be limited by nutrients.

Eppley (1972) has summarized the effect of temperature on algal growth rate. By plotting growth rate data, obtained by different people, vs. temperature, he found an empirical relation for the maximum growth rate of algae over the temperature range between 0 and 40°C, under conditions of continuous illumination:

$$\log_{10} \mu = 0{\cdot}0275\, t - 0{\cdot}070 \qquad (34)$$

where μ is the maximum possible growth rate in doublings per day (see Section 3.1.5) and t is

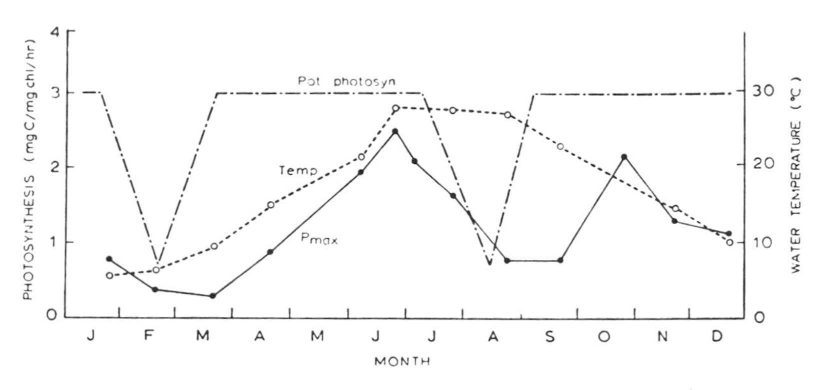

FIG. 39. Seasonal variations in temperature, P_{max} and potential photosynthetic rate in Tokyo Bay (redrawn from Ichimura and Aruga, 1964).

temperature in degrees Celsius. The equation was deduced from data on algal cultures which included a wide variety of taxonomic groups, cells with different complements of photosynthetic pigments, and diverse morphologies. The growth rate of each algal species (or clone) can be fitted to the equation over a limited temperature range; growth ceases at temperatures above a supraoptimal point, which is a characteristic of the species or clone (Eppley, 1972). The basic temperature dependency of the algal growth, shown by Eppley (1972) using batch cultures, has been confirmed by Goldman and Carpenter (1974) using various algal species from freshwater and marine origins in continuous cultures. However, there is a discrepancy between those two experiments in Q_{10}, enhancement ratio of biological reactions for each 10°C rise in temperature; 1·88 for Eppley's and 2·08 for Goldman and Carpenter's. This discrepancy seems not to be simply due to differences in culture conditions and a further evaluation is required.

Algae require certain elements for their growth. Some of these elements, such as C, H, O, N, Si, P, Mg, K, and Ca are needed in relatively large amounts and are often known as 'macro-nutrients'. Other elements are required in very small amounts and are referred to as 'micro-nutrients' or 'trace elements'; Hewitt (1957) has listed ten trace elements required by green plants as follows: Fe, Mn, Cu, Zn, B, Na, Mo, Cl, V, and Co. Most of these elements are contained in sufficient abundance for algal growth in sea water; however, nitrogen and phosphorus may often limit plant growth during the summer or in tropical and subtropical latitudes throughout the year. Thomas (1968) made comprehensive tables on nutrient requirements and utilization of many different species of algae, in which he defined requirements for vitamins, organic carbons (sugars, alcohols, fatty acids, and other organic acids), amino acids, and other nitrogenous substances, including different inorganic forms of nitrogen compounds. His summary indicates which species of algae can grow on what kind of nutrients. Considering the possible different nutrient requirement of each algal species, Takahashi and Fukazawa (1982) have tested various macro- and micro-nutrients (at several different concentrations within natural levels) in single or combination additions with other nutrient(s) using natural phytoplankton assemblages, where a variety of phytoplankton species were originally included. They performed the experiment in a semi-continuous culture system using dialysis culture tubes in order to maintain the nutrient conditions which were consistent on a diel basis. Phytoplankton assemblage developed various different dominant phytoplankton groups depending upon the differences both in nutrients and concentrations. This gives evidence that a given natural phytoplankton assemblage contains various phytoplankton species, which have the potentiality to react differently, depending upon changes in the nutrient environment.

As an inorganic carbon source, algae can use free carbon dioxide, bicarbonate, and carbonate. These three are measured as 'total carbon dioxide' and the amount is expected to be *ca.* 90 mg CO_2/l in offshore pelagic waters. At this concentration phytoplankton photosynthesis is not limited by the total amount of carbon dioxide. For example, Talling (1960) did not find any stimulation or reduction in photosynthesis following carbonate addition to a culture of *Chaetoceros affinis* in natural sea water. When phytoplankton grows vigorously, such as in a 'bloom', the total carbon dioxide content of sea water (determined as the partial pressure) may show a negative correlation with chlorophyll concentration in temperate waters (Gordon *et al.*, 1971).

Nitrogen and phosphorus are of major importance as metabolites, and their concentration should always be considered first in determining possible limitations in primary production. The uptake of carbon, nitrogen, and phosphorus by marine phytoplankton is generally found to be in the ratio of 106:16:1 (Redfield, 1934) (see also Table 10), which is commonly called the 'Redfield number'. In a recent discussion of this ratio, Goldman *et al.* (1979) have pointed out that this ratio is often maintained even in areas of apparent low phytoplankton productivity due to nutrient limitation. The most important feature is the supply of nutrients in balanced atomic ratio, and this is particularly essential at low nutrient conditions.

Under conditions in which there is an imbalance of the ratio in the water (such as due to sewage outfalls, faster regeneration rate of phosphorus than nitrogen, or under some conditions of laboratory cultures), the cellular atomic composition may change several hundred per cent (Banse, 1974 a and b).

In the oceans, nitrogen exists mainly as molecular nitrogen and as inorganic salts, such as nitrate, nitrite, and ammonia, and some organic nitrogen compounds of amino acids and urea. The usual range of concentration for these compounds in sea water is 0·01 to 50 μg at/l for nitrate, 0·01 to 5 μg at/l for nitrite and 0·1 to 5 μg at/l for ammonia 0·2 to 2 μg at N/l for amino acids (Clark *et al.*, 1972; Riley and Segar, 1970), and 0·1 to 5·0 μg at N/l for urea (Remsen, 1971). Saturation concentrations of dissolved nitrogen gas are in the range 370 to 800 μg at/l, depending upon the salinity of the seawater (Riley and Chester, 1971). Molecular nitrogen is fixed by certain blue-green algae [e.g. *Calothrix* sp. (Allen, 1963) and *Trichodesmium* spp. (Dugdale *et al.*, 1964; Goering *et al.*, 1966; Dugdale and Goering, 1967)], yeasts [e.g. *Rhodotorula* sp. (Allen, 1963)] and Bacteria [e.g. *Azotobacter* and *Closteridium* (Pshenin, 1963)]. However, compared with freshwater micro-organisms, nitrogen fixers in the marine environment have not been clearly defined. *Trichodesmium* can also actively utilize both ammonia and nitrate but since the tropical habitat of *Trichodesmium* is usually relatively poor in nitrate, nitrogen fixation, together with ammonia utilization, are probably the most important sources of algal nitrogen in such environments (Goering *et al.*, 1966). Most other algae have no ability to fix molecular nitrogen and must utilize inorganic nitrogen salts and organic forms of nitrogen compounds. Algae generally show a preferential utilization for nitrate, nitrite, and ammonia (e.g. Guillard, 1963); some exceptions are found among the green algae (Braarud and Føyn, 1931) and a few flagellates (e.g. Strickland, 1965). Of those different forms of nitrogen compounds, ammonium is generally utilized in preference to nitrate (Eppley *et al.*, 1969a) and urea (McCarthy and Eppley, 1972) in different species of phytoplankton cultures and in natural phytoplankton populations. In some areas the principal source of nitrogen for phytoplankton may differ with depth in the same water column. For example, in the eastern subtropical Pacific Ocean off the continental shelf of Mexico, a pronounced discontinuity layer often exists at between 40 and 60 m within the euphotic zone, and phytoplankton in this layer use nitrate as their major nitrogen source; those in the overlying nitrate-impoverished water 0–40 m use ammonia which is derived largely from nitrogen recycled through zooplankton (Goering *et al.*, 1970). Natural living phytoplankton populations are known to utilize amino acids as a N-source even at low concentrations of natural levels (North, 1975). However, it seems that most phytoplankton have some preference to use inorganic forms of nitrogen compounds and urea, and do not start using amino acids effectively until other forms of nitrogen are depleted or reduced down to a certain level.

The processes involved in the uptake of nitrate by phytoplankton appear to involve two steps: the first step is the uptake of nitrate from outside the cell, and the second involves the utilization of nitrate within the cell (Fig. 40). These two processes involve different enzyme systems. The first involves the active uptake of NO_3^- from the water and the translocation across the cell membrane. Falkowski (1975) showed that in the marine diatom, *Skeletonema costatum*, an enzyme (a NO_3^-, Cl^--activated ATPase) required ATP and had a K_s of 0.9×10^{-6} M which was close to the K_s generally found for nitrate uptake by the whole cell; however, no evidence was presented which demonstrated that this enzyme was specifically associated with the cell membrane or plasmalemma. Diffusion of nitrate across the cell membrane is generally discounted because it cannot maintain the intracellular nitrate levels required to insure the effective action of nitrate reductase (e.g. ~100 μM; Packard, 1979).

Within the cell, the enzyme, nitrate reductase, reduces the nitrate to nitrite, a process which also requires energy in the form of a photosynthetic reductant, NADH. The synthesis of nitrate reductase is inhibited by the presence of ammonium at concentrations above 0.5 to 1 μM (Eppley *et al.*, 1969). Thus the observed preferential of nitrogen is inversely proportional to the degree to which the nitrogen is oxidized, with ammonium being taken up approximately twice as fast as nitrite or nitrate, depending on the nutrient past history of the cell (Conway, 1977). In the presence of nitrite, only nitrite reductase is synthesized, while the presence of nitrate requires the synthesis of both nitrate and nitrite reductase (Hattori, 1962 a and b; Lui and Roels, 1972).

This high requirement for ATP in the nitrate-assimilation process accounts for a number of other observations regarding nitrate uptake by phytoplankton in the marine environment. These include: the observed uptake of nitrate in the dark (since ATP can be made available by oxidative phosphorylation), temporary inhibition of carbon dioxide uptake through competition for ATP, and increased photosynthetic pigment synthesis, together with enhanced longer term carbon dioxide uptake through the supply of more energy and eventual protein synthesis from nitrate incorporation (for discussion see Falkowski and Stone, 1975).

Ammonium, like nitrate and nitrite, appears to be transported across the cell membrane by an active process, although no specific enzyme(s) has been associated with this process. After ammonium is taken up by the cell or produced from nitrate reduction described above, there are two possible pathways by which ammonium can be assimilated into amino acids (Fig. 40). The major route of ammonium assimilation has been considered for many years to occur via the reductive animation of α-ketoglutarase, catalysed by the enzyme, glutamate dehydrogenase (GDH) (Ahmed *et al.*, 1977). However, recent work has shown the incorporation of ammonium into the amino acid pool probably occurs via glutamate synthetase from glutamate with the use of ATP (Falkowski and Rivkin, 1976; Turpin and Harrison, 1978; Dortch and Ahmed, 1978).

Phosphorus occurs in sea water in three principal phases: dissolved inorganic phosphorus, dissolved organic phosphorus, and particulate phosphorus. However, the existence of these forms is quite complex and special analytical procedures are needed to show the different forms of phosphorus in sea water (e.g. Strickland and Parsons, 1972). Phytoplankton normally satisfy their requirement for this element by direct assimilation of dissolved inorganic phosphorus (orthophosphate ion) and sometimes by utilizing dissolved organic phosphorus (Provasoli and McLaughlin, 1963; McLaughlin and Zahl, 1966). In polluted water, polyphosphate, which is inorganic, and organic soluble phosphorus detected as orthophosphate after acid hydrolysis, may be present in appreciable amounts. Some coastal algae, such as *Skeletonema costatum* and *Amphidinium carteri,* can use polyphosphate as a phosphorus source in the presence of excess nitrate. These species appear to be able to hydrolyse an external supply of polyphosphate more rapidly than it is utilized by the cells with the result that orthophosphate may accumulate in the surrrounding water (Sōlorzano and Strickland, 1968). Phytoplankton are generally unable to directly take up even simple dissolved phosphate esters. Instead they obtain the phosphorus from these compounds by producing a membrane-bound or extracellular enzyme, alkaline

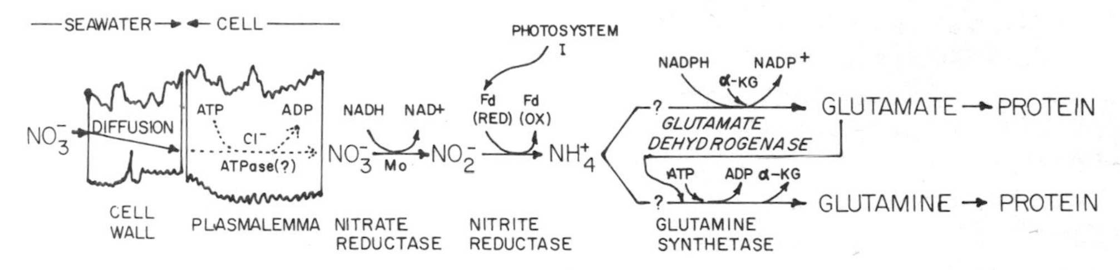

FIG. 40. Pathway for nitrate uptake and assimilation in phytoplankton. Abbreviations used: ATP, adenosine triphosphate; ADP, adenosine diphosphate; NAD^+ and NADH, nicotinamide adenine dinucleotide (oxidized and reduced forms); Fd, ferredoxin; $NADH^+$ and NADPH, nicotinamide adenine dinucleotide phosphate (oxidized and reduced forms); and α-KG, alpha ketoglutaric acid (from Packard, 1979).

phosphatase. This enzyme cleaves off the phosphate group on the compound and then the free phosphate ion can be taken in by the regular phosphate transport system. Alkaline phosphatase may be induced in some phytoplankton at orthophosphate concentrations < 0.1 μg at.l^{-1}, particularly when the organic phosphate concentration is appreciably higher than the inorganic. This enzyme has been detected in oligotrophic waters such as the Sargasso Sea and the central north Pacific gyre (Kuenzler and Perras, 1965; Perry, 1976). Phosphorus absorbed into a cell becomes part of the structural component of a cell (e.g. in poly-P-RNA) and is in part continually turned over in the energetic processes of organisms (e.g. as adenosine di- and tri-phosphate). In this sense the role of phosphorus as a metabolite is quite different from nitrogen because nitrogen is used primarily as a structural component of cells and not directly in the energy cycle of the cell.

Diatoms and silicoflagellates take up a great amount of dissolved silicon and deposit it as hydrated silica to form their elaborately patterned valves. According to Davis *et al.* (1978), steps in the uptake of silicate are governed by the external substrate concentrations, the internal pool of dissolved silicate and the amount of silica laid down in the frustule. Since growth rate was related to frustule formation, which was not directly related to the amount of dissolved silicate in the medium, it is apparent that a single equation describing silicate-dependent growth might not be an appropriate description of this process. The concentrations of dissolved silicon are generally high in coastal and deep pelagic waters and low in surface waters away from the influence of estuaries. Diatoms grown in a medium low in silicate become silicon deficient but may remain viable for several weeks in the dark. However, on exposure to bright light they photosynthesize for a limited period and soon die (Riley and Chester, 1971). Diatom blooms deplete the surface waters of silicate and Menzel *et al.* (1963) have suggested that this is the chief mechanism leading to a species succession of non-silicious flagellates following the spring diatom bloom in sub-tropical waters.

Lewin (1966) found that the silicic acid uptake of diatom was interrupted by the existence of germanium, because the element germanium is chemically related to silicon. As a result, diatom growth was specifically inhibited but not that of other algae. This has led to practical use of germanium, practically germanic acid, $Ge(OH)_4$, in inhibiting diatom over growth of mixed algal population. Thomas and Dodson (1974), by using germanic acid, have tried to separate the photosynthetic productivity of diatoms from total $^{14}CO_2$ uptake in routine marine primary productivity measurements.

Ferric iron as a hemin complex known as cytochrome plays an important role in cellular respiratory processes; in addition, the iron complex called ferredoxin is an essential part of the light reaction in photosynthesis (Section 3.1.1). Natural sea water contains from 1 to 60 μg/l of iron but only some of the iron detected analytically seems to be available for algal metabolism. A shortage of iron for phytoplankton growth has been demonstrated in the pelagic waters of the Sargasso Sea (Menzel and Ryther, 1961; Menzel *et al.*, 1963) and off the coast of Australia (Tranter and Newell, 1963). High concentrations of iron in sea water are generally associated with river runoff (see, for example, Williams and Chan, 1966). Ryther and Kramer (1961) compared the iron requirements of a number of coastal and oceanic species of phytoplankton and found that coastal species could not grow at the low iron concentrations at which oceanic species would grow. In a reversal of this situation Strickland (1965) has suggested that the growth of some oceanic species of phytoplankton in coastal waters may be inhibited by relatively high trace-element concentrations. In some instances differences in the micro-nutrient concentration may be governed more by their availability through chelating agents than through their absolute concentration. The ability of sea water samples to support the growth of phytoplankton has been observed by a number of authors to be quite variable (e.g. Johnston, 1963; Smayda, 1964 and 1970b; Barber *et al.*, 1971). These studies have led to a general conclusion that subtle differences in the micro-nutrient composition of sea water, including the presence of organic substances, may play an important role in determining the total productivity of the water and the diversity of species present. The chemical identity of these substances remains obscure although some work on the origin of soluble organic substances (e.g. Prakash and Rashid, 1968; Pratt, 1966) has shown that humic acids, and the exudates from other phytoplankton, may inhibit or enhance the growth of algae.

The total spectrum of inorganic nutrients needed for phytoplankton growth can be illustrated from the work of Provasoli and others (e.g. Provasoli *et al.*, 1957), who developed several types of artificial sea-water media for the growth of phytoplankton; one of these media which has been used extensively in phytoplankton culture experiments is given in Table 18. The ingredients illustrate the need for trace elements and organic compounds, such as the vitamins and a chelating agent (Na_2EDTA). The vitamin requirement of some organisms should be considered as a special case of autotrophy; the name 'auxotroph' is used to describe organisms which have a physiological growth requirement for one or more organic compounds but which derive their carbon from CO_2 and energy from light. The need for such growth factors should not be confused with the use of organic materials as an energy source, which is concerned in the next section under 'heterotrophy'. Provasoli (1958) reviewed some of the growth factor requirements of the algae and showed that among the commonest groups of phytoplankton, the Dinophyceae exhibited the most extensive auxotrophic requirements which often included a need for thiamine, biotin, and vitamin B_{12}. Many phytoplankton do not require any of the organic growth supplements shown in Table 18. In studies on truly autotrophic organisms it should be necessary to add the inorganic compounds shown in Table 18, together with a buffer (Tris) and the chelating agent (Na_2EDTA), neither of which are used directly by the cells. Both substances may be necessary in cultures, however, since the small volumes of culture vessels are unrepresentative of the natural sea-water environment, and the pH and chelating properties in culture vessels can change much more drastically than in the sea. In addition, for large-standing stocks of phytoplankton it may be necessary to bubble CO_2 enriched air through a culture medium.

TABLE 18. SYNTHETIC SEA WATER MEDIUM FOR THE GROWTH OF PHYTOPLANKTON (ASP 2 FROM PROVASOLI *et al.*, 1957)

Compound	wt./100 ml	Compound	wt./100 ml
NaCl	1·8 g	Tris[b]	0·1 g
$MgSO_4 \cdot 7H_2O$	0·5 g	B_{12}	0·2 μg
		Vitamins[a]	1·0 ml
KCl	0·06 g	Na_2EDTA[c]	3·0 mg
Ca (as Cl^-)	10 mg	Fe (as Cl^-)	0·08 mg
		Zn (as Cl^-)	15·0 μg
$NaNO_3$	5 mg	Mn (as Cl^-)	0·12 mg
		Co (as Cl^-)	0·3 μg
K_2HPO_4	0·5 mg	Cu (as Cl^-)	0·12 μg
$Na_2SiO_3 \cdot 9H_2O$	15 mg	B (as H_3BO_3)	0·6 mg)

[a]Vitamins: 1 ml contains 0·05 mg thiamine HCl, 0·01 mg nicotinic acid, 0·01 mg pantothenate, 1·0 μg *p*-aminobenzoic acid, 1·0 mg biotin, 0·5 mg inositol, 0·2 μg folic acid, 0·1 μg thymine.
[b]Tris buffer (2-amino-2-hydroxymethyl-propane-1,3 diol).
[c]EDTA chelating agent (ethylenediaminetetra-acetic acid disodium salt).

3.1.5 Photosynthesis and Growth of Phytoplankton in the Sea

Under natural conditions, photosynthesis of phytoplankton is regulated spatially and temporally by several environmental factors. Since the general effect of these environmental factors on photosynthesis is known, *in situ* photosynthesis can either be measured directly, or indirectly by a mathematical combination of data on environmental changes and physiological responses of the phytoplankton.

The direct measurement of photosynthesis is used extensively. In this method, samples are collected from various depths in the euphotic zone and each sample is used to fill at least one clear glass bottle and one opaque glass bottle, replicates at each depth being filled as desired. The bottles are inoculated with radioactive bicarbonate (Steemann Nielsen, 1952) and a set of light and dark bottles are then returned to the same depth from which the sample was taken. The vertical spacing of the samples to be collected and exposed depends mainly on the depth to which the water column is illuminated. For example, the sampling depths may be chosen as follows: 100, 50, 25, 10, 5 and 1% of the surface illumination. These depths can either be determined by using a known value for the extinction coefficient [eqn. (29) in Section 3.1.2], or approximated from the Secchi disc depth using the relationship, $k_e = 1{\cdot}7/T$ where k_e is the extinction coefficient (m^{-1}) and T is the depth of the Secchi disc in meters.

The light and dark bottles remain suspended from a buoy during the period of the incubation and the carbon dioxide taken up in the light bottle minus the same value for the dark bottle is considered to be the amount of particulate organic carbon produced by photosynthesis. This value is probably somewhere between a measure of net and gross photosynthesis, but it is often referred to as net photosynthesis if factors, such as the exudation of soluble organic carbon, are ignored.

The suspension of bottles in the water column is conveniently carried out during half a day (i.e. from dawn to mid-day, or mid-day to dusk). The fraction of the daily radiation occurring during the period of the incubation can either be measured with a pyranometer or calculated using a formula, such as that proposed by Vollenweider (1965) or Ikusima (1967). The transformation of short incubation periods into daily rates (Platt, 1971) does not, however, take into account that photosynthesis may decline during incubation. This may be due to a diel photosynthetic rhythm (Vollenweider and Nauwerck, 1961), or to qualitative and quantitative changes in the enclosed population (Ichimura and Saijo, 1958). Physiological reasons for the latter have not been studied extensively, but losses due to damage during manipulations could well be a contributing factor (e.g. to sensitive flagellates). Changes in the daily photosynthetic rhythm have been studied more extensively and it appears (Lorenzen, 1963) that maximum photosynthetic rate (per unit of chlorphyll *a*) usually occurs before noon and about 70% of the total daily photosynthesis at the surface is carried out in the morning (i.e. between dawn and mid-day). However, at different depths in the water column the amount of light will be a function of sun angle; consequently, the period for which light is available for photosynthesis is partially a function of depth and this tends to reduce the diurnal rhythm in the water column as compared with observed changes in surface samples. For the whole water column, Vollenweider and Nauwerck (1961) found that photosynthesis in the morning was 54 to 62% of the daily photosynthesis per unit area.

Another disadvantage to short exposures is that it ignores the influence of temporary changes in weather conditions as well as diurnal changes in illumination. Thus in spite of the physiological difficulties of long exposures mentioned above, half or full day incubations are usually performed, and correction factors are employed for day-to-day and seasonal differences in illumination and day length.

In many cases it is inconvenient to keep bottles suspended *in situ* for half- or one-day periods. Such situations arise due to the limited availability of ship time at one location or in some near-shore environments which are not readily accessible. In order to avoid this difficulty, extensive use has been made of various simulated *in situ* methods involving an experimental approach and a mathematical interpretation of the results. In this technique, instead of suspending bottles in the sea, each light and dark bottle is put under a suitable light filter adjusted to the light transmission at the depth at which the sample was taken. The sample is exposed to sunlight in a temperature-controlled incubator (Jitts, 1963; Vollenweider, 1969; Kiefer and Strickland, 1970).

The indirect method involves measuring the photosynthetic response of phytoplankton to different environmental factors; all results are then combined and integrated mathematically (Ichimura, 1956b; Talling, 1957b), or graphically (Ryther, 1956) in order to determine the total amount of carbon dioxide fixed in the water column. Since light intensity is the most changeable parameter under field conditions, spatial and temporal changes in radiation, and the response of phytoplankton, are the most important experimental results. For this technique a water sample is usually taken from just under the sea surface (or from specific light depths if time permits) and then photosynthesis is measured at the *in situ* temperature and under a series of sunlight illuminations (e.g. 100, 50, 25, 10, 5, and 1% of the surface illumination). Exposure to sunlight is made for a few hours (e.g. 3 hr) around local apparent noon. Other light sources, such as an incandescent lamp showing high colour temperature (e.g. 3200 K), or a daylight-type fluorescent tube, may be used conveniently.

Photosynthetic response of phytoplankton to light intensities (e.g. P vs I curves) has already been discussed in Section 3.1.2 (Fig. 34). Basic parameters describing a P vs. I curve, being generally understood by ecologists and physiologists, are (1) maximum rate of photosynthesis and (2) initial slope of the P vs. I curve. Based on these two parameters, the P vs. I curve can be expressed in several different mathematical forms. The simplest form is a hyperbolic function as follows (Tamiya *et al.*, 1953; Tominaga and Ichimura, 1966):

$$P_g = P_{max} \cdot \frac{I/I_k}{1 + I/I_k} \quad \text{or } P_g = \frac{P_{max} I}{I_k + I} \qquad (35)$$

where P_g is gross photosynthesis (usually expressed as mg C/mg Chl *a*/hr), P_{max} is the maximum photosynthesis (the same units as P_g), I is the light intensity in suitable energy or illumination units, I_k is the light intensity at the junction of initial slope and P_{max} (see Fig. 34); this is the point where P_g is 0·7 P_{max}. I_k is $2I'_k$, in which I'_k is the light intensity when $P_g = P_{max}/2$; it is similar to the half-saturation constant in the Michaelis-Menten equation. The second function in eqn. (35) is in a form similar to the Michaelis-Menten expression for enzyme reactions (see Section 3.1.6). The initial slope of eqn. (35) ($\Delta P/\Delta I = P_{max}/I_k$), however, falls off too fast compared to actual curves measured. Correspondence with data can be improved by rewriting eqn. (35) as follows:

$$P_g = P_{max} \cdot \frac{I/I_k}{\{1 + (I/I_k)^2\}^{1/2}} \cdot \qquad (36)$$

This formulation, first suggested for photosynthesis by Smith (1936), was applied to both culture and natural populations by Talling (1957 a, b).

Under high light intensities photosynthesis is inhibited by photoinhibition (see Figs. 34 and 35) and this is not apparent in either eqns. (35) or (36). The mechanism of photoinhibition, however, has not been fully understood yet; empirical approximation is therefore the only way to formulate a P vs. I curve with photoinhibition. Several authors have proposed solutions to this difficulty (Steele, 1962; Vollenweider, 1965; Parker, 1974). In the solution proposed by Steele (1962), P vs. I curve was expressed as an exponential function:

$$P_g = P_{max} \cdot I/I_{opt} \cdot \exp\,(1 - I/I_{opt}) \qquad (37)$$

where P_g,* P_{max}, and I are the same specified above, and I_{opt} is the light intensity maximizing P_g. The initial slope of this equation is

$$\frac{\Delta P}{\Delta I} = P_{max} \cdot e/I_{opt}. \qquad (38)$$

In practice it is difficult to fit this equation to actual data obtained experimentally, because the shape of the curve is essentially fixed, particularly at high light intensity. However, this handicap may be overcome to some extent by introducing eI_k for the first I_{opt} in eqn (37) (Vollenweider, 1965).

Vollenweider (1965) and Parker (1974) have made further modification on eqns (36) and (37), respectively, by introducing more parameters, which give more flexibility in fitting experimental data. In Parker's modification, he raised the entire right-hand side of eqn. (37) to a power. The altered function can then be written as:

$$P_g = P_{max}\{I/I_{opt} \cdot \exp(1 - I/I_{opt})\}^{\beta} \qquad (39)$$

where β is a dimensionless constant, and others are the same as specified previously. The initial slope ($\Delta P/\Delta I$) of this curve depends on β; $\Delta P/\Delta I = (P_{max}/I_{opt})\exp(\beta)$. On the other hand, Vollenweider (1965) made his modification by introducing two more parameters into eqn (36):

$$P_g = P_{max} \frac{I/I_k}{\{1 + (I/I_k)^2\}^{1/2}} \cdot \frac{1}{\{1 + (bI/I_k)^2\}^{n/2}} \qquad (40)$$

*Steele (1962) expressed P_g in terms of actual growth rate, g C/g plant C/hr instead of g C/g Chl *a*/hr.

where P_{max} is the maximum rate of photosynthesis when $b = 0$ or $n = 0$, b is a constant with dimension of (light intensity)$^{-1}$, and n is a dimensionless constant. In practice the parameters P_{max} and I_k are unmeasurable in eqn. (40), when there is any kind of inhibition. Actual methods required to evaluate these parameters, which requires a complicated treatment, have been discussed by Fee (1969). Both eqns. (39) and (40) are capable of fitting a large variety of P vs. I curves with and without photoinhibition, and are quite successful in describing experimental data. This is not surprising in view of the number of free parameters (three or four) which are available for fitting. However, there are some drawbacks accompanying this flexibility; these equations are cumbersome to integrate over depth and time, and there are some difficulties involved in evaluating the parameters as mentioned above. It should also be mentioned that there is one more approach, which is rather different from those mentioned above, but which is used to approximate any shape of P vs. I curve; this is the use of polynomial equation which has the advantages of simpler parameter estimation and analytic integrability as well as a probable low loss of fidelity in describing experimental data (cf. Takahashi and Nash, 1973).

The derivative for each equation when it is evaluated for $I = 0$, which derives the initial slope of the curve, always includes P_{max} as one of the parameters. In other words, the derivative can also be rewritten for P_{max}. In the case of eqn. (35), for example, the initial slope (ϕ will be used instead of $\Delta P/\Delta I$ for simplicity) can be expressed as,

$$\frac{\Delta P}{\Delta I} = \phi = P_{max}/I_k, \tag{41}$$

and then the whole equation can be rearranged for P_{max} as follows:

$$P_{max} = \phi I_k. \tag{42}$$

Generally, P_{max} varies widely depending on changes in environmental conditions and the physiological state of algae; however, the initial slope is more nearly constant. Considering these fundamental characters for P_{max} and ϕ, eqn (42) indicates that I_k is entirely proportional to P_{max} because ϕ is constant. By replacing P_{max} in eqn. (35) with ϕI_k, the equation becomes simple without losing any of the capabilities of the original equation:

$$P_g = \phi I_k \cdot \frac{I/I_k}{1 + I/I_k} = \phi \frac{I}{1 + I/I_k} \tag{43}$$

where ϕ must have the same light intensity unit as I and I_k. Actual values for ϕ can be obtained either from direct measurement or data already reported (cf. Table 16). This advancement for empirical approximation using ϕ was first suggested by Bannister (1974), and can be made as well for other eqns. (36), (37), (39), and (40) without any difficulty (see details in Bannister, 1974). However, as pointed out by Platt and Jassby (1976), the initial slope actually determined in a given area throughout the year was not constant but varied about 5-fold. This indicates that the initial slope determined under natural conditions is considerably lowered from the potential photosynthetic yield due to possible internal and external stresses. The model approach proposed by Bannister, therefore, is considered to give a potential rate for a given environment.

Other constants in equations mentioned above have also eco-physiolgocal meanings; β, b, and n are related to photoinhibition and the effects of changes in these constants on curve approximation have been shown by Vollenweider (1965), Fee (1969), and Parker (1974). Actual evaluation for these constants for photoinhibition has to wait until fundamental mechanisms for photoinhibition become clearer from physiological and biochemical laboratory experiments. Knowledge of how these constants, including ϕ, k, β, b, and n, actually change with environmental conditions and algal populations will have to be based on many P vs. I curves taken from different areas and at different seasons; such data may eventually provide suitable P vs. I curves, without having to make actual measurements of photosynthesis.

As has already been mentioned in this section, photosynthetic rate changes diurnally (see also Section 3.1.2). For example, evening rates at optimum light intensity may drop as low as one-tenth of that measured during the early morning hours (e.g. Doty, 1959). In some waters this decrease may be a function of diurnal nutrient depletion (e.g. in the tropics) or of saturation of photosynthetic dark reactions.

Apart from the physiological depression in photosynthetic response, the overall diurnal response of photosynthesis is dependent on changes in the light intensity. In general, the diurnal change of incident light intensity on a horizontal plane can be given by empirical equations (Vollenweider, 1965; Ikusima, 1967). According to Ikusima, for a fine day

$$I_t = I_{max} \cdot \sin^3(\pi/D)t \tag{44}$$

where I_t (expressed as a unit of photosynthetically available light energy or illumination) is the light intensity at a given time, t, on a given day and I_{max} is the light intensity when the altitude of the sun is highest at local apparent noon; D (hr) is the day length. If the relationship of light penetration in a water column (Section 3.1.2) is put into the I_{max} of eqn. (44), light intensity at a given depth and time can be calculated for a fine day as

$$I_{t \cdot d} = I_{0\text{-}max} \cdot \sin^3(\pi/D)t \cdot e^{-kd} \tag{45}$$

where $I_{o\text{-}max}$ is the light intensity at the surface for the highest altitude of the sun, k is the average extinction coefficient of photosynthetically available light (m^{-1}), and d is the depth in meters. For overcast days the equations given above should be modified by taking the $\sin^2$ instead of the $\sin^3$; if I_{max} is reduced to account for cloud cover this will effectively reduce the total radiation while keeping the same general shape of the diurnal curves.

For combining a P vs. I curve (physiological response) and light conditions (environmental response), eqn. (45) is inserted into the I of eqn. (35) or (36) and an equation for calculating gross photosynthesis at a given depth (d) and time (t), $P_{g \cdot t \cdot d}$ is obtained. For example, from eqns (35) and (45), it is possible to write

$$P_{g \cdot t \cdot d} = P_{g \cdot t} \frac{a \cdot I_{0\text{-}max} \cdot \sin^3(\pi/D)t \cdot e^{-kd}}{1 + a \cdot I_{0\text{-}max} \cdot \sin^3(\pi/D)t \cdot e^{-kd}} \tag{46}$$

Instead of using empirical equations [eqns. (44) and (45)], actual measurements of solar radiation can be put directly into an equation [e.g. eqn. (36)] for a P vs. I curve if they are available (Section 3.1.3), and then the following calculations can be continued.

By integrating eqn. (46) with time ($t = 0 \rightarrow D$), daily gross photosynthesis (usually in mg C/mg Chl a/day), at a given depth (d) is calculated as

$$P_{g \cdot d} = \int_0^D P_{g \cdot t \cdot d} \cdot dt \tag{47}$$

By integrating eqn. (47) with depth ($d = 0 \rightarrow \infty$), daily gross photosynthesis (usually mg C/day/m^2) is calculated as

$$P_g = \int_0^\infty P_{g \cdot d} \cdot dd\varphi. \tag{48}$$

For determining the growth of phytoplankton under natural conditions, net photosynthesis should be considered instead of gross photosynthesis. Net photosynthesis can be calculated by subtracting respiration, R(mg C/mg Chl a/hr), from gross photosynthesis as follows:

Hourly net photosynthesis, $P_{n \cdot t \cdot d}$;

$$P_{n \cdot td} = P_{g \cdot t \cdot d} - R. \tag{49}$$

Diel net photosynthesis, $P_{n \cdot d}$;

$$P_{n \cdot d} = P_{g \cdot d} - 24R. \tag{50}$$

where respiration rate of phytoplankton is assumed to be constant throughout the day. The depth at which $P_{n \cdot t \cdot d} = 0(P_{g \cdot t \cdot d} = R)$ or $P_{n \cdot d} = 0(P_{g \cdot d} = 24\,R)$ is called the hourly and diel compensation depth, respectively. The hourly compensation depth will change during the day and be maximum at noon and zero during darkness; the 24-hr compensation depth will change with season. Unfortunately, it is not often clear in the literature which compensation depth is being reported.

Diel photosynthesis obtained above is the diel rate when 1 mg Chl a/m³ is distributed homogeneously in the euphotic zone; this is not the actual diel photosynthesis of the water column since the amount of chlorophyll will vary in time and space. In order to obtain the actual photosynthesis, $\mathbb{P}_{g \cdot d}$, the $P_{g \cdot d}$ must be multiplied by the amount of phytoplankton biomass, S (usually expressed as mg Chl a/m³):

$$\mathbb{P}_{g \cdot d} = P_{g \cdot d} \times S, \qquad (51)$$

$$\mathbb{P}_{n \cdot d} = P_{n \cdot d} \times S. \qquad (52)$$

$\mathbb{P}_{n \cdot d}$ is sometimes referred to as ΔN, which represents the diel net increase in standing stock, N. Hourly photosynthesis in a P vs. I curve is sometimes expressed on the basis of a unit volume of water, such as mg C/m³/hr. In such a case, however, the curve can only be applied to the depth from which the sample for the measurement of P vs. I was taken, and at all depths, only in the season when the chlorophyll a is distributed homogeneously, such as during active vertical mixing. When chlorophyll a is not uniformly distributed, the expression for photosynthesis per unit of chlorophyll a must be obtained by multiplying the photosynthesis by the actual amount of chlorophyll a at different depths in the water column. Finally the depth integration of $\mathbb{P}_{g \cdot d}$ can usually be done with a planimeter, or by graphic integration.

In determining an average specific growth rate, u, for the phytoplankton over short intervals, the biomass (usually expressed as mg C/m³) is taken initially as N_0. The measured diel increase of phytoplankton, assumed to represent diel net photosynthesis (ΔN) on a unit carbon base instead of chlorophyll a, is then added to the biomass after a day's growth (t). For transforming diel net photosynthesis $P_{n \cdot d}$ (mg C/mg Chl a) into ΔN, $P_{n \cdot d}$ must first be converted to cellular carbon by multiplying by the amount of chlorophyll a in the water [eqn. (52)]. Then

$$\Delta N = N_t - N_0 \qquad (53)$$

where N_t is the biomass after t days and the average specific growth rate is calculated as

$$\mu = 1/t \cdot \ln [(N_0 + \Delta N)/N_0]. \qquad (54)$$

The average specific growth rate is often expressed in terms of 'doublings of algal biomass per day' which is called 'doubling time' and expressed as μ_2. For this purpose the base of the logarithm in eqn. (54) should be replaced by 2 instead of e (2.7183)*.

Average specific growth rates of phytoplankton in the euphotic zone for various regions are shown in Table 19. The growth rates, measured in an experimental bottle, represent the potential rate under given environmental conditions on a particular day. The actual growth rate *in situ* may be smaller or larger than the potential rate. Smaller *in situ* growth rates may result from losses of phytoplankton cells in the water column due to sinking, drifting, dying, zooplankton feeding, vertical mixing, and the movement of phytoplankton themselves. Larger *in situ* growth rates may result in nutrient-depleted environments since the exclusion of animals from the incubation bottles may decrease nutrient regeneration.

The time increase of phytoplankton population

*Cell number increase of unicellular organisms can be expressed as follows when they grow by binary cell division:

$$(N_0 + \Delta N) = N_0 \cdot 2^n \qquad \text{(i)}$$

where $(N_0 + \Delta N)$ and N_0 are cell numbers at the initial stage and after n generations. The total time duration for n generations, t, can be expressed by

$$t = n \cdot g \qquad \text{(ii)}$$

where g is the time duration for each generation (generation time, time⁻¹). Then the eqn. (i) can be written as

$$(N_0 + \Delta N) = N_0 \cdot 2^{t/g} \qquad \text{(iii)}$$

or

$$g = \frac{t \ln 2}{\ln(N_0 + \Delta N) - \ln N_0}$$

'doubling time' can be obtained from the reciprocal of 'generation time'.

TABLE 19. AVERAGE SPECIFIC GROWTH RATE OF PHYTOPLANKTON IN THE EUPHOTIC ZONE FOR VARIOUS REGIONS. TEMPERATURES INDICATED ARE FOR THE SURFACE OR THE AVERAGE IN THE MIXED LAYER (AFTER EPPLEY, 1972)

Location	Temp. (°C)	Growth rate (doublings/day) measured	max. expected*	Reference
Nutrient-poor waters				
Sargasso Sea	—	0·26	—	Riley *et al.* (1949)
Florida Strait	—	0·45	—	Riley *et al.* (1949)
Off the Carolinas	—	0·37	—	Riley *et al.* (1949)
Off Montauk Pt.	—	0·35	—	Riley *et al.* (1949)
Off Southern California				
July 1970	20	0·25–0·4	1·5	Eppley *et al.* (1972)
Apr.–Sept. 1967	12–21	0·7 av.	0·9–1·6	Eppley *et al.* (1972)
Nutrient-rich waters				
Peru Current				
Apr. 1966	17–20	0·67 av.	1·5	Strickland *et al.* (1969)
June 1969	18–19	0.73 av.	1·4	Beers *et al.* (1971)
Off S. W. Africa	—	1·0 av.	—	Calculated from Hobson (1971)
Western Arabian Sea	27–28	>1·0 av.	2·4	Calculated from Ryther and Menzel (1965)

*From eqn. (34) (Section 3.1.4) assuming the maximum specific growth rate, μ, will be one-half the value calculated as expected if daylength is 12 hr and μ is directly proportional to the number of hours of light per day.

can then be estimated under the assumption that the logarithmic growth is maintained as follows:

$$N_2 = N_1 \exp [\mu(t_2 - t_1)] \qquad (55)$$

in which N_1 and N_2 are phytoplankton biomass on volume basis ($/m^3$) at times t_1 and t_2, generally in 'days', respectively. Calculation is practically done on each day basis. By integrating the phytoplankton biomass over the depth, the phytoplankton population in the entire water column can be estimated. Results obtained in the present estimation represent the potential *in situ* increase or production of phytoplankton mentioned above if the specific growth rate determined by bottle incubation experiment is applied. Actual estimation of the *in situ* growth of phytoplankton will be discussed in Section 4.2.3.

Of these effects, the most pronounced in temperate latitudes is the depth to which the water column is mixed. Thus, under conditions of storm activity, or tidal mixing in coastal areas, it is apparent that phytoplankton cells may be mixed down below the euphotic zone and that the time that the cells spend in the disphotic zone before being mixed back to the euphotic zone, may be sufficient to result in no net production; the period of darkness below the disphotic zone having resulted in a loss through respiration of the carbon gained through photosynthesis in the euphotic zone. In this sense the euphotic zone is defined strictly as the depth to which the photosynthesis of a plant cell is equal or greater than its respiration; the depth of the euphotic zone is known as the compensation depth (D_c) and the light intensity at this depth is the compensation light intensity (I_c).

The depth to which plants can be mixed and at which the total photosynthesis for the water column is equal to the total respiration (of primary producers) is known as the 'critical depth'. The concept of a critical depth was first suggested by Gran and Braarud (1935) and developed into a mathematical model by Sverdrup (1953). The following description of the model is taken from Sverdrup's original paper, with a few modifications. In particular, Sverdrup reduced the incident solar radiation (I_0) by a factor of 0·2 to allow for absorption of the longer and shorter wavelengths of light in the first meter of sea water. In shallow water columns this may be too large a factor and it is suggested instead that the incident solar radiation should be reduced by a factor of 0·5 to allow for the absorption of non-photosynthetic ultraviolet and infrared radiation in the first few centimeters of water. An average extinction coefficient for light (400 to 700 nm) penetrating the rest of the water column should then be used; the latter value may be rather larger than the value used by Sverdrup, especially if the model is used in coastal waters.

The model is illustrated in Fig. 41. If I_0 is the surface radiation and k is the average extinction coefficient, then the compensation light intensity (I_c) is related to the compensation depth (D_c) as

$$I_c = 0{\cdot}5{\cdot}I_0{\cdot}e^{-kD_c} \qquad (56)$$

where a constant, 0.5, is a conversion factor for PAR in the total solar radiation for sunny days (see in Section 3.1.2.).

At the compensation depth the photosynthesis of a cell (P_c) is equal to its respiration R_c; above this depth there is a net gain from photosynthesis ($P_c > R_c$) and below it there is a net loss ($P_c < R_c$).

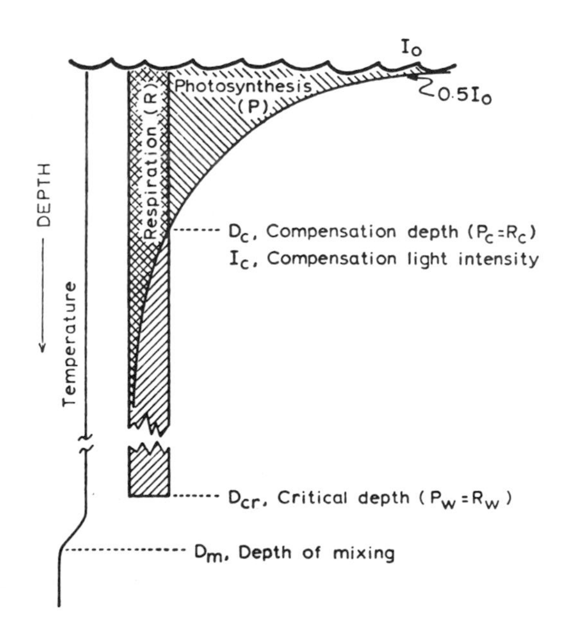

FIG. 41. Diagram showing the relationship between the compensation and critical depths, and the depth of mixing. (For explanation, see text.)

However, as the phytoplankton cells are mixed above and below the compensation depth they will experience an average light intensity ($\bar{I}$). The depth at which the average light intensity for the water column equals the compensation light intensity is known as the critical depth, D_{cr} [i.e. the depth at which photosynthesis for the water column (P_w) equals the respiration for the water column (R_w)]. The relationship of the critical depth to the compensation light intensity can be obtained by integrating eqn. (56) and dividing by the critical depth to obtain the average (compensation) light intensity for the water column. Thus

$$\bar{I}_c = 0{\cdot}5\, I_0 \int_0^{D_{cr}} \frac{e^{-kD}{\cdot}dD}{D_{cr}}$$

$$\bar{I}_c = \frac{0{\cdot}5\, I_0}{kD_{cr}}(1 - e^{-kD_{cr}}). \qquad (57)$$

From eqn. (57) it is possible to determine the critical depth for any water column, knowing the extinction coefficient (k), the solar radiation (I_0) and assuming some value for the compensation light intensity (see Section 3.1.2). Where D_{cr} is large the equation reduces to

$$D_{cr} = \frac{0{\cdot}5\, I_0}{\bar{I}_c k}. \qquad (58)$$

From the explanation given above, it is apparent that if the critical depth is less than the depth of mixing, no net production can take place since $P_w < R_w$. This is illustrated as the condition in Fig. 41 where the depth of mixing (D_m), measured to the depth to the bottom of the principal thermocline, has been drawn as being greater than the critical depth. However, if the critical depth is greater than the depth of mixing, a net positive production will occur in the water column ($P_w > R_w$) and conditions for the onset of phytoplankton growth will have been established.

Several conditions are attached to the use of the critical depth model; these are (i) that plants are uniformly distributed in the mixed layer, (ii) there is no lack of plant nutrients, (iii) the extinction coefficient of the water column is constant (this in fact has to be determined as an average value), (iv) the production of the plants is proportional to the amount of radiation, and (v) respiration is constant with depth. The last of these assumptions is made as a matter of convenience and there are in fact no ecological data on the constancy of respiration with depth. Furthermore, some difficulties may be encountered in establishing the depth of mixing; in Fig. 41 this has been simplified as being the depth to the bottom of the principal thermocline. However, in other environments the depth of the principal halocline or the depth at which there is a maximum change of density with depth ($\partial\sigma/\partial z \times 10^3$, Sverdrup *et al.,* 1946) may be better employed to determine the depth of mixing. A similar approach to the critical depth model was used by Riley (1946) and this is discussed in Section 4.2.2.

Once the mixing zone moves up above the critical depth, the standing stock of phytoplankton increases and a number of conditions for the critical-depth model may be no longer valid. Further, as the biomass of phytoplankton increases, light conditions in the water column will be changed and the water column will become stratified. Empirical relationships between light penetration and chlorophyll *a* concentrations have been given by several authors (e.g. Riley, 1956a; Aruga and Ichimura, 1968). In the former reference an equation is given relating the average extinction coefficient (k'_e, m^{-1}) to the chlorophyll *a* concentration (C, mg Chl a/m^3) for natural phytoplankton community as follows.

$$k' = 0{\cdot}04 + 0{\cdot}0088\ C + 0{\cdot}054{\cdot}C^{2/3}. \qquad (59)$$

The constant (0.04) in the equation is close to the average extinction coefficient for visible light in clean water (0·0384) (see Table 20). The original equation was derived from the actual field observations carried out in the western North Atlantic. The equation does not specifically distinguish between phytoplankton and other materials and it may also include some extinction due to particulate and dissolved materials which are not directly related to chlorophyll (Riley, 1975). Nevertheless this is still a useful empirical formula which can be used to estimate the average extinction coefficient in natural water.

Figure 42 shows diagrammatic examples of phytoplankton growth in a stratified water column. At time, t_0, just after active vertical mixing of water, phytoplankton is distributed homogeneously in the water column but the *in situ* photosynthesis per unit phytoplankton biomass (mg C/mg Chl *a*/day) at each depth is different, being inhibited at the near surface by high light intensities and then decreasing from a subsurface maximum due to light attenuation. Assuming that the losses of phytoplankton at a given depth (from sinking, zooplankton grazing, etc.) are similar, the biomass of phytoplankton after a certain period can be estimated approximately by

TABLE 20. AVERAGE LIGHT EXTINCTION COEFFICIENTS (400 TO 700 nm) OF CULTURED ALGAE AND SEA WATER (FROM TAKAHASHI AND PARSONS, 1972)

Materials	k'_2 (mg Chl $a/m^2)^{-1}$	Chl a·max (mg Chl $a{\cdot}m^{-2}$)	References
	k'_2		
Chlorella sp.	0·0115	400	Tominaga and Ichimura (1966)
Scenedesmus sp.	0·0051	900	Aruga (1966)
Cyclotella meneghiniana	0·0048	960	Steemann Nielsen (1962)
Skeletonema costatum	0·0071	650	Tominaga and Ichimura (1966)
	$k'_2 + k'_3$		
Natural phytoplankton populations (including suspended particles and dissolved matter, Kuroshio)	0·0184	250	Aruga and Ichimura (1968)
	k'_2 (m^{-1})		
Sea water (clear ocean)	0.0384		Jerlov (1968)

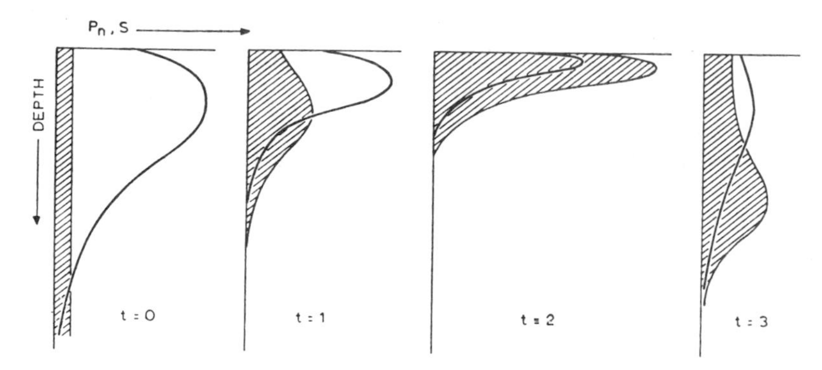

FIG. 42. Schematic changes in phytoplankton biomass (S) and daily net photosynthetic rate (P_n) after three time intervals (t) in stratified water (S or ▨, usually expressed in mg Chl a/m^3; P_n or ——, usually expressed as mgC/mgChl a/day.

using eqn. (55). Because of the difference in the *in situ* photosynthesis per unit phytoplankton biomass at $t = 0$, the *in situ* growth rate of phytoplankton, μ will be different at each depth. Thus the profile of phytoplankton biomass and photosynthesis (called the 'productive structure' — Ichimura, 1956b) will change with time $t = 1$, $t = 2$, etc. As the phytoplankton biomass increases at the sub-surface photosynthetic maximum, the average extinction coefficient of light will also increase, and self-shading will occur (see Aruga, 1966). Accordingly, the daily compensation depth and the maximum in phytoplankton growth becomes shallower (Ichimura, 1956a). Providing there are sufficient nutrients, the pattern of photosynthesis and the standing stock of phytoplankton will finally maximize at the surface as a thin layer, and a bloom will occur ($t = 2$). As nutrients become exhausted in the surface layers, however, the depth of the maxima in the phytoplankton biomass and the primary productivity deepen ($t = 3$); the latter conditions occur in stratified water columns in temperate latitudes during the summer or may be found to prevail generally in tropical and subtropical waters throughout the year. However, other factors can modify the overall effect of light penetration on the production of the water column. Thus the feeding activity of zooplankton strongly affects the standing stock of phytoplankton and more detailed information of this subject is given in Section 4.2.2. In other cases, organic substances stimulate growth of phytoplankton (e.g. humic substances in sea water are reported to stimulate the growth of dinoflagellates — Prakash and Rashid, 1968). In some tropical areas motile phytoplankton, such as dinoflagellates (Eppley *et al.,* 1968), can use nutrients from the aphotic zone by diel migrations between near surface and deeper waters. A further method of overcoming nutrient limitation in tropical waters is through nitrogen fixation (e.g. Goering *et al.*, 1966), which sometimes gives rise to large surface blooms of *Trichodesmium*.

The quantity and quality of light in a water column is affected by the depth of the water column, the presence of dissolved coloured substances, and scattering due to suspended particles, including the phytoplankton. The light attenuation in a given water column is approximated from a logarithmic equation [eqn. (29) in Section 3.1.2] which is governed by the extinction, $k'D$. The extinction is a function of the total phytoplankton biomass ($\bar{C}$), the depth of water (D) and the total amount of other suspended and dissolved matter ($\overline{M}$), as follows (Sakamoto, 1966):

$$k'D = k'_1D + k'_2\bar{C} + k'_3\overline{M} \qquad (60)$$

in which k'_1, k'_2 and k'_3 are average coefficients (400 to 700 nm) of a *unit* amount or thickness of phytoplankton, water and other matter, respectively. Then $k'D$, k'_1D_c, $k'_2\bar{C}$, and $k'_3\overline{M}$ correspond to respective k', k'_w, k'_p and k'_s in the equation. Assuming the diel compensation depth is the depth at which there is 1% of the surface illumination,* then the relation between chlorophyll a in the euphotic zone ($\bar{C}$, mg Chl a/m^2), the depth of euphotic zone (D_c in meters) and suspended and dissolved matter ($\overline{M}$, expressed in units of thickness, e.g. meters) can be expressed by the following equation:

$$0{\cdot}01 = e^{-(k_1'D_c + k_2'\bar{C} + k_3'\overline{M})} \qquad (61)$$

0·01 can be approximately replaced by $e^{-4{\cdot}6}$, which gives:

$$4{\cdot}6 = k_1'D_c + k_2'\bar{C} + k_3'\overline{M} \qquad (62)$$

then

$$\bar{C} = \frac{4{\cdot}6 - (k_1'D_c + k_3'\overline{M})}{k_2'} \qquad (63)$$

where $\bar{C}$ in eqn. (63) represents the total chlorophyll a in the euphotic zone of a given water column. If the euphotic depth is close to zero (i.e. the light absorption due to water and suspended and dissolved matter is negligible, $k'D_C - k_3'\overline{M} \simeq 0$), $\bar{C}_{max} \simeq 4{\cdot}6/k_2'$; where $\bar{C}_{max}$ represents the maximum amount of chlorophyll a in the euphotic zone of a given water column. Some values for the maximum amount of chlorophyll a in cultures are shown in Table 20. *Skeletonema* culture shows a maximum of 650 mg Chl a/m^2. Under field conditions, the maximum amount of chlorophyll a is usually less than that of pure cultures because of the effects of $k_3'\overline{M}$ and sometimes $k_1'\,D_c$ on the increased $k'\,D_c$. For example, in the Kuroshio current, the maximum amount of chlorophyll a in the euphotic zone has been computed to be 250 mg Chl a/m^2. Lorenzen (1972) found a highly significant correlation between the thickness of the euphotic zone (using the 1% light level for the euphotic depth) and the total chlorophyll content. From extensive field observations in the Pacific and Atlantic Oceans (including upwelling regions and offshore areas) he derived an equation as follows:

$$\ln \bar{C} = 8{\cdot}85 - 1.57 \ln D_c \qquad (64)$$

in which $\bar{C}$ is the total chlorophyll a in the euphotic zone (mg Chl a/m^2) and D_c the 1% light depth (m). The data used in determining this equation covered a wide range of values, i.e. D_c, 10–91 m; $\bar{C}$, 7–277 mg Chl a/m_2.

An assessment of light extinction by the three factors water, phytoplankton, and particulate and dissolved matter can be made using eqns. (62) and (64). This is done by adding the attenuation due to water ($k_1'D_c$) and the phytoplankton [$k_2'\bar{C}$, where $\bar{C}$ is estimated from eqn. (64)], and subtracting this from the total attenuation of the euphotic zone, which by definition is the natural logarithm of 0·01 ($\simeq$ 4·605). The quantity left is equal to the extinction due to dissolved and particulate substances ($k_3'\overline{M}$). The relative effect of these three factors in euphotic zones of different thickness (Fig. 43) is reflected in the ratio $k_x'\,X/kD_c$ where $k_x'\,X$ is the attenuation of each factor with the thickness of the euphotic zone D_c as determined above, and $k'D_c$ is total attenuation throughout the euphotic zone. From any set of data, the thickness of the euphotic zone (D_c) and average extinction coefficient in the euphotic zone (k', m^{-1}), one can estimate the possible light extinction due to water ($k_1'D_c$), phytoplankton ($k_2'\bar{C}$), and other materials ($k_3'\overline{M}$) from Fig. 43. In springtime in the subarctic Pacific, for example, D_c and k' are 55 m and 0·080 (m^{-1}), respectively, then respective $k_1'D_c$, $k_2'\bar{C}$, and $k_3'\overline{M}$ can be estimated as 0·040, 0·037, and 0·003 (all these have a dimension of m^{-1}).

It is clear that $k_x'\bar{C}/k'D_c$ due to phytoplankton shows a logarithmic decrease with increasing thickness of the euphotic zone. The fraction due to particulate and dissolved matter ($k_x'\overline{M}/k'D_c$). which includes 'yellow substance', has a maximum around 20–30 m. This indicates that in high productivity areas generally characterized by thin euphotic

*Compensation light intensities for marine phytoplankton are generally found to be in the approximate range of 0·002 to 0·009 ly/min; since these values are very approximately two orders of magnitude lower than averge surface radiation values, the compensation depth is sometimes simply estimated from the depth at which there is 1% of the surface radiation. A further approximate estimate of this depth is based on the relationship between the depth of the Secchi disc and the extinction coefficient. If this is assumed to be $k_e' = 1{\cdot}7/T$ (p. 71), then $0{\cdot}01 = e^{-k'Dc}$ and $k'D_c = 4{\cdot}6$ substituting for k' with the Secchi depth (T), $D_c = (4{\cdot}6/1{\cdot}7)$. $T \simeq 2{\cdot}7\,T$. Thus three times the Secchi depth is sometimes employed as an estimate of the euphotic depth.

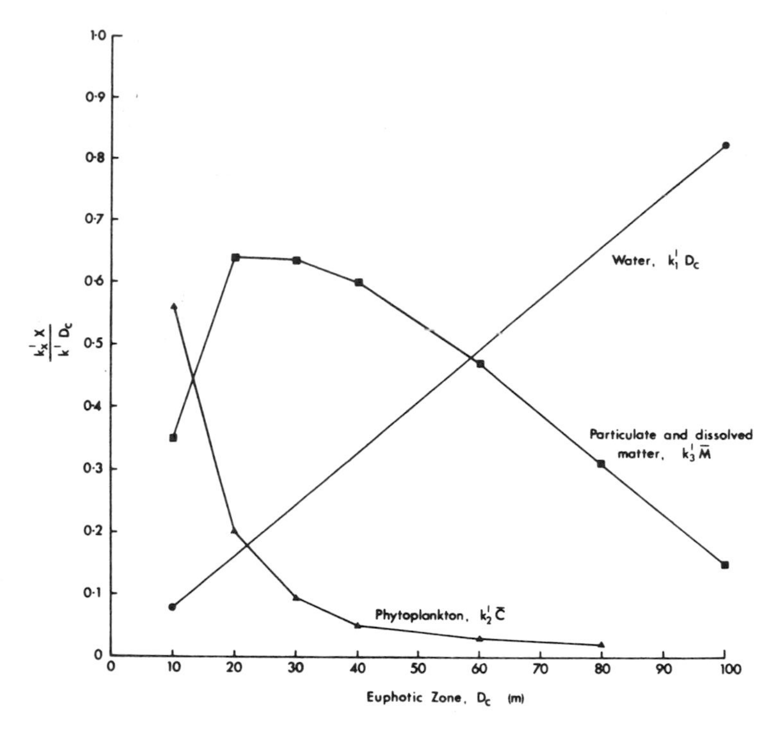

FIG. 43. Partitioning of total light attenuation ($k'D_c$), into the fractions attributable to water (k'_1D_c), phytoplankton ($k'_2\bar{C}$), and particulate and dissolved matter ($k'_3\bar{M}$). See eqn. (60) in text (from Lorenzen, 1972, 1976).

zones, and if not influenced by land runoff or tidal mixing, most of the light is attenuated by phytoplankton ($t = 2$ in Fig. 42). After a period of time, the phytoplankton concentration is decreased by grazing or a decline of available nutrients, etc., and the euphotic zone deepens. It is at intermediate euphotic zone depths (20–30 m) that an increase in the attenuation of light by particulate and dissolved matter occurs, probably as a residue to a past plankton bloom ($t = 3$ in Fig. 42). Later, the euphotic zone attains maximum depths (>60–70 m). At this time, most of the light is attenuated by water, with only a very small fraction attributed to the phytoplankton. The decline in the fraction of attenuated light attributed to particulate and dissolved materials probably results from the euphotic layer. For example, particulates may be ingested by zooplankton, decomposed by bacteria, or sink into deeper layers. A similar treatment of light attenuation in the sea will be found in Tyler (1975); in this reference the author distinguishes between light absorbed by chlorophyll *a* and phaeophytin.

It is predictable that the diel photosynthesis in a given water column will change with sustained changes in the radiant energy reaching the sea surface, even if the amount of chlorophyll *a* does not change. As an example, differences in diel net photosynthesis under different amounts of solar radiation are shown in Table 21. For the calculation of results shown in this table, the photosynthesis vs. light curve shown in Fig. 35B (in Section 3.1.3) was used, the P_{max} was assumed to be 1 mg C/mg Chl *a*/hr and the relation between the euphotic depth and the amount of chlorophyll *a* in the euphotic zone was obtained from Fig. 43. If data on the total surface radiation and the total amount of chlorophyll *a* in the euphotic zone (or the depth of the euphotic zone) can be obtained, one can estimate the approximate net photosynthesis from Table 21. In calculating production in Table 21, P_{max} was assumed to be 1 mg C/mg Chl *a*/hr, which falls

TABLE 21. DIEL NET PHOTOSYNTHESIS FOR VARIOUS AMOUNTS OF CHLOROPHYLL *a* UNDER DIFFERENT LEVELS OF IRRADIANCE (FROM TAKAHASHI AND PARSONS, 1972)

Radiation (ly/day)		Net photosynthesis (mg C/m^2/day)								
	Euphotic depth (*m*)	175	150	100	50	30	20	10	5	<1
	Chl *a* amount (mg/m^2)	1	2	10	52	100	135	190	220	260
100		1·6	3	16	83	160	220	300	350	420
200		2·6	5	26	140	260	350	490	570	680
300		3·3	7	33	170	330	450	630	740	860
400		4·0	8	40	210	400	540	760	880	1000
500		4·7	9	47	240	470	630	890	1000	1200
600		5·3	11	53	280	530	720	1000	1200	1400
700		5·8	12	58	300	580	780	1100	1300	1500
800		6·3	13	63	330	630	850	1200	1400	1600
900		6·8	14	68	350	680	920	1300	1500	1800

in the range of values for temperate pelagic waters (see Table 17). It would be better to measure the actual P_{max} in any environment in order to obtain a better prediction of the *in situ* productivity; in such a case the values for net photosynthesis in Table 21 have to be multiplied by the value of the actual P_{max}. In highly productive areas, P_{max} may be greater than 5 mg C/mg Chl *a*/hr. Thus the calculated net photosynthesis from Table 21 could be at least 1800 × 5 = 9000 mg C/m^2/day. This value is in general agreement with maximum *in situ* primary productivity values of 5 to 10 g C/m^2/day. Primary productivity values under culture conditions may be two or three times greater than *in situ* (see also Ryther, 1963).

Based on the photosynthetic production, radiation utilization efficiency in natural water can be estimated taking a ratio of photosynthetic production and the total incoming solar radiation. This is called 'light utilization efficiency' and 'ecological efficiency' and is often expressed by using a symbol of 'ϵ'. The light utilization efficiency should not be confused with the quantum (or photosynthetic) efficiency which has been described in Section 3.1.2. Considering the total incoming solar radiation, there are two treatments: one is the total radiation energy reached, and the other only considers the photosynthetically available radiation, PAR. Because only about half of the solar radiation reaching earth is PAR, it should be stated whether efficiencies are calculated relative to total energy or PAR. It is obvious that the efficiency based on the total energy is smaller than that on PAR. In order to obtain a dimensionless figure, the photosynthetic production is generally converted into energy units from carbon content and this is defined as PSR in Section 3.1.2. Since 1 mole of CO_2 fixed into carbohydrate (CH_2O) contains 112 kcal·mole^{-1}, 1 mg of carbon fixed is equivalent to an energy storage of 9.33 g cal (= 39 joules). Platt and Irwin (1973) obtained a similar value of 11·4 g cal·mg C^{-1} experimentally from calorimetric determinations of marine particulate matter, which included phytoplankton carbon as well as other organic carbon from various sources. The dimensionless utilization efficiency, ϵ, can be obtained as follows:

$$\epsilon = \frac{9.33\ (\text{g cal·mg C}^{-1}) \times \text{Photosynthetic production (mg C·m}^{-2}\text{·day}^{-1})}{\text{Incoming solar radiation (g cal·m}^{-2}\text{·day}^{-1})}. \qquad (65)$$

PAR determined by quantum units has to be converted into energy units. It has been shown (Morel and Smith, 1974) that in spite of the spectral variability that exists in the underwater light field, the conversion of quanta into energy can be made using the equivalence 1 joule = $2{\cdot}5 \times 10^{18}$ quanta for the domain 350 to 700 nm (or 40 joules = 10^{20} quanta), this approximation being valid with an accuracy better than 10% (Morel, 1978). According to the extensive study in St. Margaret's Bay, Nova

Scotia, throughout a year (Platt, 1977), the light-utilization efficiencies fluctuated between 0·02 and 0·9% being an average of 0·26% (per total radiant energy) in which chlorophyll *a* in the water column varied from 10 to 340 mg Chl $a \cdot m^{-2}$, being an average of 30 mg Chl $a \cdot m^{-2}$. For several marine and lacuserine situations estimated efficiencies on the total energy ranged from less than 0·02% in the oligotrophic Sargasso Sea to over 5% in Lake Manito, Canada, and Eniwetok Atoll (Dubinsky, 1980). Dubinsky and Berman (1976) reported an average, ϵ, on PAR of 0.48% in Lake Kinneret (Table 15).

Platt (1969) has proposed an index k_b^d in (m^{-1}) which represents that contribution to the absorption (extinction) coefficient of light energy at any depth (*d*) which is due to photosynthetic production. This should be equal to the absorption of light by photosynthetic pigments multiplied by a factor measuring the efficiency with which light is utilized in the photosynthetic process. It was shown that this index, k_b^d, was very nearly equal to the photosynthetic efficiency.

3.1.6 The Kinetics of Nutrient Uptake

The two earliest ecological considerations in the uptake of nutrients by phytoplankton were firstly, that at low nutrient concentrations, the rate of nutrient uptake was found to be concentration dependent, and secondly, that the total yield of phytoplankton was directly proportional to the initial concentration of limiting nutrient and independent of the growth rate of phytoplankton. Ketchum (1939a) demonstrated the first of these results; using the diatom, *Nitzschia closterium*, he showed that the uptake of nitrate was concentration dependent over an approximate range from 1 to 7 μg at N/l. Spencer (1954), using the same organisms, demonstrated the second result when he showed a linear relationship between the initial concentration of nitrate and total yield of cells over nitrate concentrations ranging from approximately 15 to 150 μg at/l. Ketchum's results also showed that another nutrient, phosphate, followed similar kinetics to the uptake kinetics of nitrate and that phosphate could be taken up in the absence of measurable nitrate; however, phosphate uptake was increased when the concentration of nitrate was increased.

Taking the original work of Ketchum a step further, Caperon (1967) and Dugdale (1967) showed that nutrient uptake could be described using Michaelis–Menten enzyme kinetics in which

$$v = \frac{V_m S}{K_s + S}, \tag{66}$$

where v is the rate of nutrient uptake, V_m is the maximum rate of nutrient uptake, K_s is the substrate concentration at which $v = V_m/2$ and S is the concentration of nutrient. By plotting S/v versus S, a straight line is obtained with an intercept on the abscissa $-K_s$. The constant K_s is believed to be an important property of a phytoplankton cell since it reflects the ability of a species to take up low concentrations of a nutrient and as such it may determine the minimum nutrient concentration at which a species can grow. Thus MacIsaac and Dugdale (1969) have shown that coastal phytoplankton communities generally have K_s values of >1 μM for nitrate uptake, while oceanic communities have lower K_s values of *ca* 0·2 μM.* In line with other biochemical studies on Michaelis–Menten constants it is also important to recognize that K_s values are temperature dependent (Eppley *et al.*, 1969b). Recently, however, it has been recognized that the changes observed in K_s strongly depended on V_m. To overcome this problem, the ratio of V_m/K_s, which is the slope of the Monod equation at low nutrient concentrations, has been offered as a simple way of emphasizing both parameters (Healey, 1980). The V_m/K_s ratio indicates the uptake efficiency of nutrients.

In the case of Si uptake by diatoms on the other hand, it was found that the apparent uptake kinetics are slightly different from those for other nutrients; the plot of Si uptake against the Si concentration does not start from zero concentrations (Fig. 45). Paasche (1973) has presented an equation with a

*0·2 μM = 0·2 μg at l^{-1}.

slight modification of eqn. (66) in order to express this type of uptake. His equation is

$$v = \frac{V_m(S - S_0)}{K_s + (S - S_0)}, \quad (67)$$

where S_0 is the threshold concentration of Si at which $v = 0$. According to laboratory batch culture experiments using five different species of centric diatoms, threshold concentrations varied between 0·3 and 1·3 μg at Si/1 being rather greater in species with a large K_s, with the exception of *Ditylum brightwellii* in which S_0 was small and the K_s was large (Paasche, 1973). K_s values also varied between 0·8 and 3·4 μg at Si/1 depending on different species. A similar level of K_s for Si* (2·93 μg at Si/1) has been reported from silicon limited natural phytoplankton populations measured by the recently developed non-radioactive mass-spectrometric technique (Goering *et al.*, 1973).

It should be noted both from eqns. (66) and (67) and Figs. 44 and 45, that the kinetic approach of nutrient uptake is only valid for a condition under which nutrients are limiting. In waters with naturally saturating nutrient concentrations, it is not possible to measure K_s.

Under conditions of no nutrient stress, nutrient uptake rate shows a hyperbola-shaped response to light intensity which is similar to the response commonly observed in photosynthesis/light relation. Such nutrient/light relation can also be described by using the Michaelis–Menten equation, but the following two cases must be considered. The first point is that the nutrient/light curve quite often shows a positive value of nutrient uptake even at zero light intensity, because some nutrients (e.g. nitrate and ammonia) are known to be taken up in the dark at low rates, with ammonium being taken up more readily (Dugdale and Goering, 1967). The second point is that nutrient uptake may be depressed under bright light intensities (e.g. above 10 or 25% of surface illumination) in the same way as are observed 'photoinhibition' in photosynthesis/light responses. MacIsaac and Dugdale

*Precisely silicic acid {$Si(OH)_4$}, in this case, which is a hydroxylated product of silicate.

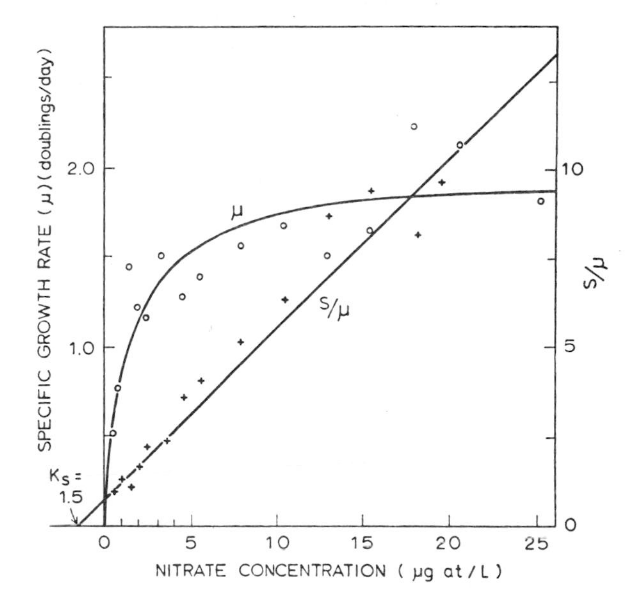

FIG. 44. Specific growth rate of *Asterionella japonica* as a function of nitrate concentration (redrawn from Eppley and Thomas, 1969).

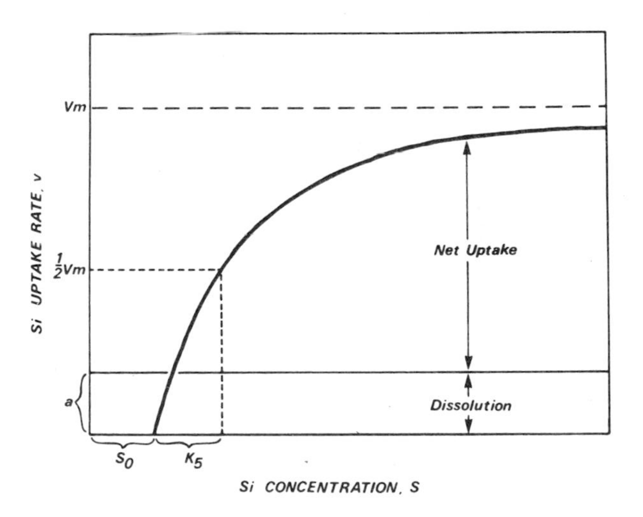

FIG. 45. Si uptake rate as a function of reactive silicate concentration in the medium (redrawn from Paasche, 1973) when the uptake experiments were carried out with silicate-depleted cultures; a correction for silica dissolution (a, generally 10 to 20% of the maximum uptake rate) is needed to drive the net uptake rates. S_0 is the threshold concentration.

(1972) have proposed an equation slightly modified from eqn. (66) in order to describe the nutrient uptake versus light intensity, in which they have taken into account the first point but not the second point mentioned above.

$$v = V_d + V'_m \left(\frac{I}{K'_I + I} \right) \quad (68)$$

where V_d is dark uptake, V'_m is the maximum velocity at nutrient and light saturation, K'_I is the constant of the equation and I is light intensity. This equation only describes uptake over the hyperbolic part of the curve but not for photoinhibition. There is also an assumption made in this equation that dark uptake is a constant at all light intensities, which has to be verified in future. Photoinhibition problems may be overcome by using Parker's (1974) expression* [eqn. (39)] for the second term on the right-hand side of eqn. (68). Values for K_I for nitrate and ammonium for some natural phytoplankton populations, for example, are within the range of 1 to 14% (*ca.* 0·004–0·06 ly/min) of the surface light intensity (MacIsaac and Dugdale, 1972), ~ 2·3% (*ca.* 0·01 ly/min) for silicic acid (Goering *et al.*, 1973), and 11·3 to 18·3% for phosphates (Reskin and Knauer, 1979).

It is anticipated that nutrient uptake is also affected by temperature. Actually there are several reports showing that increases in K_s for nitrate uptake followed with increasing temperature within a certain temperature range; K_s for nitrate uptake of *Skeletonema costatum* were 0·0 and 1·0 μg at N/1 at 8 and 28°C, respectively (Eppley *et al.*, 1969b), and those for *Gymnodinium splendens* were 1·0 and 6·6 μg at N/1 at 18 and 28°C, respectively (Thomas and Dodson, 1974). However, there is still not enough information available to make a generalization of temperature effect on nutrient uptake in a manner similar to effects by ambient nutrient concentration and light intensity as described above.

It is now apparent that uptake of a nutrient (e.g. K_s) is affected by light intensity (and possibly quality also) and temperature, at least within a certain range of changes, as well as physiological properties of cell membranes. Furthermore, other nutrient(s) not considered here may also affect the uptake of the rate-limiting nutrients. Most past experiments using algal cultures done by many investigators were carried out under optimal or suboptimal environmental conditions. Under such circumstances (with no environmental limitation except for the nutrient being considered), K_s values obtained may reflect the physiological potential of the species or clones being used, and they can be compared with each other in that sense. However, in nature the situation may be rather different and it is hard to anticipate that there is only one parameter limiting nutrient uptake and that all others are optimal, even supposing that every different species of phytoplankton reacts the same way. For example, if a nutrient, which is the primary limiting factor, is introduced into a natural population in order to estimate K_s, some other parameter might start causing a limitation, other than the nutrient being studied. Nutrient uptake measured under such conditions then reflects the uptake characteristics under possible environmental stress, which does not necessarily reflect the potential capability of the population themselves in the water.

As mentioned on the heterotrophic uptake of phosphate in the natural water by Reskin and Knauer (1979), many other inorganic nutrients are also utilized by heterotrophic organisms. This makes a possible competition for nutrient uptake occurring between phytoplankton and heterotrophic organisms in the water. With the existence of a large quantity of organic carbon compounds in the water, such as a polluted area caused by waste water from some food industries, heterotrophs utilize phosphate and inorganic nitrogen compounds at faster rates than phytoplankton because of the rapid growth rate of heterotrophs. There is no convenient way to separate the uptake of each group of organisms at the time of writing. We should be careful for the further evaluation on the nutrient uptake kinetics in natural waters.

An important question to resolve is whether the K_s of a phytoplankton species in the sea can be determined only from the rate of nutrient uptake or whether the same value can be determined from the

* $$v = V_d + V_m[I/K_I \exp(1 - I/K_I)]^{\alpha} \quad (68a)$$
where α is a dimensionless constant. This equation also describes the saturated curve expressed by eqn. (68).

phytoplankton growth rate. It has been demonstrated experimentally that nutrient uptake can occur without cell division (Fitzgerald, 1968); ecologically, however, it is important to know whether this is a general phenomenon. From studies conducted by Eppley and Thomas (1969) on two species of diatom it was concluded that the half-saturation constants for growth and nitrate uptake are very similar so that the expression given above, for the velocity of nutrient uptake can also be given in the form

$$\mu = \mu_{max} \frac{S}{K_s + S'} \tag{69}$$

where μ and μ_{max} are the growth rate and maximum growth rate, respectively, in units of time^{-1}. Algal cells are under steady-state conditions of nutrient limitation. A plot of the above expression is given in Fig. 44 together with a plot of S/μ vs. S which gives the intercept value on the abscissa of K_s. A summary of some K_s values is given in Table 22. From these data it was suggested by Eppley *et al.* (1969b) that there was a clear trend in K_s, being large in large cells and small in small cells; thus *Coccolithus huxleyi*, which is *ca.* 5 μm diameter, had a K_s of 0·1 μg at NO_3^-/l while *Gonyaulax polyedra*, which is *ca.* 45 μm diameter, had a K_s of > 5 μg at NO_3^-/l. Similar trends were also found in silicic acid uptake by diatoms (Goering *et al.*, 1973). It was also apparent from additional experiments conducted by Eppley *et al.* (1969b) that the ability to take up nitrate and ammonium ions differed between species and that some species could take up nitrate at lower concentrations than they could take up ammonia, and vice versa. Carpenter and Guillard (1971) showed that differences in K_s values were not confined to species but that clones of the same species had different K_s values depending on their environment. Clones isolated from low nutrient oceanic water had K_s values of < 0·75 μg at/l while the same species taken from an estuarine region had a K_s of > 1·5 μg at/l.

Differences in K_s values coupled with differences in the ability of species to reach their maximum growth rate at different light intensities, have been

TABLE 22. HALF-SATURATION CONSTANTS FOR NITRATE AND AMMONIA UPTAKE

Phytoplankton species clone or area	K_s (μgat/l) Nitrate	K_s (μgat/l) Ammonia	Reference
Oligotrophic, tropical Pacific	0·04	0·10	MacIsaac and Dugdale (1969)
	0·21	0·55	
	0·01	0·62	
	0·03		
	0·14		
Eutrophic, tropical Pacific	0·98		
Eutrophic, subarctic Pacific	4·21	1·30	
Oceanic species	0·1 to 0·7	0·1 to 0·4	Eppley *et al.* (1969b)
Neritic diatoms	0·4 to 5·1	0·5 to 9·3	
Neritic or littoral flagellates	0·1 to 10·3	0·1 to 5·7	
Thalassiosira pseudonana			
Clone 3H	1·87		Carpenter and Guillard (1971)
Clone 7–15	1·19		
Clone 13–1	0·38		
Fragilaria pinnata			
Clone 13–3	0·62		
Clone 0–12	1·64		

suggested by Dugdale (1967) and Eppley *et al.* (1969b) as being important factors in determining species succession in phytoplankton blooms. A schematic example is given in Fig. 46. In the first figure it may be seen that *Coccolithus huxleyi* reached its maximum growth at a low light intensity using ammonium ions and that its growth was approximately double that of the other species at a concentration of 0·5 μg at NH_4^+/l. However, in the second figure, at a higher light intensity, both *Skeletonema costatum* and *Ditylum brightwellii* grew faster than *C. huxleyi* at concentrations above 2 μg at NO_3^-/l. Since specific growth rates are exponents it is quite apparent that even small differences, over a period of time, would lead to very large differences in the standing stock of a particular species. The validity of this approach to field studies has been demonstrated by Thomas (1970b) and Thomas and Owen (1971), who showed that it was possible to obtain good agreement between growth rates calculated from $^{14}CO_2$ uptake and from ammonium concentrations in tropical Pacific waters. However, this empirical justification for the determination of K_s and phytoplankton growth constants from field data is not wholly supported by experimental and theoretical considerations advanced by several authors. The relationship between growth rate and nutrient concentration described in the above equation depends on μ_{max} and K_s remaining constant; this condition is generally met during exponential growth when the nutrient content per cell remains relatively constant, such that

$$\frac{\text{uptake rate}}{\text{cell division rate}} = Q\text{, the nutrient content per cell.} \quad (70)$$

However, Kuenzler and Ketchum (1962) showed that the uptake of phosphorus by *Phaeodactylum* was not determined by the phosphate concentration in the medium but by the past history of the cells and their internal phosphorus content. The cell content of phosphorus varied between 2 and 66 × 10^{-15} moles/cell; Caperon (1968) showed a similar variability in the nitrogen content of *Isochrysis galbana*, from 2 to 40 × 10^{-15} moles/cell. Further it has been known from some of the earliest studies (e.g. Ketchum, 1939a) that nutrient uptake can take place in the dark when the cell division rate is zero. Eppley and Strickland (1968) have suggested that phytoplankton cells require a certain minimum content of nitrogen (or phosphorus) before cell division can proceed; for *Dunaliella tertiolecta* the minimal nitrogen per cell (Q_0) varied with light intensity between 0·7 × 10^{-15} and 1·4 × 10^{-15} moles/cell. Caperon (1968) showed that there was a hyperbolic relationship between the reserve nitrate content per cell (Q') and growth rate (μ), such that

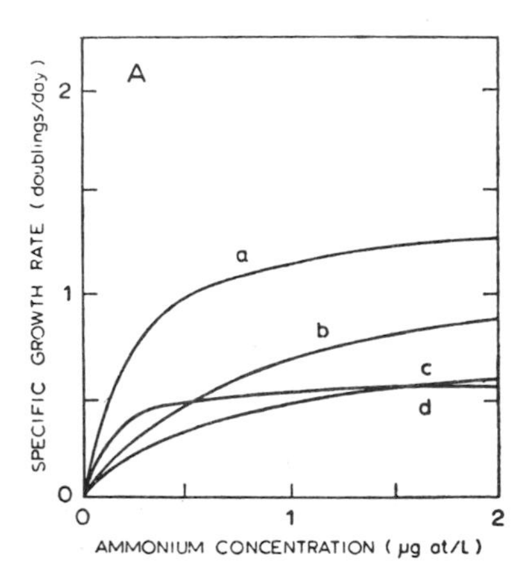

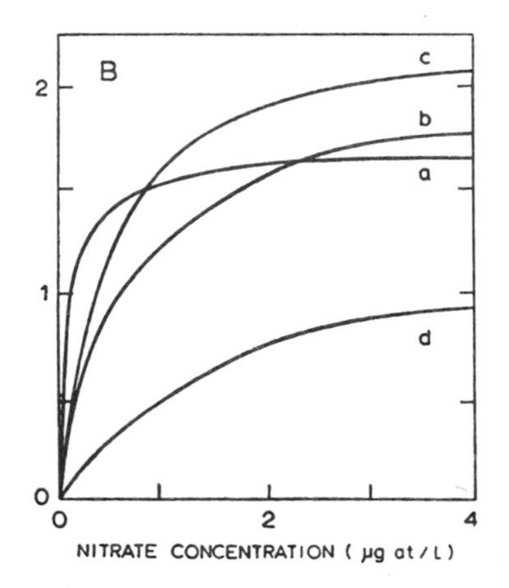

FIG. 46. Specific growth rate vs. ammonium and nitrate concentration at two light intensities, (B) approx. 4 times (A). (a) *Coccolithus huxleyi*, (b) *Ditylum brightwellii*, (c) *Skeletonema costatum*, (d) *Dunaliella tertiolecta* (redrawn from Eppley *et al.*, 1969b).

$$\mu = \mu_{max} \cdot \frac{Q'}{A + Q'}, \qquad (71)$$

where A is a growth constant. The value Q' used in the above equation represents the nitrate in the cell reservoir which is in addition to Q_0, the minimum amount of nitrate required for a normal cell at zero growth rate. Thus A represents the concentration of reservoir nitrate within the cell which is necessary to obtain a growth rate of $\mu_{max}/2$. In chemostat studies on vitamin B_{12} requirements, Droop (1970) showed that the growth rate could be expressed in terms of Q_0 and Q as

$$\mu = \mu_{max} \left(1 - \frac{Q_0}{Q}\right). \qquad (72)$$

From these findings it is apparent that the ecological use of K_s values (Table 22) should be approached with caution. As pointed out by Eppley *et al.* (1969b), for example, the dinoflagellate, *Gonyaulax polyedra,* has a low growth rate and an apparently high K_s. However, this organism displays a diel migration in which it absorbs nutrients from 10 to 15 m during the night and swims to the surface during the day where light is available for photosynthesis and growth, but where the nutrients are much lower than at 10 to 15 m. Thus the uptake of nutrients can be independent of growth and the phenomenon of 'luxury consumption' (see Eppley and Strickland, 1968) may result in growth being proportional to the nutrient content per cell (e.g. Droop, 1968 and 1970) rather than the nutrient concentration *in situ* (e.g. Thomas, 1970b).

Nutrient-dependent differences in growth rates may account for the dominance of one species over another on the basis of their respective growth rates but as Hulbert (1970) has pointed out, there is little or no possibility of an abundant species monopolizing the nutrient supply and forcing a less abundant species to extinction. Each algal cell in a phytoplankton community may be visualized as the centre of a volume of water in which nutrient depletion decreases from the cell outwards; this volume may be referred to as the cell's nutrient zone. Hulbert (1970) based his conclusions on the fact that the nutrient zone surrounding a cell would have to overlap with that of another cell if one growth was to affect another. For cellular nutrient zones to overlap, Hulbert calculated that there would have to be at least 3×10^8 phytoplankton cells per litre; since most natural populations rarely exceed 10^6 cells/litre in open and coastal waters, there is generally no possibility of one nutrient zone overlapping with another. However, at cell concentrations of 10^9/litre, such as may occur in some estuarine and eutrophic environments, there is a possibility that dominant species will lead to the elimination of residual species and hence decrease the diversity of the system.

Parsons and Harrison (1983) have made an extensive review of the competition among phytoplankton species under different nutrient regimes. The following is a quotation from their work.

Hutchinson (1961) suggested that the simultaneous existence of many species of phytoplankton in an apparently uniform body of water presented a 'paradox' because the continued coexistence of closely competing species of phytoplankton for this has been referred to as 'temporal succession' in which it was supposed that the nutrient regime of aquatic habitats was changing too rapidly for equilibrium to be established (i.e. all species were in different states of reaching or declining from their maximum abundance, depending on the exact conditions at any moment of time). Another suggestion was made by Richardson *et al.* (1970) in which it was supposed that microhabitats develop within relatively small volumes of water, providing turbulence is low. Each microhabitat then favours the growth of a distinct patch of plankton. This theory is known as 'contemporaneous disequilibrium' because of the temporary division of an otherwise uniform body of water into microhabitats separated in space. In discussing both of the above concepts, Peterson (1975) advanced a third hypothesis in which it is supposed that an assemblage of coexisting phytoplankton may be limited by the availability of several nutrients, with two or more species coexisting providing each is limited by a different nutrient. Good evidence for this concept has been provided (Titman, 1976;

Tilman, 1977) and the following discussion is taken from the two references cited.

At some point where a species has equal growth rates (μ), regardless of whether it is being limited by one of two nutrients, it follows that:

$$\mu^1 = \mu^2 \tag{72a}$$

or from eqn. (69).

$$\frac{[S_1]}{[S_2]} = \frac{K_1}{K_2} \tag{73}$$

where S_1 and S_2 are the steady-state concentrations of resources 1 and 2, and K_1 and K_2 are the half-saturation constants for growth limited by resources 1 and 2, respectively. It follows that if $[S_1]/[S_2] < K_1/K_2$, the species is limited by substrate 1. If two species are now considered, such that K_i of species A is less than K_1 of species B, but that K_2 of species A is greater than K_2 of species B, then if $[S_1]$ is limiting, species A will dominate, while if $[S_2]$ is limiting, species B will dominate. Under these conditions, the nutrient ratio $[S_1]/[S_2]$ will determine which species dominates, while the difference in the K_1/K_2 ratio will determine the range over which each species can coexist in the same water mass. Thus, if K_1/K_2 for species A is 100 and for species B is 10, then coexistence of the two species, each limited by different substrates, will occur between $[S_1]/[S_2]$ ratios of 100 and 10. This is illustrated in Fig. 47. The model suggested here was confirmed in practice by Tilman (1977) using two species of lake diatoms under conditions of silicic acid and phosphate growth limitation.

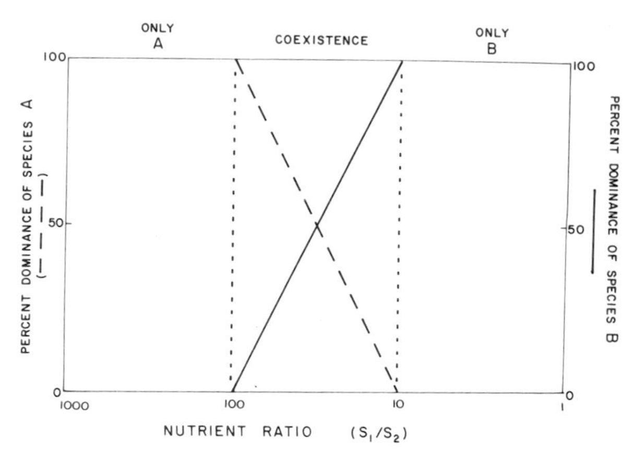

FIG. 47. Relative abundance of two phytoplankton species, A and B, plotted against the log of the ratio of limiting nutrients, S_1 and S_2 (after Parsons and Harrison, 1983).

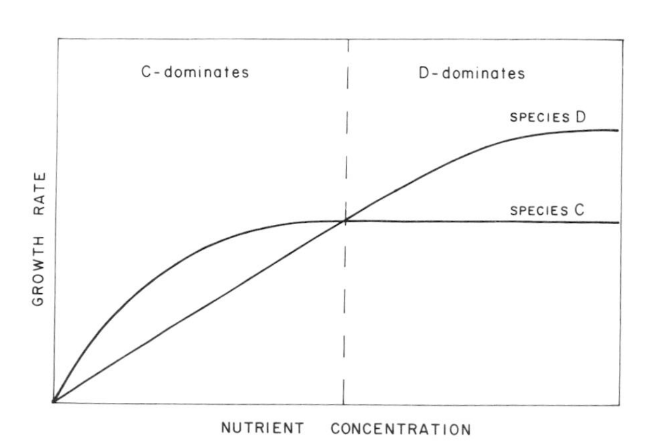

FIG. 48. Species domination based on differences in K_s and μ_{max} of two phytoplankton species, C and D (after Parsons and Harrison, 1983).

Turpin and Harrison (1979) have further extended our concept of species coexistence from the above consideration of two or more limiting nutrients to the case in which a single nutrient could give rise to coexistence under a varying temporal or special distribution of the limiting nutrient. In an earlier paper, it was shown by Eppley *et al.* (1969) that a single nutrient concentration which changed over time, could lead to a shift in phytoplankton species dominance, as indicated in Fig. 48. In an extension of this concept, Turpin and Harrison (1979) considered what would happen if the nutrient supply was provided, (1) at low or high rates, but continual addition (Harrison and Davis, 1979), or (2) at the same total concentration per unit time, but pulsed by a single large addition followed by the absence of nutrient over the same unit time as in (1). The results of their experiments are summarized in Fig. 49. From these observations it was generally concluded that cells having a high V_{max} would outcompete cells with a low V_{max} under conditions of high specific nutrient flux (time^{-1}). However, under conditions of low specific nutrient

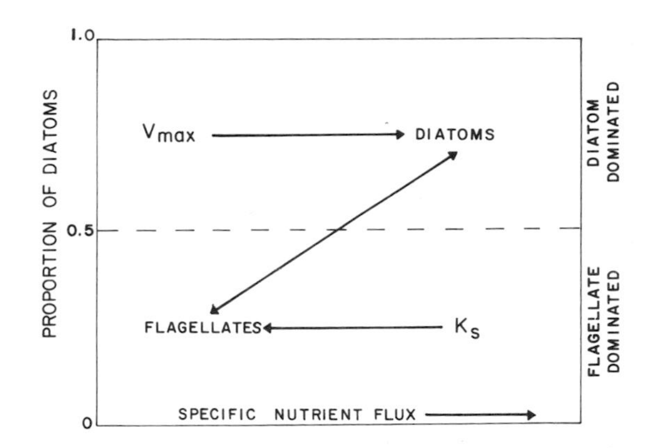

FIG. 49. A schematic representation of the possible relationship between specific nutrient flux (nitrogen) and community structure (from Turpin and Harrison, 1979).

flux (i.e. low steady-state nutrient concentration) cells having a low K_s for the limiting nutrient would dominate. A high specific nutrient flux (providing nutrients in pulses, rather than a continual addition) tended to select for different types of centric diatoms. Therefore, the specific nutrient flux appears to be a coarse tuning variable, while the patchiness of the nutrient is more of a fine tuning variable. Since it can be shown that diatoms in general have the highest growth rates but that small flagellates may have the lowest K_s, it can be postulated that a high specific nutrient flux will lead to diatoms-dominated phytoplankton, while oligotrophic waters with a low but relatively steady nutrient supply would tend to be dominated by flagellates (Turpin and Harrison, 1979; Parsons and Takahashi, 1973; Parsons, 1979, and references cited therein).

It appears that various theories on the uptake of nutrients by phytoplankton, including earlier concepts of 'temporal succession' and 'contemporaneous disequilibrium', are sufficient to illustrate the important role of nutrients in maintaining phytoplankton species diversity. In addition, other factors such as selective zooplankton grazing, temperature and salinity tolerance and inhibition of certain phytoplankton by antimetabolites (allelopathy) would serve to further broaden the number of species which could coexist at any given time.

For a further consideration of nutrient effects on the growth of phytoplankton under various environmental stresses other than nutrients, two approaches can be mentioned; the Blackman type model and the multiplicative model. In the Blackman model, which is described in Section 3.1.4, a single environmental parameter only affects the growth at a given time, whereas simultaneous effects of various environmental parameters are allowed in the multiplicative model. Rhee and Gotham (1981 a,b) pointed out from mono-species culture experiments that the effects on the phytoplankton by environmental parameters were not multiplicative but the Blackman type. Further evidence was given by Tilman *et al.* (1981), who suggested (based on laboratory culture experiments using two species of freshwater diatoms under silicic acid limitation) that the half-saturation constant K_s, was temperature-independent for temperatures within or below the optimal temperature range, but increased sharply for temperatures above it. It appears that environmental growth regulation follows the Blackman type in a *single species* condition, although this subject still requires further evaluation.

The Blackman type model has been applied to nutrient effects upon phytoplankton growth under various environmental stresses. One application was done by MacIsaac and Dugdale (1972) and the other was by Takahashi *et al.* (1973). Both of these applications were made for multispecies natural phytoplankton populations. The approach can be extended to analyse the various environmental growth control of a single species population and a further extension for multispecies assemblages can also be made by the use of multiple separate equations.

An example of the multiplicative model has been given by Goldman and Carpenter (1974). In the model, they pointed out that μ_{max} and K_s were dependent on temperature and light intensity, and suggested the following equation:

$$\mu = \mu_{max}(T,I) \frac{S}{K_s(T) + S} \tag{74}$$

in which $\mu_{max}(T,I)$ and $K_s(T)$ indicate the maximum specific growth rate controlled by temperature and light intensity, and the half-saturation constant controlled by temperature, respectively. For the possible relation between the maximum specific growth rate and temperature, the Arrhenius equation was evaluated by using a set of literature values on continuous culture experiments with freshwater and marine phytoplankton, and found to fit nicely. Equation obtained was

$$\mu_{max} = A\ e^{-E(I)/RT} \qquad (75)$$

where A is constant (d^{-1}), E is the activation energy ($cal \cdot mole^{-1}$), R is the universal gas constant ($cal \cdot {}^{\circ}K^{-1}$ $mole^{-1}$) and T is temperature in Kelvin scale (°K). Light intensity gives a strong effect on the activation energy, E, then shown by $E(I)$ representing light-dependent activation energy. However, actual relations cannot be formulated due to incomplete data available at this stage. This is also the case for the temperature effect on K_s; $K_s(T)$.

Once functional relations between two variables, μ_{max} and K_s, and the environmental parameters, nutrient concentrations, temperature and light intensity, are known for each phytoplankton species, the specific growth rate for each species, μ_x, can be estimated at a given condition controlled by the environmental parameters mentioned above. The actual growth of a given phytoplankton species population can then be estimated by the application of the growth rate obtained into a growth equation such as eqn. (55). With the aid of a computer, growth calculation can be made with minimum efforts even though actual calculation is highly complicated. However, care must be taken to over-generalize and indiscriminately apply models, because computer calculations often go further than they should.

3.1.7 Chemosynthesis

Some micro-organisms can satisfy their primary energy requirements by utilizing simple inorganic compounds, such as ammonia, methane, or nitrite, or elements, such as ferrous iron, hydrogen gas, or water-insoluble amorphous sulfur. All of the known organisms which comprise this (chemosynthetic) group are bacteria. Most of their inorganic substrates are derived from the decay of organic matter which is itself primarily formed through photosynthesis. Consequently, if one follows the origin of the energy used in chemosynthesis, chemosynthetic processes may not be considered as primary production. However, chemosynthesis usually involves carbon dioxide fixation and the primary formation of new particulate material. Thus chemosynthesis may be considered as a special case of primary production on the grounds of its trophic position in the marine food chain.

In the past it was believed that the inorganic substrates reacted directly with molecular oxygen to form oxidized end-products. However, Bunker (1936) showed that molecular oxygen was not directly involved in sulfur oxidation by thiobacilli. The fact that oxygen in the end-products originated from water resulted in a new concept of chemosynthesis, although in some cases it is still questionable whether the concept is adaptable to all cases of chemosynthetic activity. Thus reactions of the type suggested by Bunker involve dehydrogenations rather than oxidations. In general terms the chemosynthetic process can be expressed conveniently in three stages according to Bunker's concept (Gundersen, 1968).

(1) As the result of dehydrogenation, high reducing power is produced as follows:

$$\underset{\text{(Inorganic substrates)}}{nAH_2} + nH_2O \xrightarrow{\text{dehydrogenase}} \underset{\substack{\text{(oxidized}\\\text{end-product)}}}{nAO} + \underset{\text{(reducing power)}}{4n[H^+ + e^-]} \qquad (76)$$

where the symbol $[H^+ + e^-]$ is used as a synonym for the reducing power.

(2) A proportion of the reducing power is then utilized for energy production (adenosine triphosphate, ATP, synthesis) by being transferred through the cytochrome system to molecular oxygen. A second part of the reducing power is transferred to NAD (nicotinamide adenine dinucleotide) in order to produce reduced NAD (or $NADH_2$, reduced nicotinamide adenine dinucleo-

tide). These relationships might be visualized as follows:

$$4[H^+ + e^-] + mADP + m\ P_i + O_2 \xrightarrow[\text{system}]{\text{cytochrome}} 2H_2O + mATP \quad (77)$$

$$2[H^+ + e^-] + NAD \longrightarrow NADH_2 \quad (78)$$

where ADP is adenosine diphosphate, and P_i is inorganic phosphate. Some anaerobes can use bound oxygen derived from inorganic compounds instead of free oxygen as shown in the above equation (e.g. sulphate-reducing bacteria).

(3) The ATP and $NADH_2$ are then used for the assimilation of carbon dioxide:

$$12NADH_2 + 18ATP + 6CO_2 \rightarrow C_6H_{12}O_6 + 6H_2O + 18ADP + 18P_i + 12NAD. \quad (79)$$

Thus the different inorganic substrates used by chemosynthetic bacteria are not merely the sole source of the organism's energy but also the sole source of their reducing power.

Depending on differences in the organic substrate (AH_2), chemosynthetic bacteria are classified into several groups, such as nitrifying, sulfur, hydrogen, methane, iron and carbon monoxide bacteria. Table 23 shows some representative chemosynthetic bacteria inhabiting marine environments. The overall efficiencies of chemosynthesis during the growth of the bacteria (e.g. nitrifying or sulfur bacteria) can be expressed as the ratio between the total energy consumed in carbon dioxide assimilation and the energy liberated by the primary inorganic compounds during oxidation; these efficiencies are 6 to 8% (Gibbs and Schiff, 1960) although the efficiency may sometimes change drastically with different stages of cultures. For example, young cultures of *Nitrosomonas* gave values approaching 50%; in older cultures, the efficiency dropped to 7% (Hofmann and Lees, 1952). Most chemosynthetic bacteria require free oxygen as the electron acceptor in the second step described above. However, facultative or obligate anaerobic bacteria, such as *Thiobacillus denitrificans* and *Desulfovibrio desulfricans*, can use bound oxygen derived from nitrate or sulfate.

Among inorganic substrates available for the chemosynthetic bacteria, nitrogen and sulfur compounds are relatively abundant and widely distributed compared with the other reduced compounds in the pelagic environment of the oceans. Among reduced nitrogen compounds, ammonia may be present in concentrations up to *ca.* 5 μg at/l and nitrite at concentrations up to *ca.* 2 μg at/l (Riley and Chester, 1971). Actual occurrences of nitrifying bacteria, expressed in colonies per litre of sea water, were found to be < 1 in the north Atlantic Ocean, < 10 in the Pacific Ocean, *ca.* 10^4 in the Indian Ocean near an island, and *ca.* 10^6 in Barbados harbor (Watson, 1965; Hattori and Wada, 1971). Some of the strains of nitrifying bacteria isolated from marine waters are the same or similar to those from fresh water or soil, but others are peculiar to marine environments (e.g. see Watson, 1971). Nitrification by marine bacteria, using either ammonia or nitrite as a substrate, is reported to be more efficient at low ($<0{\cdot}1$ ml/l) oxygen concentrations (Carlucci and McNally, 1969).

The qualities of reduced sulfur compounds in the marine environments are generally much greater than the quantities of reduced nitrogen compounds. Thiosulfate and polythionates may sometimes be present at levels from 0 to 100 μg at/l, the highest concentrations of these compounds being detected near to shore (Tilton, 1968). Sulphides are generally not detected within the analytical limit of < 1 μg at/l in open ocean waters (Tilton, 1968). The number of colonies of sulfur bacteria (*Thiobacillus* spp.) have been found to range from 0 to 275 per 100 ml (Tilton *et al.*, 1967). These values for colony counts are several orders of magnitude smaller than those (10^3 to 10^4 per 100 ml) which were calculated by Tilton *et al.* (1962) on the basis of the amount of reduced sulfur compounds.

The reduced inorganic substrates are mainly produced through anaerobic metabolic processes. Consequently, anaerobic environments, which sometimes develop in fjords and estuaries, create favourable habitats for chemosynthetic bacteria. Under such conditions vigorous growth of certain species of chemosynthetic bacteria (usually sulfur, hydrogen, or methane bacteria) has been observed

TABLE 23. SUMMARY OF CHEMOSYNTHETIC BACTERIA GROWING IN THE OCEAN AND THEIR INORGANIC SUBSTRATES

	Inorganic substrate	Oxidized end-product	Oxidizer	ΔF* (K cal)	Ability to grow heterotrophically	Habitat
(1) Nitrifying bacteria						
Nitrobacter spp.	NO_2	$N0_3$	O_2	18		Soil, fresh, and sea waters
Nitrococcus mobilis WATSON & WATERBURG	NO_2	NO_3	O_2	18		Sea water
Nitrospina gracilis WATSON & WATERBURG	NO_2	NO_3	O_2	18		Sea water
Nitrosomonas spp.	NH_3	NO_2	O_2	85		Soil, fresh, and sea waters
Nitrosococcus oceanus (WATSON) *comb. nov.*	H_3 or NH_2OH	NO_2	O_2	85		Sea water
(2) Sulfur bacteria						
Thiovulum majas HINZE	H_2S	S		50	?	Sea water
Beggiatoa spp.	H_2S	S		50	+	Fresh and sea waters
Thiospira bipunctata MOLISCH	H_2S	S		50	?	Sea water
Thiothrix spp.	S	SO_4^-	O_2	119	?	Soil, fresh, and sea waters
Thiobacillus thioparus BEIJERINCK	5/4 $S_2O_3^-$	3/2 SO_4^{--} + s	O_2	112	+	Soil, fresh, and sea waters
Thiobacillus denitrificans BEIJERINCK	5S	$5SO_4^{--}$	O_2	112	+	Soil, fresh, and sea waters
(3) Hydrogen bacteria						
Hydrogenomonas spp.	H_2	H_2O	O_2	56	+	Soil, fresh, and sea waters
Desulfovibrio desulfricans (BEIJERINCK) KLUYVER & VAN NIEL	H_2	H_2O	SO_4	56	+	Soil, fresh, and sea waters
(4) Methane bacteria						
Methanomonas spp.	CH_4	CO_2	O_2		+	Soil, fresh, and sea waters
(5) Iron bacteria						
Grallionella spp.	$4FeCO^3$	$4Fe(OH)_3$	O_2	81	+	
(6) Carbon monoxide bacteria						
Sarcina bakerii SCHNELLEN	CO	CH_4	H_2		+	Soil, fresh, and sea waters

*Free energies per number of moles of electron donor indicated. Values reviewed by Gibbs and Schiff (1960) are mainly quoted here.

(Kuznetsov, 1959; Sokolova and Karavaiko, 1964).

It should be mentioned that an obligate anaerobic strain, *Desulfovibrio desulfricans*, has been isolated even from oxic ocean waters (Kimata *et al.*, 1955; Wood, 1958). In order to explain this phenomenon, Baas Becking and Wood (1955) proposed the idea of 'metabiosis' in which they suggested that several kinds of micro-organisms co-exist on or in bacterial aggregates and conditions for the production of reduced substrates could occur within the aggregates. Thus the obligate anaerobic bacteria existing in oxic environments are probably present within the microcosm of a bacterial aggregate (Seki, 1972).

In the ocean, chemosynthetic activity can be estimated from carbon dioxide uptake in the dark using $NaH^{14}CO_3$. Compared with carbon fixation through photosynthesis, dark CO_2 uptake is usually small (i.e. less than 5% of the photosynthesis on a daily basis within the euphotic zone in pelagic areas). Dark CO_2 uptake is not entirely carried out by chemosynthesis, but heterotrophic processes by bacteria and algae (e.g. Wood and Werkman, 1935, 1936, and 1940) may result in the uptake of small amounts of CO_2 in the absence of light. Algal dependency on dark fixation of CO_2 is low (i.e. 3 to 5% of the total CO_2 fixed in the light) and this is about the same proportion of CO_2 fixed by aerobic heterotrophic bacteria when they are growing on an organic substrate (Sorokin, 1961 and 1966). However, among facultative autotrophic bacteria which belong to an intermediate metabolic type between obligate heterotrophs and chemoautotrophs, there is a requirement of between 20 and 90% CO_2 during the oxidation of low molecular weight organic compounds, such as methane and formic acid; for obligate chemoautotrophs the requirement for CO_2 is close to 100%. Since in natural environments, micro-organisms which depend on different kinds of substrates exist together, it is practically impossible to separate out and estimate the actual activity of chemosynthetic organisms except at the time of vigorous growth of any one species. However, a measure of dark CO_2 uptake is a useful measure of chemosynthetic activity in any environment. As an example, Seki (1968) studied seasonal changes in dark uptake of CO_2 in a small bay and showed that dark uptake of CO_2 near the bottom of the bay could be as high as 20% of the photosynthesis throughout the year and that during the spring dark uptake for the water column was sometimes 50% greater than photosynthesis.

Active chemosynthesis occurs in waters and sediments in which both aerobic and anaerobic environments exist in the same column. Sorokin (1964a and b) has studied the importance of chemosynthetic bacteria in the Black Sea, where a thick anaerobic zone exists below the depths of *ca.* 150 m in the central part of the sea. A summary of his finding is shown in Fig. 50. At depths shallower than 50 m, inorganic carbon is fixed through photosynthesis by algae, and the daily photosynthesis is reported to be *ca.* 350 mg $C/m^2/day$ during October. At the transition zone between 100 and 250 m, where environmental conditions are changing from aerobic to anaerobic, chemosynthetic fixation of inorganic carbon is predominant. An *in situ* maximum value for daily chemosynthesis was *ca.* 9 mg $C/m^3/day$; chemosynthesis for the water column amounted to approximately 200 mg $C/m^2/day$. This active chemosynthesis is carried out mainly by sulfur bacteria (aerobic and anaerobic *Thiobacillus*), and it is supported by a continual supply of reduced sulfur compounds, such as H_2S, S, and S_2O_3, from the anaerobic zone. The potential activity of thiobacilli in water samples taken from the depth of their maximum activity is very high compared with the *in situ* activity; by determining the oxidation rate of thiosulfate added to a water sample, a potential activity of 200 mg $C/m^3/day$ has been obtained.

Although the studies described above may appear to have limited geographical significance, it is probable in fact that similar microzonation in environmental factors can occur in many marine environments, especially in bays and estuaries. Also in all nearshore sediments where wave action does not disturb the benthos, a high chemosynthetic activity will occur in a depth zone of a few cm just below the sediment surface. If the transient zone of aerobic to anaerobic conditions comes up above the depth receiving a few per cent of the surface

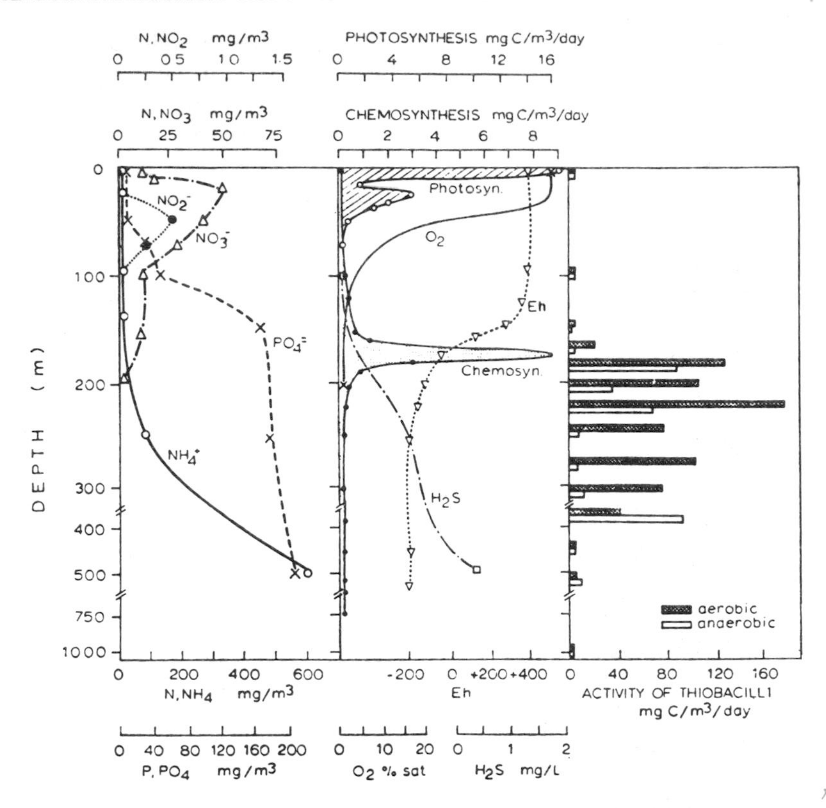

FIG. 50. Daily chemosynthetic production and some factors influencing it in the Black Sea (after Sorokin, 1964c).

illumination, anaerobic photosynthetic sulfur bacteria, which requires reductants (e.g. H_2S, S) for the hydrogen donor, grow vigorously. These organisms impart a purple or deep green colour to the sediment or water column in which they exist.

3.2 HETEROTROPHIC PROCESSES

3.2.1 The Origin of Organic Substrates

Heterotrophic organisms are dependent on an organic carbon source to provide energy for growth. In the sea, most of the organic carbon utilized by heterotrophic organisms originates from the marine biota, and only in near-shore coastal areas is there any appreciable contribution of organic materials from the land. Exceptions to this generalization are found where major rivers may influence the oceanic environment for a considerable distance off shore; in particular, the Amazon River, with an annual discharge of 6×10^{12} m^3 (or *ca.* 20% of the entire world wide river runoff), may contain sufficient organic carbon (2 to 10 mg/l) to influence heterotrophic activity over an oceanic area approximately 10^7 km^2 during maximum runoff (Williams, 1968). Duursma (1965) recognized four main groups of organic compounds which occur as dissolved substances in sea water. These are (1) nitrogen-free organic matter, including carbohydrates, (2) nitrogenous substances, including amino acids and peptides, (3) fat-like substances, and (4) complex substances, including humic acids, derived from groups (1) and (2). In addition to these dissolved materials, particulate organic materials also serve as a substrate for heterotrophic organisms. The nature and chemical composition of these substances are

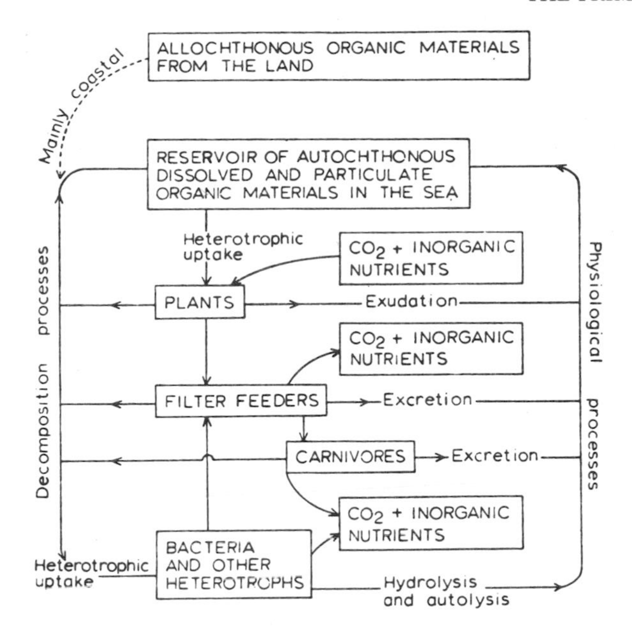

FIG. 51. Pathways of transfer and regeneration of organic substrates in an aquatic ecosystem (modified from Johannes, 1968).

discussed in Section 1.4; their origin is illustrated in Fig. 51, which is adapted from Johannes (1968). The figure illustrates that the two principal pathways of organic materials which act as substrates for heterotrophic organisms result from the food chain of the sea, firstly through the physiological release of materials and secondly through the decomposition of plants and animals themselves. Autochthonous organic materials derived by physiological processes include the release of dissolved organic materials by phytoplankton; this subject has been reviewed by Fogg (1966), who records that in some cases, up to 50% of the CO_2 photosynthetically fixed by phytoplankton may be released as soluble carbon (Allen, 1956; Fogg, 1952). This process is sometimes called 'excretion' but should probably be called 'exudation'. The latter term is used by Sieburth (1969) and Sieburth and Jensen (1969) in referring to the loss of soluble organic material from seaweeds; the authors report that this loss may amount to 40% of the net carbon fixed daily. Further studies on the exudation of soluble organic carbon by phytoplankton (Eppley and Sloan, 1965; Hellebust, 1965) have shown that in general the release from healthy cells amounts to 15% or less of the total carbon fixed; Watt (1966) has suggested a maximum loss of extracellular products of up to 30%. Two factors which appear to effect the release of soluble organics are the age of the cells and the light intensity. Hellebust (1965) and Anderson and Zeutzschel (1970) have shown that exudation is greatest at high light intensities and among phytoplankton cells collected at the end of a plankton bloom.

The relationship between algal exudation and bacterial growth has been discussed by Bell and Mitchell (1972). From their own experimental results and from literature reports the authors conclude that bacterial growth may show a small increase during the logarithmic phase of algal growth but that a much greater maximum in the number of bacterial cells occurs when the algae reach a stationary growth phase and when there is some cellular lysis. The authors introduce the concept of a 'phycosphere' which is defined as a zone extending outward from an algal cell (or bloom) in which bacterial cells are stimulated by extracellular products of the algae. It appears that bacterial chemotaxis towards this zone can occur under experimental conditions but only when the organic material released by algae has reached relatively high concentrations from lysing algal cells.

It is quite apparent from these studies that since plants are the major producers of organic matter in the sea, any fraction lost through exudation will constitute an appreciable input to the organic reservoir in the environment. The type of compounds released by algae are known to include small amounts of amino acids and generally larger amounts of short chain acids (e.g. glycollic acid), glycerol, carbohydrates, and polysaccharides (Fogg *et al.*, 1965; Watt, 1966).

The excretion of organic materials resulting from digestive processes of animals forms a second physiological pathway for the input of organic materials available for heterotrophic growth. In addition to the process of true excretion, animals may also release organic material directly from their bodies (e.g. Johannes and Webb, 1970), and this may in part be coupled with the moulting process, which results in the specific production of chitinous debris; for example, Lasker (1966) estimated that the loss of chitin by euphausiids amounted to 10% of their body weight every 5 days.

The natural mortality of plants and animals in the food chain forms the main pathway for the production of organic substrates by decomposition processes. Post-mortem changes in the permeability of cell membranes and the effect of autolytic enzymes (Krause *et al.*, 1961; Krause, 1961 and 1962) may result in an initial loss of 15 to 50% of the total biomass of an organism. Soluble organic compounds released from dead organisms include amino acids, peptides, carbohydrates, and fatty acids.

Heterotrophs which utilize the organic particulate and soluble substrates produced by the food chain, themselves contribute to the organic reservoir of the sea, either by serving as food for filter-feeding organisms or by hydrolysing substrates, and self-autolysis, to provide soluble organic materials. Williams (1970) found that soluble organic materials were assimilated by bacteria with an average efficiency of 67% for glucose and 78% for amino acids. Sorokin (1970) found that algal hydrolysates were utilized by natural populations of marine heterotrophs with an efficiency of *ca.* 45%. These efficiencies are much higher than are found in culture experiments (e.g. Parsons and Seki, 1970) where only about one-third of the organic substrate taken up is retained as new cellular material. Williams (1970) suggests that higher conversion efficiencies result with natural populations utilizing a variety of organic compounds, in contrast to the usual laboratory studies, where all organic carbon must be derived from a single organic compound. The largest fraction of organic carbon lost during heterotrophic activity is respired back to carbon dioxide, a process which also occurs throughout the food chain as organic carbon is transferred to higher trophic levels.

While information summarized in Fig. 51 shows the major pathways for the release of organic compounds in the heterotrophic cycle of the sea, there are in addition very specific organic compounds released in sea water which can modify the overall balance of organic materials in any one environment. Included among specific organic substances are growth factors and antimetabolites. As an example of the former, Burkholder and Burkholder (1956) and Starr (1956) showed that in coastal areas, vitamin B_{12} is produced by the bacteria; secretion of vitamins by some phytoplankton has also been demonstrated (Carlucci and Bowes, 1970). Since other species of phytoplankton require vitamins for growth, their presence can be assumed to affect the organic food chain. Antimetabolites have been detected in sea water, particularly in respect to the production of antibacterial substances by phytoplankton (e.g. Sieburth, 1964). More detailed studies on the release of antibacterial materials by phytoplankton (Duff *et al.*, 1966) have revealed species specific differences and in at least one case (Antia and Bilinski, 1967) this has led to the identification of a lecithinase as being an antibacterial agent in the chrysomonad, *Monochrysis lutheri*.

3.2.2 Heterotrophic Uptake

The two pathways of heterotrophic uptake shown in Fig. 51 are firstly through the plants (facultative heterotrophs) and secondly through the bacteria and other obligate heterotrophs, such as yeasts and moulds. The most important of these two pathways is undoubtedly the second, and the role of plants as autotrophic (or sometimes auxotrophic organisms) far exceeds their role as heterotrophic organisms.

The ability of marine bacteria to decompose a wide range of naturally occurring organic substrates has been widely demonstrated. These substances include the decomposition of chitin (Hock, 1940; Seki, 1965 a and b), cellulose (Waksman *et al.*, 1933), protein (Wood, 1953), and alginates (Meland, 1962). Jannasch (1958) showed that the presence of large numbers of bacteria in sea water was dependent on the concentration of dissolved organic matter; in the absence of soluble substrates, bacteria are found attached to particulate materials, such as inorganic particles and chitin. The heterotrophic activity of bacteria results in a number of complex transformations which are not implied in Fig. 51; these have been reviewed by ZoBell (1962) and include dissolution and precipitation of organic and inorganic compounds as well as the large-scale production of inorganic energy reserves (such as methane and hydrogen sulphide), when heterotrophic decomposition occurs under anaerobic conditions.

The heterotrophic activity of phytoplankton has been demonstrated in a number of specific cases. Out of forty-four pure cultures of littoral diatoms, Lewin and Lewin (1960) found that twenty-eight were capable of heterotrophic growth on glucose, acetate, or lactate media. Kuenzler (1965) found that only one out of thirteen species of phytoplankton could assimilate glucose for growth in the darkness. Antia *et al.* (1969) demonstrated that a photosynthesis cyptomonad, *Chroomonas salina*, could grow heterotrophically in the dark at high concentrations of glycerol but not at low concentrations. However, a variety of substrates including acetate, glutamate, and glucose stimulated growth of the organism in the light. Sloan and Strickland (1966) used radioactive carbon to determine the heterotrophic uptake of glucose, acetate, and glutamate by four species of marine phytoplankton. While some uptake was detectable, it was generally less than the dark uptake of radioactive carbon dioxide; the latter can be fixed non-photosynthetically in the dark through biochemical mechanisms, such as the Wood–Werkman reaction (Wood and Werkman, 1936 and 1940). At times the dark uptake of carbon dioxide may exceed the amount taken up by photosynthesis; this indicates that in such environments heterotrophic activity, resulting in the mineralization of organic matter, is in excess of autotrophic activity resulting in the production of organic matter (e.g. Seki, 1968, and Section 3.1.6).

The uptake of organic substrates at low concentrations has been demonstrated among a few species of coastal phytoplankton. North and Stephens (1971 and 1972) showed that *Platymonas* and *Nitzschia ovalis* could take up amino acids. In the case of *Platymonas*, for example, arginine, glycine, and glutamate were taken up at concentrations of *ca.* 5×10^{-6} moles/litre; the rate of uptake was increased 10-fold in nitrogen-starved cells. As a general conclusion, however, it appears that heterotrophic activity among the phytoplankton is limited to a number of species and that among these, some species may require very high concentrations of substrate in order to demonstrate heterotrophic activity.

Parsons and Strickland (1962b) introduced the use of radioactive organic substrates to a study of the kinetics of heterotrophic uptake in sea. The author's original experiments showed that the uptake of radioactive glucose and acetate by natural microbial populations in the sea could be fitted to the equation

$$v = \frac{V_m(S + A)}{K + (S + A)}, \tag{80}$$

where v was the rate of uptake of the substrate, S was the natural *in situ* concentration of substrate, A was the concentration of the same substrate added to the water, V_m was the maximum rate of uptake and K was a constant, similar to the Michaelis-Menten constant for enzyme reactions (see Appendix to this section). Quantitatively v, in units of mg/m^3/hr, was obtained from the expression

$$v = \frac{cf(S + A)}{C\mu t}, \tag{81}$$

where c was the radioactivity of the filtered organisms (cpm), S and A were as defined above, C was the activity of 1 μCi of ^{14}C in the counting assembly used, μ was the number of microcuries added to the sample bottle, f was a factor for any isotope discrimination between the ^{14}C isotope and normal carbon, and t was the incubation time in hours. By employing a reciprocal plot of eqn. (70), to obtain a straight line, the negative intercept on the x-axis becomes $-1/(K + S)$. If one could assume that $K \ll S$, the method would give a measure of the amount of natural substrate in sea water; alternatively, if $K \gg S$ the method would give a measure of the 'relative heterotrophic potential' based on K in mg C/m^3, which expresses the ability of the microbial population to take up the substrate. Wright and Hobbie (1965 and 1966) discussed the use of this method as applied to lake water and from their work it has been possible to draw a number of conclusions. By combining eqns. (80) and (81) and rearranging, the following form can be obtained:

$$\frac{C\mu t}{fc} = \frac{(K + S)}{V_m} + \frac{A}{V_m}. \quad (82)$$

When $C\mu t/fc$ is plotted against different concentrations of A (assuming A to be much larger than S), a straight line is obtained with a slope $1/V_m$, an intercept on the ordinate of $(K + S)/V_m$ and an intercept on the abscissa at $-(K + S)$. If an independent determination is made of S (e.g. by chemical analysis), the rate of uptake of the substrate under natural conditions can be determined, providing the organisms are complying with the relationship in eqn. (80). Three experimental qualifications governing the use of this equation are (i) that there must be no appreciable change in the microbial population during the experiment, (ii) that there should be no appreciable change in substrate concentration, and (iii) that careful temperature control is maintained. A plot of $C\mu t/fc$ versus A is shown in Fig. 52 from Wright and Hobbie (1965). From these results with natural lake water it is apparent that eqn. (80) applied over concentration range from 0 to 580 mg/m³ of added acetate, but that above this concentration the relationship did not apply. In additional experiments, it was further shown by Wright and Hobbie (1965) that at high substrate concentrations a second uptake mechanism was involved which accounted for the departure from an asymptotic value of V_m predicted by eqn. (80). This was explained in terms of a passive transport of substrate due to diffusion at high substrate concentrations and the process could be expressed in terms of a diffusion coefficient (K_d) as

$$K_d = \frac{V_1 - V_2}{A_1 - A_2}, \quad (83)$$

where V_1, V_2 and A_1, A_2 are two rates of uptake and substrate concentrations, respectively. The difference in these two uptake mechanisms is illustrated in Fig. 53, curves A and B.

From these studies it is apparent that there are at least two transport systems for heterotrophic uptake: one showing dependence on Michaelis–Menten kinetics at low substrate concentrations and the other showing dependence on diffusion at high concentrations. Some indication of how low substrate concentrations can be in nature for the

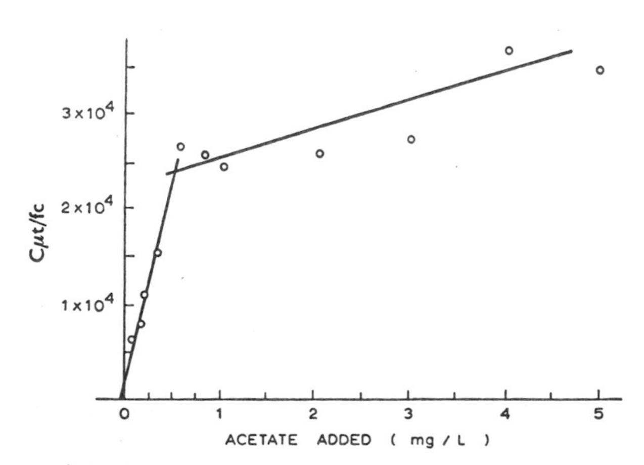

FIG. 52. Kinetics of acetate uptake by natural lake water samples at different substrate concentrations plotted according to eqn. (82) (redrawn from Wright and Hobbie, 1965).

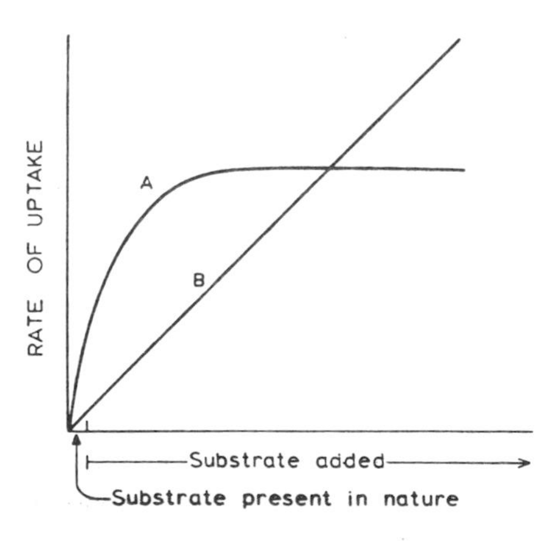

FIG. 53. Theoretical relationships between substrate addition and velocity of heterotrophic uptake showing (A) active transport, in which the transport system becomes saturated and (B) transport by diffusion in which the uptake velocity increases with substrate concentration. Note: added substrate is greatly in excess of natural levels.

first mechanism, is given by the K values for two bacterial isolates obtained by Wright and Hobbie; these values were $5 \cdot 0 \text{ mg/m}^3$ for acetate and 7 mg/m^3 for glucose. Since 90% of the maximum uptake velocity is obtained at *ca.* $10 \times K$, the active uptake mechanisms may be assumed to operate maximally at substrate concentrations below 100 mg/m^3. The second mechanism for heterotrophic uptake appears to operate about 10 times the maximum concentration for active uptake (i.e. at *ca.* 1.0 g/m^3) and from some additional experiments it was shown by Wright and Hobbie (1965) that phytoplankton were largely responsible for the uptake of organics by the diffusion mechanism. From these findings it was concluded that, in general, bacterial activity in aquatic environments would effectively maintain concentrations of substrates which would be too low for algal heterotrophy, if the latter was dependent only on the diffusion of substrates at high concentrations. However, several authors have found that Michaelis–Menten kinetics may apply to the uptake of organic substrates by some species of coastal phytoplankton. For examples, Hellebust (1970) found that one species of marine diatom, *Melosira nummuloides*, showed an ability to take up amino acids through an active transport system which compared favourably (K as low as $7 \cdot 7 \times 10^{-6}$ m) with transport constants of many bacteria (see also North and Stephens, 1971 and 1972). At present, however, these organisms do not appear to account for heterotrophic activity in oceanic environments. This is indicated from field data collected by Williams (1970), who found that 80% of the heterotrophic activity in sea-water samples from the Mediterranean and Atlantic Ocean would pass an 8-μm filter. This would eliminate a large part of the phytoplankton biomass. Particles of $<8 \ \mu$m would include all bacteria and most bacterial aggregates, other than those attached to large particles of detritus.

The hyperbolic curve (A) in Fig. 53 is shown passing through zero rate at zero substrate concentration. In practice it has been found (Jannasch, 1970) that zero growth may occur at some low but finite, or threshold, concentration. This phenomenon has been explained as a population effect in which a certain initial population of a particular species has to be present in order to modify the environment (e.g. redox potential) for the cells to multiply. The effect has been observed in chemostat studies but it may also be a common phenomenon under natural conditions in the sea. Kinetics accounting for a small departure from a zero intercept have been given by Jannasch (1970).

In some cases it has been reported to be difficult to establish any meaningful relationships between substrate concentration and uptake (e.g. Vaccaro and Jannasch, 1967; Hamilton and Preslan, 1970): these difficulties may be due to the existence of several competing heterotrophic populations or to the occurrence of more than one uptake mechanism operative over the same range of concentration. From Hellebust's (1970) data it is also apparent that a form of competitive inhibition of substrate uptake may be caused by the presence of similar substrates.

Sorokin (1970) has investigated the uptake of radioactive algal hydrolysates in water samples taken at different depths from the surface to 5000 m in the Pacific Ocean. The author found very little heterotrophic activity below 800 m and the maximum activity in tropical waters was between 400 and 600 m, which coincides with the depth of the oxygen minimum. From fairly extensive studies on the uptake of glucose and amino acids in the English Channel, Andrews and Williams (1971) concluded that heterotrophic processes in the area account for an uptake of organic material equivalent to 50% of the measured phytoplankton production. If this is converted into bacterial biomass with an efficiency of between 30 and 60%, then the contribution of heterotrophs to the formation of particulate organic material in the ocean would amount to between 15 and 30% of the production by autotrophic photosynthetic processes.

(*Appendix to Section 3.2.2.*)

In a number of sections throughout this text reference is made to Michaelis–Menten kinetics. Caperon (1967) discussed the general applicability of this approach to the uptake of food materials by micro-organisms. From his review of the subject it is possible to write a general expression for the growth of a population of micro-organisms limited by food supply. The approach follows the original expression given by Michaelis and Menten for the action of enzymes on substrates.

The process of a population feeding may be described by the following equation in which C is the acquisition site for food, B is a food particle or unit quantity of food, $C \cdot B$ is an acquisition site filled by B, and P is a particle, or unit amount, of ingested food.

$$C + B \underset{k_2}{\overset{k_1}{\rightleftharpoons}} C \cdot B \xrightarrow{k_3} C + P. \tag{84}$$

If $(C + B)$ to $(C \cdot B)$ is a reversible process it can be defined by constant K_m such that

$$K_m = \frac{[(C) - (CB)](B)}{(CB)} \tag{85}$$

or

$$(CB) = \frac{(C)(B)}{K_m + B}$$

If the breakdown of (CB) to $(C + P)$ is largely an irreversible process, then the rate (v) for this process is given by

$$v = k_3(CB) \tag{86}$$

and

$$v = \frac{k_3(C)(B)}{K_m + (B)}. \tag{87}$$

The maximum rate (V_{mc}) will be attained when the concentration of CB is maximized, which is when all the acquisition sites are filled [i.e. $(CB) = (C)$].

$$V_m = k_3(CB) = k_3(C). \tag{88}$$

By combining eqns. (87) and (88), the general Michaelis–Menten expression is obtained:

$$v = \frac{V_m(B)}{K_m + (B)}. \tag{89}$$

Under these conditions it is possible to relate growth to the rate of food intake. These conditions are (1) that the concentration of only one food is limiting population growth, (2) that a given amount of ingested food always results in the production of a fixed number of new individuals, (3) that there is no time lag in response of growth rate to change in food concentration, and (4) that food once absorbed is ingested and not returned in any appreciable quantity as a food item. Under these conditions, growth, measured as production dn/dt per unit biomass (n), can be expressed as μ, the growth constant, where

$$\mu = \frac{1}{n}\frac{dn}{dt} \tag{90}$$

and μ and μ_{max} substituted for v and V_m in eqn. (89)

$$\mu = \frac{\mu_{max}(B)}{K_m + (B)}. \tag{91}$$

CHAPTER 4

PLANKTON FEEDING AND PRODUCTION

4.1 FEEDING PROCESSES

Among the zooplankton, methods of feeding may be broadly divided into filter feeding and raptorial feeding. In the former process a number of different mechanisms are employed to induce a flow of water against a screen which removes most of the particulate material in suspension. In the second process, animals actually seize individual prey items; the latter may either be large plants, such as individual diatom cells, or planktonic animals generally smaller in size than the predator. The two processes are not mutually exclusive and examples of both filter feeding and raptorial feeding can often be found in one species, especially among the planktonic crustaceans. On the basis of diet, zooplankton feeding may be herbivorous, omnivorous, or carnivorous; other divisions used by some authors are to describe the dietary requirements as being phytophagous, zoophagous, euryphagous, or detritivorous.

A general account of suspension feeding has been written by Jørgensen (1966), and an extensive study of the nutrition of abyssal plankton has been reported by Chindonova (1959); specific reviews on planktonic crustacean feeding have been written by a number of authors including Marshall and Orr (1955), Wickstead (1962), and Gauld (1966).

PROTOZOA

Among the planktonic protozoans, the feeding of the foraminifera and radiolaria is generally accomplished through the extrusion of pseudopodia from within the main body of the animal. The size of particles captured efficiently appears to be generally in the range 1 to $<20\ \mu m$ diameter (Rassoulzadegan and Etienne, 1981). The grazing rate of tintinnids was measured in the waters of Long Island Sound where population densities of these microzooplankton were *ca.* 10^3 to 10^4 animals/litre. The maximum filtering rates of these animals was estimated to be *ca.* 85 μC/animal/hr which resulted in the consumption of up to 41% of the phytoplankton standing stock per day (Capriulo and Carpenter, 1980). The role of microflagellates as part of the protozoan ecology is an important subject which has been neglected until quite recently. Haas and Webb (1979) showed that a number of microflagellates were heterotrophic organisms feeding on bacteria. In a series of papers (e.g. Fenchel, 1982) it has been shown that the relatively low but constant (*ca.* 10^6/ml) number of bacteria in sea water is probably controlled by the microflagellate population. Among these organisms are chrysomonads and choanoflagellates in the size range *ca.* 1 to 20 μm; numbers are reported to be about 10^3/ml and to increase in response to bacterial blooms. On a depth scale in different oceans Sorokin (1981) has observed that the maximum number of microflagellates and other protozoa are closely associated with the deep chlorophyll and bacteria maxima. Larger heterotrophic flagellates, such as the dinoflagellate *Noctiluca*, may also form abundant blooms at certain seasons and consume larger autotrophic phytoplankton in competition with metazoan herbivores.

CTENOPHORES

Ctenophores become very abundant in many temperate waters from *ca.* July to September in the northern hemisphere. In tropical waters they are a ubiquitous member of the plankton community.

There are two classes of Ctenophores—the Tentaculata and the Nuda. The Tentaculata are by far the most abundant and include those ctenophores with well-developed tentacles, such as *Pleurobrachia*, as well as those with large lobes, such as *Bolonopsis.* Both orders are believed to be entirely carnivorous and to feed largely on small crustaceans. The class Nuda are predatory ctenophores which feed on the tentaculate ctenophores and some other soft-bodied zooplankton—an example is the genus *Beröe*, different species of which generally feed rather specifically on a single species of tentaculate ctenophore. A review of the nutritional ecology of ctenophores is given by Reeve and Walters (1978).

Tentaculate ctenophores are voracious predators of copepods and may consume up to 1000% of their body weight per day at high prey densities (*ca.* 10^6 copepods of the genus *Acartia* or *Pseudocalanus* per m^3). At this rate of feeding the organisms (*Pleurobrachia* and *Mnemiopsis*) could grow at a rate of just over a doubling in the size of the population per day. Under such conditions ctenophore population 'explosions' occur, decimating copepod standing stocks, and releasing nutrients for the production of phytoplankton. Assimilation efficiencies of ctenophores are *ca.* 80% but growth efficiency is generally low, *ca.* 10% (Reeve and Walter, 1978). Since many of the ctenophores are quite fragile, their abundance in the ocean has never been greatly appreciated. Recently Harbison *et al.* (1978) have made visual observations on ctenophores in open ocean waters and report numerous sightings. The food of oceanic ctenophores ranges from small copepods to large euphausiids and even small fish. Among the predators of ctenophores are amphipods, medusae, heteropods, and other predatory ctenophores (e.g. *Beröe*).

CHAETOGNATHS

In temperate waters, chaetognaths become abundant in the latter part of the summer along with ctenophores. The number of generations per year varies from *ca.* 5 in subtropical waters to one or less in polar waters. Since chaetognaths have an ability to sense their prey and hunt, there is some evidence that they are successful predators on small crustaceans at prey densities below those required by ctenophores (i.e. $<$ *ca.* 10^3 copedods/m^3). The prey of chaetognaths is not limited to crustaceans and there are many observations on larval fish being caught by these predators. As in the case of the relationship between ctenophores and copepods (Fig. 63), young chaetognaths are also known to be eaten by the copepod *Acartia* (Davis, 1977). The life history of an abundant temperate chaetognath, *Sagitta elegans*, is given by King (1979) and food-size selection has been reported by Sullivan (1980) and Pearre (1980) as well as in references cited by these authors.

UROCHORDATES

The Urochordata (or tunicates) consist of three classes: the Ascidiacea, the Thaliacea, and the Larvacea. The Ascideans are sessile and commonly referred to as sea squirts. The two other classes are pelagic; the Thaliaceans, or salps, have six genera and are often most abundant in sub-tropical and tropical seas. The Larvaceans (sometimes called the Appendicularians) have neotenic adults which retain some larval characteristics as well as having a remarkable feature — the larvacean 'house' which is used to filter particulate material.

Salps and Larvaceans are extremely abundant in some ocean areas (e.g. several thousand/m^3 in some aggregated populations). Both types of organisms are fine filter feeders and largely consume phytoplankton and bacteria. In sub-tropical waters, under conditions of a favorable food supply, the salp, *Thalia democratica*, was observed to grow at 10% per hour and to complete a generation in *ca.* 46 hr (Heron, 1972). Growth of large numbers of these organisms gives rise to accumulations of flocculant fecal material which becomes the medium for populations of bacteria and protozoa (Pomeroy and Deibel, 1980). Grazing and ingestion rates for two species of salps are discussed by Deibel (1982). Daily rations range from *ca.* 60 to 130% of body weight per day, depending on species. Concentrations of Larvaceans may reach several thousand per m^3—with filtering rates of between 100 and 1000 ml/animal/day, such populations can rapidly decimate the smaller (<10 μm) sized phytoplankton and bacteria. Large blooms of Larvaceans also result in much detrital material from the cast-off houses of these organisms (Alldredge, 1981).

COELENTERATES

Among the Coelenterates, two classes, the Hydrozoa and the Scyphozoa, form important groups in the plankton. The medusoid stage of the hydrozoans is represented by such common genera as *Obelia, Pennaria* and the large pelagic hydroid, *Velella.* The colonial hydroid, *Physalia* (a siphonophore), is also a conspicuous member of this group when present in large numbers, particularly on bathing beaches. In the class Scyphozoa, the dominant form of the animal is the medusa. The class includes some of the most poisonous jellyfish in the sea including *Cyanea* and *Chironex fleckeri* (the sea wasp), as well as more abundant genera such as *Aurelia.* In practically all cases the coelenterates feed as carnivores using nematocysts on their tentacles to paralyze their prey. While most pelagic coelenterates are relatively small and feed on other plankton (e.g. crustaceans) some such as *Cyanea* may have a medusoid bell of *ca.* 1 m and tentacles greater than 10 m in length. These animals are capable of capturing small fish and other nekton. The feeding of a siphonophore, *Rhizophysa eysenhardti*, on fish larvae is described by Purcell (1981a). The author has also described the feeding of twenty-four species of siphonophores from different oceans (Purcell, 1981b). She found that each sub-order among the species studied had a characteristic diet. These diets varied from being primarily composed of fish larvae to those in which small copepods dominated. The maximum size of prey was generally correlated with gastrozoid length, food intake was proportional to prey density and more prey were consumed at night among most siphonophore species than during the day.

ANNELIDA (Class Polychaeta)

There are several species of polychaete worms which occur in the plankton. The best known of these is the genus *Tomopteris.* The pelagic polychaetes are generally regarded as raptorial carnivores, preying off smaller zooplankton, including larval herring.

MOLLUSCA

The mollusca consist of several classes of which the cephalopoda (squids) are important among the nekton and the sub-classes Prosobranchia and Opisthobranchia are important among the plankton. The prosobranchian molluscs include the pelagic heteropods; these are exclusively carnivorous, locating prey by sight and capturing it with the large grasping teeth of the radula located at the tip of a proboscis. There are also prosobranchs which are members of the neuston community, such as *Janthina*, which feeds from a raft of bubbles on jellyfish. Among the opisthobranchian molluscs, two orders, the Gymnosomata and the Thecosomata, are often abundant among the plankton. The thecosomatous, or shelled pteropods, feed primarily on phytoplankton, although larger species also capture planktonic protozoa including foraminifera and radiolarians (Chindonova, 1959; Gilmer, 1974). Food particles are collected on mucus produced in the mantle cavity or on the foot, and the consolidated food particles are transferred to the mouth by cilia. Gilmer (1974) reported that the size of particles retained by different species of thecosomes depended upon the size and species of pteropod and varied from very small (<5 μm, including bacteria and detritus) to large particles up to 800 μm (including crustacean larvae). The specialized pseudothecosomes (*Corolla, Gleba*) employ a unique method of feeding, capturing small food particles in large (up to 2 m diameter), external, mucus webs produced by special glands in the wings (Gilmer, 1972). The food and feeding of gymnosomatous (unshelled) pteropods has been reviewed and studied by Lalli (1970, 1972) who indicates that most of these molluscs feed on particular thecosome species. The complex mouth parts of gymnosomes appear to be specifically adapted toward only one or two species of thecosome prey and are employed in prey capture and in removal of the prey body from its shell. Another order of opisthobranch which is sometimes important in the plankton community is Nudibranchia; while many are largely benthic, the genus *Glaucus* is pelagic and feeds on jellyfish from which it may capture and retain nematocysts.

CRUSTACEANS

The position of the zooplanktonic crustaceans, as by far the largest group of suspension feeders in the ocean, together with their ability to consume a great variety of prey, warrants special consideration of the feeding habits of these animals in the marine food chain. The major groups of planktonic crustaceans are the calanoid copepods, amphipods, and euphausiids. In some areas ostracods and cladocerans may be relatively abundant, while in coastal regions mysids, isopods, cumaceans, and harpacticoid copepods also occur as part of the bentho-pelagic community.

The type and arrangement of feeding appendages among the planktonic crustaceans is complex compared with many of the zooplankton discussed above. The appendages of copepods, amphipods, and euphausiids are quite different in form but are similar in function in that generally at least two forms of feeding are displayed, either within the group of animals, or even by a particular species of animal depending on the availability of prey type. These two forms of feeding have been referred to in the past as 'filter' feeding and 'raptorial' feeding. In the former case it is known that some crustaceans can remove very small particles from sea water, although the exact mechanism by which they do this is controversial. In the latter case it is apparent that the appendages of some crustaceans are better adapted to seizing their prey. In general the former mode of feeding is suitable for collection of phytoplankton (herbivores) while the latter is suitable for grasping small animals (carnivores). Since both modes of feeding are displayed by some crustaceans, such animals are omnivorous.

The complexity of appendages on a marine calanoid copepod (e.g. *Eucalanus pileatus*) is shown in a stylized form in Fig. 54A. It is the movement and function of these appendages which is the subject of alternate explanations. In earlier versions of the feeding process it was assumed that because many of the appendages (particularly the two maxilla) were composed of chitinous fibres (setae) and even finer fibrils (setules) they must act as a filtering basket. Thus the mechanics of feeding was assumed to be similar to that of any other filter, or screen, which removed particles if a current of water was passed through the screen (e.g. as illustrated in Fig. 54B). This in turn gave rise to explanations on size selection which essentially explained prey 'preference' in terms of the pore size of the filtering basket as decided by the spacing between setae and setules (e.g. Boyd, 1976). In some cases such a mechanism may exist and there is certainly some size selection mechanism as shown later in this discussion. The alternate explanation on 'filter feeding' is that it should not be regarded as a filtering process since although the setae of the appendages appear as rakes, they act more like paddles because of viscous forces which dominate in the small world of copepods. Observations made with high-speed, close-up cinematography are illustrated diagrammatically in Fig. 54C,D and E. In this sequence the flapping of appendages produces a stream of water past the copepod. When an alga occurs in the vicinity of the copepod, the feeding appendages beat asymmetrically, so as to draw water preferentially from the direction of the alga. As the alga nears the 'filters', they fling apart creating a gap that is filled by inrushing water (Fig. 54D). The water carries the alga within the basket,

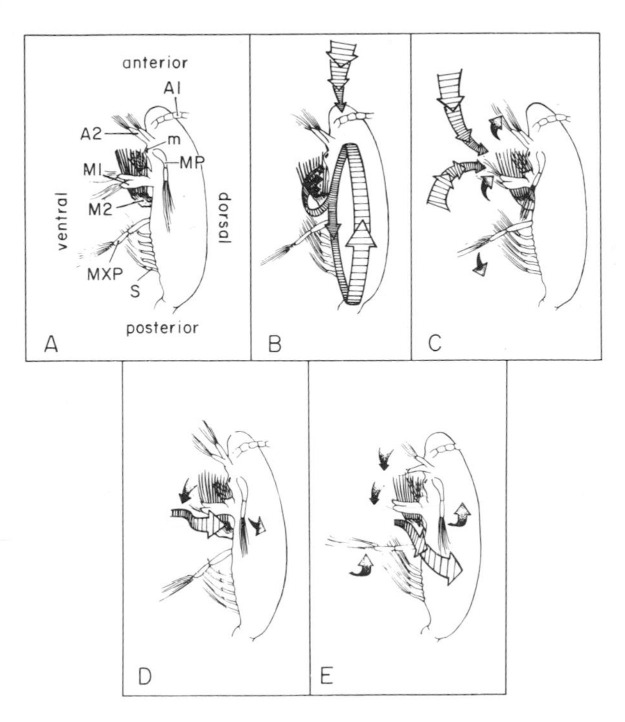

FIG. 54. *Eucalanus pileatus*. (A) Diagram of an animal in typical feeding position viewed from its left side. Only the left appendage of each pair is shown. Feeding appendages are: A2—second antenna; MP—mandibular palp; M1—first maxilla; M2—second maxilla; MXP—maxilliped. Other structures labeled are: A1—first antenna; S—swimming legs; m—mouth. Structure of appendages has been grossly simplified for clarity. (B) Diagram of 'textbook version' of copepod feeding currents (wide arrows) drawn from information presented elsewhere (Cannon, 1928; Lowndes, 1935; Marshall and Orr, 1955; Russell-Hunter, 1979; Barnes, 1980). (C–E) Diagrams of feeding appendage movements (stippled arrows) and water currents (striped arrows) they produce as revealed by our films. An arrow with a narrow shaft and wide head indicates lateral movement out of the plane of the page toward the reader; an arrow with a wide shaft and narrow head indicates medial movement away from the reader. (C) Outward movements of second antennae and maxillipeds sucks water toward copepod's maxillae. (D) Postero-medial movement of the first maxillae and dorso-lateral movement of mandibular palps sucks water laterally. (E) Inward movements of second antennae and maxillipeds coupled with dorso-lateral movement of mandibular palps shoves water posterolaterally. (With permission from Koehl and Strickler, 1981).

and as the appendages close, water is squeezed out between the setae. In this description there has to be some prior detection mechanism of the presence of an algal cell, as opposed to water being continuously pumped through a filter screen, as suggested in earlier descriptions (Fig. 54B). Independent support for these observations made by cinematography is found in the fact that the intersetule distance in many crustaceans is a poor indicator of the size of particles retained by these organisms (e.g. Vanderploeg and Fallscheer, 1982). On the other hand, the presence of chemoreceptors on various appendages, including the (1) antennae, gives some credence to the ability of copepods to detect the presence of a phytoplankton prey item (Strickler and Bal, 1973; Friedman and Strickler, 1975). Among cyclopoid copepods (which are largely carnivorous) visual detection of prey has also been reported (Gophen and Harris, 1981). While this explanation of 'filter feeding' appears satisfactory for small copepods, there are also recent descriptions on euphausiid feeding which follow more traditional explanations of 'sieving' particles. Kils* (personal communication), for example, has made detailed studies on two feeding modes of *Euphausia superba* in the Antarctic. In the filtering mode, this animal appears to actually exclude very large phytoplankton cells with one group of setae while collecting smaller phytoplankton with another, finer group of setae and setules. Raptorial feeding by the same animal on large particles did not involve the filtering mechanism.

The adaptation of crustacean appendages for filter feeding, raptorial feeding or a combination of both, has been discussed by a number of authors. Anraku and Omori (1963) showed that the (2) antennae, mandibles, (1) maxillae and maxillipeds of a predominantly herbivorous filter feeder were all well developed in terms of setae which increased the surface area of these appendages and therefore assured their efficient use in producing water currents through the filter chamber; setae of the (2) maxillae were also well developed in order to assure an efficient particle filter. In contrast, in predatory species of copepod, *Tortanus discaudatus*, the appendages have few setae and instead there are modifications in structure which aid in the use of these appendages for seizing and holding a prey item. An illustration of two extremes in structural

*A detailed report on *E. superba* feeding appeared at the time this book was in press (see McClatchie, S. and Boyd, C. M. (1983) *Can. J. Fish. Aquat. Sci.* **40**, 955–967).

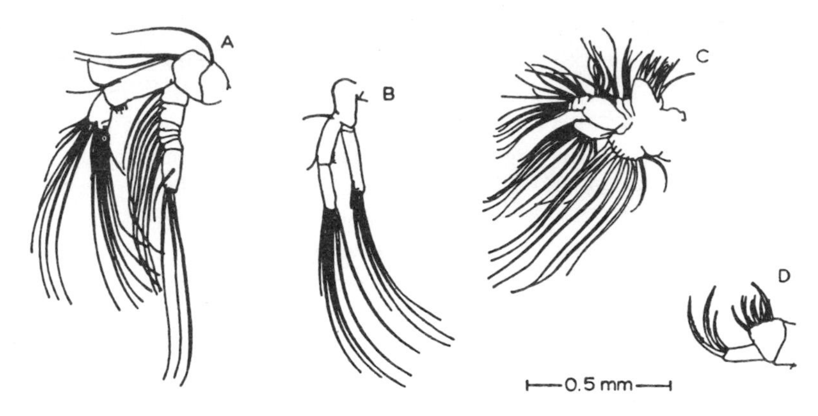

FIG. 55. Adaptation of crustacean appendages showing (2) antenna of (A) *Calanus finmarchicus* and (B) *Tortanus discaudatus* and (1) maxilla (C) and (D) of the same species, respectively (redrawn from Anraku and Omori, 1963).

development in *C. finmarchicus* and *T. discaudatus* are shown in Fig. 55. Anraku and Omori (1963) also noted differences in the cutting edges of mandibles in different copepods and they describe a typical filter feeder as having grinding teeth, while those of a typical predator have very sharp teeth. Between these two extremes, there are a variety of structures which enable some copepods to be omnivorous. Itoh (1970) has been able to summarize the difference between herbivores, omnivores, and carnivores on the basis of an 'Edge Index' (EI) derived from measurements of the cutting edges of the mandible. Calculation of the 'Edge Index' is illustrated in Fig. 56 and a plot of this value versus the number of cutting edges is illustrated in Fig. 56B. From the latter figure it is apparent that pelagic copepods may be divided into three groups, viz. Group 1, with EI $\leqslant 500$ consisting of the Families Calanidae, Eucalanidae, Paracalanidae, and Pseudocalanidae, which are largely herbivorous filter feeders; Group II, with $500 < \text{EI} \leqslant 900$, consisting of generally omnivorous feeders of the Families Euchaetidae, Centropagidae, Temoridae, Lucicutiidae, and Acartiidae, and Group III EI >900 consisting of predominantly raptorial feeders of the Families Heterorhabdidae, Augaptilidae, Candaciidae, Pontellidae, and Tortanidae. From earlier work on stomach contents of these animals it

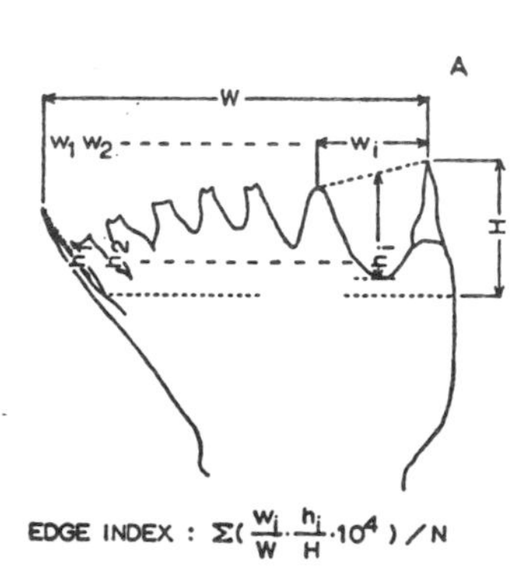

FIG. 56A. Schematic representation of the 'Edge Index' on the cutting edges of the mandible, $w_1, w_2, \ldots, w_i$ represent the width of individual cutting edges and W the total width; similarly $h_1, h_2, \ldots, h_i$ and H represent the edge heights, respectively; N represents the number of edges (redrawn from Itoh, 1970).

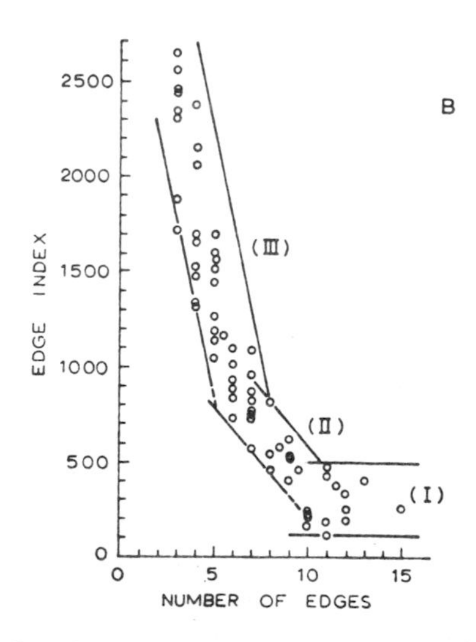

FIG. 56B. Relation between 'Edge Index' and number of edges of the mandible (redrawn from Itoh, 1970).

is sometimes possible to identify the raptorial feeders as being carnivores although the inclusion of stomach contents of a herbivorous prey makes it difficult to diagnose whether an animal is exclusively carnivorous.

In addition to extensive studies on the feeding of copepods, Nemoto (1967) has reported on differentiation in the body characters of euphausiids. Among euphausiids, raptorial feeders usually have very long 2nd or 3rd thoracic legs which terminate in grasping spines or small chelae; predominantly filter-feeding euphausiids have thoracic legs of similar length and well-developed setae. In addition to modifications in the appendages of these animals, Nemoto examined differences in stomach structures and reported that herbivorous euphausiids had a cluster of spines along the posterior wall of the stomach while the same structures were largely absent from deep water carnivorous species.

4.1.1 Rate of Filtering

The rate of filtering among microcrustaceans is broadly related to body size but can vary in any individual depending on such factors as temperature and food concentration. Gilmer (1974) showed that filtering rate of thecosomatus pteropods was an approximate log/log relationship with size from 4·3 to 55·7 ml/hr for animals between 0·03 and 15 mg dry weight. Such a relationship would appear reasonable if one considers that the weight of an animal will increase as the cube and the filtering surface will increase as the square of an animal's linear dimensions. From a summary of measurements made by Marshall and Orr (1955) and Jørgensen (1966) for copepods, it appears that the volume of water filtered by an animal may range from *ca.* $<1\cdot0$ to 200 ml/day. Cushing and Vucetic (1963) determined that maximum filtering rates for adult copepods could be as high as 1 litre/day. The latter figure appears high, however, when compared to the filtering rates of the much larger euphausiid crustaceans; these have been measured by Raymont and Conover (1961) and Lasker (1966) who report a range of values for *Meganyctiphanes norvegica, Thysanoëssa* sp., and *Euphausia pacifica* from *ca.* 1 to 25 ml/hr. The filtering rate of planktonic crustaceans may decline as food and concentration is increased. In experiments conducted by Mullin (1963), adult *Calanus hyperboreus* appeared to filter between 200 and 300 ml/day, decreasing linearly to less than *ca.* 50 ml/day as the concentration of food increased. Frost (1972) also has shown a decrease in filtering rate of *Calanus pacificus* from *ca.* 10 to 2 ml/copepod/hr over increasing phytoplankton concentrations from 200 to 600 μg C/l. In addition, however, Frost (1972) has shown that filtering rate is also proportional to the size of the phytoplankton cells; a semi-log relationship is demonstrated showing an increase in filtering rate from 4 to 9 ml/copepod/hr over a cell size range of *ca.* 10^3 to 10^5 μm^3. At very low phytoplankton cell concentrations it appears (Frost, 1975) that filtering rate may decline. On the other hand, hunger can greatly stimulate the filtering rate; for the copepod, *Calanus pacificus*, Runge (1980) found, for example, that starved females filtered at *ca.* 3 times the rate of well-fed animals.

A relationship between filtering rate and body size of a single species of copepod is shown in Fig. 57. In general, filtering rate increases with body size both within and between species, as summarized in Table 24. Most authors have observed among

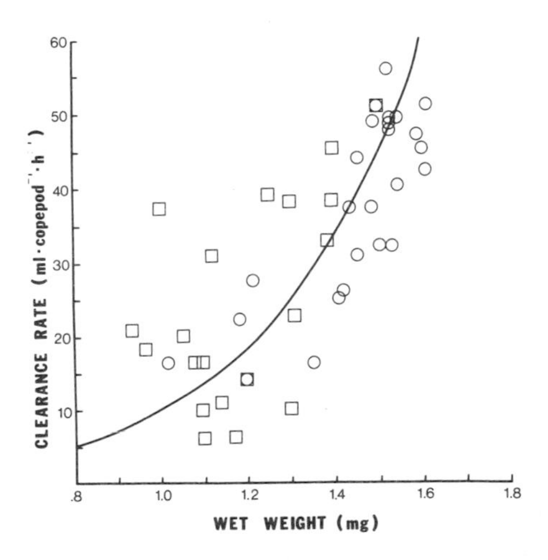

FIG. 57. The relationship between body size and clearance rate for *Calanus pacificus* (different symbols for points from 1975 and 1976—redrawn from Runge, 1980).

TABLE 24. EXAMPLES OF FILTERING (OR CLEARANCE) RATES OF SOME PLANKTONIC ANIMALS

Organism	Approximate size or stage	Filtering rate per animal	Reference
Copepods:			
Calanus helgolandicus	Nauplius adult	4 to 21 ml/day 286–773 ml/day	Paffenhöffer (1971)
Calanus hyperboreus (1) high, (2) low food concentration	(1) Adult (2) Adult	200 ml/day <50 ml/day	Mullin (1963)
Larvacean:			
Oikopleura dioica	0·6 to 1·5 mm trunk length	4 to 15 ml/hr	Alldredge (1981)
Tintinnid:			
Mixed population, Long Island Sound	<200 μm	*ca.* 1 to 85 μl/hr	Capriulo and Carpenter (1980)
Tunicates (salps):			
Pegea confederata	15 mm 50 mm	212 ml/day 53 l/day	Harbison and Gilmer (1976)
Thalia democratica	4 mm	192 ml/day	Diebel (1982)
Ctenophore:			
Mnemiopsis	Larva adult	*ca.* 1 l/day *ca.* 20 l/day	Reeve *et al.* (1978)
Thecosomatus pteropods	0.03 to 15 mg dry wt.	*ca.* 4 to 55 ml/day	Gilmer (1974)

herbivorous copepods that there is a decrease in volume filtered with an increase in prey concentration. However, this is disputed by Conover and Huntley (1980) who consider that copepods can adapt over time to higher prey concentrations and in fact show no decrease in filtration rate over longer time periods. Reeve (1964) found that in the case of the chaetognath, *Sagitta hispidia*, a maximum capacity of fifty *Artemia* nauplii were consumed per day regardless of food concentration. Filtering rates generally increase with temperature to some optimum, and then decline. Kibby (1971) has shown that within a species, the optimum temperature for maximum filtration rate depended on the temperature to which the animal had become adapted (Fig. 58).

The filtering rate (F) of a zooplankter over a given time period (t) can be measured as the decrease in phytoplankton cell concentration at the beginning (C_o) and end (C_t) of the experiment (Gauld, 1951). If the assumption is made that the concentration of cells decreases exponentially with time, then

$$F = \frac{v(\log C_o - \log C_t)2.303}{t},$$

where v is the volume of water per animal. A possible error in the use of this expression is that it assumes that F is constant over the period of the experiment. However, since the filtering rate is known to vary with the concentration of cells, the assumption that F is constant is only an approximation which will be true for small changes in $C_o - C_t$. Control flasks, in which changes in C_o can be measured, and gently stirred or rotated incubation chambers which assure an even distribution of cells, are necessary refinements in most experiments. A discussion of the methodology of measuring filtering rates has been given by Rigler (1971).

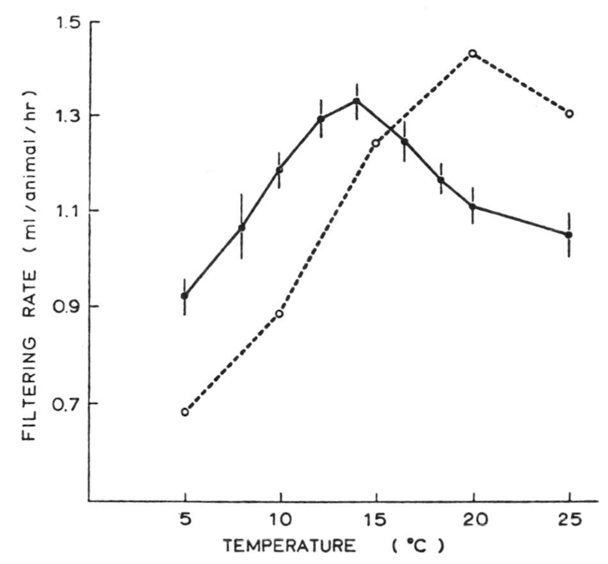

FIG. 58. Filtering rates of *Daphnia rosea* grown at 12°C (solid line) and 20°C (broken line) (redrawn from Kirby, 1971).

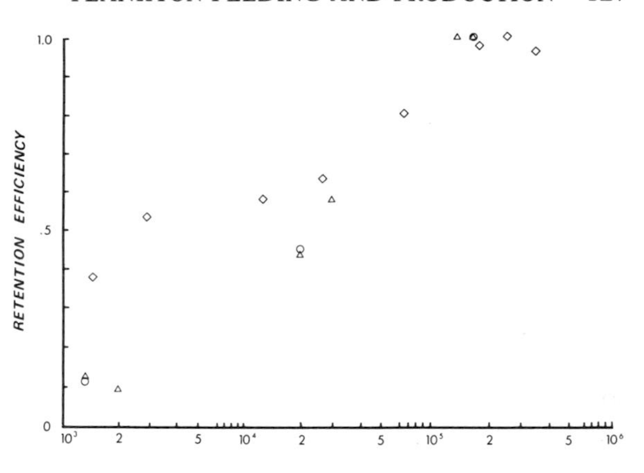

FIG. 59. Comparison of retention efficiencies of *Calanus pacificus* under various experimental conditions (different symbols represent different feeding conditions—data accumulated by Runge, 1980).

4.1.2 Prey Selection

The size of prey items is probably the single most important factor governing prey selection among various organisms in the zooplankton community. This is true for herbivorous feeding (e.g. Frost, 1972 and 1974) and carnivorous feeding (e.g. Reeve and Walter, 1974). This size-selection hypothesis has two properties: these are firstly that predators are generally larger than their prey, and secondly, within the prey size range of a particular predator, the largest prey items will be selected when available (e.g. Mullin, 1963; Hargrave and Geen, 1970). The first of these properties has received general discussion by Sheldon *et al.* (1977) who showed that if predator and prey items of a whole variety of planktonic and nektonic animals were summarized, in relation to the size ratio of predator to prey, then this ratio was on the average about 14:1. The second property, on size selection by a single species, is illustrated in Fig. 59, where it is shown that the ability to retain food items by *Calanus pacificus* is a log. linear function of their size, the largest particles (up to some maximum value) being the best retained.

Additional factors which determine prey selection are very numerous and it is perhaps sufficient to mention a few of these as examples. For some carnivores (e.g. herring larvae, chaetognaths) prey items must be alive and moving in order to provoke attack. Bioluminescence has been reported by Esaias and Curl (1972) to discourage copepod feeding on some species of dinoflagellates. A difference in prey selection towards male and female copepods by two freshwater carnivores (Haly, 1970) has also been observed. The role of large spines on some species of diatoms (e.g. genus *Thalassiosira*) has been investigated by Schnack (1979) and Gifford *et al.* (1981). In the case of *Calanus helgolandicus* and *C. finmarchicus* it was observed that the filtration rate was higher on the spinose plankton. On the other hand, an example of the preferential survival of spined prey has been reported for freshwater plankton communities (Kerfoot, 1977). Food selection may change in relation to the age of predators (as discussed also in respect to fish, Chapter 7). Using an elegant technique of ^{3}H and ^{14}C differential labelling, Lamport (1974)was able to show with *Daphnia* that the smallest animals (<1·5 mm) showed a preference for bacteria, but that larger animals (>2 mm) could eat diatoms (28 μm) or bacteria (1·5 μm). Reversals of the size dominance of predators over prey are apparent in the feeding of some pelagic molluscs (e.g. *Glaucus*) on jellyfish, as well as among some small crustaceans which feed off large salps (e.g. Heron, 1973).

The role of chemosensory mechanisms in deciding the prey of some predators has become recognized among zooplankton as well as the nekton. In experiments in which the swarming and feeding of two copepods were observed, Poulet and Ouellet (1982) found that copepod activity was most sensitive to dicarboxylic amino acids. In other work cited by these authors it is noted that such compounds as carbohydrates, polypeptides, and urea were non-stimulatory. The release of compounds which are inhibitory to zooplankton grazing may also occur; for example, an abundant alga in some temperate waters is *Phaeocystis* which is known for its unpleasant odour due to the compound, acrylic acid.

The problem of feeding in deep-water environments where food is scarce has been reviewed by Harding (1974) who suggests that three pathways are available for the transport of food to deep-water animals. These are zooplankton vertical migrations (Vinogradov, 1955), the 'rain' of detritus from the surface (Vinogradov, 1962), and the transfer of dissolved organic carbon into particulate matter (e.g. Riley, 1970; Parsons and Seki, 1970). While the first of these pathways is an active transport of animals to different levels in the upper water column, it is probable that it is not an effective mechanism below 100 m where there is no evidence for diurnal migrations (Angel and Fasham, 1973). The question with both the other two mechanisms is that even if they are effective in producing particles at great depths, are these concentrations sufficient to support the feeding requirements of deep-sea zooplankton? From metabolic requirements for growth and maintenance, the minimum particulate carbon concentration required for a copepod would be *ca.* 25–50 μg C/l (Parsons and Seki, 1970) which is below the general background level for deep-water particulate carbon (e.g. Menzel, 1967). However, it has been suggested (Parsons and Seki, 1970) that microlayers with high concentrations of particulate matter may exist in the deep ocean. Some evidence for this has been given by Wangersky (1974) who found anomalously high particulate organic carbon values in about 8% of his samples. These samples were also found to contain small zooplankton.

The suggestion contained in Harding (1974) that deep-water particulate material is formed *in situ* from soluble organic carbon has been further examined by Williams (1975). Using data from Williams *et al.* (1970) on the ^{14}C enrichment of deep water in the aftermath of atomic bomb testing, Williams (1975) has pointed out that both surface plankton samples and bathypelagic organisms are enriched with ^{14}C. In contrast, deep-water soluble organic carbon shows no such enrichment. This is taken as indicating a relatively rapid transportation of surface organic carbon to the bathypelagic community. Recent evidence for a rapid flux of particulate material from the surface to deep water is discussed in Section 1.3.3 and illustrated in Fig. 15. From these data it appears that the most likely source of new particulate material, below the euphotic zone, is from the detrital material which sinks rapidly in the water column.

In conclusion, the process of feeding in the plankton community can result in the removal of very fine particles (*ca.* 0·1 μm) by animals possessing a mucus net; larger particles such as bacteria (*ca.* 1–2 μm) are removed by flagellates and some metazoans; phytoplankton are captured along with microzooplankton (*ca.* 2–50 μm) by some form of sieving action, while the macroplankton are generally removed by raptorial predators.

4.2 SOME TROPHODYNAMIC RELATIONSHIPS AFFECTING THE PLANKTON COMMUNITY

Comprehensive reviews on the trophodynamics of zooplankton have been written by several authors (e.g. Steele and Mullin, 1977; Conover and Huntley, 1980). In addition there are a number of papers which deal extensively with the energy budget of a single species of zooplankton (e.g. for the copepod, *Pseudocalanus*, see Corkett and McLaren, 1978, or for the amphipod, *Calliopius laeviuscules*, see Dagg, 1976).

4.2.1 Food Requirements of Zooplankton

The distributional needs for food among marine zooplankton are similar to any other representative of the animal kingdom. This is shown in Fig. 60 as a flow diagram in which the energy in the ration consumed is distributed as a flow chart into the various life requirements of the animal. The major divisions of food energy (R) are first between how much is assimilated (A) and how it is excreted (E) and secondly, of the assimilated ration (AR), how much is used for various metabolic needs (T) and how much is used for growth (G). Growth is generally defined as an increase in body weight but among crustaceans a certain amount of growth tissue will be lost as moults, while the female of a species will also lose an appreciable fraction of her body weight in egg production. Among metabolic needs, movement, body maintenance and the metabolic cost of the extra effort required in digestion and growth are clearly identifiable. Further, the waste products of these metabolic activities must then be excreted; in high animals this excrement would be identifiable with urine but in small zooplankton it is extremely difficult to identify this loss.

An important point to recognize in Fig. 60 is that the fractions of energy found in any one section are not always the same for different species or for the same species, either under different feeding regimes or at different ages in the organism's life cycle. Thus, while we shall attempt to give values to some of these life requirements, there are actually 'feed-back' relationships which can alter the absolute fraction assigned to any one process [e.g. the fraction of food used for growth (G) is not a simple fraction of the amount of food consumed (R)]. These 'feed-back' relationships will also be discussed.

The food requirement of a zooplankter can be expressed as the sum of the major requirements for growth, metabolism, and feces (Richman, 1958).

$$R = G + T + E. \quad (92)$$

Since E can be expressed as $(R - AR)$ where A is the assimilation efficiency of the food, eqn. (92) can be written as

$$G = AR - T. \quad (93)$$

The extent to which food is assimilated by a zooplankter is difficult to measure experimentally since there are technical difficulties in making quantitative collections of feces from zooplankton, even under laboratory conditions. By assuming that the ash content of the food material was largely unabsorbed in the digestive process, Conover

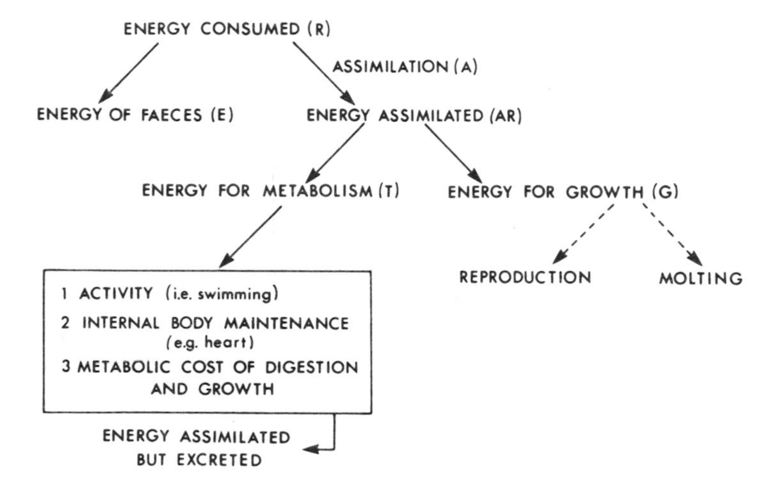

FIG. 60. Distribution of food energy in a marine zooplankter.

(1966a) derived a formula which allows for an estimation of the assimilation efficiency (A) in terms of (J), the ash-free dry weight to dry weight ratio in the ingested food and (L), the same ratio in a sample of feces.

$$A = \frac{(J - L)}{(A - L)(J)} \times 100(\%). \tag{94}$$

Use of this formula avoids the difficulty of quantitatively collecting feces and also permits measurement of assimilation efficiencies on samples of particulate material collected *in situ.* Conover (1966b) examined factors which affect the assimilation efficiency and found it to be largely independent of temperature, the amount of food offered, or the amount of food consumed. The assimilation efficiency was affected by the ash content of the phytoplankton food, however, and a simpler empirical relationship than the one described above was established, such that

$$A = 87{\cdot}8 - 0{\cdot}73X, \tag{95}$$

where (X) is the percentage ash per unit dry weight of food.

The determination of assimilation efficiencies based on the ash content of food and feces may be a useful method for determining organic carbon assimilation but it appears that the assimilation efficiency of elements, such as nitrogen and phosphorus, cannot be determined with any accuracy by this method (Kaczynski, personal communication). The quantities of these elements absorbed are much smaller than the quantities of organic carbon and small changes in the ash content of the food and feces can greatly affect the calculated elemental assimilation efficiency. An alternative method which relies on the ratio of elements to each other (e.g. N to P) in the food and feces is given by Butler *et al.* (1970).

Conover (1968) summarized his own data and a number of experimental reports in the literature and concluded that herbivorous zooplankton assimilated phytoplankton with 60 to 95% efficiency. This may reflect a maximum range of assimilation efficiencies since Gaudy (1974) has clearly shown that assimilation decreases with food intake. His results are given in Fig. 61 and show that herbivore assimilation efficiencies are much lower (10–20%) at high food intake than indicated by Conover (1968). For carnivorous zooplankton, assimilation is generally higher than for herbivores; this is explained by assuming that a carnivore can better digest an animal prey that is biochemically more similar to itself compared with a plant prey item. Thus, Conover and Lalli (1974) found that the carnivorous pteropod, *Clione limacina*, had an assimilation efficiency of >90%; Cosper and Reeve (1975) showed by direct measurement that the chaetognath, *Sagitta hispida*, had an assimilation efficiency of 80%. Assimilation efficiencies may also vary with age; Mootz and Epifanio (1974) found that the pelagic larvae of the stone crab, *Menippe mercenaria*, showed an increase in assimilation efficiency from 40 to 85% during maturation from Stage I to megalopa, when fed on *Artemia* nauplii. Direct measurements of fecal material excreted by zooplankton (E) have been

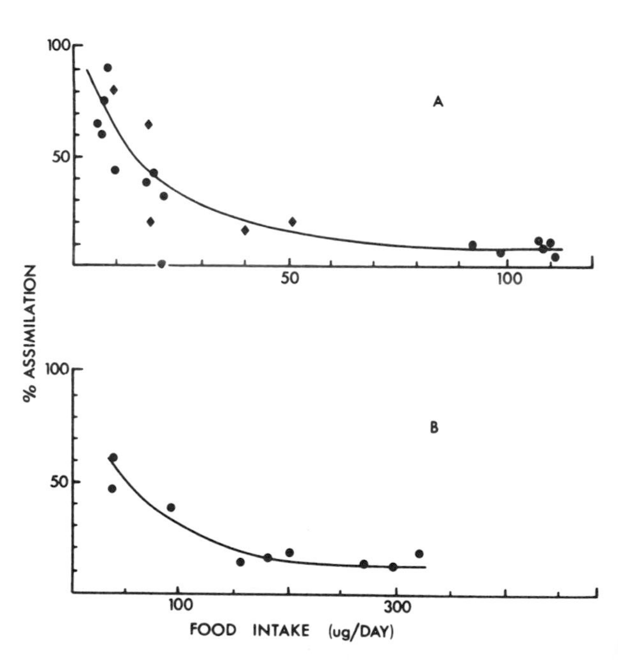

FIG. 61. Relationship between assimilation rate and daily food intake in (A) *Centropages typicus* (diamonds) and *Temora stylifera* (circles), (B) *Calanus helgolandicus* (redrawn from Gaudy, 1974).

made in a few cases. For example, Hargrave (1970) measured the feces produced by a deposit feeding freshwater amphipod and found that the fecal material represented *ca.* 83% of the ration; this gives an assimilation efficiency of less than 20%.

Carbon may be lost by predators due to breakage of prey items before consumption. This has been shown to occur with euphausiids and copepods feeding on phytoplankton (Parsons and Seki, 1970; Copping and Lorenzen, 1980) and with amphipods feeding on copepods (Dagg, 1974). The loss is reported to range from *ca.* <10% up to *ca.* 30% which would lead to an over-estimation of the assimilation efficiency (A).

Welch (1968) has derived an interesting relationship between assimilation efficiency and growth efficiency [K_2 eqn. (105)] for aquatic organisms, including copepods and fish. His relationship shows that assimilation decreases as growth efficiency increases. The interpretation of this result is that high assimilation must be associated with high energy expenditure and therefore low growth efficiency. For example, a carnivore spends more energy searching for and catching a prey than a herbivore spends in grazing. However, once having caught the prey, carnivore assimilation efficiencies are high—higher that is than herbivore efficiencies for reasons indicated in the paragraph above. It is important to recognize, however, that the relationship between assimilation and net growth efficiency (K_2) does not hold for gross growth efficiency (K_1); only ration (R) and not assimilated ration (AR), is considered in the latter.

Food used for metabolism within an animal (T) can be expressed as a function of the animal's body weight (W) such that

$$T = \alpha W^{\gamma} \tag{96}$$

or $$\log T = \log \alpha + \gamma \log W.$$

T can be measured in terms of an animal's respiration (e.g. $\mu l\ O_2$/animal/hr) and the value of α will be a function of the units which are used to express the animal's weight (e.g. wet weight, dry weight, etc.) as well as being influenced by environmental factors such as temperature. γ, on the other hand, has been found to be relatively constant (Zeuthen, 1970) and to reflect the internal metabolism of the organism; since there are certain overall similarities in cellular metabolism of a large variety of species, the relative constancy of γ appears logical. However, from a very extensive survey of different zooplankton organisms, including crustaceans, pteropods, chaetognaths, coelenterates, and polychaetes, Ikeda (1970) established that there was a significant difference between the value of γ for tropical, temperate, and boreal species taken from the tropical Pacific, the temperate Pacific southeast of Hokkaido, and the Bering Sea, respectively. The results of Ikeda's studies are shown in Table 25 and Fig. 62. These

TABLE 25. REGRESSION EQUATIONS OF LOG SPECIFIC RESPIRATION RATE, R' ($\mu l\ O_2$/mg BODY WT./hr) AND LOG BODY WEIGHT EXPRESSED AS THE DRY WEIGHT OF THE ANIMAL (mg/ANIMAL) DERIVED FROM DATA IN FIG. 51 (IKEDA, 1970)

Species	Regression equation
Boreal	$R' = -0{\cdot}169W + 0{\cdot}023$
Temperate	$R' = -0{\cdot}309W + 0{\cdot}357$
Tropical	$R' = -0{\cdot}464W + 0{\cdot}874$

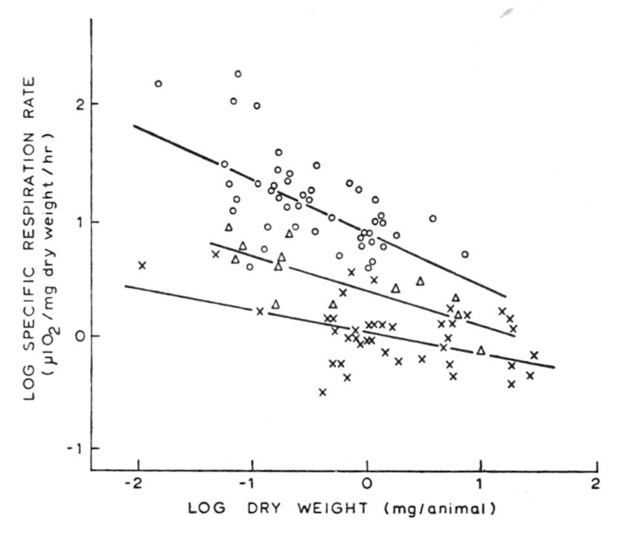

FIG. 62. Regression lines for planktonic animals from boreal (×), temperate (△), and tropical (○) waters drawn for log specific respiration rate ($\mu l\ O_2$/mg dry wt./hr) against log dry weight/animal (mg)—(redrawn from Ikeda, 1970).

results, in which metabolism is expressed per unit weight of animal, clearly show that the smallest animals have the highest metabolic rate and that there are geographic differences, not only in the value of α, but also in the value of γ, which reflect an environmental difference in basic metabolic processes of animals from different areas.

For the same species of animal at different temperatures (i.e. within species variation), Comita (1968) determined a multiple regression equation for *Diaptomus* which gave the oxygen uptake (T in μl/mg wt./hr) in terms of the temperature (t in °C) and the dry weight (W in mg) as follows:

$$T = 0{\cdot}0364t - 0{\cdot}3418 \log W + 0{\cdot}6182. \quad (97)$$

The conversion of metabolism, which is generally measured as respiration in units of oxygen consumed, to biomass of food in terms of organic carbon is made by the following equations:

$$\text{mg } O_2 \text{ consumed per unit time} \times \frac{12}{32} \times RQ$$

$$= \text{mg C utilized per unit time;}$$

or

$$\text{ml } O_2 \text{ (NTP) consumed per unit time} \times \frac{12}{22{\cdot}4} \times RQ$$

$$= \text{mg C utilized per unit time}$$

where RQ is the respiratory quotient, $+\Delta CO_2/-\Delta O_2$, which may range in animals from 0·7 to 1·0 depending on whether fats or carbohydrates are being utilized for energy. Thus, in a copepod living off fat reserves, an RQ of 0·7 may be recommended, while a copepod feeding off phytoplankton may have an RQ closer to 1·0.

For a more precise evaluation of metabolism, the total metabolic cost (T_t) to an animal can be divided into three components, viz.

$$T_t = T_1 + T_2 + T_3 \quad (98)$$

where T_1 is the basal or resting metabolism, T_2 is the energy used in capturing food, and T_3 is the energy used in biochemically transforming the food. Since T_2 and T_3 are difficult to separate, they are usually measured together and the total metabolism at rest (T_1) is compared with the total metabolism of an actively feeding animal ($T_1+T_2+T_3$). This difference is reported to result in an approximate 2- to 3-fold increase in metabolism between resting and active feeding (e.g. Heyraud, 1979; Vidal, 1980b). However, since food type and concentration are both highly variable, it is apparent that the metabolic feeding needs (T_2+T_3) of any animal will vary with time and location.

The food consumed by a zooplankter (R) can be expressed in terms of the amount of food available. Using a variety of fish, Ivlev (1945) determined that the quantity of food eaten increased with the concentration of food offered, up to some maximum ration which could not be further increased by increasing the concentration of food. From this observation it was stated that if the maximum ration is taken as R_{max}, then the relation between the size of the actual ration, R, and the concentration of prey, p, must be proportional to the difference between the actual and maximal ration such that

$$\frac{dR}{dp} = k(R_{max} - R) \quad (99)$$

where k represents a proportionality constant. Integrating, this expression gives

$$R = R_{max}(1 - e^{-kp}). \quad (100)$$

Sufficient evidence exists in the experimental data from a number of authors (e.g. Reeve, 1963; Mullin, 1963; Parsons *et al.*, 1967; Sushchenya, 1970, and references cited therein) to show that the general form of this relationship can be applied to zooplankton grazing. Two modifications to this expression have been proposed on the basis of additional experimental results. The first of these was shown by Ivlev (1961) to apply to the feeding of fish under conditions in which the effect of prey aggregation on feeding was to increase the ration obtained when the prey concentration was held constant. It is difficult to establish on the microscale of zooplankton whether this effect with fish is applicable to zooplankton feeding. It is apparent, however, that natural mechanisms which tend to concentrate prey cause patchy distributions of both plants and animals (see Section 1.3), and that from the observation given above, the occurrence of patchy distributions may increase the food available when the average prey concentration is quite low (Paffenhöfer, 1970). The second modification to eqn. (100) concerns the prey concentration at which feeding starts. From observations on zooplankton grazing conducted at sea (Adams and Steele, 1966; Parsons *et al.*, 1967), as well as from laboratory experiments (e.g. Nassogne, 1970), it appears that grazing ceases at some minimum or threshold prey concentration, p_0. Equation (100) can be modified to include this value as

$$R = R_{max}(1 - e^{k(p_0 - p)}). \qquad (101)$$

Threshold values for p_0 have been reported to range from 40 to 130 μg C/l depending on species of microcrustacea and food (Parsons and LeBrasseur, 1970). In other experiments, however, it appears that filtering continues down to zero prey concentration (e.g. Mullin, 1963; Paffenhöfer, 1970). The question of a prey threshold concentration has been further examined by Frost (1975) who found that as prey concentrations decreased, the copepod *Calanus pacificus* fed at a constant maximal rate until food intake fell to below about 15% of the maximal hourly ration. At this point, a significantly depressed feeding rate was observed. The actual concentration of phytoplankton at which this occurred (presumably a function of p_0) depended on the size of the phytoplankton species studied.

There is some discussion of the actual mathematical form of the feeding relationship given in eqn. (100) (Frost, 1974; Mullin *et al.*, 1975). From the latter reference it is apparent that a rectilinear equation statistically best fitted experimental data. However, both the curvilinear eqn. (100) and a Michaelis–Menten equation [eqn (89)] have been used to describe experimental results and neither could be statistically rejected in the study by Mullin *et. al.* (1975).

The concentration of phytoplankton at which herbivorous copepods reach their maximum ration is almost certainly dependent both on the species of zooplankton and the species (or size of phytoplankton). Parsons and LeBrasseur (1970) generally found concentrations of food organisms greater than 300 μg C/l for copepods and up to 1000 μg C/l for euphausiids. Paffenhöfer (1970) showed that the copepod *Calanus helgolandicus* could attain its maximum ration, as determined from the largest sized animals, at food concentrations of between 100 and 200 μg C/l. Frost (1974) showed that the maximum ration for female *Calanus pacificus* was attained at between 100 and 300 μg C/l depending on the size of phytoplankton. It was found that small phytoplankton cells (11 μm) had to be approximately three times more concentrated in terms of biomass (μg C/l) than large cells (87 μm) for the animal to obtain the same ration (R_{max}).

A more specialized predator/prey relationship than that given by eqn. (100) has been described by Greeve (1972). In the feeding relationship between copepods and ctenophores, Greeve (1972) noted that adult copepods tended to destroy larval ctenophores but that adult ctenophores ate copepods. A further complication in this feeding relationship was that if there were too many copepods, then adult ctenophores themselves suffered a certain mortality because their tentacles were eventually broken by too frequent encounters with large numbers of copepods. Greeve's (1972) representation of these observations could be modelled as shown in Fig. 63. Here, the probability of ctenophore survival among young ctenophores is shown to decline with an increase in copepods. As ctenophores become older the probability of ctenophore survival increases with an increase in copepod abundance, but decreases again at high copepod abundance.

The actual quantity of food eaten by zooplankter (R) depends on the type of prey as well as its concentration; thus *E. pacifica* obtained 15% of its body weight per day off a bloom of *Chaetoceros* but at equivalent concentrations of nanoplankton, *E. pacifica* could only obtain approximately one-third of this amount of food per unit time (Parsons *et al.*, 1967). In this example it may be assumed that the

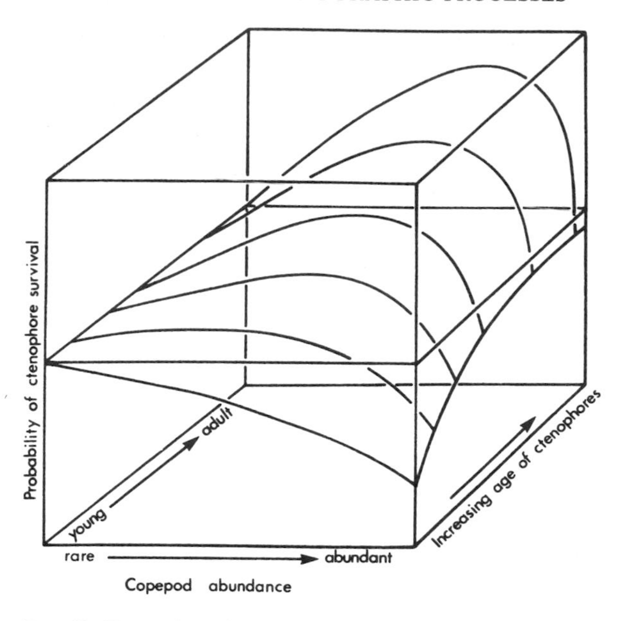

FIG. 63. Three-dimensional model showing the probability of ctenophore survival with increasing abundance of copepods and age of ctenophores (redrawn from Greve, 1972).

small nanoplankters (8 μm diameter) were much less efficiently filtered than the larger diatoms (32 μm diameter). Prey selectivity may also determine the amount of food eaten. Conover (1966c) and others have observed, for example, that some food particles will be rejected by zooplankton even after they have been seized during raptorial feeding. Also it has been shown that within the same species of phytoplankton, larger food particles, consisting of paired cells, may be selectively grazed over smaller single cells (Richman and Rogers, 1969). In addition, although some phytoplankton cells may be eaten, their digestion by crustaceans may be ineffectual and they may pass through the animal as viable cells. This was observed by Porter (1973) who showed that gelatinous green algae remained alive after passage through the guts of *Daphnia*. The degree to which an animal selects a prey over the natural abundance of different prey in the environment can be expressed as an electivity index, such as that given by Ivlev (1961) where

$$E = \frac{r_i - p_i}{r_i + p_i} \tag{102}$$

and r_i is the relative proportion (e.g. percent) of a prey in the ration and p_i is the relative proportion of the same prey in the water. Prey concentration may be expressed as units of biomass or numbers of organisms and the choice of units will affect the value of E. The value of E in the above expression ranges from -1 to $+1$, higher values denoting a greater selection of a prey item over lower values. Although Ivlev's selectivity index has been widely used it has been criticized by Berg (1979) for showing poor comparability of values obtained from different samples because the digested food is included in the divisor, r_i+p_i. An alternate index is suggested by Berg (1979) such that

$$\text{Selectivity} = \log_{10}\frac{\%N_i \text{ in the ingested food}}{\%N_i \text{ in the potentially available food}} \tag{103}$$

where $\%N_i$ is the numerical percentage of an item.

In general it appears that food ingested (as indicated by R in experiments) has been found to range from a few percent up to *ca.* 100% of the body weight of zooplankton per day, with lower average values of 10 to 20% of the body weight per day for the larger crustacean zooplankton and 40 to 60% of the body weight per day for the smaller crustacean zooplankton (Bell and Ward, 1970; Sushchenya, 1970; Mullin, 1963; Parsons and LeBrasseur, 1970). A summary of food intake, as a percent body weight, versus the body weight of different species of zooplankton, has been presented from various literature references by Ikeda (1977) in Fig. 64. From these references it is also apparent that prey densities at which zooplankton reach the asymptotic value of R_{max} appear to be in the range 200 to 1000 μg C/l, depending (as in the case of p_0) on the particular combination of predator and prey. Inhibition of zooplankton food intake at high prey concentrations appears in some experimental data. From Mullin's (1963) results it was shown that the number of diatom cells (*Thalassiosira fluviatilis*) ingested by *C. hyperboreus* increased over the range *ca.* 200 to 4000 cells/ml but then decreased to about 25% of their maximum intake over the range 4000 to 8000 cells/ml. Diel periodicity in zooplankton grazing has been observed by Duval and Geen

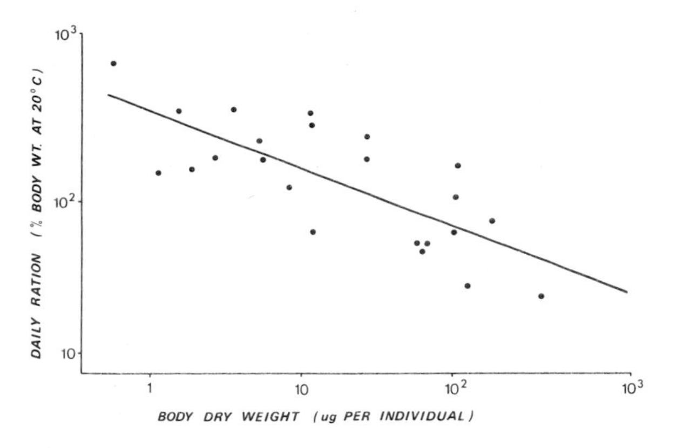

FIG. 64. Relation between daily ration and body weight of various marine planktonic copepods (data accumulated by Ikeda, 1977).

(1976). These authors found that several species of freshwater crustaceans exhibited bimodal maxima in feeding rates at dawn and dusk, with the lowest feeding at midday. The diel increase in feeding rates amounted to 6 times the midday minimum; a corresponding diel periodicity in respiration was on the average twice as great at dawn and dusk compared with midday.

Apart from the obvious need for major metabolites such as proteins, carbohydrates and fats, there are other properties of diets, as well as properties of the sea water in which animals reside, which may influence an animal's growth and survival. These properties are at present poorly defined but they are apparent from a number of examples. Thus Wilson (1951) showed that *Echinus* larvae survived better in some waters than others; Provasoli *et al.* (1959) showed that only one or two algae were satisfactory as food for *Artemia* and *Tigriopus* while a combination of several algae generally proved a much more adequate diet. Lewis (1967) showed that a mixture of trace metals improved the survival of *Euchaeta japonica* and that periodicity in the occurrence of natural chelators in sea water was a possible factor affecting *E. japonica* survival, *in situ* (Lewis *et al.*, 1971).

In eqn. (93) the food available for growth (G) is achieved when the gain in food (AR) exceeds the loss (T). This indicates that for a constant basal metabolism (T) a linear increase in growth can be achieved by increasing the ingested ration (AR). However, in a review of published data on the feeding of fishes, Paloheimo and Dickie (1965, 1966 a and b) found that increasing the ration (R) resulted in a decreased efficiency of food utilization for growth and that the relationship between gross growth efficiency (K_1) and ration was log-linear, such that

$$\ln K_1 = -a - bR \tag{104}$$

where a and b are constants, characteristic of the predator and its prey, and gross (K_1) or net (K_2) growth efficiency are defined as

$$K_1 = \frac{\Delta W}{R\Delta t} \quad \text{and} \quad K_2 = \frac{\Delta W}{AR\Delta t} \tag{105}$$

Since $\Delta W/\Delta t$ represents growth per unit time it follows that

$$\frac{\Delta W}{R\Delta t} = e^{-a-bR} \tag{106}$$

or

$$\frac{\Delta W}{\Delta t} = Re^{-a-bR} \tag{107}$$

which is an expression of growth (G) in terms of ration (R). Substituting this into eqn. (93),

$$Re^{-a-bR} = AR - T$$

or

$$T = R(A - e^{-a-bR}). \tag{108}$$

The application of Paloheimo–Dickie equations to zooplankton feeding has been discussed by Conover (1968) and Sushchenya (1970) who have concluded from experimental data on zooplankton that the approach may have general application to the study of food requirements of zooplankton. Thus Sushchenya (1970) showed from Richman's data that there was a linear relationship between the logarithm of the gross growth efficiency and ration in the feeding of *Daphnia pulex* and that the maximum gross growth efficiency was 60%. Further it was possible to derive an equation [eqn. (108)] for *Daphnia* feeding on *Chlamydomonas* which closely fitted the experimental growth data. Sushchenya (1970) accumulated various values for the gross and net growth efficiency, and the food used for respiration (T) as a percentage of the food assimilated (AR); the values are reproduced in Table 26. From these results it is apparent that gross efficiencies vary over a wide range but the K_1 for most animals lies in the range of *ca.* 10 to 40%; K_2 values are less variable and show a range of *ca.* 25 to 55%, while the expenditure of food on metabolism generally lies in the range 40 to 85% of the food assimilated.

The significance of the values (a) and (b) in eqn. (104) has been discussed by Paloheimo and Dickie (1966b). Neither value was affected by temperature which is known to affect the level of metabolism. This was interpreted as meaning that temperature affected the total rate of turnover of food but not the distribution of food among the various metabolic components. By contrast, changes in factors such as salinity and type of food (especially particle size) influenced both parameters.

From Fig. 55 it is apparent that the specific metabolism of a zooplankter decreases with body size. This gives rise to another general relationship if we consider also that the amount of energy used for growth will be large in young (and therefore small animals) compared with older (and larger) animals; as the latter become adults, the amount of energy used for growth reaches zero. Thus, at some point in an animal's life cycle there must be a period of maximum growth efficiency when the specific metabolism per unit body weight is lowest for a maximum in the growth rate. Experimentally, this was shown to be true by Makarova and Zaika (1971) who found that for *Acartia* maximum growth efficiency (K_2) of *ca.* 30% was achieved at approximately one-third of the animal's adult body weight (Fig. 65). While these data apply to growth efficiency changes within the life cycle of a particular animal, it must also be considered that the same general relationship may apply to different species of different size. Thus, Parsons (1976) calculated that for animals growing at 7% of their body weight per day, growth efficiency (K_1) of a small (0.05 mg) copepod would be about half that of a large (5 mg) copepod, if all other factors were equal. This would appear to give a distinct advantage to the growth of large zooplankters.

TABLE 26. VALUES OF ENERGY COEFFICIENTS K_1 AND K_2, AND PERCENTAGE OF ENERGY EXPENDITURE FOR RESPIRATION (T), IN % OF AR (FROM SUSHCHENYA, 1970)

Species	t °C	K_1	K_2	T	Author
Artemia salina	25	18·5	23·6	76·4	Sushchenya (1962)
Artemia salina	25	13·0	26·5	73·5	Sushchenya (1962)
Artemia salina	25	9·0	27·4	72·6	Sushchenya (1962)
Daphnia pulex	20	13·2	55·4	44·6	Richman (1958)
Daphnia pulex	20	9·1	57·6	42·4	Richman (1958)
Daphnia pulex	20	4·8	56·9	43·1	Richman (1958)
Daphnia pulex	20	3·9	58·7	41·3	Richman (1958)
Calanus helgolandicus	10	48·9	52·5	47·5	Corner (1961)
Calanus helgolandicus	10	37·2	47·0	53·0	Corner (1961)
Calanus helgolandicus	10	41·9	46·2	53·8	Corner (1961)
Asellus aquaticus	14–19	28·3	40·4	59·6	Levanidov (1949)
Asellus aquaticus	14–19	23·3	33·3	63·7	Levanidov (1949)
Asellus aquaticus	14–19	20·6	30·0	70·0	Levanidov (1949)
Asellus aquaticus	14–19	18·7	26·7	73·3	Levanidov (1949)
Aseilus aquaticus	14–19	16·3	23·3	76·7	Levanidov (1949)
Euphausia pacifica	10	7·1	7·4	92·6	Lasker (1960)
Euphausia pacifica	10	31·8	39·2	60·8	Lasker (1960)
Euphausia pacifica	10	14·2	14·9	85·1	Lasker (1960)
Euphausia pacifica	10	28·0	29·0	70·0	Lasker (1960)

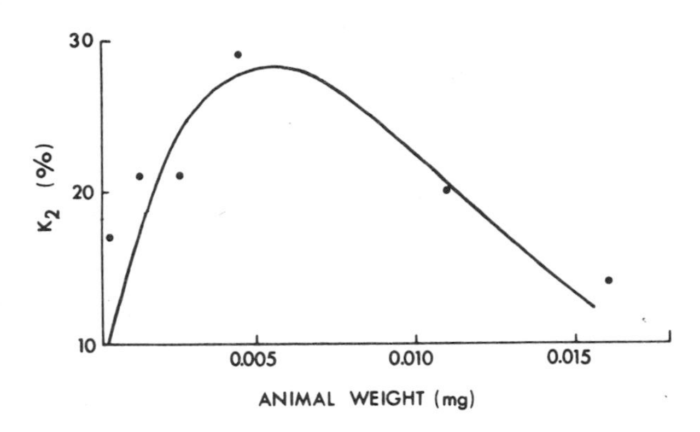

FIG. 65. Changes in the growth efficiency (K_2) during the growth (mg dry wt.) of *Acartia* (from Marakova and Taika, 1971).

However, such an assumption is made without consideration to life cycles, food supply, reproduction, and a number of other factors. For example, as Taniguchi (1973) and Fenchel (1974) have pointed out, small zooplankton grow faster than large zooplankton and thus reach maturity sooner. This gives small zooplankton an advantage, especially, for example, in tropical waters where increases in food supply may be short-lived as a result of temporary upwelling in warm waters. Such a growth regime gives an advantage to small zooplankton maturation in short time periods because of their inherently more rapid growth rate. This growth rate is further accentuated in warm waters because the Q_{10} for growth exceeds that for metabolism by a factor of *ca.* 2 (e.g. Ivleva, 1970; Chang and Parsons, 1975). Another advantage in the life-cycle strategy of a small zooplankter is that more energy is devoted to reproductive tissue in small crustaceans compared with large crustaceans (Khmeleva, 1972). In contrast to these advantages for small animals in tropical environments, an advantage for large zooplankters in colder waters would be to grow slower but at the same time to be able to carry greater energy reserves through periods when food is virtually absent (i.e. during winter). As Taniguchi (1973) has pointed out, the size distribution of small zooplankton in the tropics and large animals in temperate waters can be explained in part by these metabolic and growth strategy considerations.

In further extensions of generalized relationships governing animal size and metabolic complexity (e.g. unicellular vs. multicellular organisms, and heterotherms vs. homiotherms), Fenchel (1974) has discussed the relative advantages and costs of different life strategies. From correlations drawn from a wide range of animal sizes at different levels of metabolic complexity, Fenchel (1974) reaches certain conclusions which may be of general interest to trophic relationships in the marine habitat. His first conclusion is similar to Khmeleva's (*loc. cit.*) conclusion that larger animals spend more of their assimilated energy on maintenance than productivity; this has the advantage of giving larger animals a more independent life style. Obviously metabolic complexity (i.e. from unicellular to multicellular organisms) is another means of achieving greater ecological independence. However, according to Fenchel this is only achieved with an approximate 8-fold increase in the specific metabolic needs of multicellular organisms. Since the growth rate of a metazoan is approximately twice as great as a protozoan of the same size, the overall cost in terms of growth efficiency is approximately 4 times lower for metazoan growth compared with protozoan growth. Moving further afield, Fenchel (1974) shows that the metabolic cost of homiothermic existence is about 28 times greater than that of a heterotherm of the same size, while productivity of the latter is about 1·7 times lower. This indicates that growth efficiency of heterotherms should be about 16 times higher than homiotherms (e.g. compare growth efficiency for a fish of *ca.* 30% with that of a cow, *ca.* 2%). While these latter generalizations have drifted away from the subject of trophodynamics within a plankton community, they serve to tie together the ultimate extremes of life strategies available both in marine and terrestrial habitats (for the latter, see MacArthur and Wilson, 1967). Thus, in one extreme a highly successful strategy for survival is in being small, numerous, highly productive, but open to severe predation; in another situation being large, scarce, relatively unproductive but highly independent and consequently generally safe from predation is also a successful strategy. One aspect of these survival strategies not considered above is the relative annual net growth efficiency of different invertebrate species, both small and large. While *within* a species, growth efficiency may change according to Fig. 65, *among* species of invertebrates it appears that there is no demonstrable significant dependence of annual net growth efficiency on species size (as measured by body weight at sexual maturity). This finding was indicated by an analysis of data reported by Banse (1979). If it is substantiated it implies that in spite of the higher metabolic cost of small animals vs. large animals there must be a compensating mechanism, possibly associated with greater feeding efficiency [i.e. T_2 and T_3 in eqn. (98)], which allows the overall growth efficiency of small species to be similar to that of large species.

Several food budgets involving the terms G, AR, and T in eqn. (93) have been derived for microcrustaceans. Some results given in Table 27 from Petipa (1966) show that the growth and rate of metabolism of *Acartia* nauplii are 2 or 3 times greater than for stage V copepods, and that consequently the food intake per unit body weight of the young animals is considerably greater than for the older animals. These results are similar to results obtained for euphausiids (Table 28) which show that faster-growing animals devoted proportionally more energy to growth than metabolism, when compared to slower-growing animals. In the case of euphausiids, however, an appreciable fraction of up to 15% of the food is devoted to moults. This latter figure may be higher in euphausiids than in copepods. Corner *et al.* (1967) estimated <1% of the food assimilated by copepods was lost to moults; Mullin and Brooks (1967) showed that *Rhincalanus* lost *ca.* 6% of its body carbon as moults during its life cycle from egg to adult. The amount of tissue devoted to reproduction (Fig. 60) may be generally related to body size. Khmeleva (1972) suggests that 'generative growth' is proportionally higher in small compared with large crustaceans. From some recent data on the percentage of assimilated food available for reproductive tissue (i.e. eggs) it appears that in harpacticoid copepods this may be as high as 23% (Harris, 1973), in mysids 19% (Clutter and Theilacker, 1971), in copepods 12% (Corner *et al.*, 1967) and in euphausiids this is seen (Table 28) to amount to less than 10%. However, in another species of euphausiid Nemoto *et al.* (1972) found that the carbon content of eggs in *Nematoscelis difficilis* was equivalent to 28% of the body carbon. Marshall and Orr (1955) showed that the total number of eggs laid by *Calanus* was related to the amount of food available. From these results it appears that while size may in part govern generative growth, other factors, such as temperature and food supply, must also influence the fecundity of zooplankton.

TABLE 27. DISTRIBUTION OF FOOD EATEN AS A % OF THE BODY WEIGHT IN *Acartia clausii* AT 17–25°C ASSUMING AN ASSIMILATION (A) OF 80% (FROM PETIPA, 1966)

	Dry weight (mg)	Daily growth (%)	Daily metabolism (%)	Daily ration (%)
Nauplii	0·000 09	20·3	98·3	148
Stage V	0·002 56	8·3	49·8	73

TABLE 28. DISTRIBUTION OF FOOD REQUIREMENTS IN A EUPHAUSIID (FROM LASKER, 1966)

Animal	Dry wt. (mg)	Growth Final wt./Initial wt.	Time (days)	% food assimilated (A)	% food for growth (G)	% food for moults (G')	% food for eggs (G'')	% food for metabolism (T)	Location of study
E. pacifica	1·19	3·95	69	86	30	8·0	0	62	Laboratory
E. pacifica	1·65	1·16	63	78	6·2	7·0	0	86·8	Laboratory
E. pacifica	0·23	39·2	580	—	9·4	15·3	8·9	66·4	Natural population

Vlymen (1970) has discussed the energy requirements for movement (swimming) among copepods. From a theoretical analysis of the problem he concludes that the energy cost of acceleration from rest to a given constant velocity is very small, requiring slightly more than 0·1% of basal metabolism. Further, that even during vertical migrations the rate of energy expenditure is slightly less than 0·3% of basal metabolic rate.

The principal product of excretion among crustacean zooplankton appears to be ammonia. However, the methodology of estimating how much ammonia is lost as assimilated excreta (Fig. 60) is extremely difficult. This is because it is difficult to separate soluble material that may come from fecal pellets as opposed to excreta from the products of metabolism. Butler *et al.* (1970) performed some experiments in which it was shown that the percent of body nitrogen excreted per day was found to vary seasonally from a high of *ca.* 10% in the spring to a low of *ca.* 2% in the winter. Phosphorus is also excreted as a product of assimilation and Butler *et al.* (1970) found that the ratio of N:P in soluble excreted material was *ca.* 5 in spring and *ca.* 6 in winter. The fraction of phosphorus excreted was a greater (*ca.* 5 to 25%) fraction of the body phosphorus compared with the body nitrogen. Some other reports in the literature indicate that much higher fractions of body nitrogen and phosphorus are excreted (e.g. Harris, 1973; Pomeroy *et al.*, 1963). While these reports may be correct it also should be considered that excessively high excretion rates are difficult to accommodate in terms of known growth efficiencies unless there is a form of wasteful or 'superfluous' feeding (see Butler *et al.*, 1970, for discussion). *In situ* measurements of nitrogen and phosphorus excretion by *Euphausia pacifica* and *Metridia pacifica* have been made by Takahashi and Ikeda (1975). They showed that compared with the excretion rate at zero phytoplankton density, the estimated increase in excretion rate was as high as 5 times at 15 μg Chl *a*/l for *M. pacifica* and almost 9 times at 70 μg Chl *a*/l for *E. pacifica*.

A summary of nitrogen excretion rates is shown in Fig. 66. The data on different species of copepod are compiled from Corner *et al.* (1975) and Christiansen (1968). On the same figure, the nitrogen excretion rate for different stages of the copepod *Pseudocalanus* have been added from Corkett and McLaren (1978).

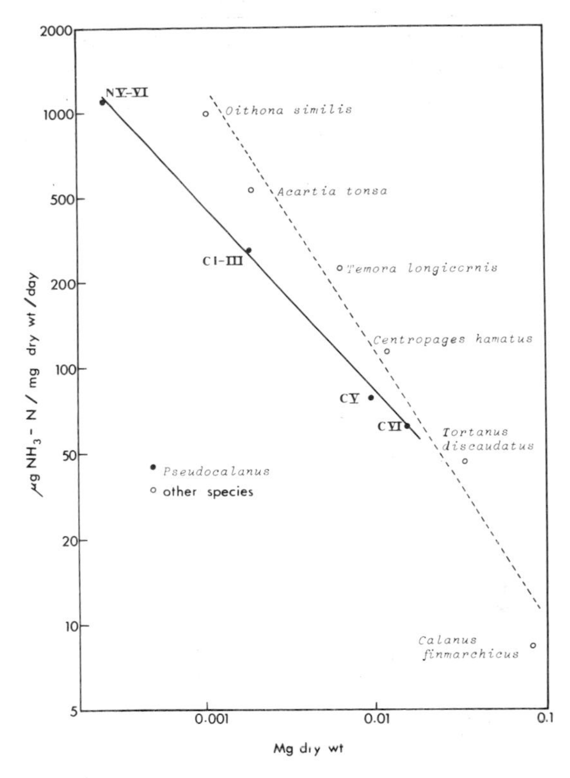

FIG. 66. Excretion of ammonia by different species of zooplankton and different stages of *Pseudocalanus* (data compiled by Corkett and McLaren, 1978).

Throughout the above discussion reference has been made to 'food' requirements and its metabolic distribution. Food may be better expressed in terms of energy units, and the following conversion factors can be used as general approximations:

1 ml of O_2 consumed ≡ 5 g cal assuming,
RQ of approximately 1·0,
1 mg of food (dry wt.) ≡ 5·5 g cal.

assuming an average dry weight composition of food as 50% protein, 20% fat, 20% carbohydrate, 10% ash, and caloric equivalents of 5·65 cal/mg for protein, 4·1 cal/mg for carbohydrate and 9·45 cal/mg for fat.

The effect of temperature on the metabolic rate (T) has been discussed earlier in this section. However, since temperature affects such overall processes as growth, reproduction, and the final size of organisms, it is appropriate to consider whether there is a general relationship between such complex biological processes and temperature.

McLaren (1963, 1965, and 1966) has discussed the general acceptability of empirical relationships between temperature and the rate of biological reactions with particular reference to the zooplankton community. From his discussions it appears that the two extremes in such approximations are firstly that biological reactions increase in rate 2- to 3-fold with a 10°C rise in temperature (the Q_{10}); this is described by McLaren as a clumsy approximation which may be assumed to have been adopted for its mathematical simplicity. At the other extreme, mathematically complex polynomial equations can be devised to fit almost any deviations in different groups of data. However, from his own data, as well as from a wide variety of published reports, McLaren concluded that the closest fit was generally obtained with Belehradek's empirical formula which is given as

$$V = a(t + \alpha)^b \tag{109}$$

where V is the rate of metabolic function, t is the temperature and a, b, and α are constants. When other factors could be assumed not to be limiting (e.g. food supply), this function gave good descriptions of development rate, metabolic rate, and size (McLaren, 1963); a number of examples are shown in Fig. 67 A, B, and C (note that metabolic rate, V, has been expressed in these figures in terms of time taken to complete a process or length achieved as adults, i.e. as $1/V$ on the ordinate). Some physiological and environmental justifications for the use of this function have been discussed by McLaren in terms of the 'constants', a, b, and α. Thus the scale correction, α, is known as the 'biological zero' and expresses the temperature at which $V = 0$. The value α appears to be positively related to environmental temperature and consequently has been shown to vary with latitude (and altitude in the case of terrestrial lake communities); the lowest values of α being found in the most northerly populations. The constant, b, reflects the degree of curvature and represents the general dependence on temperature of all metabolic processes leading up to changes in the measured parameter, V. Thus when the same metabolic function (e.g. egg hatching) is measured among similar groups of organisms, the curvilinear response, b, should be the same for all groups. The proportionality coefficient, a, is determined by the units in which V is measured; however, it is also apparent the value of a can be related to size in such a way as to indicate that it is an expression of a surface to volume restriction governing the exchange of gases across a membrane. This was illustrated by McLaren (1966) who showed a positive correlation between a and egg diameter for a variety of copepod eggs, corrected for differences in their yolk content.

Vidal (1980 a, b and c) has examined the effect of temperature on body size, growth rate, net production efficiency and metabolism for different stages of *Calanus pacificus* and *Pseudocalanus* sp. Because of the complexity of the response of different species to the effect of temperature on the above physiological functions, the author concluded that the overall effect of temperature could not be extrapolated from one species of zooplankton to another on the basis of similarity of size. Rather, it appears from his data that zooplankton are spatially distributed to take maximum advantage of differences in temperature coefficients of the above processes, as well as differences in the abundance of prey. This generalization also appears to be true for stages, within a species. For example, early developmental stages of *C. pacificus* optimized growth and food-conversion efficiency at relatively high temperatures, under a wide range of food abundance; late stages, however, grew best and transformed food more efficiently at low temperatures, particularly at low food concentrations.

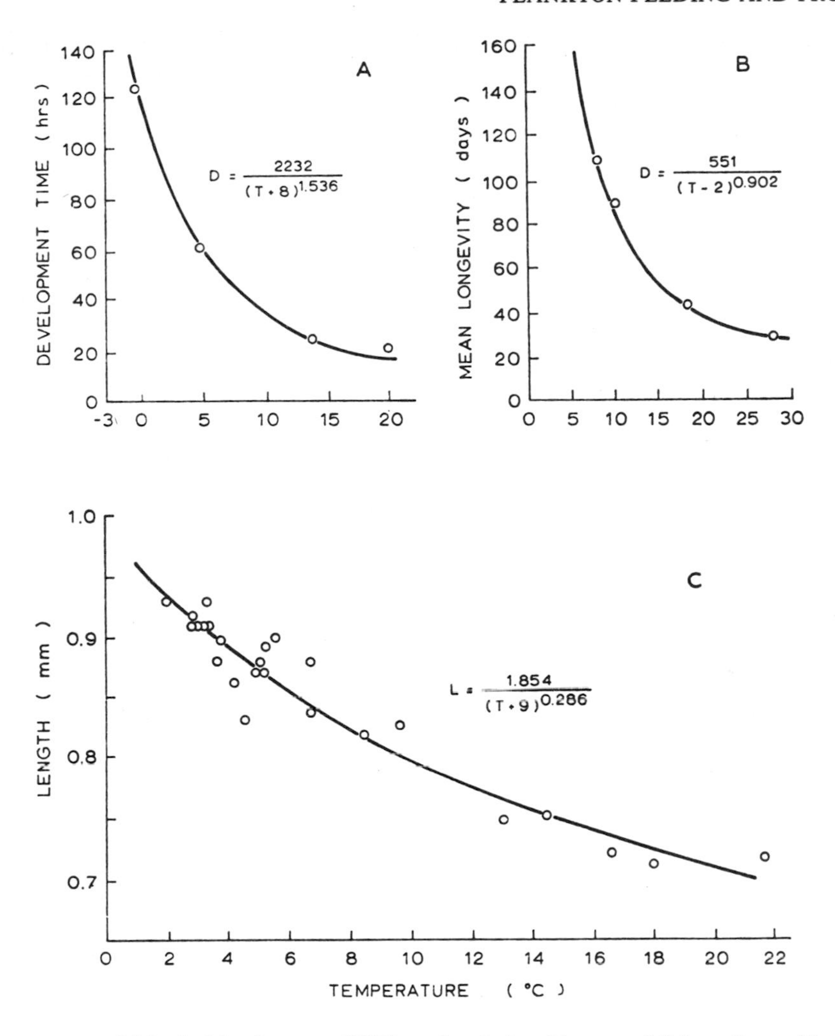

FIG. 67. Effect of temperature on biological development. (A) Time taken for hatching eggs of *Calanus finmarchicus*. (B) Mean longevity of *Daphnia magna*. (C) Mean cephalothorax length in female *Acartia clausi* (redrawn from McLaren, 1963).

4.2.2 The Measurement of Production

Production is defined as the total elaboration of new body substance in a stock during a unit time, irrespective of whether or not it survives to the end of that time (Ricker, 1958). Production depends on the time interval over which it is measured, the presence or absence of predators, and the growth and natural death rate of the population. The combination of these factors is very difficult to measure *in situ* and consequently production estimates are only approximate.* The simplest relationship for the estimation of production is to consider the *average* standing stock of the population and multiply this by an estimate of the generation rate, both determined over short time intervals. For example, if the average standing stock of zooplankton is 1 g/m^2 and the average generation rate (or doubling time) is 1 month, then

*Note: Primary productivity estimates as measured by the ^{14}C– or oxygen techniques actually represent the *potential* production in the absence of sinking, grazing, etc.

the annual production is 12 g/m^2/yr. The obvious difficulty in making this determination is in obtaining figures for the average standing stock and generation rate of a zooplankton population *in situ*.

A direct experimental measurement of potential production can be made in the case of primary producers because their growth is generally rapid so that a change in the population over a short interval (e.g. several hours) can be made in an isolated sample using a variety of sensitive techniques, such as the uptake of radioactive $^{14}CO_2$. In the case of animals, however, direct measurement of production is difficult although *growth rate* of fish, or zooplankton can be measured over extended periods of time in the laboratory and the information applied to field conditions.

Where a discrete population (or cohort) of animals can be enumerated and the average weight of an individual determined, then the production (P_t) of the cohort over a short time period (t) is given by Mann (1969):

$$P_t = (N - N_t) \times \frac{(\overline{W} + \overline{W}_t)}{2} + (B_t - B) \qquad (110)$$

where B and B_t are the respective biomasses at the beginning and at time t, and where N and N_t are the number of animals alive at the beginning and at time t, and $\overline{W}$ and $\overline{W}_t$ are the respective mean weights of the animals.

In this expression the choice of a *short* time period over which to measure changes in W and N really depends on the growth and mortality rates of the population since over long periods growth tends to be exponential. As a very approximate guide for animal populations, a short time interval over which to measure a growth stanza should be 10% or less of the total generation time of the animal. For a zooplankton population maturing from egg to adult in 3 months, growth observations using eqn. (110) could be made over *short* time intervals of about a week to 10 days.

Equation (110) represents the production of the total population. That is the production removed by predators (and natural mortality) is the number of individuals lost per unit time, multiplied by the average weight of those lost, plus the increase in biomass of the surviving population. If the production removed by predators is determined over short time intervals and the loss totalled for the year, then the total production can be determined as the sum of the surviving population at the end of the year plus the sum of the total lost to predators, (i.e. $P_{\text{Total}} = P_{t_1} + P_{t_2} + \cdots + P_{ti}$).

For a population of copepods having one generation per year the growth rate can be determined *in situ* from changes in the weight of two different stages and the time interval between the maximum numbers of each stage (Cushing, 1964; Parsons *et al.*, 1969). This is illustrated in Fig. 68

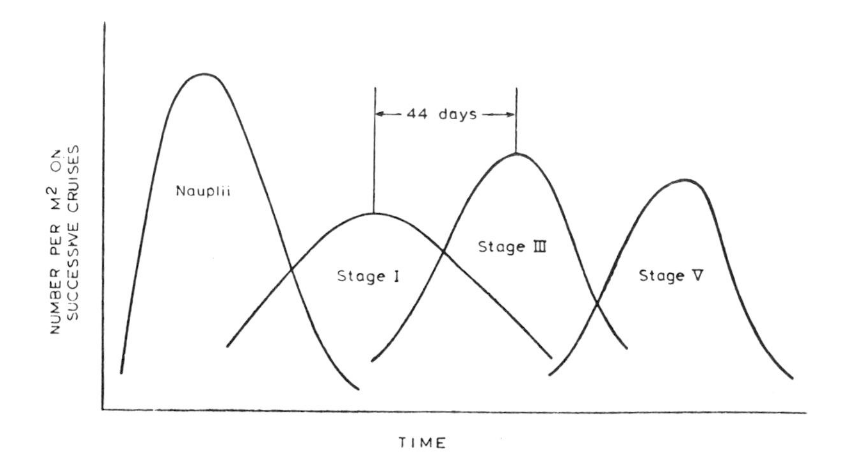

FIG. 68. Estimation of the time intervals in the maximum occurrence of different stages of a copepod having one generation (schematic representation of data from Parsons *et al.*, 1969).

where the time interval between the maximum number of Stage I and Stage III copepodites is given as 44 days; the mean weight of the respective stages was 0·15 and 0·60 mg. Since growth over an extended period tends to be a logarithmic function of time, the growth rate per day can be approximated from the expression

$$\%\Delta W = 100[10^{1/t(\log W_2 - \log W_1)} - 1] \quad (111)$$

where %ΔW is the percent increase in weight per day, W_1 and W_2 are the weights of Stage I and III copepodites respectively and t is the time interval. For the values given above, $\%\Delta W = 3{\cdot}5$ (Parsons *et al.*, 1969). From this growth rate the potential production of stage I copepodites can be determined using $N_1 \times W_1$ (as the total biomass of the cohort) and determining $N_2 \times W_2$ from eqn. (100). The difference between the mean *in situ* biomass of copepods (N_2W_2) and the calculated value of N_2W_2, will represent an estimate of the production lost to predators and natural mortality.

In a population of zooplankton which has more than one generation per year it is more difficult to determine production since several generations of the species may be present and the time of development for a single stage cannot be determined as in Fig. 68. However, if the development time of a stage (or the individual daily growth) is found experimentally (e.g. by isolating stages and incubating them under *in situ* conditions of temperature and food supply), then the production of the population can be calculated as the sum of the individual stages. In this technique it may not be necessary to deal with the species in stages, and weight groups are in fact more satisfactory for the purposes of the calculation. An example of this determination was made by Greze and Baldina (1964) and is reported in Table 29 from Mann (1969).

TABLE 29. PRODUCTION OF *Acartia clausii* DURING THE SUMMER (FROM MANN, 1969)

Size groups (mg × 10^{-3})	Number per m^3	Individual daily growth (mg × 10^{-3})	Daily production (mg × 10^{-3})
0–2·5	850	0·09	76·5
2·5–10	250	0·45	108·0
10–20	87	1·20	104·0
20–30	35	1·68	58·8
30–50	47	1·20	56·4
50–70	40	0·47	18·8
		Total	422·5

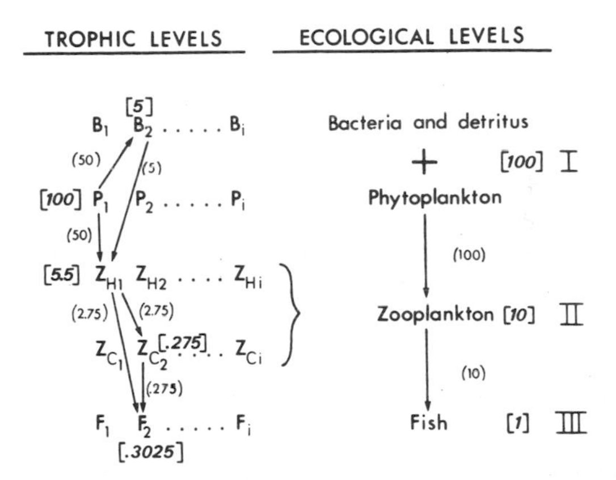

FIG. 69. Diagram of a pelagic food pyramid involving three ecological levels and five trophic levels (adapted from Ricker, 1968). $B_1, B_2, \ldots, P_1, P_2, \ldots$, etc., represent different organisms at each ecological level. Z_H represents herbivorous zooplankton and Z_C represents carnivorous zooplankton. The amount of energy available to be transferred is represented by the standing stock [5] and (2·75) represents the amount of energy transferred by a particular route assuming that all energy at one trophic level is available to the next but that it is transferred with only 10% efficiency. In this schematic food web it is also assumed that where a food resource is shared, 50% goes to each predator.

4.2.3 The Pelagic Food Pyramid and Factors Affecting its Production and Stability

The flow of energy up the pelagic food pyramid is illustrated in Fig. 69. When the energy flow is considered from a point of view of groups of animals of a generally similar habitat, then these are referred to as *ecological levels* and the system is referred to as a *food chain* (right-hand side of Fig. 69). If successive steps in the food pyramid are considered, then these are referred to as *trophic levels* and the system is called a *food web.* The amount of energy transferred up a food chain is not the same as that transferred by a food web. This is also illustrated in Fig. 69 where an arbitrary 100 units of primary production has been moved up the food pyramid through both a food chain and a food web. The details of this example calculation are given in the figure caption and the result obtained shows that approximately 3 times more energy reaches the fish community if one considers a model based on ecological levels compared with one based on trophic levels. Furthermore, the position of animals within a food web is not fixed throughout their lifetime; for example, copepodite stage *Oithona* are considered to be largely herbivores (Z_H) while Stage V *Oithona*, becoming adult, transfers to the level of primary carnivores (Z_C) (Petipa *et al.*, 1970).

From the above discussion it is apparent that the representation of the food pyramid in the sea as a food web in which there are many trophic levels is probably a more accurate representation of what is actually going on than the simpler representation of a food chain. However, for comparative purposes between different environments it is also apparent that the representation of a community as a food web could become highly complex. Thus some approximate comparisons between different communities may be made in terms of food chains. In this event it becomes important only to know how much energy is being produced by the primary producers, the number of ecological levels and the *ecological efficiency* (E); the last is defined as

$$E = \frac{\text{Amount of energy extracted from a trophic level}}{\text{Amount of energy supplied to a trophic level}}. \tag{112}$$

Slobodkin (1961) indicated that this value was about 10%. However, it is very important to know the correctness of this value since it can greatly influence final estimates of production in which man is interested. For example, Schaefer (1965) considered the effect of ecological efficiencies ranging from 10 to 20% on the production of fish at the fifth trophic level. The general form of this production (P) can be given as

$$P = BE^n \tag{113}$$

where B is the annual production at the primary trophic level, E is the ecological efficiency and n is the number of trophic levels. Starting with an annual production of phytoplankton of $1{\cdot}9 \times 10^{10}$ metric tons of carbon, the production of fish at the 5th trophic level (3rd carnivore; $n = 4$) was $1{\cdot}9 \times 10^6$ and $30{\cdot}4 \times 10^6$ metric tons of carbon at $E = 10\%$ and $E = 20\%$, respectively. Thus a doubling of the ecological efficiency can result in an order of magnitude increase in the production of a terminal resource.

However, the ecological efficiency as defined above does not take into account recycling processes which take place in nature, but which would not be apparent in a single predator/prey relationship. Thus a 10% transfer of material between trophic levels actually includes a 90% loss to the system. Part of this loss will be real, in that organic carbon will be respired back to CO_2 at each trophic level. However, another fraction of the 90% loss will appear as organic debris and be recycled through the food chain (e.g. as indicated in Fig. 51 and discussed in Section 3.2.1). Further, it is apparent that at least at the herbivore level, experimental data have recently been presented to show that the gross growth efficiency of copepods [K_1 see eqn. (105)] from nauplius to adult may be as high as 30 to 45% (e.g. Mullin and Brooks, 1970). Ecological efficiencies will not approach these values for various reasons connected with life cycles, the variety of organisms included in natural trophic levels, predator/prey electivity and the need for wild organisms to spend more energy in hunting their prey. Nevertheless it is apparent that the value of 10% is probably too low for all trophic levels and that it is especially low as an ecological efficiency involving transfers from the first trophic level. For overall estimates it may be assumed that the ecological efficiency at the herbivore level is probably no less than 20% and that efficiencies at higher trophic levels are probably between 10 and 15%. From eqn. (113) it is also apparent that the number of trophic levels (n) occurs as an exponent and as such it can greatly affect production. In an examination of differences in fish production throughout the world, Ryther (1969) considered the number of trophic levels in three communities which may be described as 'oceanic', 'continental shelf', and 'upwelled'. Over large areas the authors suggested that oceanic communities have long food chains with low ecological efficiencies, the latter being determined by the three or four levels of carnivorous feeding. As an example of such a community the author gave the following food chain which is summarized here as a diagram:

Nanoplankton (small flagellates) → Microzooplankton (Herbivorous protozoa) → Macrozooplankton (Carnivorous crustacean zooplankton) →

Megazooplankton (e.g. Chaetognaths & euphausiids) → Planktivores (e.g. lantern fish & saury) → Piscivores (e.g. squid, salmon & tuna)

The food chain is shown as a continuous flow of biomass from phytoplankton to fish but in fact salmon, for example, may also feed on euphausiids and squid; thus their position at the 6th trophic level would actually be better represented as being between the 5th and 6th trophic levels. For purposes of comparison with other environments, however, it is apparent that there are approximately five trophic levels leading to the production of a commercial species of fish. Further it was determined from the literature that the annual primary production of oceanic areas was generally low and 50 g C/m^2/yr was considered to be an average value.

The second food chain given by Ryther was described as 'coastal' but should be more properly called 'continental shelf' since it may sometimes exist at considerable distance from the coast, such as in the Atlantic on the Grand Banks. This food chain may be represented diagrammatically as follows:

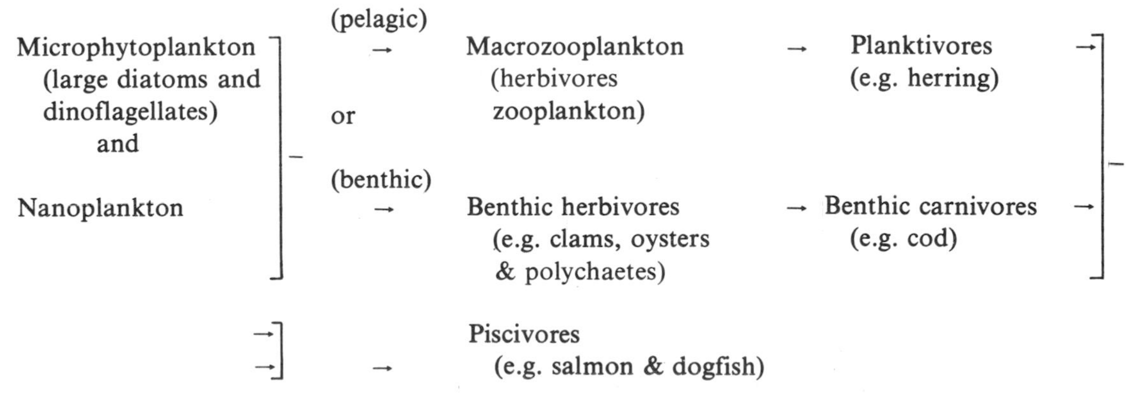

As in the oceanic food chains, the flow of food represented above is an average picture and may include such steps as carnivorous zooplankton in addition to those shown. Essentially, however, the food chain can be represented by three trophic levels, whether this is via the benthic or pelagic community. The total annual primary production of these areas was assessed as *ca.* 100 g C/m^2/yr.

Finally the food chain of areas in which there is persistent upwelling, such as off the coast of Peru or in Antarctic seas, could be represented by the diagram shown at the top of p. 146. In this food chain it is known that certain fish, such as adult anchovy, may feed directly off phytoplankton, while another relatively short food chain may exist between euphausiids and whales. The total number of trophic levels in such environments was represented by Ryther as one and a half. The total primary productivity of this community type was assessed at 300 g C/m^2/yr since the process of upwelling leads to a continual supply of nutrients.

In these three communities ecological efficiencies at each trophic level were assumed to be highest, when governed largely by phytoplankton/herbivore associations and lowest for communities in which there were secondary and tertiary carnivores. Consequently Ryther assigned a 10% overall

Macrophytoplankton (large diatoms and dinoflagellates including chain-forming species) — → Planktivores (e.g. anchovy)
or
→ Megazooplankton (e.g. *Euphausia superba*) → Planktivores (e.g. whales)

efficiency to the oceanic food chain, a 15% efficiency to the continental shelf food chain, and a 20% efficiency to the food chain in upwelled areas. The resulting potential fish production is shown in Table 30. From these results it may be readily concluded that the upwelled areas should produce the largest fisheries and that the continental shelf areas and oceanic areas are of decreasing importance. This is borne out in part by the fact that the anchovy fishery of Peru is the largest fishery in the world, exceeding the total catch of an intensive fishing nation such as Japan; in contrast very few fisheries are conducted in the open ocean waters except where there is a continental shelf, such as in the northern hemisphere off the Grand Banks. The description of food chains given by Ruther (1969) may be open to certain criticisms as to details of his calculations (e.g. Alverson *et al.*, 1970) but the general approach is illustrative of the importance of trophic relationships in assessing the productivity of the ocean.

In the previous paragraphs it will be noted that the size of phytoplankton cells appears to be a critical factor in determining the length of the pelagic food chain and the ultimate yield of fish to man. The next question to ask is, what factors govern the size of phytoplankton cells? Malone (1971 a, b) suggested that large phytoplankton cells only predominate during periods of positive vertical advection and when nitrate concentrations are above 1–3 μM. Further, he found some observational data supporting earlier experimental results that zooplankton selectively grazed large cells as opposed to the nanoplankton. Semina (1972) suggested on the basis of environmental data that the size of phytoplankton cells was governed by at least three factors, viz. (1) the direction and velocity of vertical water movement, (2) the value of the density gradient of the main pycnocline, and (3) the phosphate concentration. Some of these and additional ideas were incorporated in an equation which attempted to describe cell size in terms of eight environmental and physiological variables (Parsons and Takahashi, 1973). While this equation appeared superficially correct, its use has been criticized (Hecky and Kilham, 1974; Malone, 1975) mainly on the basis of a lack of terms governing alternate possibilities, including grazing and horizontal advection. Laws (1975) has suggested that cell size is determined largely by three factors: phytoplankton growth rate, respiration, and sinking. By showing that large phytoplankton cells catabolize a smaller fraction of their biomass than small cells, Laws (1975) uses a model to demonstrate that the net growth rate of large cells may exceed that of small cells when the mixed layer is relatively deep. From these reports it appears that phytoplankton cell size must be linked ultimately to

TABLE 30. ESTIMATED FISH PRODUCTION IN THREE OCEAN COMMUNITIES (ADAPTED FROM RYTHER, 1969)

Marine environment	Mean primary productivity (g C/m²/yr)	Trophic levels	Efficiency (%)	Fish production (mg C/m²/yr)
Oceanic	50	5	10	0·5
Continental shelf	100	3	15	340
Upwelled	300	1·5	20	36 000

the physical, chemical, and biological processes but that it may still be too early to clearly define the causative agents. Sournia (1982) has provided a good review of various concepts which need to be considered as affecting both phytoplankton cell size and shape.

Cushing (1971 and 1973) determined the ecological efficiency from environmental data by assuming that the amount of energy extracted (yield) from a trophic level was proportional to the production at that level. Consequently the ratio of production at one ecological level to the production at the next is an estimate of the ecological efficiency which Cushing (1971) called the *transfer efficiency.* By plotting the transfer efficiency between primary and secondary producers against the primary production, Cushing (1971) showed that ecological efficiencies probably declined with production (Fig. 70). Experimental support for this assumption is found in Paloheimo and Dickie (1966b) who showed that growth efficiency [K_1, eqn. (105)] declined with ration. The connection between growth efficiency and ecological efficiency is established through yet another type of efficiency used by biological oceanographers and that is the *ecotrophic efficiency* (E_c). This is defined as the fraction of the annual production of a trophic level which is consumed by predators (Ricker, 1968). In Fig. 69 it was assumed for simplicity that all of one trophic level was available for transfer to the next trophic level; in fact a certain fraction must remain in order to maintain the production. For example, if we assume that 10% of the primary production is needed to maintain production against other losses, then 90% of the primary production is available for consumption by secondary producers. Further, if this is consumed by the secondary producers with a growth efficiency (K_1) of 20% then the ecological efficiency is 18%. The relationship between these three efficiencies can be summarized as

$$E = E_c \times K_1. \qquad (114)$$

Two further corollaries of eqn. (114) are that ecological efficiency (E) should decline with successively higher steps in the trophic pyramid and, secondly, that ecological efficiency should decline with the age of a population. The first of these corollaries comes from the fact that P/B ratios decrease with trophic position (Fig. 73) and therefore less production can be made available to the next trophic level (i.e. E decreases); the second corollary is indicated by the fact that K, as a measure of growth, must decrease with age.

Riley (1946) first considered quantitative relationships governing productivity at the primary and

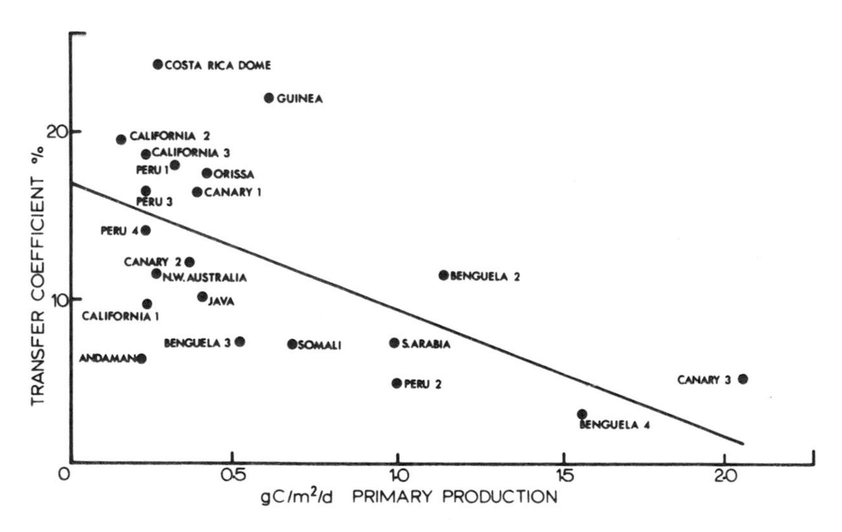

FIG. 70. Dependence of transfer coefficients on the intensity of primary production for different upwelling areas in different seasons (redrawn from Cushing, 1971).

secondary levels in the marine food chain. Using data from Georges Bank, Riley (1947) considered two approaches; in the first several quantities were measured simultaneously and their interrelationships derived by some statistical method, such as by multiple correlation. In the second approach, a number of simplified assumptions were made, based on experimental data, and these were synthesized into a mathematical description of the events.

As an example of the first approach, Riley (1946) determined multiple regression equations between plant pigments, the depth of the water, temperature, the amount of phosphate, the amount of nitrate, and the abundance of zooplankton. Proportionality 'constants' in the equations varied with season and the effect of changing the mean value for each component by one standard deviation caused different seasonal effects which indicated the temporary importance of each factor. This is illustrated in Table 31 which shows that major changes in the phytoplankton crop were caused by increasing the nitrate concentration during March, phosphate concentration during April, and zooplankton during May. Thus it is apparent that no one factor exercises complete control over phytoplankton production and that the relationship between the few major components in Table 31 indicates a complex system in which one factor after another gains momentary dominance. Combining data from different seasons Riley (1946) developed a multiple regression equation in which

$$PP = -153t - 120P - 7{\cdot}3N - 9{\cdot}1Z + 6713 \qquad (115)$$

where PP was the phytoplankton crop in Harvey pigment units (not used today, but proportional to chlorophyll a), t was the temperature, P and N were phosphate and nitrate concentrations, respectively, and Z was the zooplankton standing stock. This equation reproduced the observed data with an average error of about 20% for all seasons.

More recent studies using various types of multiple-component analysis have been used to show the principal components governing plankton production in different environments. Williamson (1961) showed that out of eleven factors in a correlation matrix, four accounted for 83% of the variance in plankton abundance in the North Sea and of these, one factor related to the extent of vertical mixing in the water column predominated by accounting for 48% of the variance. Walsh (1971) showed that 66% of phytoplankton variation across the Antarctic Convergence could be predicted in terms of water mass structure, turbulence, light, silicate, and an index of heterotrophic conditions. However, as the latter author has concluded, definite statements about functional relationships between the biological and habitat variables cannot be made because of the inherent non-causal nature of correlation and regression analyses. A similar conclusion led Riley (1946) to consider a second approach to describing changes in a plankton community. In this approach he assumed that the rate of change in a phytoplankton community could be determined as the difference in reaction rates between processes of accumulation and loss of biomass (or energy) in the plankton population. Riley considered that the most important reactions could be summarized by

TABLE 31. PERCENTAGE CHANGE IN THE PHYTOPLANKTON CROP PRODUCED BY INCREASING THE VALUE OF EACH ENVIRONMENTAL FACTOR FROM ITS MEAN TO THE LIMIT OF ITS STANDARD DEVIATION (FROM RILEY, 1946)

	Sept.	Jan.	Mar.	Apr.	May	June
Depth	-1	60	-20	11	0	-1
Temperature	-9	-37	-31	-5	10	-26
P	7	74	-23	-24	6	-28
N	7	41	-57	1	-21	1
Zooplankton	-6	5	-1	-9	-31	-10

an equation which can be given in a modified form as

$$\frac{dN}{dt} = N(P_h - R) - G \tag{116}$$

in which dN/dt was the rate of change in the phytoplankton population (N) having a photosynthetic rate (P_h), a respiration rate (R), and a rate of grazing by zooplankton (G). Riley (1947) considered that each term in the above equation would be subject to environmental influence but that a description of their effect could be given in some mathematical form. Thus the average light intensity ($\bar{I}$) to the depth of the mixed layer, D_m can be given by the equation

$$\bar{I} = \frac{I_0}{kD_m}(1 - e^{-kD_m}), \tag{117}$$

I_0 is the photosynthetic radiation at the surface and k is the extinction coefficient (for the derivation of this equation see Section 3.1.5). The photosynthetic response to this light intensity can then be obtained from a P vs. I curve (e.g. Fig. 34) which might be expressed in the form

$$P_h = P_{max}(1 - e^{\alpha(I_c - \bar{I})}) \tag{118}$$

where P_h is the photosynthetic rate at $\bar{I}$ and P_{max} is the maximum photosynthetic rate for the phytoplankton, and α is a constant.

Since in eqn. (118) the compensation light intensity (I_c) is determined by the respiration (see Fig. 34), the term R in the eqn. (116) does not have to be treated separately. However, temperature (which was included in Riley's original equation as only affecting respiration) should be included as an effect on the photosynthetic rate. This has been discussed in Section 3.1.4 and using Eppley's empirical relationship between temperature and photosynthetic rate, a temperature adjustment can be included having the form

$$P_h = P_{max}\ e^{(at - b)} \tag{119}$$

where a and b are constants and t is the temperature.

On the basis of more recent findings (see Chapter 3) the nutrient limitation on phytoplankton growth can now be expressed in the form of Michaelis–Menten kinetics as

$$P_h = P_{max}\frac{[S]}{[S] + K_m} \tag{120}$$

where P_{max} is the maximum rate of photosynthesis at which an increase in the rate limiting nutrient concentration [S] no longer results in an increase in P_h, and K_m is the Michaelis constant for the uptake of the nutrient.

The grazing component, G, can also be expressed more appropriately than in Riley's original equation. Thus if $G = HR'$ where H is the standing stock of zooplankton and R' is the ration per unit time per animal, then from Section 4.2.1

$$R' = R'_{max}(1 - e^{-k'N}) \tag{121}$$

where N is the concentration of the phytoplankton population. Thus one attempt to construct a determinate model of phytoplankton productivity on the basis of Riley's classical paper might be written as

$$\frac{dN}{dt} = N\left[P_{max}(1 - e^{\alpha(I_c - \bar{I})})\left(\frac{[S]}{[S] + K_m}\right)(e)^{(at - b)}\right] - HR'_{max}(1 - e^{-k'N}) \tag{122}$$

where $\bar{I}$ is defined by eqn. (117) and all the variables affect dN/dt independently.

Further embellishments of this type of equation can be made by adding factors, such as might be required to account for the sinking rate of phytoplankton, or by using other expressions for the photosynthetic light response, such as are discussed in Section 3.1.5. In Riley's (1946) original model, which differs from eqn. (122), he was able to reproduce the essential features of the phytoplankton standing stock with approximately the same precision as was obtained by the statistical estimate given in his first approach eqn. (115). Riley

(1947) extended his determinate equations to express changes in the herbivorous zooplankton population (dH/dt) as

$$\frac{dH}{dt} = H(gP - R'' - aC - D), \qquad (123)$$

where H was the herbivorous zooplankton population, gP was the input from grazing on phytoplankton, R'' was loss from zooplankton respiration, aC was loss by carnivorous predation, and D was natural mortality.

Vinogradov and his colleagues (Vinogradov *et al.*, 1972, 1973) have developed an ecosystem model in which they have tried to simulate the time changes in biomass of organisms at different trophic levels, inorganic biogenous elements (i.e. nutrients), and detritus. The model is intended to represent a natural water column and the possible co-reaction between each parameter (e.g. Fig. 71). In this model the animal community was classified into several different trophic categories as follows: protozoa (F_1), nauplii (F_2), small and large herbivores (F_3, F_4), cyclopods (S_1), carnivorous calanoids (S_2), and chaetognaths and polychaetes (S_3). This treatment makes the model more complicated than others (e.g. Riley's), but the final result is probably more realistic for the description of a natural ecosystem. In the Vinogradov model, the rate changes of nutrients (N), phytoplankton (P), and bacteria (B) were described in a manner similar to eqn. (116) but including also several other terms such as mortality, sinking and input and output due to physical water-mass transportation. The changing rate of zooplankton biomass was then described as follows:

$$\frac{\partial X_i}{\partial t} = u_i \Sigma C_p - r_i X_i - \mu_i X_i - \Sigma X_q \qquad (124)$$

$$Cp = P, B, D, F_1 \ldots F_4, S_1 \ldots S_3$$

$$Xq = F_1 \ldots F_4, S_1 \ldots S_3$$

where X_i is the biomass of animal in the ith category of the trophic food web ($F_1 \ldots F_4$, $S_1 \ldots S_3$), in the water column, u_i food consumption efficiency of the animal, ΣC_p total food eaten, r_i respiration rate of the animal, μ_i mortality of the animal, and ΣX_q biomass loss of the animal due to feeding by other animals. The rate of detritus production was described as follows:

$$\frac{\partial D}{\partial t} = \Sigma(h_i + \mu_i) X_i - \Sigma d_i X_i + k\frac{\partial^2 D}{\partial Z^2} - w\frac{\partial D}{\partial Z} \qquad (125)$$

where D is the concentration of detritus in the water column, h_i is undigested food of the ith zooplankters ($i = F_1 \ldots F_4, S_1 \ldots S_3$), d is the removal rate of

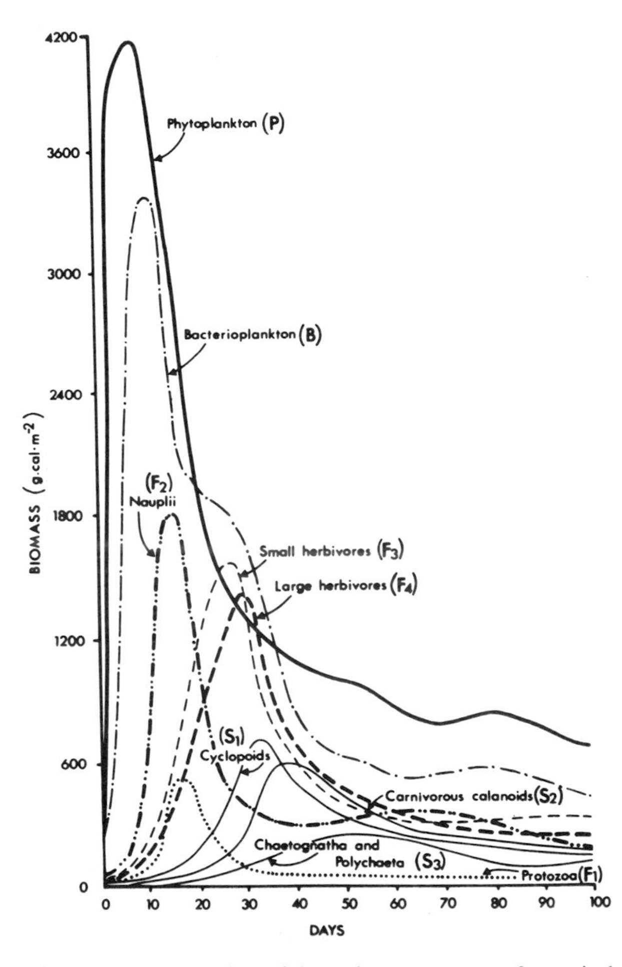

FIG. 71. Changes in time of the major components of a tropical upwelling ecosystem in the upper 200 m, predicted by a simulation model; for explanation see text (redrawn from Vinogradov *et al.*, 1973).

detritus by ith zooplankters ($i = F_1 \ldots F_4$), k is the turbulent diffusion constant, w is the average sinking rate of detritus, and Z is the depth.

This approach has been applied to a tropical upwelling community (Fig. 71), in which the X-axis can also be read as the distance from the upwelling area as well as days. The biomass of both phyto- and bacterio-plankton increases rapidly, followed by a development of herbivorous animals. Then phyto- and bacterio-plankton decrease in their biomass due to heavy grazing by herbivores, as well as possible nutrient deficiency. Predators develop their biomass even later. The peak height of organisms in each category generally diminishes at higher trophic levels. Fifty to sixty days after starting, the system comes to a semi-stationary state, characterized by low concentration of all live elements. The time and spatial changes predicted are supported by actual observations (e.g. Vinogradov *et al.*, 1970). Menshutkin *et al.* (1974) expanded Vinogradov's original approach, using data from the Sea of Japan, and taking into account the nekton (i.e. fishes and squid).

Many additional descriptive models using both approaches have been developed since Riley's (1946 and 1947) papers. Cassie's (1963) paper serves as a more elaborate example of recent developments in statistical analyses of plankton communities while Patten (1968) has reviewed models which synthesize empirical relationships between the major components governing the production of plankton. Recent extensive use of various types of computers has made the mathematical modelling approach convenient and easy. However, the effective use of models is not always made. Platt *et al.* (1981) have reported various mathematical methods in marine ecology for the design of research programmes in the open sea and the near-shore waters, and various experiments for the treatment of biological-data collections with particular reference to the development of mathematical models. The book includes mention of the overall difficulties and cautions, as well as the superior character of different models, based upon discussions of a workshop organized particularly for the evaluation of the modelling approach by the Scientific Committee on Oceanic Research (SCOR).

The structure of a plankton community is often not so readily resolved into definite trophic levels as indicated by the use of Riley's models. More often there is a complex food web in which a species of plant or animal may belong to more than one trophic group at any time (e.g. as illustrated in Fig. 69). There are very few quantitative descriptions of such food webs but one has been attempted for the planktonic communities of the Black Sea (Petipa *et al.*, 1970); unfortunately the data presented in this study contain several inconsistencies but the approach serves as an example of a planktonic food web. There are also one or two similar studies in which authors have considered the distribution and flow of energy (or biomass) in other aquatic environments (e.g. the Thames river study by Mann, 1965; the Georgia salt marsh study by Teal, 1962).

In order to understand the energetics of the Black Sea planktonic communities it is necessary first to describe the trophic levels involved, and particular attention is given here to Petipa's description of the epiplanktonic community. The phytoplankton of this community consisted of autotrophic organisms (such as chlorophyll containing diatoms and dinoflagellates) and saprophagous organisms (principally *Noctiluca* which feeds by engulfing particulate food, 70 to 90% of which was reported to be detrital in origin). Thus the authors chose to consider these two groups of organisms as a single trophic level, based on the fact that both groups of organisms were involved in the primary formation of particulate material for filter feeding animals. The distribution of biomass (standing stock) within the phytoplankton community was over 90% in favour of the saprophagous *Noctiluca*. It appears from the data, however, that this organism had a very high death rate (42% per day) and that this was the chief source of detritus being utilized by filter feeders. In contrast the smaller autotrophic phytoplankton were producing one or two generations per day with a natural mortality of less than 5% per day. Herbivorous organisms formed the second trophic level and these included nauplii of copepods, some copepodite stages, *Oikopleura*, and the larvae of some benthic molluscs and polychaetes. Omnivorous organisms, which consumed both plant and animal plankton, formed the third level and consisted mostly of the later stages of the copepods, *Acartia, Oithona*, and *Centropages.* One primary carnivore, adult *Oithona*, was recognized as the fourth trophic level, while the fifth and sixth trophic levels were recognized as secondary carnivores (mainly *Sagitta*, which ate both herbivores and primary carnivores) and a tertiary carnivore (*Pleurobrachia*) which fed off all other zooplankton; for most calculations the two latter levels were considered together as one. A diagram of this food web is shown in Fig. 72. While the figure represents fixed relationships described above it was also considered by Petipa *et al.* (1970) that certain organisms transferred from one level to another during their lifetime so that *Oithona*, for example, transferred to the primary carnivore level on becoming an adult; in addition, a number of early copepodite stages transferred from the herbivore to the omnivore level as they matured.

Petipa *et al.* (1970) considered that the diel rate of production of matter (P') at one trophic level was determined as the number of organisms eaten (G), the number of organisms dying per day (M), the number of organisms transferring to the next level per day (L) and ($B_1 - B_0$) the difference in standing stock at the end of the day. The best units in which to compare the activity of different levels in such a food web are units of energy (e.g. calories/m^2/day) although actual measurements are usually made in terms of numbers or biomass of organisms which are assigned caloric equivalents. Experimental data on the six trophic levels described above were collected from an area of high stability in the Black Sea. Food composition and daily rations were

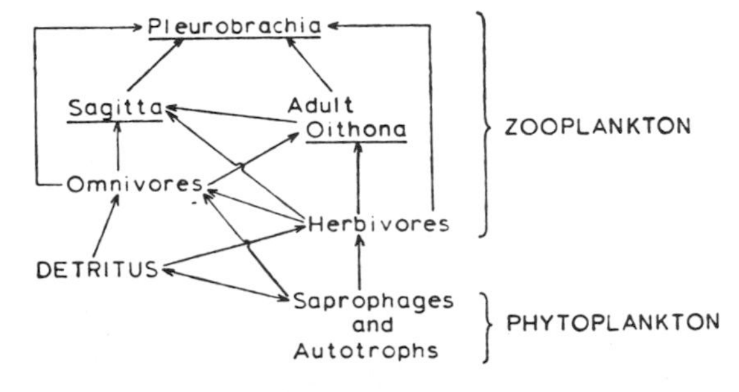

FIG. 72. Food web of the epiplankton community in the Black Sea during early summer (redrawn from Petipa *et al.*, 1970).

obtained from gut contents; reproductive rate of the algae and the respiration of animals were obtained from samples incubated *in situ* and the total weight increment of animals was determined from the duration of stages at different temperatures and the known weight of each stage. The results of these measurements are shown in Table 32. From these data it is apparent that the phytoplankton community is predominated by the rather unusual death rate which has been attributed to the saprophage, *Noctiluca.* Further, it is apparent that both biomass and production show a general decrease going from the lowest trophic level towards the highest. Production per unit biomass (P'/B) would have been highest at the primary level if data on *Noctiluca* had been excluded. There is some evidence in the data that P'/B ratios decrease with increasing trophic levels, as might be expected for slower-growing organisms which have to spend an increasing amount of energy on capturing their prey.

Dickie (1972) has attempted to summarize P'/B ratios for different trophic levels and their results are reproduced in Fig. 73. P'/B ratios in Table 32 are not directly comparable to data in Fig. 73 since the former are based on daily estimates and do not allow for an organism's life cycle which may impose a natural restriction on its growth for part of a year. Thus phytoplankton may continue to multiply throughout the year while more complex organisms have to grow to maturity, mate, and reproduce; these processes may require several years in higher organisms, including many fish. P'/B ratios on an annual basis indicate that at the three principal trophic levels, the respective ratios are *ca.* 300 for phytoplankton, 10 to 40 for most zooplankton, and 1 or less for most fishes depending on how long they take to complete their life cycle. The data also indicate that the annual average standing stock at different trophic levels may be very similar, but that low ecological efficiencies reduce the transfer of high productivity at low trophic levels, as material is moved up the food chain.

Banse and Mosher (1980) studied the annual P'/B ratio with respect to size of organisms (i.e. body weight, m_s, in kcal equivalents). For thirty-three invertebrates living in the temperature range 5° to

TABLE 32. PRODUCTION OF THE TROPHIC LEVELS OF BLACK SEA EPIPLANKTON COMMUNITY (mg/m²/DAY) FROM PETIPA *et al.* (1970)

Trophic level	Amount of food eaten from a given level (*G*)	Death rate (*M*)	Difference between final and initial standing stock ($B_1 - B_0$)	Transport to the following level (*L*)	Production (*P*)	Standing stock of living organisms (*B*)	*P'/B* (%)
Primary producers and saprophages	478·5	12593·8	5397·2	—	18 469·5	29 249·6	63
Herbivores	361·2	4·0	−6·3	156·0	515·0	792·0	65
Omnivores	67·0	1·2	20·0	82·0	170·4	193·5	88
Primary carnivores	53·5	1·3	27·4	—	82·2	172·2	47
Secondary–tertiary	33·0	2·8	−1·1	—	34·7	560·5	6·2

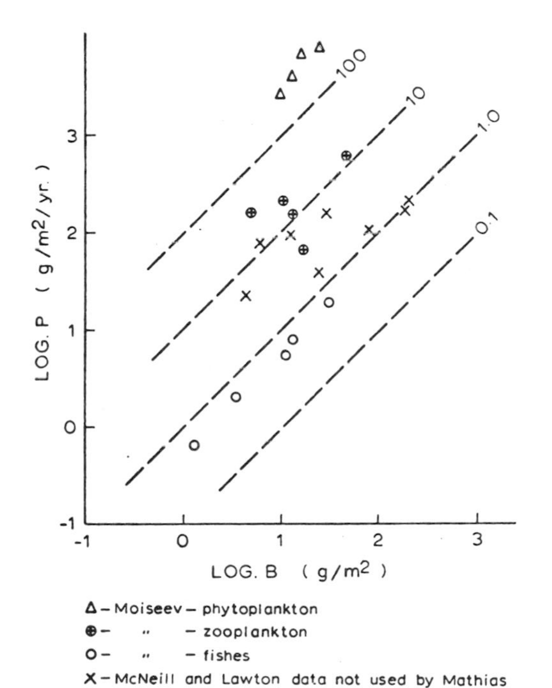

FIG. 73. Production (*P*) and biomass (*B*) relationships (data accumulated by Dickie, 1972 from various sources. Dashed lines indicate log intervals of constant *P'/B*).

20°C the authors found a significant decline in *P'/B* (years⁻¹) such that

$$P'/B = 0{\cdot}65M_s^{-0{\cdot}37}. \qquad (126)$$

For mammals and fishes, the exponent 0·37 was not significantly different but the proportionality coefficients were different indicating that for the same mass of animal, the *P'/B* ratio would be *ca.* 4 times higher for fishes and 20 times higher for mammals. This is illustrated in Fig. 74. Since the previous statement taken from Fig. 74 appears to contradict data in Fig. 73 it should be made clear that the higher *P'/B* ratio of invertebrate populations in the latter figure (i.e. zooplankton *ca.* 10) is for *populations* and reflects annual turnover. Data in Fig. 74 refer to *animals* of the same weight, but different phylogenetic origin (e.g. compare the *P'/B* ratio of a lobster and a fish of similar weight). Further, an exception to the use of the equation is noted by Banse and Mosher (1980) for most protozoa which appear to have lower *P'/B* ratios than would be predicted by the equation. These authors also note that the *P'/B* ratio is in fact an expression of the mortality coefficient, since high *P'/B* ratios must be accompanied by high mortality.

An important property of a food web is its stability. The simplest case is to consider a single predator and a single prey over the course of time (Fig. 75). If there are no other restrictions on this system the predator will increase to a point where it runs out of food and then declines and becomes extinguished. Such a system lacks stability. Under natural conditions, however, various restrictions are placed on predator/prey associations which

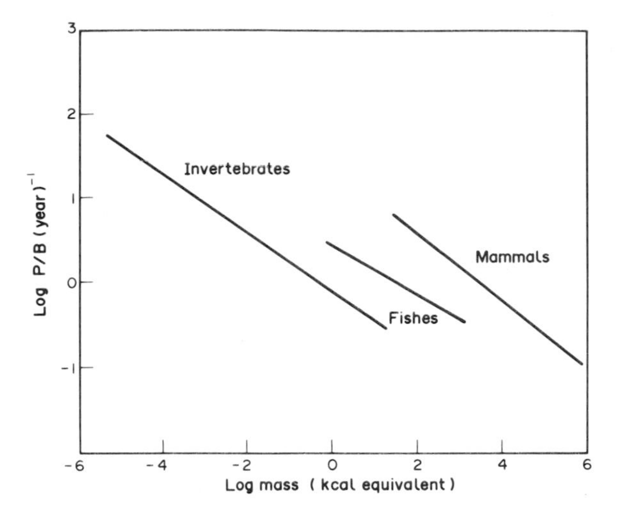

FIG. 74. Adult body mass and annual production/biomass relationships from field populations (redrawn from Banse and Mosher, 1980).

tend to dampen out violent oscillations. Dunbar (1960) suggested that in general oscillations are bad for any system and that violent oscillations may be lethal. This has led to a definition (Hurd *et al.*, 1971) in which it can be stated that stability is the ability of a system to maintain itself after a small external perturbation.

Factors which tend to stabilize natural systems have been discussed by several authors (e.g. Rosenzweig and MacArthur, 1963; McAllister *et al.*, 1972). MacArthur (1955) and others have postulated that development of many species (i.e. a high diversity) is the principal component in establishing community stability. Since more energy may be required in establishing a more complex food web (e.g. see Ryther's three trophic relationships) it is also apparent that where stability has been acquired through diversity, productivity per unit biomass will be low. Thus in general, tropical plankton communities and food webs tend to have a high diversity, low productivity, and high stability. In contrast a temperate plankton community has a high productivity caused by one or two species of zooplankton (e.g. *Calanus plumchrus* and *cristatus* in the subarctic Pacific) and a low stability in which there are large seasonal fluctuations in production. Some other important factors leading to the stabilization of a system are the imposition of a limitation on the predator, other than its supply of prey; a hiding place for the prey; periodic migration of the predator away from its food source; a threshold concentration below which the prey is not consumed by the predator; and the patchiness of prey distributions. Examples of these stabilizing influences can be found in the aquatic environment. Thus the diel migration of zooplankton into the euphotic zone may be interpreted as adding to community stability (as well as being of some physiological advantages to the zooplankton population, e.g. see McLaren, 1963). The possible occurrence of a threshold prey concentration in phytoplankton/zooplankton relationships (e.g. Parsons *et al.*, 1967) and the ability of zooplankton, such as euphausiids, to change from being carnivores to herbivores (Parsons and LeBrasseur, 1970) are further examples of ways in which stability of a community may be maintained. On the other hand, instability may be imparted on a

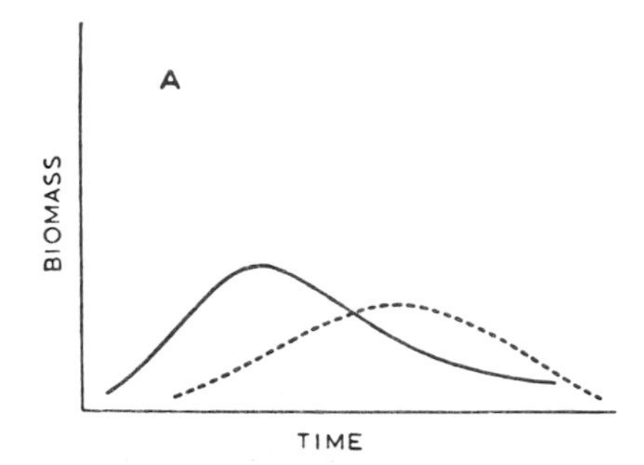

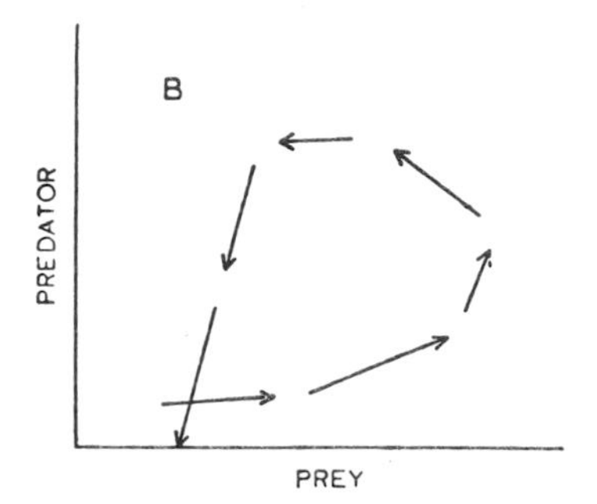

FIG. 75. (A) Predator (----) prey (——) association with time in which (B) predator exhausts its food supply and the population is extinguished (redrawn from Maly, 1969).

plankton community by time delays (Dickie and Mann, 1972); one example of this is the maintenance of the barnacle community in the Firth of Clyde which is highly dependent on the timing of *Skeletonema* bloom in association with the release of barnacle larvae (Barnes, 1956).

An important observation regarding the stability of a system is based on the natural tendency for components of a system to be more variable than the system itself. Thus it is apparent from nearly all ecological oceanographic data (e.g. Riley's 1947 data on George's Bank) that individual components of a food web leading to production at a particular trophic level are more variable *in toto* than the variability in the production level itself. This has been expressed by Weiss (1969) and given by Dickie and Mann (1972) as a definition: viz. a system exhibits stability when the variance of the whole is less than the variance of the parts. This definition has some merit in that it can be expressed mathematically as

$$S = \frac{N_i V}{\Sigma(v_1 + v_2 + v_3 \ldots v_i)} \qquad (127)$$

where S is an *index* of stability of a system having a variance V and a number of components (N_i) making up the system with variances of v_1, v_2, $v_3 \ldots v_i$. An advantage to the use of the above ratio as an index of stability is that it removes the problem of deciding on the relative size of perturbations as destabilizing effects on a system. Thus one system which shows regular oscillations over time of several hundred percent may be as stable as a system which shows practically no variation with time.

Patten (1961 and 1962b) considered the stability of a plankton community and its environmental parameters. The stability index evolved by Patten (1962b) was expressed as

$$S' = \frac{\sum_{j=1}^{m} \det P_j}{\sum_{j=1}^{m} (s/\bar{x})_j} \qquad (128)$$

where

$$P_j = \begin{bmatrix} P_{id} & P_{ii} \\ P_{dd} & P_{di} \end{bmatrix} \qquad (129)$$

which is the matrix of transition probabilities for the jth of m variables, P_{id} being the probability for a decrease in value of the variable following an increase, P_{ii} that of an increase following an increase and so on. det P_j is then determined as

$$\det P_j = P_{id}P_{di} - P_{ii}P_{dd} \qquad (129a)$$

and $(s/\bar{x})_j$ is the standard deviation divided by the mean of the jth variable. Equation (128) takes into account both the direction of changes and their amplitude. Thus if a parameter tends to show more of an increase than a decrease (or vice versa) the stability index will be negative; a large $s/\bar{x}$ ratio will further increase the value S' indicating a lack of stability. In comparing S' for components of the physical environment and plankton community, Patten found that the plankton (S'_p) was 6·7 times more stable than the environment (S'_e); this is not unreasonable since the plankton must to some extent absorb the 'shock' of environmental change and if $S'_p < S'_e$ the plankton community would collapse.

From these observations it is apparent that variability in the components of a system contributes to the system's stability which is the opposite process to that contributing to the instability of a machine (Weiss, 1969), or a highly controlled ecosystem, such as is attempted in aquacultural projects. In controlled operations, such as machines, the variances of each of the components add up to describe the variance in the performance of the total unit. Only by imposing maximum control over the components is it possible to predict the operation of the machine. In contrast, in systems the components vary widely but the system itself (if stable) may show only small (or if large, regular) oscillations. This contrast is brought out when attempts are made to simulate a system using a computer model which inadequately describes the buffering action of the natural environment. Thus small changes in a component of such models can be shown to produce practically any change in a population, however unrealistic such changes may be in a real ecosystem (e.g. McAllister, 1970). In contrast it appears that a natural aquatic ecosystem is made up of highly variable components which at some point in their oscillating states interlock to allow for a flow of energy up the food chain; the exact position of the interlocking point in any process being maintained by fluctuations about a mean, rather than by rigid control of an absolute value. One corollary of these observations is that a perturbation applied to the top of a food chain should have more effect on the nature of the food chain than if a perturbation of similar magnitude is applied to the bottom of the food chain. This may be seen in part by two large-scale perturbations applied to sockeye-producing lakes in British Columbia. In the first experiments (Foerster and Ricker, 1941; Foerster, 1968) 90% of the sockeye predators were removed from a lake with a resulting *ca.* 300% increase in the biomass of sockeye produced in the lake. In a second experiment (Parsons *et al.*, 1972; Barraclough and Robinson, 1972) nutrients added to a large sockeye-producing lake were increased by 100% which resulted in only about a 30% increase in the biomass of sockeye, due to depensatory components (particularly a temperature effect). While these two experiments are only very crudely comparable they serve to illustrate that the proportional increase in fish production was at least twice as effective when a perturbation was applied to the top, in contrast to the bottom, of an aquatic food chain. However, while production is the only consideration in this section, the removal of a predator as a means of raising production may not be economically sound if the same predator is the focal point of another interest, such as the sport fisherman.

Other work on lake communities has been helpful in obtaining an understanding of marine pelagic food chains. Since lakes are relatively discrete bodies of water, it is often easier to establish cause and effect relationships in the aquatic food web than in the sea where the exchange of water may remove organisms from the study area. Density-dependent relationships between the growth of planktivorous fish, and phytoplankton and zooplankton growth rates and abundance are particularly difficult to establish in the marine environment. In lakes, however, it can be demonstrated (e.g. Brocksen *et al.*, 1970) that the abundance of fish is generally related to the abundance of plankton. However, depensatory mechanisms are often apparent; thus in the case of sockeye-salmon-producing lakes, there is a general relationship between zooplankton production and sockeye-salmon production *between different* lakes. However, *within* a single lake, a large number of young fish may compete so heavily for a limited amount of food that the entire population is weakened and the number of fish surviving to become adults may be very much smaller than if fewer young fish were present initially. Another depensatory mechanism has been discussed by Kerr and Martin (1970). In a study of trout populations in Ontario lakes, these authors found that while there was a general relationship between phytoplankton productivity and trout production there was no relationship between trout production and the length of the food chain in different lakes, as implied by the general relationship in eqn. (113). Thus the biomass of large trout feeding as piscivores was similar to the biomass of small trout feeding as planktivores (other factors being equal, between lakes). This was explained by the fact that the large piscivorous trout expended less energy on feeding and were more efficient feeders than the smaller, planktivorous trout. While this depensatory mechanism may be valid in a relatively confined environment, it is probably less valid in marine ecosystems where herbivorous zooplankton are much larger and often more densely aggregated than in lakes.

Predation by planktivorous fish on the plankton community may also affect the structure of an aquatic food web. This has been illustrated by Brooks and Dobson (1965) in the case of a number of lake communities in the eastern United States. Under conditions in which there were few planktivorous fish, it was found that large zooplankton generally predominated. It was assumed that this was because large herbivorous zooplankton are generally more efficient phytoplankton feeders since they filter both large and small phytoplankton. However, in lakes where planktivorous fish were predominant, large zooplankton were selectively grazed by the fish; this markedly decreased the dominance of large zooplankton allowing smaller zooplankton to flourish. Further since the smaller zooplankton could not feed off large phytoplankton, the latter also tended to flourish. Thus the whole size structure of planktonic organisms in similar lakes was largely determined by the abundance of planktivorous fish.

CHAPTER 5

BIOLOGICAL CYCLES

In the preceding section the relationship between different organisms in the food web of the sea has been expressed without regard to life cycles. For unicellular organisms, such as phytoplankton and bacteria, this may be a useful simplification but it should not be interpreted as meaning that unicellular organisms do not have life cycles; many standard texts will describe life cycles of unicellular organisms including resting spore and auxospore formation, sexual reproduction, etc. (e.g. Fritsch, 1956). Rather, it means that the growth and decay of unicellular populations can often be accurately represented by a simple exponential function [e.g. eqn. (54)]. In the case of more complex organisms, however, the life cycle of the organism over a period of one or more years may not permit the use of simple kinetics. This is because life cycles of multicellular organisms often involve different strategies for which there is no simple mathematical expression, other than for the period in which the population shows maximum growth. In order to gain some appreciation of these strategies, examples of life cycles of a number of zooplanktonic organisms are discussed below.

In addition to life cycles, it should also be apparent from the preceding sections that organic and inorganic materials are distributed throughout the food chain and that the mineralization of organic matter back to carbon dioxide and inorganic radicals serves, in itself, to assure a 'food' supply for autotrophic and chemosynthetic organisms. Thus the distribution of carbon and other elements tends to follow a cyclic pattern in which a single element may be present in several different phases (e.g. as a solid, gas or in solution). In contrast to the recycling of materials, energy can only be used once. Mann (1969) has commented that while materials circulate through the biosphere, energy flows through the system in one direction, entering as light during photosynthesis and being lost as heat during respiration. In one sense this is true but the passage of energy through organic compounds into inorganic compounds (e.g. H_2S, NH_3^+, etc.) and back to organic compounds and the food chain represents a redistribution of energy which is cyclic.

Many attempts have been made to depict organic and inorganic cycles in aquatic systems (e.g. Strickland, 1965; Kuznetsov, 1968; Nakajima and Nishizawa, 1968; Riley and Chester, 1971). The closer these come to representing all the intricate possibilities involved in the flow and feedback of materials, the more complex they become. Unfortunately, this leads to difficulties in understanding their significance. In an attempt to clarify this situation, we have represented degrees of cyclic complexity by separating overall processes from some of the more detailed interdependent reactions.

5.1 PLANKTON LIFE CYCLES

Details of life cycles of some species of plankton will generally be found in textbooks on invertebrate zoology. Such details are mostly concerned with morphological and physiological changes; the fact that life cycles also represent different strategies for survival in the pelagic environment of the sea is not generally emphasized and consequently a number of different examples are given here.

While diatoms are generally regarded as undergoing asexual division as their dominant form of reproduction during exponential growth, sexual reproduction and spore formation may also occur. This is illustrated in Fig. 76. The figure shows a decrease in cell size during asexual (vegetative) division (1 to 2), sexual reproduction with the formation of flagellated gametes (2 to 3), and auxospore formation (4). Resting spore formation may also occur (5) directly from vegetative cells. The figure serves to illustrate the complexity of phytoplankton cycles in which only the step 1 to 2 is generally included in trophodynamic relationships.

Reproduction among zooplanktonic crustaceans is generally unisexual, involving both male and female animals, although parthenogenesis is known among the cladocera and ostracods. The typical life cycle of a pelagic copepod has been described by Marshall and Orr (1955). From egg to adult the genus *Calanus* passes through six naupliar and six copepodite stages. Metamorphosis in the first few naupliar stages may take place in a matter of days and may not involve feeding. The six copepodite stages may be completed in less than 30 days (depending on food supply and temperature) and several generations of the same species may occur in the same year (i.e. an 'ephemeral' life cycle). The organism generally overwinters as a late copepodite stage in deep water and starts to breed again as an adult in the spring. The total number of eggs per

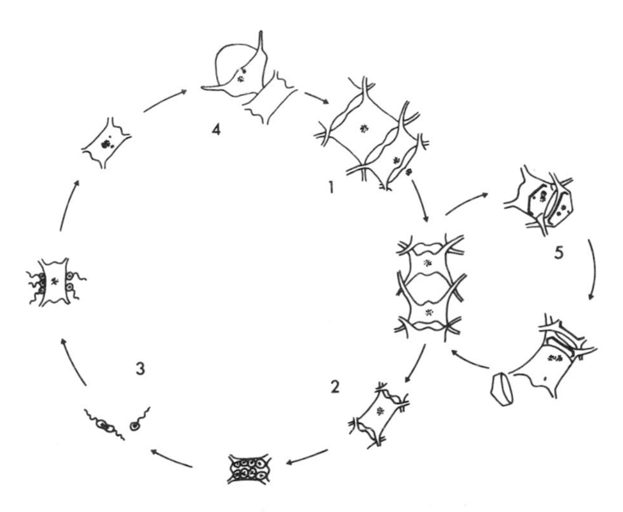

FIG. 76. Life cycle of a marine diatom. *Chaetoceros didymum*, from Harrison and Turpin (1982) as modified from V. Stosh *et al.* (1973).

female (or fecundity) depends on the food supply and may vary from <10 to several hundred. McLaren (1965) showed that for *Pseudocalanus* both egg size and body size of adults decrease with an increase in temperature so that late summer generations of the same species are smaller than early spring generations. Typical common species following this life cycle are *Calanus finmarchicus, C. pacificus,* and *Pseudocalanus minutus*. Because these species take advantage of new supplies of food to create new generations, they are often described as 'opportunistic'.

A distinct variation in the above life cycle among common pelagic copepods is that exhibited by such copepods as *Calanus plumchrus, C. cristatus,* and *C. hyperboreus.* As an example, the life cycle of *Calanus plumchrus* is illustrated in Fig. 77 from Fulton (1973). In this species the development of naupliar stages through to stage V copepodite takes place in the near-surface waters, *ca.* February to June. Copepodite stage V then migrates to a depth of *ca.* 200–400 m where it matures to an adult form, mates and lays eggs. The eggs rise in the water column starting from January and may still be found as late as March. The eggs hatch into nauplii during their passive drift towards the surface and the first naupliar stages do not feed prior to arriving at the surface layers where phytoplankton is abundant. This vertical change in the distribution of the species during its life cycle is known as 'ontogenetic migration'.

Another large group of zooplanktonic crustaceans of ecological importance in the total biomass of the oceans are the euphausiids. The number of developmental stages in the life cycle of euphausiids is greater than in copepods and consists of nauplius, metanauplius, calyptopsis, furcilia, cyrtopia, and adult forms. Breeding populations may occur within 1 year (e.g. *Euphausia pacifica*) or animals may continue to mature and breed in their second to third years (e.g. *Euphausia superba* or krill). With the latter species eggs appear from November to March (southern hemisphere) and sink to 500–2000 m. Larvae start to develop soon after the eggs are laid and ascent to the photic zone occurs during three stages from nauplius I, nauplius II to metanauplius. Independent feeding starts at

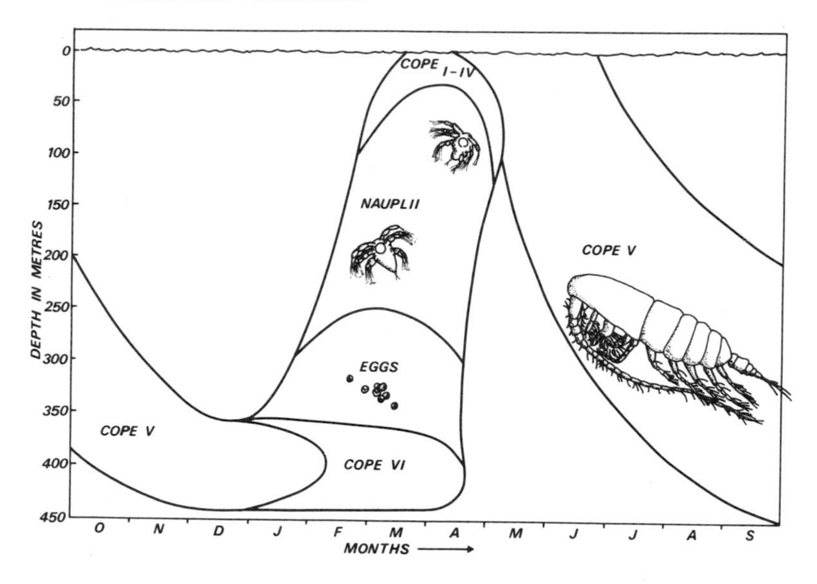

FIG. 77. Schematic diagram of the life cycle of *C. plumchrus* in the Strait of Georgia (from Fulton, 1973).

the calyptopsis stage. Juvenile euphausiids overwinter at 150 to 200 m and mature further the following summer: adults occur after two full years and eggs are laid during the following Antarctic summer (November–March); thus, the complete life cycle of the krill involves a period of three summers and two winters (Marr, 1962).

Some species of zooplankton are 'oviparous' in that their eggs are released into the water column where they are free-floating and beyond the control of the parents (e.g. *E. superba*). In some other zooplankton (e.g. among mysids, isopods, and amphipods) eggs are laid into a modified broad chamber where they hatch to be liberated from the adult as young stages (i.e. the species are 'ovoviviparous'). Grice and Gibson (1975) showed that among some oviparous crustacean zooplankton, eggs which were laid in the fall remained viable on the bottom during the winter; the eggs hatched as nauplii to re-enter the pelagic community in the spring.

Mauchline (1973) showed that among mysids, egg production amounted to about 10% of the adult body weight regardless of the size of adult. However, meso- and bathypelagic species of mysids had fewer but larger eggs. Mauchline (1972) was unable to diagnose specific seasonal breeding cycles among a number of bathypelagic organisms but the adaptation to longer-lived species would indicate a slow, irregular breeding cycle which would maximize conservation of resources in a nutritionally sparse environment. In addition it appears that with deep-water zooplankton, brooding of young tends to be more common than among shallow-water animals. This extra care for the offspring may be another adaptation towards a general scarcity of food at great depths.

Among chaetognaths, sexual hermaphroditic reproduction occurs giving rise to free-floating eggs which may hatch in several days. Young chaetognaths are referred to as larvae but are essentially similar in form to the adults. The number of generations per year may range from several in tropical waters to a biennial life cycle in arctic waters. As with many other species, size is inversely correlated with temperature (Reeve and Cosper, in press; McLaren, 1966; Dunbar, 1962).

Among the coelenterates reproduction is commonly carried out by alternate generations (metagenesis) involving a sexual budding from a polyp stage and sexual reproduction in a free-floating jellyfish stage. In the true jellyfish (Scyphozoa) the polyp stage is often suppressed and although metagenesis occurs, members spend most of their life as free-floating jellyfish. In the Hydrozoa, only the medusan stage is planktonic and thus the hydrozoans may contribute only seasonally to the meroplankton. One or several generations of coelenterates may occur per year but where there is a hydroid (or polyp) stage, this is used for hibernation.

Pteropods reproduce as protandric hermaphrodites by sexual reproduction. Eggs are usually released as an egg mass or ribbon in which the eggs are embedded in mucus. As molluscs, the first larval stage is known as a veliger and these are generally reported from laboratory experiment to be herbivores. Veliger metamorphosis may occur in 10 to 20 days giving rise to larval forms similar in structure to the adults. Among gymnosomatous pteropods, carnivorous feeding starts after metamorphosis of the veliger stage (Lalli and Conover, 1973).

In addition to the examples given above, in continental shelf areas the life cycle of many benthic organisms will contribute larval forms to the plankton. Such larval forms will commonly include echinoderm and polychaete larvae, Veligers and cypris larvae of barnacles.

Figure 78 contains a summary of some of the different pathways in the life cycle of marine planktonic organisms. This is not intended as an exhaustive survey of various options but as an introduction to the subject. As Cole (1954) has commented, "the number of conceivable life history patterns is essentially infinite, if we judge by the possible combinations of features which have been observed. Every existing pattern may be presumed to have survival value under certain environmental conditions." From this, one must conclude that there is a certain danger in employing over-simplified descriptions of growth processes, especially within higher trophic levels. Further, the deployment of different life cycle strategies is not unrelated to what has been called 'stability' in the previous chapter. Obviously there remains a wide gap between the real world biology of the sea and our ability to view it as a series of coupled processes that respond to physical and chemical forces.

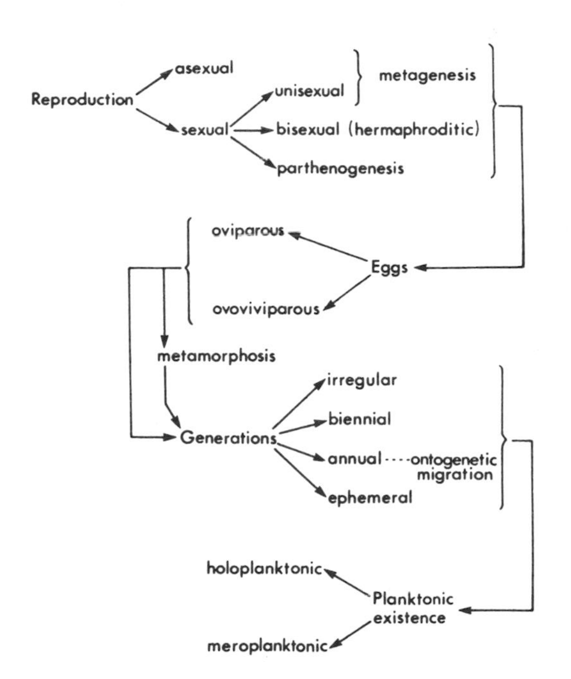

FIG. 78. A summary of some life cycles strategies among the zooplankton illustrating various options employed by different organisms in the pelagic environment of the sea.

5.2 ORGANIC CARBON AND ENERGY CYCLES

In a preceding section it has been noted (Section 2.4) that the largest fraction of organic carbon in the oceans exists as debris, either as organic compounds dissolved in sea water or as particulate organic detritus. It is also apparent from the utilization of organic substrates by heterotrophic organisms, as well as from the utilization of particulate detritus by zooplankton (e.g. Chindonova, 1959), that the reservoir of organic debris in the sea feeds back into the food chain. From these observations Riley (1963) has concluded that there is a flexible system of reversible reactions allowing organisms to draw upon the reservoir of organic carbon and replenish it in a variety of ways. This in turn has tended to stabilize the aquatic environment by providing a food source for living organisms over a longer period of time than that in which a single phytoplankton bloom can be sustained by the environment.

This system can be illustrated by Fig. 79, which

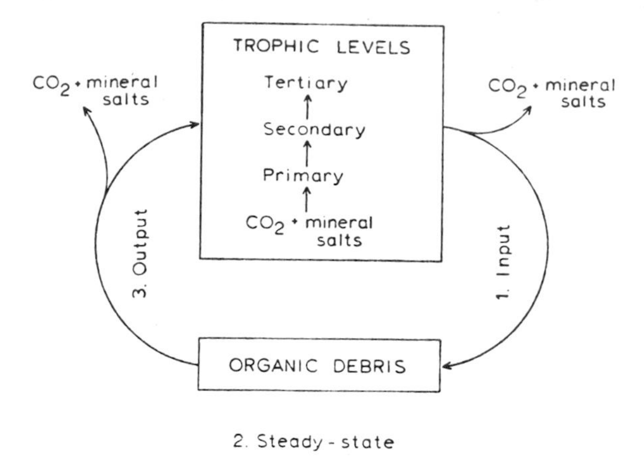

FIG. 79. A generalized organic carbon showing the exchange of carbon between the food chain and the organic debris of the sea.

has been numbered to show (1) the input into the system from the food chain, (2) the reservoir of organic debris which is assumed to have reached a steady state over a long period of time, and (3) the output from the reservoir which includes both the mineralization of organic matter to carbon dioxide and inorganic radicals, as well as the utilization of organic matter by the food chain. Olson (1963) has discussed the kinetics governing the input of organic material (L) to a standing stock of organic carbon (X). In the simplest case of a continual input and with a constant decomposition rate, $k(\text{time}^{-1})$, changes in X with time can be given by the equation

$$X = \frac{L}{k}(1 - e^{-kt}). \tag{130}$$

In a situation where sufficient time has been allowed for the system to come to equilibrium (t large), a steady-state value of X_{ss} will be reached such that

$$X_{ss} = \frac{L}{k}. \tag{131}$$

Minderman (1968) has criticized the use of this approach in terrestrial environments on the grounds that the initial decay of fresh organic material is much faster than can be accounted for by use of the kinetics given above (see also Section 2.4). In aquatic environments, however, decomposition starts as the debris descends towards its final place of accumulation; in shallow seas this will generally be the sediments while in the deep oceans the largest accumulation of organic material is in the deep-water column. Thus the initial rapid loss of organic material will probably have already occurred when the input term (L) is assessed. Skopintsev (1966) using the above kinetics [eqn. (130)] estimated that the one or two parts per million of soluble organic carbon in deep ocean waters had accumulated over a period of several thousand years; a value which is similar to the residence time of 3400 years found by radio-carbon dating (Williams *et al.*, 1969). Using the same approach, the turnover time of organic carbon in a near-shore sediment was found to be about 30 years (Seki *et al.*, 1968).

The need to consider biological systems in terms of energy units instead of weight has already been expressed at several points in this text. In practice, however, the expression of results in terms of weight per unit area is more easily conceived and compared with everyday problems of agriculture, fisheries, and nutrition. However, the absolute need to use energy units when discussing biological cycles emerges whenever a consideration is given to autotrophic and chemosynthetic processes in an environment. This is illustrated in Fig. 80, which is modified from Sorokin (1969).

The left-hand side of the figure is divided into aerobic and anaerobic environments; since oxygen disappears in anaerobic environments, the best division of the two environments is in terms of the 'oxidizing potential', or E_h, measured in millivolts. Generally most of the environments discussed in this text have been aerobic environments in which the oxygen concentration was adequate to supply all trophic levels. In the Black Sea, and in some fjords and lakes, however, anaerobic environments exist below the surface and it is at the interface of these two zones that a number of biological reactions may occur. The areas mentioned above are not large compared with the hydrosphere but it should be considered that certain near-shore sediments, which are not mixed by wave action, may contain very similar aerobic/anaerobic zones. The important aspect of an anaerobic zone

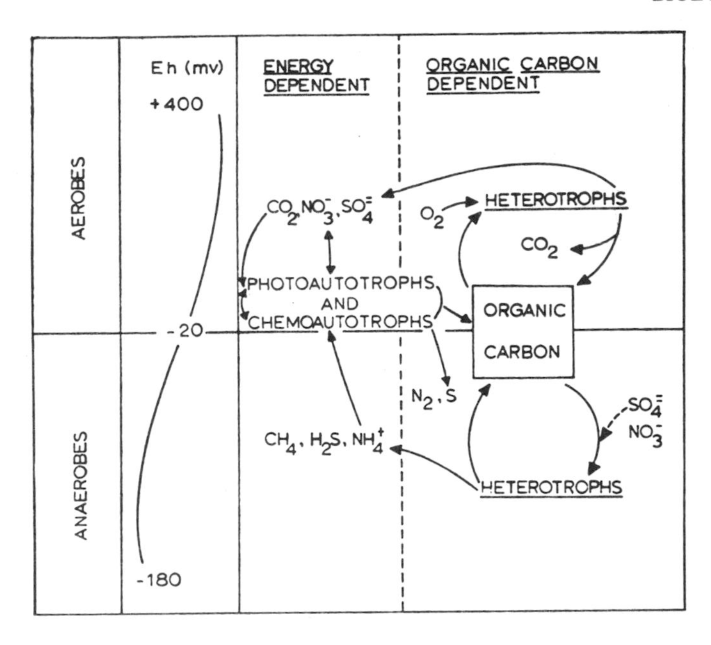

FIG. 80. Organic carbon and energy-dependent cycles in marine aerobic and anaerobic environments (modifed from Sorokin, 1969).

wherever it occurs is that it represents a storehouse of chemical energy which may have somc ability to feed back into the food chain; certainly this has already been demonstrated in meromictic lakes and the Black Sea (Sorokin, 1969). In these environments different groups of bacteria exist which can decompose organic material by using sulphate and nitrate as a source of oxygen, and in the process form reduced substances, such as CH_4, H_2S, and NH_4^+. The latter compounds can be utilized by other bacteria, some of which are strictly chemoautotrophs since they can use CO_2 as a source of carbon, and inorganic compounds as a source of energy. The extent to which CO_2 is taken up independently of photosynthesis has been studied by a number of authors (e.g. Romanenko, 1964 a, b; Sorokin, 1969) and a discussion of these processes is given in Section 3.1.7.

The role of chemoautotrophs in the primary production of deep sea volcanic vents has been described by a number of authors (e.g. Jannasch and Wirsen, 1979). In these communities the only original source of energy is reduced compounds which are given off by the earth's interior. Thus unlike the reduced compounds derived from excessive photoautotrophic production, the bacterial communities in deep sea vents are entirely independent of sunlight. Substantial communities of animals exist in close association with the bacterial production of these volcanic communities.

The top of Fig. 80 is divided into energy-dependent reactions and organic carbon-dependent reactions. In moving from either photoautotrophic or chemoautotrophic reactions, the cycle can either continue to be described in energy units, or quantities of organic carbon can be substituted (the latter are generally easier to measure in small amounts). Thus a photosynthetic organism receives its energy to grow from light, but once new organic material is formed, it can be expressed as so many calories of organic matter in terms of dry weight of organic carbon, wet weight, and so on. All other steps in the food chain, represented by the 'organic carbon' box in Fig. 80, may also be expressed in terms of biomass. However, due to the very different inorganic chemical composition of marine organisms (including the water and ash content of plankton, e.g. Table 11), it is often advisable to retain units which express either organic carbon or a growth-limiting element, such as nitrogen, and to avoid units such as total wet or dry weight. Some authors recommended the exclusive use of energy units but it is apparent from Table 10 that the primary producers may at times be rich in energy compounds (i.e. carbohydrates and lipids) and low in nitrogen compounds (i.e. proteins) and vice versa depending on the availability of inorganic nitrogen. Thus two different metabolic cycles or food webs may exist in nitrogen limited (e.g. tropical) and nitrogen available (e.g. temperate) environments; the former being characterized by excessive amounts of energy rich compounds and the latter being an energy-starved but nitrogen-rich system.

Figure 51, in Section 3.2.1, is another representation of a carbon cycle; in this figure emphasis is given to the loss of organic material through various processes in the marine food chain. A large component which could be added to this cycle is the loss of materials to the benthic community by sinking. In certain areas, such as the Grand Banks, an important pathway for organic material is through sedimentation of phytodetritus, benthic filter feeders and from thence to demersal fish which inhabit such shallow areas in great abundance.

5.3 INORGANIC CYCLES

Many of the earliest studies on the inorganic micronutrients of the sea were concerned with the study of the phosphorus cycle (e.g. Harvey, 1957, and references cited therein). The occurrence of phosphorus in three different forms (i.e. particulate phosphorus, soluble organic phosphorus, and inorganic phosphate) over a period of a year in a coastal environment is shown in Fig. 24, Section 2.1. From this figure it is apparent that the biological phosphorus cycle of the sea involves the uptake of inorganic phosphate by phytoplankton during the summer in temperate latitudes. Phosphorus is then redistributed as particulate and soluble organic phosphorus, the latter resulting from the breakdown of cellular material, as well as from the release of organic phosphorus from living plants and animals. In tropical and subtropical latitudes seasonal cycles are far less well defined and the distribution of phosphorus may be assumed to be in a state of flux in which phosphorus from decomposition processes is utilized as it becomes available.

Pomeroy *et al.* (1963) and Johannes (1965) showed that zooplankton, and particularly marine Protozoa, could excrete phosphorus as organic compounds and as inorganic phosphate, in daily amounts in excess of their body phosphorus content. The excretion of phosphorus by zooplankton was shown by Martin (1968) to be inversely proportional to food abundance; the explanation for this was that phosphorus was used for storage products or egg production when food was abundant, but excreted due to a relative increase in body metabolism when food was scarce. A distribution of dietary phosphorus for *Calanus* during active feeding (April) has been given by Butler *et al.* (1970) as 17·2% retained for growth, 23·0% excreted as fecal pellets, and 59·8% excreted as soluble phosphorus. Thus the grazing activity of *Calanus* effectively returns more than 80% of the phytoplankton phosphorus to the environment. This process of phosphorus remineralization has been observed in nature. For example, Cushing (1964) showed that during 10 weeks of the spring phytoplankton/zooplankton bloom in the North Sea, inorganic phosphate did not decrease below 0·6 μg at/l. However, Antia *et al.* (1963) observed a decrease in phosphate to 0·1 μg at/l after 2 weeks of phytoplankton growth in the absence of zooplankton.

Both the excretion of organic phosphorus by phytoplankton and the uptake of phosphorus from organic phosphorus compounds for phytoplankton growth have been demonstrated by a number of authors (e.g. Kuenzler, 1965 and 1970). Studies on the remineralization of phosphorus from decaying phytoplankton in the absence of light (Antia *et al.*, 1963) showed that half the organic phosphorus could be released as reactive inorganic phosphate in a period of 2 weeks. Phosphorus lost to sediments through sinking as particulate phosphorus may be partly converted into soluble mineral phosphates or recycled by the benthic animals and microflora.

A general summary of the major pathways in the phosphorus cycle of the sea is shown in Fig. 81.

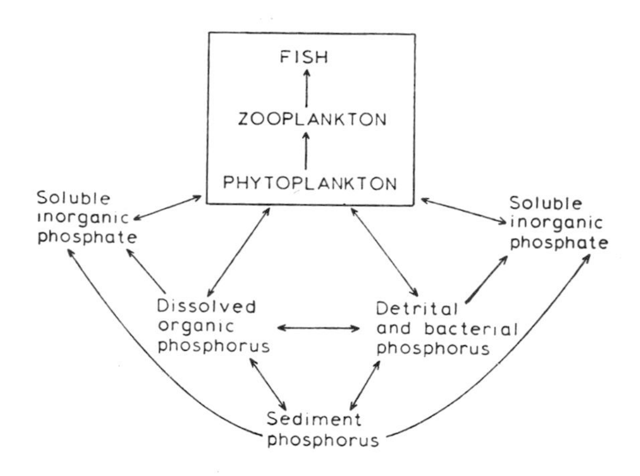

FIG. 81. Major pathways in the phosphorus cycle of the sea.

From this cycle and the observations given above it is apparent that the various forms of phosphorus are often readily exchangeable as metabolites. The rate of turnover, especially in the presence of zooplankton, indicates that phosphorus is generally available in the marine environment although the absolute concentration of phosphate may sometimes be sufficiently low as to determine the actual growth rate of the phytoplankton species.

Phosphorus is readily hydrolysed from organic compounds, either by hydrolysis at the alkaline pH of sea water or by phosphatases, which are hydrolytic enzymes present in many bacteria and on the surface of some phytoplankton, particularly those from environments low in inorganic phosphate. In contrast, organic nitrogen is cycled with much greater difficulty since its fixation, reduction, and oxidation back to nitrate requires the exchange of energy. Thus in the same experiment in which half the phytoplankton phosphorus was mineralized in 14 days, no mineralization of organic nitrogen was observed in a period of up to 75 days (Antia *et al.*, 1963). However, as in the case of phosphorus, zooplankton feeding greatly enhances the recycling of nitrogen, either as ammonia (e.g. Harris, 1959; Corner and Newell, 1967) or through the release of soluble organic nitrogen compounds (Johannes and Webb, 1965; Webb and Johannes, 1967 and 1969). A dietary nitrogen budget for actively feeding *Calanus* has been given by Butler *et al.* (1970) as 26·8% retained for growth, 37·5% excreted as fecal pellets, and 35·7% excreted as soluble nitrogen compounds.

In temperate waters the uptake of nitrate nitrogen follows a seasonal cycle similar to that for phosphate with the exception that the supply of nitrate in the surface layers often becomes completely exhausted during the summer months, while phosphate may continue to be present in low concentrations. In tropical and subtropical waters nitrogen availability is often regarded as the rate-limiting nutrient throughout the year. Dugdale and Goering (1967) have shown that ammonia is taken up by phytoplankton more rapidly than nitrate; this was particularly noticeable in phytoplankton from subtropical compared with temperate regions. Urea may also serve as a nitrogen source for some phytoplankton and significant concentrations (*ca.* to 5 μg at/l) of this compound have been found in coastal and oceanic waters (Newell, 1967; McCarthy, 1970; Remsen, 1971). In tropical environments the fixation of molecular nitrogen by certain bacteria and more especially by blue-green algae, may be very important. Goering *et al.* (1966) measured the uptake of molecular nitrogen in waters containing *Trichodesmium* sp. and found a maximum rate of 0·32 μg N/l/hr. The loss of nitrogen through denitrification of nitrate appears to be an anaerobic reaction. However, in some aerobic environments, dentrification may occur in the presence of detrital particles; this is probably due to the formation of anaerobic microzones of bacterial activity within the particles (Jannasch, 1960).

From the comments given above it should be apparent that the nitrogen cycle is much more complex than the phosphorus cycle in the sea. This is because unlike phosphorus, nitrogen occurs in both oxidized and reduced forms, it can be fixed from, or lost, as gaseous nitrogen and different forms of nitrogen can be utilized with various degrees of success by both phytoplankton and by bacteria. Some indication of these complex reactions is given in Fig. 82 which was modified from previous diagrams (e.g. Riley and Chester, 1971) by Turpin (1980). In this diagram, the food chain (living particulate nitrogen) is the primary

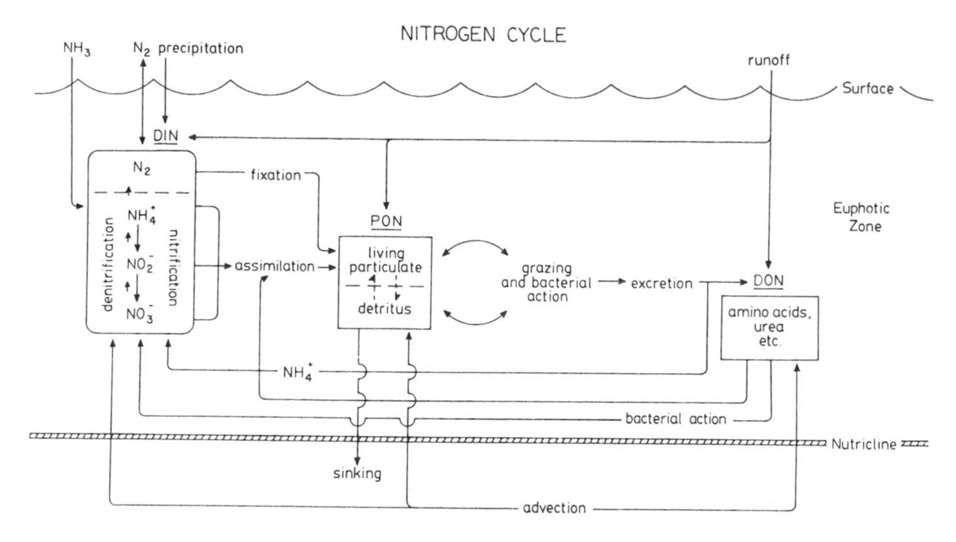

FIG. 82. Nitrogen cycle of the sea (from Turpin, 1980).

consumer of inorganic nitrogen which then can be recycled by zooplankton and bacteria (e.g. Takahashi and Ikeda, 1975). In addition, organic forms of nitrogen from the dissolved nitrogen (DON) pool may also drive the food chain. Antia *et al.* (1975) found that among twenty-six species of phytoplankton, 88% showed good growth on urea, 69% could grow on hypoxanthine, and 50% could grow on glycine. The sources for nitrogen in the ocean are the atmosphere, river run off, and deep water nitrate from below the nutricline. Losses occur by denitrification and sinking of particulate material out of the euphotic zone, a small part of which may be retained in the sediment.

The concentrations of nitrate, ammonia, urea, nitrite, and DON in any one body of water are a function of the biological history of the water mass. Ammonia is generally high following a phytoplankton bloom and in reduced sediment waters. Nitrite generally shows a small maximum at depth in thermally stratified waters (for review, see Wada and Hattori, 1971).

The cycling of other micronutrients in sea water is generally less well defined than in the case of nitrogen and phosphorus. Silicon, which is required in diatom metabolism for the formation of cell walls, follows a seasonal cycle of abundance in temperate waters. The lack of soluble silicate may to some extent determine species succession from a diatom to a flagellate community. Many biologically essential materials in sea water are present in very large amounts so that any cyclic process involved in their distribution is not important to the rate control of metabolic processes in the food chain. These nutrients include sodium, magnesium, potassium, calcium, sulphate, water, and carbon dioxide. In some cases elements may become limiting to phytoplankton growth due to their availability and not to their absolute concentration. Thus iron, particularly in sub-tropical and tropical oceanic waters, may become growth limiting in the absence of organic chelators, which makes it available for uptake by plants.

5.4 TRANSFER OF ORGANIC COMPOUNDS WITHIN THE FOOD CHAIN

Certain organic compounds are transferred within the food chain of the sea; this may also involve a cyclic process in which the compounds are recycled between different trophic levels but at present this has not been a subject of investigation. In the following section some examples are given of compounds which are transferred intact between different trophic levels.

The importance of preformed compounds in the nutrition of lower organisms has been well recognized in the vitamin requirements of certain algae. In a review of this subject Provasoli (1958) showed that thiamine, biotin, and vitamin B_{12} are often required by marine algae. Since these compounds are also required in a preformed state by higher organisms it is reasonable to consider their transfer up the food chain as a necessary part of food chain stability. The exact state of the preformed compound has been investigated in the case of vitamin B_{12} by Droop *et al.* (1959), who showed that various analogues and precursors of vitamin B_{12} could also serve to meet the requirements for this vitamin in some algae.

The transfer of an antibacterial substance in the food chain of penguins has been investigated by Sieburth (1961). The compound which was found to have strong antibacterial properties was acrylic acid; this is present in relatively large amounts in a common antarctic phytoplankton, *Phaeocystis* (a genus which is also found in the northern hemisphere). In the Antarctic this organism forms substantial blooms which are consumed by *Euphausia superba* or krill; the euphausiids in turn are one of the principal food organisms of penguins. Acrylic acid is transferred to the penguin where it causes bacteriological sterility in the anterior segments of the gastrointestinal tract.

The transfer of fatty acids up the food chain from phytoplankton to fish has been indicated by Williams (1965) and studied experimentally by Kayama (1964) and Lee *et al.* (1971). Kayama (1964) fed a diatom, *Chaetoceros* sp., to brine shrimp which were then fed to guppies. Fatty acid compositions of the lipids from the three trophic levels were then examined and the results for C_{12} to C_{22} fatty acids are shown in Table 33. These results

TABLE 33. FATTY ACID COMPOSITION OF *Chaetoceros, Artemia,* AND GUPPY OILS (WEIGHT PERCENT OF TOTAL ESTER) [FROM KAYAMA (1964)]

			Guppy	
Fatty acid	*Chaetoceros*	*Artemia*	17 ± 1°C	24 ± 1·5°C
Shorter chain	trace	0·4	trace	
12:0	0·4	trace	0·2	trace
13:0	0·7	trace	trace	trace
14:0	13·0	4·8	1·5	0·9
15:0	1·8	1·5	trace	0·2
14:2	0·6	trace	0·6	0·5
16:0	18·1	11·6	22·9	36·0
16:1	47·9	44·9	15·9	8·9
16:2	2·7	trace	0·2	0·2
16:3?	4·0	1·7		0·6
16:4?	trace			0·5
18:0	0·5	1·9	8·2	9·8
18:1	8·7	18·4	18·3	15·0
18:2	1·7	0·7	trace	trace
18:3	trace	0·5	1·4	0·8
20:1		0·9		
18:4 & 20:2		0·8	0·3	trace
20:3			0·2	trace
20:4		trace	2·0	2·0
20:5		12·0	4·8	4·6
22:4			1·3	1·0
22:5			6·1	7·3
22:6			16·5	11·5

show a transfer of 14:0, 16:0, 16:1, and 18:1 fatty acids, but an apparent synthesis of polyunsaturated fatty acids by *Artemia* (20:5) and guppies (22:5 and 22:6). However, from Table 33 it is also apparent that C_{20} and C_{22} polyunsaturated fatty acids occur among some phytoplankton, in which case they are probably transferred intact up the food chain (e.g. Lee *et al.*, 1971). Hinchcliffe and Riley (1972) fed different phytoplankton diets to brine shrimp (*Artemia*) and found that the levels of saturated acids were comparatively constant regardless of the food organism used. They showed also that polyunsaturated fatty acids occurred in their different phytoplankton diets and that they were transferred to the brine shrimp. However, some fatty acids occurred in higher proportions in the brine shrimp than in the algal food (e.g. oleic acid, as shown also by Kayama, 1964). Ackman *et al.* (1970) studied the fatty acids and lipids in north Atlantic euphausiids and showed differences between samples which could be explained in terms of diet. Jeffries (1975) made use of the transfer of fatty acids to the gut of juvenile Atlantic menhaden (*Brevoortia tyrannus*) as an assay procedure for the determination of the proportion of different foods eaten by the fish. Since the gut contents of many fish may be difficult to recognize, the distribution of fatty acids in the gut contents, when compared with food items in three habitats (bay, river, and marsh), gave an elegant way of chemically identifying the proportion of food items eaten from each location. The fatty acid spectrum was also used by Tanoe and Handa (1980) to determine the origin of particulate matter transferred to deep water using sediment traps deployed to a total depth of 5250 m. From their results it was apparent that the fatty acid composition of deep water particulate material was very different from material collected in traps; this indicated that the trapped material at great depths was part of a rapidly sinking surface flux (see Section 1.3.3) while the *in situ* particles at depth were part of the more refractory standing stock of detritus.

From studies on hydrocarbons in algae it appears that an important biological substance found in many algae is a 21:6 hydrocarbon, all-*cis*-3,6,9,12,15,18-heneicosahexaene or 'HEH'. This substance is present to the extent of a few thousandths of a percent of the cell weight of algae and is also found in small quantities in zooplankton and higher organisms including oysters, herring, and the basking shark, among animals investigated (Blumer *et al.*, 1970). It has been suggested (Youngblood *et al.*, 1971) that this compound is involved in some way in a biochemical function of the reproductive cycle in both marine plants and animals. The evidence for this is at present circumstantial but it appears that polyunsaturated hydrocarbons are highest in algae during exponential growth, they are concentrated in the reproductive structure of at least one benthic algae, and in the specific case of HEH there is some correlation between the predominance of HEH to other hydrocarbons in algal foods and the relative proportions of male animals hatched from *Calanus helgolandicus* when fed different algae. The general chemistry of hydrocarbons in the marine food chain also appears interesting from the fact that different species of algae appear to have quite specific hydrocarbon ratios and that these differ from those synthesized by copepods and present in mineral oils. Zooplankton contain complex mixtures of C_{19} and C_{20} isoprenoid alkanes and alkenes while mineral oils contain many isomers including large amounts of cyclic compounds, but no olefins. Since hydrocarbons are relatively indestructible in the food chain of the sea, it is apparent that transfer routes within the food web as well as the source of oil pollutants might be discovered through hydrocarbon chemistry (Blumer *et al.*, 1971).

CHAPTER 6

BENTHIC COMMUNITIES

6.1 DISTRIBUTIONS OF BENTHIC ORGANISMS

6.1.1 Taxonomic, Habitat, and Size Groups of Benthos

Organisms which live in, on or are occasionally associated with, the bottom collectively form the 'benthos'. Species, such as bottom-living fish, may also be considered to be part of the benthos. Bacteria, plants, and animals from all phyla are represented, but, unlike the plankton, sizes of bottom-living organisms span several orders of magnitude. Truly benthic organisms, as opposed to demersal species which temporarily come to the bottom, are considered to be sessile and relatively inactive. Organisms such as seaweeds, encrusting sponges, corals, barnacles, and some echinoderms are firmly anchored to solid surfaces for the duration of their life. Burrowing worms and molluscs demonstrate sedentary behaviour but movement within and over the bottom also occurs. In contrast, some benthic animals such as certain molluscs, crustaceans, and demersal fish species often undergo extensive vertical and horizontal movements away from the bottom. The distinction between a pelagic and benthic mode of life is arbitrary for such organisms.

Benthic organisms differ from planktonic ones in their adaptation for an association with a substrate. While a planktonic existence necessitates small size and a specific gravity close to that of sea water, a sedentary life permits a variety of size, shape, and density. Seventy-five per cent of the total number of marine species live on firm substrates (rocks, coral reefs), 20% occur on sandy and muddy bottoms, and only 5% of the total are planktonic (Thorson, 1957). Calcareous shells, elongated stalked and branched body forms and the development of appendages (cilia, bristles) and body musculature which enable movement over, into, and through sediment are characteristic of benthic organisms.

A characteristic feature of all sediments is their vertical zonation into a surface oxidized layer and a subsurface reduced zone where dissolved oxygen is depleted (see Section 6.4.1.1). Vertical distributions of sediment fauna and flora clearly reflect these sharp vertical gradients and organisms themselves produce and utilize metabolites which maintain chemical profiles. Bacteria are most important in this regard. They are the most numerous and ubiquitous organisms present in sediments and their metabolic activities largely control the chemical nature of sedimentary environments. Groups of bacteria isolated from marine sediments are generally similar to those in water in antibiotic sensitivity, degradative, and fermentative capability and cation requirements (Wood, 1965). Besides the common genera of bacteria which occur in water (Section 1.1), *Flavobacterium* and *Achromobacter* are often present in sediments. Both gram negative and gram positive cell types can be present and cell size has been observed to be larger than in bacteria isolated from water (Stevenson *et al.*, 1974). Most motile sediment bacteria possess polar flagella and rod and coccoid cell forms predominate.

Bacteria using oxygen as a hydrogen acceptor to oxidize organic compounds (aerobic heterotrophs) occur in surface sediment layers to depths determined by the penetration of dissolved oxygen. Sediments are always anoxic below some depth and anaerobic bacteria use various organic and reduced inorganic compounds as hydrogen acceptors. Bacteria capable of anaerobic respiration (fermentation) and chemosynthesis characterize these sediment layers. Sulfurbacteria can occur in patches when reducing conditions extend to the sediment surface where light is present. This type of microbial community is described as a 'sulfuretum' since energy transfers which involve sulfur predominate (Fenchel, 1969).

Fungi have not been thought to be numerically important in marine sediments but microbial flora may be dominated by fungi at certain times of the year (Litchfield and Floodgate, 1975). Marine waters subject to fresh water input may receive fungal spores through drainage and humic materials in sediments could provide an available carbon source for these micro-organisms. Burnett (1981) estimated that yeast-like cells predominated the microbiota (2 to 50 μm size class) in sediments from the San Diego Trough.

Benthic microalgae and macrophytes grow, often in great abundance, where water is sufficiently shallow for light to reach the sediment surface. Benthic microflora can be very diverse with all major classes present in the phytoplankton represented. Besides photosynthetic bacteria, cyanophyceans, cryptomonads, euglenoids (green and colourless), dinoflagellates, and diatoms are often found in well-structured vertical distributions corresponding to the presence of light and oxidation-reduction conditions. Attached macrothytes (Thallophytes—seaweeds) and emergent vascular plants (sea grasses) are often abundant in shallow-water littoral and intertidal areas provided exposure to water movement is not too excessive.

Benthic fauna, unlike the zooplankton, is not dominated by any particular class of animals. Representatives of all phyla occur but a particular habitat may be characterized by a few dominant species. Thus, species associations described below have been used to identify bottom community types. This approach has arisen from numerous studies of soft sand or silt bottom deposits which are sampled by dredge or grab. There have been far fewer investigations of rock and gravel substrates where quantitative collection methods are less easily applied. Bottom communities in such areas are often dominated by a few large species (Mills, 1975). An obvious example of dominance exists in coral reefs, oyster banks, and mussel beds. Species diversity may also be low in intertidal mudflats and sand deposits exposed to excessive wave motion. A diverse and heterogeneous fauna, both in species abundance and size, usually exists where substrate particle size is not homogeneous.

It has been apparent since the voyage of the *Challenger* that trawl and dredge samples from the deep sea contain numbers of benthic animals in inverse proportion to depth. Faunal changes associated with water depth have been classified by a system of vertical faunal regions based on the distribution of particular genera and species (Menzies *et al.*, 1973). These authors calculated an 'index of distinctiveness' (the percentage of genera or species not held in common by any two sampling points along a depth gradient) on the basis of isopod crustacea. 'Intertidal', 'shelf', 'archibenthal' (zone of transition), and 'abyssal' faunal groups are terms usually used to indicate vertical zonations in the sea (Fig. 83). The zone of transition (continental slope and rise) is also referred as the 'bathyal' zone. Trenches and canyons in the abyssal region which may drop to below 6000 m form the 'hadal' zone. These regions have distinctive faunal assemblages perhaps reflecting reproductive isolation which exists between populations at great depths. Animals such as decapod crustaceans, echinoderms, and fish living on or near the bottom also show distinct faunal assemblages which may be zoned according to depth (Haedrich *et al.*, 1980). Depth zones of rapid faunal change were observed to be separated by regions of relative faunal homogeneity. While the causes of zonation are not known, these assemblages are of recognized importance as geographic units.

Benthic animals are described on the basis of their position in the sediment relative to the surface and their size. 'Infauna' are animals of any size which live within the sediment. They move freely through interstitial spaces between sedimentary particles or they build burrows or tubes. 'Epifauna' which may be attached (sessile) or capable of movement live at or on the sediment surface. Many of these species which are mobile frequently enter the water column. Amphipods, shrimp and scallops, for example, make extensive movements

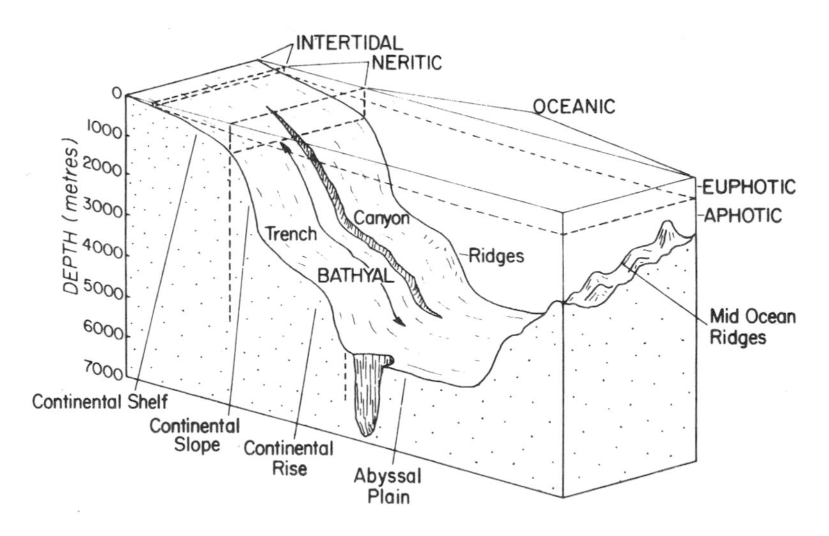

FIG. 83. Horizontal and vertical zonation in a hypothetical cross-section of a transect across the western Atlantic. Geological features of the ocean floor are illustrated in conjunction with terms commonly used to differentiate water column depth layers. Depth and distance relations are not drawn to scale (redrawn from Heezen *et al.,* 1959).

by swimming off the bottom. Gametes are often released freely into the overlying water during reproduction, or migration of sexually mature animals occurs. Tube-dwelling polychaetes may leave the sediment during such spawning migrations.

Size separation of benthic organisms, like that for plankton, is operationally useful. Mare (1942) separated 'microfauna' (1 to 100 μm) (bacteria, protophytes and protozoans excluding foraminifera), 'meiofauna' (100–1000 μm) (foraminifera and small metozoans-nematodes, turbellarians and juvenile macro-invertebrates), and 'macro or mega fauna' (>1000 μm) (Figs. 84, 85, and 86). Metazoans passing through a 500–μm sieve are usually considered as meiofauna with a lower size limit near 50 μm, although the actual limit is somewhat arbitrary. Thiel (1981) has referred to micro-organisms of a size between that of bacteria and meiofauna (2–50 μm) as 'nanobiota'. These organisms may include yeast-like cells, large prokaryotes, flagellates, amoebae, and testate cells (Burnett, 1981). All feeding types, from selective particle feeders to omnivores and carnivores, are reprented in each size category. 'Demersal' (bottom-living) fish which prey on various sizes of benthic organisms enter the benthos as larvae. Most demersal fish species remain at or just above the bottom but some bury themselves in surface sediment.

The separation of benthic organisms based on Mare's size classification could have biological meaning. Schwinghamer (1981) pointed out that the three major size groups (bacteria, meiofauna, macrofauna) correspond to three habitat types (on sediment grains, between particles or on the surface of deposits) which exist for benthic organisms in particulate substrates. When biomass present in each size category is plotted against organism size on a logarithmic scale as described for planktonic particles (Section 1.1), characteristic spectra arise (Fig. 87). Schwinghamer compared his data from six intertidal locations with published data from estuarine and abyssal habitats to show that similar size distributions are characteristic of a variety of benthic communities. The absolute amount of biomass over all size categories in one area may depend on the level of supply of organic matter. The maximum size of organism present may depend on physical factors which limit successful colonization of a habitat by large-bodied animals.

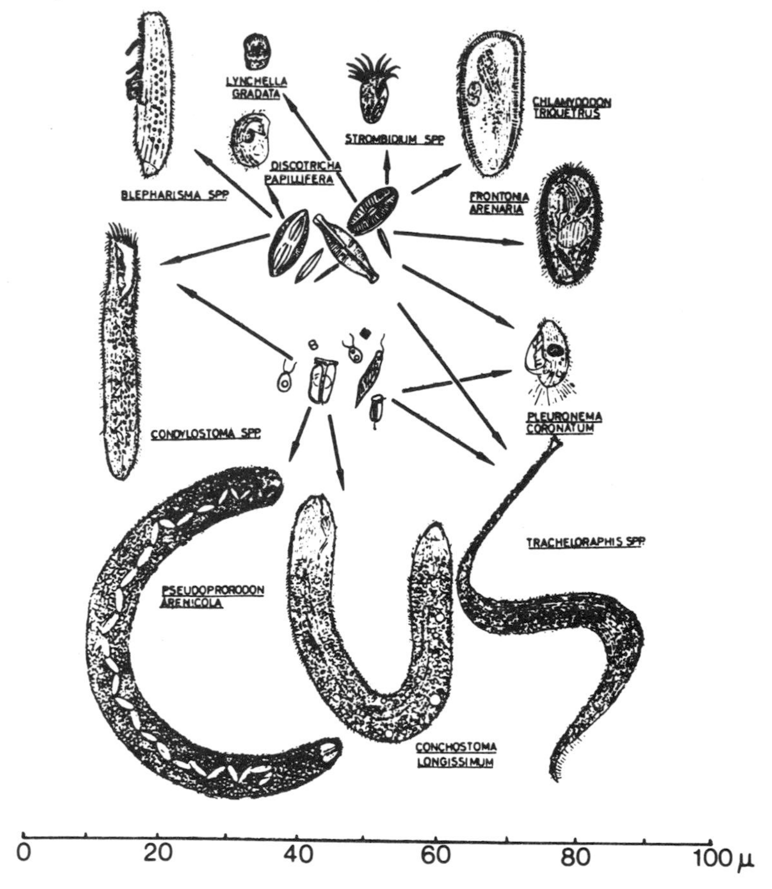

FIG. 84. Examples of microbenthos: bacteria, flagellates, diatoms and herbivorous ciliates in surface sublittoral sands (from Fenchel, 1969).

6.1.2 Diversity and Species Associations in Benthic Communities

6.1.2.1 Indicator species, community associations and guilds. Descriptions of benthic communities are of three major types. Broad sampling over large geographical areas has been used to define species composition and large-scale distribution in terms of dominant (usually macrofauna) species. Studies by Petersen (1913), Thorson (1957), and Nesis (1965) are examples of this type. The regular recurrence of a few common macrofauna species over a wide area is used to name a community in terms of one or two of the most abundant species. Alternatively, studies such as those by Sanders (1960) and Parker (1975) describe and enumerate all major benthic macrofauna species within one localized area. A third approach has been to arrange fauna into functionally similar species groups (guilds) based on feeding, mobility, or ability to modify sediment (Fauchald and Jumars, 1979). This classification scheme emphasizes interactions between an organism and its environment on a functional basis.

Thorson (1957) presented the concept of 'parallel communities' which stated that similar animals will be found in association wherever similar environmental conditions exist. He believed that, where similar selective forces and responses (predation and settlement) occurred, similar groups of more or less closely related species would always occur. The accumulation of evidence for latitudinal

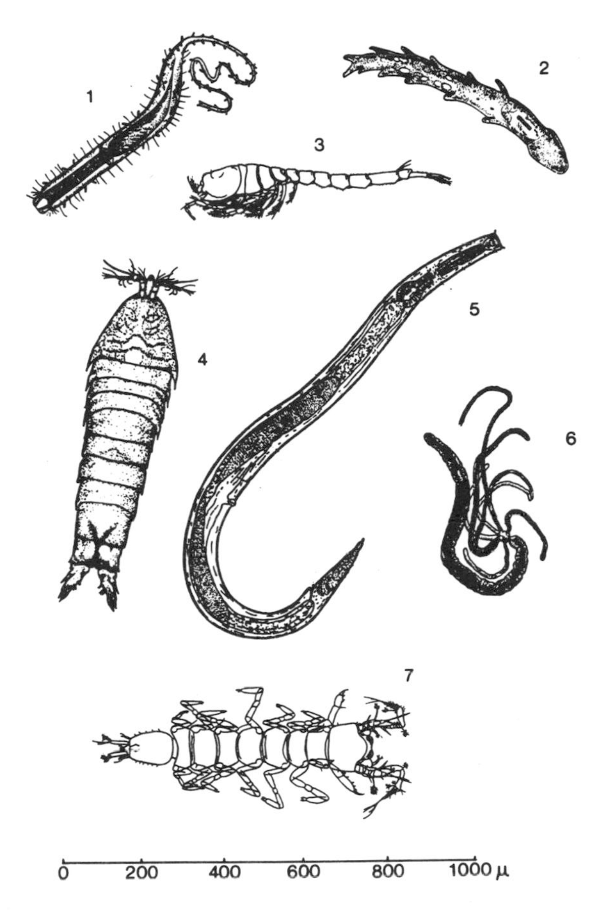

FIG. 85. Examples of meiobenthos: gastrotrich [*Urodasys* (1)], mollusc [*Pseudovermis* (2)], cumacean [*Eudorella* (3)], harpacticoid copepod [*Paramphiacella* (4)], nematode [*Ethmolaimus* (5)], coelenterate [*Halammohydra* (6)], isopod [*Microjaera* (7)]. From Eltringham (1971), Hyman (1959 a,b), Smith (1964), and Marcotte (1974).

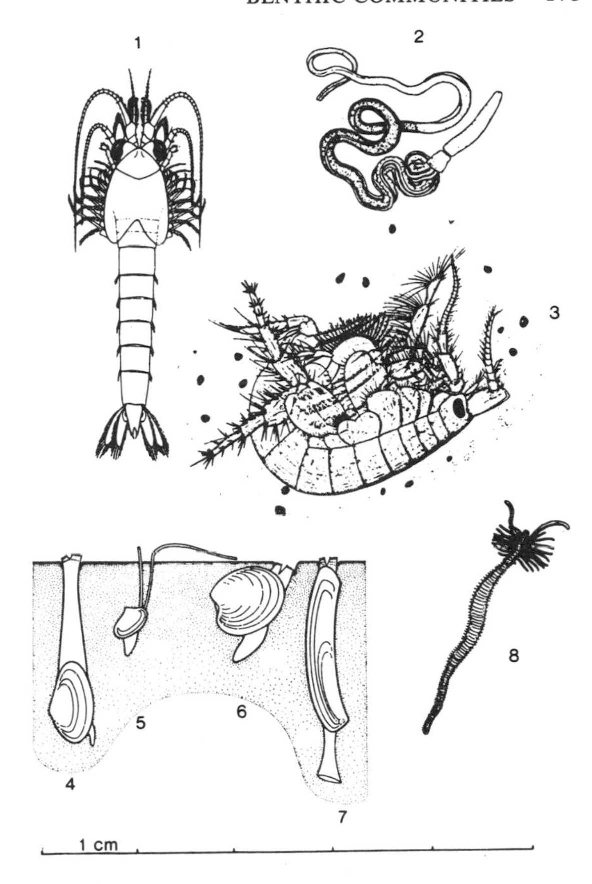

FIG. 86. Examples of macrobenthos: mysid [*Heteromysis* (1)], hemichordate [*Saccoglossus* (2)], amphipod [*Bathyporeia* (3)], molluscs [*Mya* (4), *Tellina* (5), *Mercenaria* (6), *Ensis* (7)], polychaete [*Dodecaceria* (8)], (From Smith, 1964 and Nicholaisen and Kanneworff, 1969).

gradients in diversity created difficulties in applying the concept, however. An increasing gradient in diversity (species number) occurs in benthic communities as one moves from northern to equatorial latitudes. Tropical and deep sea benthos are highly diverse, and usually have no clear dominant species (Sanders and Hessler, 1969).

While some benthic communities may be characterized in terms of a few dominant species, the concept of a community may be an abstraction arising from a series of species distributions along environmental gradients (Mills, 1969). Mills demonstrated that the abundance of smaller less numerous species along a transect of an intertidal sandflat was not correlated with peak numbers of abundant species. There was thus little justification for assuming that dominant species were functionally related to other less abundant species. This interpretation corresponds to Jones' (1950) classification of benthic communities as ecological units separable by physical (temperature, oxygen, sediment type), and biological interactions. The scheme assumes that groupings of organisms are caused more by physical than by biological factors.

The role of physical factors in controlling the formation of species associations is particularly evident in shallow water where abiotic factors

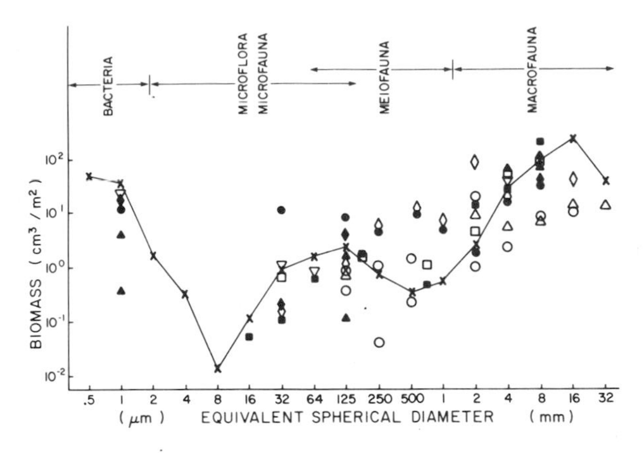

FIG. 87. Size classifications, spectrum, and ranges of biomass of benthic organisms in different size classes of samples from various intertidal and subtidal habitats. Organism volumes (V) were converted to equivalent spherical diameter (D) by the formula

$$D = 2\sqrt[3]{\frac{V}{4\pi}}$$

with total biomass of each $\log_2$ size class expressed as volume/sediment surface area (modified from Schwinghamer, 1981).

(substrate type, water motion) vary considerably in space and time (Warwick and Uncles, 1980). The physical control of benthic community structure is thought to decrease in the deep sea, however, where organisms themselves provide a source of disturbance (Jumars, 1975a). Activities of a single predatory species may also greatly modify the physical and biological nature of some benthic communities. Paine (1969) has termed these 'keystone species' since their activities and numbers determine community structure, integrity, and stability. Starfish which feed on a variety of prey, especially those capable of monopolizing rocky shores, demonstrate that through predation one species can determine overall community composition and structure. Fager (1964) and Mills (1967) describe the stabilizing effect which tube-building polychaetes (*Owenia fusiformis*) and amphipods (*Ampelisca abdita*) can have on fine-grained sediment. The structure imposed was temporary and could have been physically disturbed by the feeding activity of rays and snails. Both numbers and species composition in benthic communities may thus be regulated by the activities of a single species.

Species other than macrofauna have been used to indicate associations of benthic organisms. Analyses of species abundance and diversity of meiobenthic copepods above and below the Atlantic continental shelf break demonstrate unique assemblages in the deep sea (Coull, 1972). Shelf samples were dominated by few species while those from below 1000 m were different and more diverse. Wieser (1960) and Warwick and Buchanan (1970) were able to differentiate communities on the basis of nematode meiofauna in Buzzards Bay and off the Scottish coast. Differences in species composition corresponded to differences in sediment grain size. Ankar and Elmgren (1976) report a similar dependence of species dominance on sediment texture. Macrofauna associations predominated sandy sediments in the northern Baltic while meiofauna (ostracods and nematodes) dominated mud bottom communities. Benthic foraminifera also exist in characteristic community associations (Schafer, 1971).

The functional classification of benthic macrofauna into guilds depends upon knowledge of organism behavioral attributes and not only on taxonomic identification. Species groups are separated by their type of feeding, movement, or sediment-modifying ability (Table 34). The advantage of such a classification scheme lies in the identification of the functional role of each organism. Some understanding of the separation which exists between major groups of benthic macrofauna such as suspension and deposit feeders, for example, may be gained by this type of classification. Rhoads (1974) has suggested that bioturbation by deposit feeding species creates conditions of high turbidity which exclude filter feeders. Wildish (1977) has offered an alternative explanation by suggesting that tidal currents which scour the bottom provide a substrate and food supply adequate for epifauna filter feeders which cannot be utilized by burrowing infauna. Both physical and biological factors appear to combine to create conditions of species dominance in a particular habitat.

TABLE 34. GUILDS OF BENTHIC MACROFAUNA BASED ON A CLASSIFICATION OF FEEDING-MOBILITY AND SEDIMENT MODIFYING ABILITY (FROM LEE AND SWARTZ, 1980) 0 INDICATES THAT A PROCESS IS USUALLY ABSENT OR OF MINOR IMPORTANCE IN MODIFYING SEDIMENT. + INDICATES THAT THE PROCESS IS PRESENT AND OF MODERATE TO MAJOR (++) SIGNIFICANCE IN MODIFYING SEDIMENT.

Guild classification	Representative taxa	Feeding zone	Particle reworking from:			Biode-position	Pellet-ization	Tubes	Fluid transport
			Feeding	Burrowing	Excavation				
I. Suspension-feeders									
IA. Filter-feeders									
Mobile infaunal	*Mya, Mercenaria*	Bottom few cm of water column	0	+	0	+	+	0	++
Stationary infaunal	sabellids, phoronids	Bottom few cm of water column	0	0	0	+	+	+	+
Stationary epifaunal	*Crassostrea, Mytilus*	Bottom to several meters into water column	0	0	0	++	++	0	0
IB. Raptorial feeders	cerianthids, corals	Bottom to several meters into water column	0	0	0	++	++	0(+)	0(+)
II. Deposit-feeders									
IIA. Surface deposit-feeder									
Mobile infaunal-vagile	*Tellina, Macoma* cirratulids, nereids	0–1 cm of sediment	+	+	0	0(+)	+	0(+)	+
Mobile infaunal-excavator	*Uca*	0–1 cm of sediment	++?	0	++	0	++?	0	++
Stationary infaunal	spionids, onuphids *Amphitrite*	0–1 cm of sediment	+	0	0	0(+)	+	+	+
Mobile epifaunal	*Hydrobia* holothurians	0–3? cm of sediment	+(++)	+	0	0	++	0	0
IIB. Subsurface deposit-feeder									
Mobil infaunal-vagile	*Yoldia, Nucula Scoloplos, Nephtys*	0– >20 cm of sediment	+	++	0	0	++	0	++
Mobile infaunal-excavator	*Callianassa, Upogebia*	0–>1 m of sediment	++	0	++	0(+)	+(++)	0	++
Funnel feeders	*Leptosynapta, Arenicola Balanoglossus*	Predominantly upper 0–1 cm, down >10 cm	++	+	0	0	0	0	++
Conveyor-belt species	*Pectinaria, Clymenella Molpadia*	Deeper sediment 3–30 cm	+(++)	0	0	0	+?	+	+(++)

6.1.2.2 Indices of species associations. Species-number relations have been used to provide a quantitative measure of benthic community structure. The approach has served as a rationale and focus for collections of benthos as for other organisms. Abundance and importance (ecological functions) need not be related, however, and calculations of indices of diversity provide only one measure of potential interactions between organisms. Two different approaches to measures of diversity have been described for planktonic organisms (Section 1.2). Sanders' (1968) rarefaction method is considered to be independent of *a priori* assumptions about the structure of the community sampled, although random spatial distribution of species is assumed. The method, dependent on the shape of the species abundance curve rather than the absolute number of specimens per sample, does not measure any biological property of the community. Also, the demonstration of differences in species numbers between regions depends on the use of standard sampling techniques and comparison of similar taxonomic groups (Abele and Walters, 1979).

Hurlbert (1971) proposed a measure of the probability of inter-specific encounters which has parameters with biological interpretation. All individuals in a community can encounter every other individual assuming random movement. Of the (N) $(N - 1)/2$ potential encounters of N individuals, $\Sigma\,(N_i)\,(N - N_i)/2$ encounters involve different species and

$$\Delta_1 = \sum^{S} \left\{\frac{N_i}{N}\right\} \left\{\frac{N - N_i}{N - 1}\right\}$$
$$= \left\{\frac{N}{N - 1}\right\} \left\{1 - \sum_{i=1} \left\{\frac{N_i}{N}\right\}^2 \right\} \qquad (132)$$

represents the proportion of potential encounters that are interspecific where N_1 = number of individuals in the *i*th species in the community (or collection), $N = \Sigma_i N_i$ = the total number of individuals in the community, and S = the number of species in the community. Δ_1 varies from 0 to 1 and it expresses the probability that any two individuals encountered at random in a community will belong to different species. If an encounter is non-lethal (that is, it does not involve predation or cannibalism), subsequent encounters are possible and the probability corresponds to

$$\Delta_2 = 1 - \sum_{i=1}^{S} \left\{\frac{N_i}{N}\right\}^2 \qquad (133)$$

Hurbert's attempts to develop diversity or 'species composition' parameters with biological interpretations also permit estimates of species richness = the number of species present in a collection of specified number of individuals.

$$E(S_n) = \sum_{i=1}^{S} \left\{1 - \left\{\frac{\binom{N-N}{n}}{\binom{N}{n}}\right\}\right\} \qquad (134)$$

This describes the expected number of species in a sample of n individuals selected at random from a collection containing N individuals and S species. Besides cumulative plots of n versus $E(S_n)$ for comparison, as in Sanders' rarefaction method, the parameter quantifies the expected encounters, for example, of a predator which enters a community and in a specified time encounters n individuals at random. The value could serve as a measure of prey availability and should be proportional to a predator's searching rate.

Although Hurlbert's parameters have biological meaning, data required to measure the probability of encounters will be difficult to obtain. Information on habitat size (ambit) and movement of organisms thought to share a common space will be required. No past studies provide such information. Measures of small-scale spatial distribution of benthic fauna (discussed in Section 6.1.3.2), however, show that intra-specific aggregations might be used to quantify the partitioning of habitat space by benthic species. Encounter probabilities might then be calculated assuming random movement of individuals within that space.

A weakness of indices of species diversity based on relative numbers of individuals per species or sample is that abundance *per se* may be related to relative energy consumption. Numerical abundance provides one measure of potential interspecific interactions but it does not specify in any way the number of energy pathways. 'Trophic diversity' was calculated by McArthur (1955) where the proportion of a community's food energy which passed through a particular pathway in a food web was used to describe community stability. Information content (H') in this usage refers to the uncertainty of predicting the pathway of a given unit of food energy in a food web. Alternatively, other measures, such as reproduction or respiration per species, may be informative. Banse *et al.* (1971) ranked macrofauna species on the basis of respiratory demand and demonstrated differences in relative importance of the same species ranked according to numbers and biomass.

Hurlbert (1971) suggested that the importance of a species can be defined by the sum, over all species, of changes (±) in production which would occur on removal of a particular species.

Importance of jth species =

$$\sum_{i-1}^{S} \left\{ P_{i,\,t=1} - P_{i,\,t=0} \right\} \qquad (135)$$

with P_i = production of the ith species before ($t = 0$) and after ($t = 1$) removal of the jth species. This calculation requires that production of all species affected by the removal of a particular species be measured — a requirement no previous study has provided.

One alternative approach to direct calculations of species abundance curves, information statistics, or other measures calculated from observed relative abundance of species in a collection involves representing aspects of species composition in terms of parameters fitted to theoretical models. The use of logarithmic functions of species abundance or rank are examples. Numerous studies have demonstrated that indices determined by these methods are dependent on sample size and the fit of actual data to theoretical distributions is seldom perfect. In addition, there is no reason why actual distributions should conform to a theoretical distribution. One advantage of using parameters of theoretical distributions as descriptive statistics, however, lies in the fact that quantitative measures of variance of predicted distributions may be derived. Some objective judgment of goodness of fit is thus provided. Also, use of certain distribution statistics permits separation of diversity into two components — 'species richness' and 'unevenness' — each of which can be quantitatively estimated (Gage and Tett, 1973).

The distribution of individuals among species can often be represented by a log-normal curve and many studies of the abundance of infauna benthic species have been based on this observation (Gray and Mirza, 1979). The method involves converting numerical abundances of species into logarithms, ranking these and estimating the median abundance ($\bar{a}_R$), the total number of species sampled (s^*) and the standard deviation of the distribution of species abundances (σ). Gage and Tett (1973) found a reasonable fit to the distribution of collections of macrobenthos in two Scottish sea lochs except for rarer species. They further interpreted the log-normal parameters in ecological terms — $\bar{a}_R$ = the abundance of the typical species in a community (those neither common nor rare), s^* = the total species number (equivalent to 'specimen richness'), and σ = 'unevenness' (the inverse of 'evenness' 'equitability').

In an earlier study, Gage (1972) analyzed the same data set using Sanders' rarefaction method. There was a significant positive correlation between stations along a Scottish sea loch ranked by rarefaction diversity and by values of s^*. Additionally, a positive correlation existed between s^* and median sediment particle diameter and values of σ increased with distance up the lochs. Tett (1973) also observed increased 'unevenness' in phytoplankton species abundance at the head of two of the same lochs using similar methods. Increased salinity fluctuations and reduced exchange at the head of the narrow fjord-like lochs were thought to contribute to unevenness among species abundance. However, values of s^* for macrobenthos did not decrease with distance up the loch as was observed for phytoplankton. Thus, on the basis of parameters of the log-normal distribution, factors affecting benthic community structure did not necessarily affect phytoplankton communities in a similar way.

Various types of numerical analyses have been applied to benthic survey data as an alternative approach to identifying species associations. All such methods use statistical criteria to evaluate the frequency of occurrence and/or abundance of species and to quantify associations as groups through various types of 'similarity analyses'. Co-occurrence is assumed to indicate interaction between species within groups and thus these methods provide an objective form of community classification. As Mills (1969) has pointed out, however, co-occurring species may be independently distributed and thus structural groupings are not necessarily functional ones. Correlation and concordance tests have also been applied to tests for similarity among species rank orders of abundance between communities. Jumars (1980) has pointed out that significance tests of correlation necessitate that all rankings are equally likely (perfect evenness of the distribution of individuals among species) which is unlikely to occur in most communities.

Numerical methods used to quantify species associations have two aims: to group stations as to faunal similarity and to compile species lists characteristic of each group. When surveys are conducted over large geographical areas, a heterogeneous collection of organisms results. Both species composition and abundance change and a single sample contains only a fraction of the total number of species. Analysis of such data on the basis of species presence or absence alone is adequate.

Two similarity coefficients have been applied to characterize species groups when samples include heterogeneous types of benthic communities. Jaccard's coefficient is the number of species common to both samples expressed as a percentage of the total number of species in both samples:

$$JC = \frac{c}{a + b + c} \cdot 100 \qquad (136)$$

where a is the number of species in sample A, b is the number of species in sample B, and c is the number of species common to both samples. Czekanowski's coefficient which includes measures of species abundance in each sample is:

$$CC = \frac{2c}{a + b} \cdot 100 \qquad (137)$$

where the same notation is used to express the number of species common to both samples as a percentage of the mean number of species in both samples. As in planktonic populations, logarithmic transformation of the abundance of common species gives a better measure of abundance than do actual numbers and coefficients are calculated on this basis. Dendrograms derived from matrices of coefficients can be used to indicate levels of similarity between groups of stations (Day *et al.*, 1971).

Fager (1957) calculated an index of affinity ('recurrent group analysis') between species in a series of collections based on the geometric mean of the proportion of joint occurrences corrected for sample size. Only presence or absence records are required. The grouping of zooplankton on this basis has been described above. The method has the advantage that levels of affinity can be specified. Connections between groups are determined in terms of the number of species in each group which show affinity with species in other groups. Data on occurrence of demersal fish along trawl stations off the coast of Africa demonstrated the existence of up to twenty-one recurrent groups which could be marshalled into major assemblages (communities) corresponding to water depth (Fager and Longhurst, 1968). Analysis based on joint occurrences of groups along transects formed two assemblages which corresponded to differences in sediment or bottom type.

Multivariate statistical methods have also been used in numerical studies of benthic organisms. Many of these methods serve as a quantitative basis for assessment of the impact of disturbance on benthic faunal communities (Green, 1979). These methods assume a multivariate normal distribution and raw data are usually logarithmically transformed. Two types of analyses have been used. Principal component analysis assigns stations or species coordinates on axes representing the major components of variance. Canonical correlation is analogous to multiple regression analysis where the maximum correlations between variates are computed for each vector (variable). Chardy *et al.* (1976) summarized relations between these and other methods which fit a set of points with given weights and distances into a subspace of defined dimension. Hughes *et al.* (1972) used these methods to demonstrate the proportion of variation in occurrence of certain benthic macrofauna attributable to various components. Approximately 50% of the variance was associated with differences in sediment type.

Numerous clustering or sorting techniques have been used to consider hierarchical association between groups of benthic species. Comparison of sum-of-square agglomeration (Field, 1971; Hughes *et al.*, 1972) separates classes into a dendrogram on a probabilistic basis. Macrofaunal groups in both studies were identified which were associated with different depth zones and bottom type. Vilks *et al.* (1970) used an index of association based on chi square (χ^2) values to construct a hierarchy of benthic foraminifera groups in an Arctic marine bay. Nearshore-shallow (50 m) and offshore-deep (250 m) water assemblages could be differentiated.

All of these analyses provide an objective method for separating species groups which often confirm subjective classification. Their value lies in the quantitative associations which they demonstrate between co-occurring species. These can be used to make statistical comparisons between areas or over time. Species groups which occur in particular habitats may be assumed to have similar environmental requirements. It is not surprising that sediment related properties are common factors upon which groups of benthic species can be differentiated. Associations identified are also dependent upon the sampling methods used and the species compared. Often, only one size or group or organism is considered in a particular study. Associations recognized in these restricted groups are seldom compared across a variety of sizes of organisms.

6.1.2.3 Colonization and successional changes. Experimental manipulation by the removal or addition of species, or the colonization of artificial substrates, allows consideration of diversity-causing mechanisms. Studies by Paine (1971a) showed that diversity (measured as species abundance) decreased when predatory starfish were removed from rocky intertidal areas. Similar simple exclusion experiments showed that the abundance of attached macroalgae could be greatly reduced by the presence of large macrofauna such as sea urchins which on an areal basis were unimportant as energy transformers. Predation can enhance coexistence between species of benthic organisms by preventing monopolization of space.

Experiments to evaluate ecological dominance of attached macroalgae are described by Dayton (1975). Competitive advantages existed for dominant species under various conditions of light and exposure to wave action. Sea urchins were the only herbivore to over-exploit most algal species. Two important carnivores were capable of clearing large areas of urchins, resulting in patches in which algal succession followed. Breen and Mann (1976) described similar interactions between a sea urchin (*Stronglyocentrotus*), a subtidal seaweed (*Laminaria*), and lobsters. The sea urchin was not only important in determining community structure, however. They were also major energy pathways within the littoral community (Miller and Mann, 1973).

The rates and patterns of species colonization and subsequent development of community structure in habitats initially devoid of species form a body of ecological research which has only recently been extended to consider marine benthic communities. Studies of marine fouling attempt to interpret patterns of colonization, species composition, and mechanisms of attachment which can be altered by chemical and physical characteristics of submerged surfaces. Dempsey (1981) and earlier studies used scanning electron microscopy to observe the effect of substrate type on the sequence of microbial colonization. Little information exists about the predictability of colonization events, however. Osman (1977) and Sutherland (1977) both concluded that stable climax communities may not form. The sequence of arrival of colonizing species has a profound effect on how space is utilized.

Changes in benthic communities with time will be discussed in the following section. However, disturbance through physical erosion during storms may expose fresh surfaces for colonization. Johnson (1972) postulated that natural disturbances of the sediment–water interface could operate to maintain a mosaic pattern of distribution of some species in localized areas. Reineck (1968) and Ewing (1973) observed that erosion of sediments at depths greater than 30 m does occur during storms. However, short-term changes in benthic species composition in small areas following such catastrophes have not been documented.

Rhoads *et al.* (1978) differentiated pioneering and equilibrium species assemblages which colonize dredge spoils dumped on the sea bed. Pioneering species tend to be small, sedentary organisms which live and feed near the sediment surface. Species which are representative of equilibrium (stable, undisturbed) benthic communities in shallow water are typically of larger size than the colonizing species, and feeding occurs below the sediment surface. This results in intensively reworked (bioturbated) deposits. Yingst and Rhoads (1980) have compared the changes in trophic structure of benthic communities which occur over time following physical disturbance of sediments. The changes are similar to those which occur over distance along a gradient of chronic organic enrichment (Fig. 88). The comparison illustrates that benthic fauna respond to and modify physical conditions in sediments through their burrowing and feeding activities.

The sequence of colonizing events which Rhoads *et al.* (1978) described in coastal marine sediments may not occur as rapidly or follow the same pattern in the deep sea where organism size is restricted. Grassle (1977) documented the slow rate of recolonization of azoic sediments at a depth of 1760 m. Desbruyères *et al.* (1980), on the other hand, observed a rapid colonization of sediment in trays exposed in the deep sea over 6 months. Abundance of fauna was nearly 5 times greater than

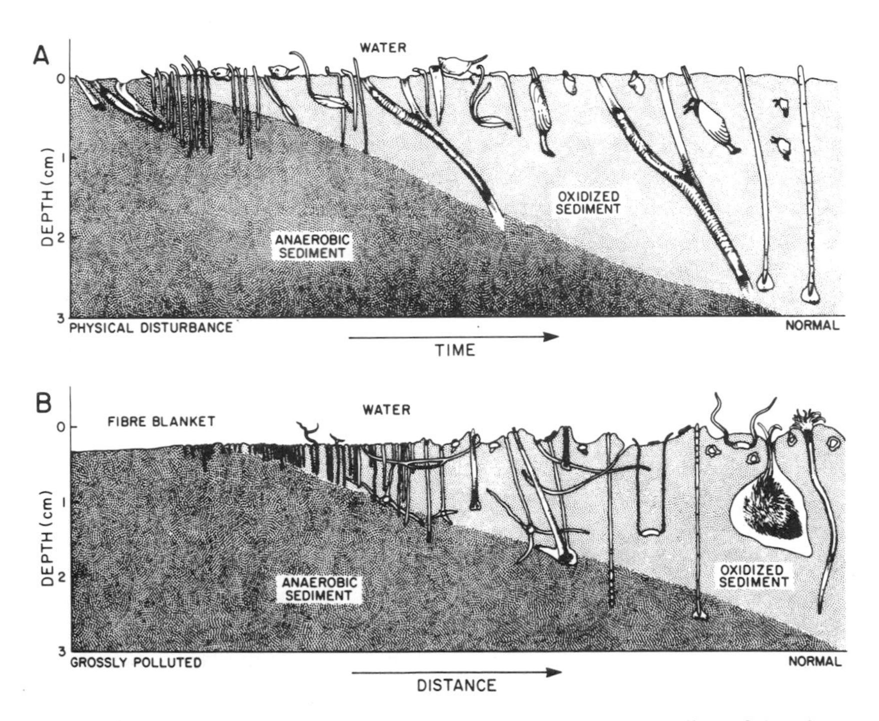

FIG. 88. Changes in the trophic structure of benthic communities over time and along a gradient of chronic pollution (redrawn from Yingst and Rhoads, 1980).

in ambient sediment taken by box coring, but the two dominant polychaete species in the trays were not present in the surrounding deposits. This demonstrates that opportunistic species which invade newly exposed sediments may not represent the fauna normally present in the community — an obvious limitation if this approach was to be used in monitoring studies.

Chemical disturbances can also induce successional changes in benthic communities. The disappearance and recolonization of benthic fauna in the Bornholm Basin of the Baltic, for example, was correlated with oxygen depletion and subsequent replacement (Leppäkoski, 1973). High numbers of many previously unrecorded species occurred 2 years after oxygen-rich water re-entered the area. Such a phenomenon could be used to follow changes in benthic communities during recolonization and thus establish quantitative relationships between species. Rosenberg (1973) has documented such a succession in benthic macrofauna in a Swedish fjord following closure of a sulphite pulp mill.

6.1.3 Abundance in Space and Time

6.1.3.1 Sampling statistics. Marine benthos, like all other organisms, are seldom distributed uniformly. Even relatively homogeneous areas of sediment can contain dispersed patches of fauna and flora. Problems of sampling such heterogeneous associations are not unique to benthic ecology. It is thus not surprising that techniques to quantify spatial distribution patterns developed by botanists, geographers, and geologists have been applied to measure certain geometrical properties of distribution patterns. The ecological relevance of geometric distribution patterns may be obscure, however, especially when organisms are compared which vary in size and mobility. Pattern recognition is also strongly dependent on the size of the sampling unit (quadrate size). It is thus impossible to quantitatively assess non-random distribution patterns for all organisms in a community by use of a single sample size.

Sampling inefficiency and variability in organism numbers due to non-random distribution are separate problems which hinder the accurate assessment of the abundance of all benthic organisms. Holme and McIntyre (1971) reviewed collection methods commonly used to sample benthic fauna. Photography and television permit enumeration of large surface-dwelling invertebrates and fish (Conan *et al.,* 1981; Rice *et al.,* 1982). These methods may also be applied to rock substrates where few other quantitative measures of abundance can be used. Scavenging mobile fauna may be attracted to baited traps. Deployment of traps with cameras attached allows observations of arrival rate of animals at bait (Thurston, 1979), but the area from which animals have been attracted has not been measured.

Although large areas may be surveyed by cameras, specimens cannot be collected. Quantitative collections are usually made by grabs, coring tubes, dredges or trawls. Whether diver or linc operated, the efficiency of all samplers is problematic. Particular substrates or groups of organisms usually require the use of a specific sampler. Once collected, sediments must be sieved, and the organisms removed for counting or otherwise enumerated. No quantitative estimates of abundance are possible without assessment of extraction or counting efficiency. Even when efficient collection methods are used, it is common to observe considerable variation in both species composition and abundance in replicate samples from the same sampling area. Explanations for aggregated distributions of benthic organisms will be considered below, but patterns of variation can often be due to spatial differences in sediment properties or hydrographic conditions. Often, however, heterogeneity occurs over a small area where these factors are thought not to vary significantly. Tests which provide information on the spatial distribution of organisms can then be used to assess quantitative differences between samples.

Elliott (1977) has summarized statistical measures commonly used to assess spatial distribution patterns of benthos. Tests for various types of frequency distribution: positive binomial ($s^2 < \bar{x}$), Poisson series ($s^2 = \bar{x}$), negative binomial ($s^2 > \bar{x}$)

and normal distribution (with transformations) can be used as models for analyzing samples from a population. Frequency distributions fitted to these models permit definition of population dispersion in mathematical terms.

A random distribution is usually an initial hypothesis and early workers assumed that this was the general pattern of distribution for most benthic species. Recent studies have not supported this assumption. The detection of non-randomness depends greatly on the size of sampling units. If these are large or small relative to the size of clumps and clumps are regularly or randomly dispersed, then dispersion appears random. Thus, even small sampling units (500 cm^2) will only detect non-randomness for certain species if there are more than a few individuals in each clump.

The simplest and most widely used test for randomness, the variance-to-mean ratio or 'index of dispersion' (I), has been described (Section 1.3.1) where:

$$I = \frac{\text{sample variance}}{\text{theoretical variance}} = \frac{s^2}{\bar{x}} = \frac{\Sigma(x - x)^2}{\bar{x}\ (n - 1)} \qquad (138)$$

where s^2 is the variance, $\bar{x}$ the arithmetic mean, and n the number of sampling units. The ratio will tend towards unity for samples from populations with a random distribution. The expression $[I(n - 1)]$ gives a good approximation to χ^2 with $n - 1$ degrees of freedom and thus the significance of departures of the ratio from unity can be tested.

The choice of sample size in analyses of distributions of benthic organisms is critical since it affects the numbers sampled. Counts with low means cannot be reliably fitted to a Poisson distribution. Thus, the smaller the mean for an aggregated species, the more nearly will numbers in samples conform to the Poisson distribution. Clark and Milne (1955) observed that evidence for aggregation may disappear in smaller sampling units because the mean per unit is reduced.

A further difficulty with the use of the dispersion index is the dependence of the variance on the mean. With larger and larger sampling area, values for both variance and mean increase but not necessarily in a linear manner. Taylor (1961) observed empirically that sample variance (s^2) was proportional to a fractional power (b) of mean population abundance (m) where

$$s^2 = am^b. \qquad (139)$$

The exponent was considered to be an index of aggregation which remains the same for each species in one environment. Downing (1979) attempted to use the relation to describe a general consistency in the spatial distribution of freshwater benthos, but there is no reason to expect that a common relationship between abundance and aggregation will exist for all species (Taylor, 1980). Despite the inappropriate assumption, slope values of regression curves have been used to compare coefficients for various species to provide a comparative measure of overall community aggregation. Gage and Geekie (1973) concluded that macrofauna from muddy sand were generally more aggregated than those from soft mud on this basis.

Angel and Angel (1967) and Rosenberg (1974) attempted to standardize for the effect of sample size (and number) by calculating coefficients of dispersion for increasing numbers of sampling units. Both studies showed that I values increased for all species when sample unit size increased. However, the former workers compared these curves with those generated due to changing values of n alone by:

$$I \pm 2 \left\{ \frac{2n}{(n - 1)^2} \right\}. \qquad (140)$$

These curves were used as baselines or limits against which the significance of aggregation could be tested. Three of the eleven common macrofauna species analyzed were aggregated at all sample unit sizes, five were randomly distributed and three had fluctuating distributions. The authors concluded that multiple random sampling was not a suitable method for quantifying distribution patterns of benthic populations unless a variety of samples size was used.

The distribution of macrobenthos on a within-sampling unit basis (i.e. by subcoring) has also been determined. Rosenberg (1974) concluded that benthic macrofauna populations were aggregated at least on a scale larger that 600 cm^2 (his minimal sample size) since values of I increased when multiple samples were combined. As discussed above, however, this increase would be expected on the basis of increased numbers alone. Cores covering 46 cm^2 were collected by Angel and Angel (1967) and the three species which demonstrated aggregation did so down to the smallest sampling block size (2 cores = 92 cm^2). For small organisms, such as meiofauna, McIntyre (1969) suggested that numbers be assessed per 10 cm^2.

Jumars (1975b) tested the assumption that macrofauna species in 0·25 m^2 box cores from 1200 m off California were randomly dispersed by subsampling twenty-five contiguous samples of 100 cm^2 each. Dispersion coefficients were calculated for each species to test for strong aggregation of individuals within species. These numbers were summed to calculate a 'total dispersion chi-square' value which expressed the proportion each species comprised of the total number of individuals per sample and a 'pooled' value reflecting the total number of individuals per sample. Since contiguous subcores were sampled, a 'joint method' was also used which considered the number of adjacent presences or absences in particular species.

Jumars observed that few species showed aggregation either between or within cores of 100 cm^2 area. If all species were considered together as replicates, intraspecific aggregation could be detected between cores, but uniform dispersion predominated within cores. Environmental structure, either physical or biological, was assumed to have a 'grain' smaller than 100 cm^2 and it was concluded that organisms themselves provide a major source of environmental variability in the deep-sea. Relations between the scale of variation and organism size will be considered below.

An alternative to testing for goodness of fit to a random distribution, when the variance exceeds the mean, is to use a negative binomial distribution as a model. This probability series is given by:

$$(q - p)^{-k} \tag{141}$$

where $p = \bar{x}/k$ and $q = 1 + p$. $\bar{x}$ is the arithmetic mean and k is not the maximum possible number of individuals possible per sampling unit, but is related to the spatial distribution of organisms in bottom samples. $1/k$ is thus a measure of clumping. As the exponent approaches infinity a Poisson series results and as it approaches zero the distribution converges to the logarithmic series. The relationship between normal, Poisson and logarithmic distribution series is illustrated in Fig. 89.

Negative binomial distributions have been found to adequately describe distribution patterns of some groups of microfauna, while other groups were normally distributed. Gärdefors and Orrhage (1968) used paired samples taken at a constant and known distance from each other to assess macrofauna aggregation. A method of calculating number and the radius of patches by 'tieline' sampling was used but results depend on the form and size of aggregates relative to the distance between double samples. Populations of ophuiroids

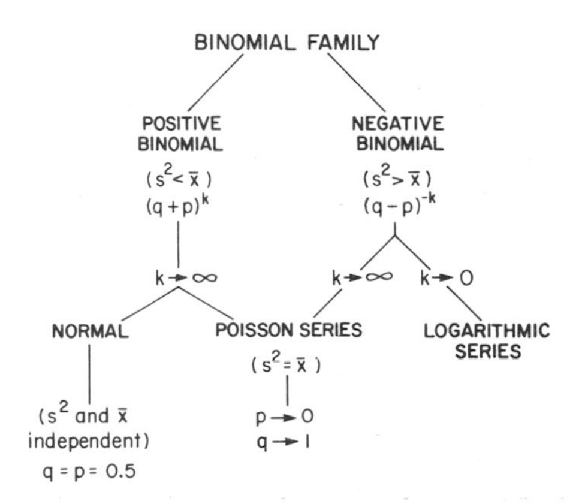

FIG. 89. Frequency distributions used as models for the three possible relationships between the variance and arithmetic mean of a population (redrawn from Elliott, 1977).

and polychaetes were not randomly distributed and negative binomial curves described observed distributions.

A final group of tests used to measure departures from random patterns of distribution in benthic organisms involve distance measurements between individuals. The emphasis is on the spatial location of variate values, not on the overall variance. Heip (1976) and Eckman (1979) quantified aggregated faunal distributions by examining abundance in contiguous cores taken along transects by these methods. The analysis provides a measure of patch size if transects are chosen with axes which intersect. Contouring of data when all subcores from a given area are sampled is also possible (Findlay, 1981). This method has the advantage that patch size and shape may be defined if significant correlations can be shown to exist between adjacent subcores. However, if scales of heterogeneity vary with organism size, the arbitrary choice of one sampling scale will not permit recognition of various patterns. Jumars (1978) used spatial autocorrelation procedures to statistically evaluate the significance of autocorrelations between an irregularly spaced network of macrofaunal samples in the San Diego Trough. Scales less than 200 m showed little spatial autocorrelation in total numbers of individuals per sample. Some species of polychaetes, however, displayed a clumped distribution in recognizable patches while other species were either not aggregated or showed negative spatial autocorrelation at the smallest inter-sample distance. Horizontal habitat segregation was thus highly species specific.

Distances from each animal to its nearest neighbour and from a randomly located point to the nearest animal may be recorded by X-ray analysis (Levinton, 1972). If organisms are randomly arranged, the mean distances of the latter measurement would equal the distance from an individual to its nearest neighbour. Levinton observed that *Nucula proxima,* a small infaunal deposit-feeding mollusc, showed no tendency to aggregate in experimental trays and random spatial patterns occurred over different elapsed times. There was a slight tendency for field samples to be aggregated, but the total area sampled was up to twice that in trays. Environmental heterogeneity could have induced non-random patterns at this scale.

Harvey *et al.* (1976) calculated cumulative frequency distributions or r^2 values, where r was measured as the distance from each individual to its nearest neighbour, to compare settlement of a bryozoan and an annelid polychaete on plastic discs. Distributions were compared with those expected assuming random spacing. At low abundance, the polychaete *(Spirorbis)* showed aggregated settlement, but at high numbers spacing apart occurred. The bryozoan *(Alcyonidium)* showed an aggregated distribution at all levels of abundance.

The spatial dispersion of a population is a fundamental and characteristic property of a species which reflects environmental pressures and behavior patterns. In addition, population abundance cannot be adequately assessed without some effort to estimate errors in population parameters (mean and variance). The choice of sample (quadrate) size is critical in this respect. A small sampling unit is usually more efficient than a larger one when contagious distributions occur. However, organism size and scales of obvious environmental heterogeneity affect the choice. In general, clumped distributions occur but the size of sampling units may affect our ability to quantify their size and shape.

Comparisons of organism abundance are desirable and numbers are calculated per square meter of sediment for macroscopic species. There are usually three main objectives in benthic surveys of faunal abundance (to assess the number of species present, the distribution of the fauna and the number or biomass per unit area). Rosenberg (1974) concluded that a quantitative evaluation of abundant macrofauna was obtained by multiple (5–10) random sampling within one area to sample a total area of 0·5 to 1 m^2. Saila *et al.* (1976) calculated that as few as three 0·1 m^2 grab samples would permit an estimate of the five to seven most abundant macrofauna species to a precision of ±50% of the mean with a probability of 90%. Replication of sampling at single stations is necessary if tests for significant difference among stations are to be carried out (Green, 1979).

6.1.3.2 Small-scale variation in horizontal and vertical distribution. Principal factors which cause local non-random aggregations of benthic organisms can be grouped as follows:

(1) Physical/chemical factors (sediment grain size, oxidation-reduction state, dissolved oxygen, organic content and light).
(2) Biological factors:
   (a) food availability and feeding activity,
   (b) predation effects and removal of certain species,
   (c) reproductive effects on dispersal and settlement,
   (d) behavioral effects which induce movement and aggregation.

Variations in numbers or biomass in benthic species which occur over small distances usually reflect effects of at least one of these factors.

Sediment which appears to be of uniform texture can be composed of a variety of microhabitats of varying size. Lackey (1961) demonstrated this for protozoa and microflora populations in restratified sediments. Aggregations occurred apparently in response to or through facilitating heterogeneity on a microscale of sediment physical and chemical characteristics. Thistle (1978) observed non-random dispersion at scales of separation from 100 m to 1 cm for samples of harpacticoid copepods in the San Diego Trough. Schafer (1971) quantified the patchy lateral distribution of living benthic Foraminifera in shallow coastal waters. Species proportions and total numbers in replicate samples were most variable in samples from shallow water where turbulence and sediment movement could occur more frequently. Significantly higher numbers of living specimens were associated with sand-ripple crests and troughs at different times which implied that hydrodynamic transport occurred across the sediment surface. Environmental gradients for these small organisms appear to occur over distances on the order of a few centimeters. The observation that positive correlations occur between the distribution of living and dead agglutinated deep-sea foraminifera (Bernstein and Meader, 1979) implies that small-scale patch structure of these organisms in abyssal sediments may persist for more than one generation. The clustering of suspension feeding invertebrates around hydrothermal vents in the deep sea (Lonsdale, 1977) also provides an extreme example of aggregated distributions. Enhanced food supply in the form of chemosynthetic bacteria, which may only be concentrated in the immediate vicinity of vents, and reduced predation may account for high biomass of macrobenthos at these sites.

Studies of benthic organism distribution which have concentrated on small localized areas and attempted to describe and enumerate as many species as possible show that different scales of variation exist for different organisms. Communities where a few species are characteristically dominant over a large area seldom exist in shallow water. Heterogeneity in habitat features, such as differences in sediment texture, and hydrodynamic factors prevent the development of well-defined community associations over broad areas. It has been possible, however, to recognize benthic species assemblages associated with specific sediment types.

Classification schemes used to describe differences in sediment texture are based on the proportion of silt, clay, sand, and water (Fig. 90). Commonly used size ranges of particles in each group are given in Table 35. The total amount (as weight, volume or number) of particles in various size classes of a deposit may also be measured and size-frequency distributions compared. The scale of size class interval ('grade scale') chosen for grouping is important for comparison of results from different studies. Logarithmic transformation of the Wentworth scale yields the ϕ grade scale which gives an arithmetic series of equal intervals. Sediment texture is given by references to the proportion of silt, clay, and sand.

Jones (1956) attempted to classify marine benthic communities on the basis of sediment texture. He demonstrated that muddy sand contained a fauna recognizably different and of higher biomass than other grades of deposit in adjacent areas. Sanders (1956) identified a soft-bottom polychaete *(Nephthys)*–mollusc *(Yoldia)* community in Long Island Sound on the basis of sediment texture. Deposit feeders were most numerous in areas of fine

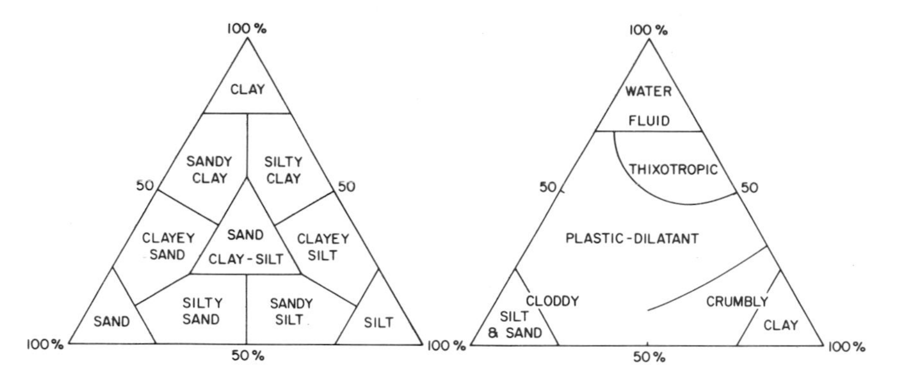

FIG. 90. Classification schemes for sediment texture according to the ratio of silt, clay, and sand (redrawn from Shepard, 1954) and geotechnical properties based on the ratio of these particle types to water content (redrawn from Boswell, 1961).

TABLE 35. COMMONLY USED SEDIMENT PARTICLE SIZE IN THE WENTWORTH GEOMETRIC SCALE.
The phi scale is based on a logarithmic transformation of particle diameter ($\phi = \log_2$ particle size in mm).

	Particle type	Size (mm)	Phi units (ϕ)
Gravel	Boulder	>256	beyond −8·0
	Cobble	256–64	−8·0 to −6·0
	Pebble	64–4	−6·0 to −2·0
	Fine gravel	4–2	−2·0 to −1·0
Sand	Very coarse sand	2–1	−1·0 to 0
	Coarse sand	1–0·5	0 to 1·0
	Medium sand	0·5–0·2	1·0 to 2·0
	Fine sand	0·25–0·125	2·0 to 3·0
	Very fine sand	0·125–0·063	3·0 to 4·0
Silt	Coarse silt	0·063–0·020	4·0 to 5·0
	Medium silt	0·020–0·005	5·0 to 7·0
	Fine silt	0·004–0·002	7·0 to 8·0
Clay	Clay	<0·004	beyond 8·0

sediments with a high organic and clay content. Filter feeders, on the other hand, reached an optimum abundance in sediments with a median grain size of 0·18 mm. Substrate movement and optimal amounts of suspended matter were thought to interact to provide optimum sediment with a texture suitable for the establishment of populations of filter feeders. Bloom *et al.* (1972) made similar comparisons and identified species assemblages of intertidal infauna in which numbers of deposit and filter-feeding species were inversely related. The predicted optimal grain size for filter feeders was observed.

Hughes *et al.* (1972) observed that 46% of the variance in occurrence of polychaetes and echinoderms in a coastal bay was associated with an area of soft mud. Numerical methods described above have also been used to identify species groups of macrofauna and demersal fish which were correlated with differences in sediment texture (Fager and Longhurst, 1968; Field, 1971). In contrast to these observations, Buchanan (1963) found that macrofaunal communities off the east coast of England were poorly correlated with sediment texture. Explanation for this discrepancy might involve complex relationships between tidal and wave-induced patterns of water and sediment movement in near-shore environments. Erosive and depositional habitats occur with characteristic macrobenthic species, but the relation between sediment grain size and species occurrence is not simple (Warwick and Uncles, 1980).

Meiofauna species might be expected to be more sensitive to changes in sediment texture because of their small size. McIntyre (1969) reviewed aspects of the ecology of marine meiobenthos which indicated that characteristic fauna occur in sand and mud deposits. Wieser (1960) observed that particular nematode species occurred only in muddy sediments. Warwick and Buchanan (1970) found that diversity decreased in nematode fauna and increased in meiobenthic copepods with an increase in the silt–clay fraction. While more ecological niches may be present for nematodes in sandy habitats, presumably the geometric diversity is greater for copepods in mud deposits. Gage and Geekie (1973) also observed that fauna on current-swept muddy sand was more aggregated than that on soft mud sediments. The presence of spatial organization could be considered indirect evidence for greater niche diversification, but the environments studies are also structurally heterogeneous in space and time.

The proportion of silt and clay (in addition to the absolute amounts of either size fraction) is of importance for the distribution of many organisms since porosity (water content) and interstitial space are directly controlled by the relative abundance of different-sized sediment particles. Driscoll and Brandon (1973) observed that the distribution of a selective deposit-feeding mollusc *(Macoma tanta)* was directly related to the silt:clay ratio. Parker (1975) noted that silty-coarse sand generally supported an order of magnitude more animals per unit area than silty-clay sediment. These deposits are stabilized, yet water exchange supplies dissolved nutrients and gases and large particles of detritus are readily trapped in spaces between sediment grains.

Sediments which differ in grain size also differ in numerous properties of significance for organisms. Food availability, discussed below, is related to sediment particle size for many deposit-feeding species. Also, sediment porosity, or 'interstitial space' (voids between sediment particles), is critical for small organisms living within the sediment. Webb (1969) and Gray (1974) discussed numerous biologically significant physical properties of marine sediments. Many of these factors are directly controlled by water movement through the sediment which depends on the size, shape, and degree of packing of particles.

Sediment water content (loss of weight by drying) or pore volume (the amount of water necessary for saturation) have been used as measures of available space within sediments. Porosity (total pore space) is not necessarily related to average grain size, however, and permeability (the rate of flow from a constant head pressure) may be a more representative measure of spatial relations of particles within sediments (Morgan, 1970). Frazer (1935) suggested that in systematically packed spheres:

$$P = kD^2 \qquad (142)$$

where permeability (P) varies directly with the square of the diameter of a sphere (D). Morgan tested this relationship and found that it adequately described permeability through various grades of natural sand except with coarse material over 2000 μm diameter.

Webb (1969) and Morgan (1970) have also pointed out that spaces theoretically exist when spheres of equal diameter are packed. Frazer (1935) calculated that the diameter ratio of smaller spheres capable of passing through 'throats' between larger spheres ('critical ratio of entry') varied between $0{\cdot}15D$ for tight packing and $0{\cdot}41D$ for loose packing (where D is the diameter of the large sphere). Similarly, a 'ratio of occupancy' existed as the ratio of the diameter of a sphere too large to pass between larger spheres but capable of existing within a void without disturbing packing and the diameter of the larger spheres. This ratio varied between $0{\cdot}44D$ and $0{\cdot}72D$ for tight and loose packing respectively. Morgan (1970) observed that maximum diameters of interstitial amphipods *(Pectenogammarus)* capable of burrowing and inhabiting various grades of substrate corresponded to calculated diameters of 'throats' between sediment particles. Amphipods had a preference for substrates which permitted free movement through voids.

Direct pore size measurements on sections of resin-impregnated sediments show that interstitial space may be partly filled by small particles. Williams (1972) observed that this occurred in four different sediment types and that the abundance of major faunal groups was related to pore-size distribution. Positive correlations existed between animal size and pore diameter for all groups of interstitial fauna. Although total biomass of interstitial fauna was the same in all grades of sediment, the estimated volume of animals/voids was higher in samples which contained silt.

The quantitative importance of benthic meio- and macrofauna in various marine sediments have been compared (Table 36). While the absolute biomass or numbers of organisms in different areas is not clearly related to grain size or organic content, certain regularities in proportions of numbers or biomass of organisms in these two major size groups appear to exist. Numbers of meiofauna may be high in sandy-silt sediments (10^9–10^{10} individuals/m^2) but macrofauna biomass generally exceeds meiofauna biomass. Numbers of meiofauna and macrofauna are also usually much more variable than biomass estimates. Sanders (1960) considered numbers a more valid estimate of population size for benthic fauna than biomass since less than 1% of fauna numbers in Buzzards Bay sediment constituted over 50% of the total weight. However, ratios of biomass of macrofauna:meiofauna are much less variable than numbers (Table 36) implying that different communities with different taxonomic composition may conform to some uniformity in faunal size composition.

Schwinghamer (1981) provided a quantitative evaluation of this by demonstrating that a uniform pattern of size spectra of biomass existed between 0·5 μm and 32 mm for benthic organisms in various locations (Fig. 87). The absolute concentration of biomass may be an integral function of organic-matter supply in any location, but the coherent pattern of the spectra in different areas implies that whatever the energy input, the resulting biomass of organisms which live on or between grains, or as macroscopic surface dwellers is distributed among organisms of different size in a conservative manner.

The presence of large-bodied macrofauna in particular benthic environments is highly variable and not strictly a function of food supply. Thiel (1975) speculated that the restricted size of macrofauna in deep sea sediments arises from inadequate levels of organic-matter supply. Schwinghamer (1981), on the other hand, who sampled in intertidal habitats, concluded that food supply was not an important limit to the maximum size of the biomass size spectrum. Winter ice scour was thought to be a major factor in removing populations of benthic fauna which could otherwise survive. The macrofauna present at a Bay of Fundy intertidal site (*Corophium* and *Aglaophamus*) are opportunistic species of small size capable of rapid colonization once ice-free conditions exist. Elmgren (1978) also observed that the relative importance of meiofauna increases in the oxygen-poor zones (below 60–80 m) in the Baltic Sea. In well-oxygenated shallow-water areas, the ratio of macro- to meiofauna biomass is typically 5:1, but in deeper water the fauna is dominated by anoxic-resistant nematodes. Thus, in oxygen-deficient environments, it is the lack of resistance to low levels of dissolved oxygen which limits the presence of large macrofauna.

Fenchel (1969) estimated metabolic rates of different size groups of benthic animals to provide an alternative comparison of relative importance. Microfauna, particularly ciliates, predominated metabolism in fine-grain sediments (Table 36). Nematodes were responsible for the bulk of animal respiration in sand finer than 100 μm, while other meiofauna (harpacticoides, ostracods, rotifers) were dominant in coarse sediment (< 250 μm).

McIntyre (1969) discussed problems which arise from the lack of standard methods in quantitative studies of benthic organisms. This applies particularly to estimates of sediment microflora (bacteria and algae) abundance. Bacteria are generally considered to predominate all benthic communities in both numbers and metabolic importance. Fluorescence staining and direct microscopic counts, colony counts on nutrient

TABLE 36. PROPORTIONS BETWEEN VARIOUS SIZE GROUPINGS OF BENTHIC FAUNA ON A NUMERICAL FRESH WEIGHT (g) AND ESTIMATED METABOLIC RATE PER SQUARE METER BASIS IN SEDIMENTS FROM DIFFERENT AREAS.

Animals weighing more than 10^{-4} g but less than 5 g each are considered as macrobenthos infauna. Ratios refer to macrobenthos:meiobenthos (and ciliates when considered separately).

Location sediment type	Proportions Numbers	Biomass	Metabolism	Reference
English Channel (coarse silt 20–50 μm)	1:60	90:1		Mare (1942)
Buzzards Bay (clayey silt 2–5 μm)	1:20	33:1		Wieser (1960)
Fladen Grounds, North Sea (coarse silt 20–50 μm)	1:40	40:1		McIntyre (1961)
Loch Nevis (clayey silt 2–20 μm)	1:100	15:1		McIntyre (1961)
Martha's Vineyard (coarse sand 1470 μm)	1:35	14:1		Wigley and McIntyre (1964)
(fine medium sand 250 μm)	1:180	18:1		Wigley and McIntyre (1964)
(fine sand 130 μm)	1:170	22:1		Wigley and McIntyre (1964)
Ålsgårde Beach (fine sand 175 μm)	1:40:1500	190:1·5:1	2·7:1:1·4	Fenchel (1969)
Helsingor Beach (fine sand 200 μm)	1:28:3000	3·9:1·6:1	1:3:8	Fenchel (1969)
Nivå Bay (medium sand 350 μm)	1:10:50	170:10:1	4:2:1	Fenchel (1969)
Subtidal silty sand (four locations 50–250 μm)	1:3	10:1 (dry weight)		Gerlach (1978)
Baltic Subtidal (above 60 m)		1–10:1		Elmgren (1978)
Nova Scotia Intertidal sand-silt locations		8:1		Schwinghamer (1981)

plates, or chemical methods specific for bacterial cell wall constituents have shown that numbers *per unit volume* usually exceed those in sea water by several orders of magnitude. Only recently, however, has abundance been calculated per unit area of attachment surface to allow comparison between different substrates. Marine silt, sand and pebbles have comparable numbers per unit area

(100–400 μm^2 per cell) (Dale, 1974). This approaches Mare's (1942) estimate of 3×10^{11} cells/m^2 in marine mud. Hargrave (1972a) replotted ZoBell's (1946) data and observed an inverse logarithmic relation (slope of −1·0) between average sediment particle diameter and bacteria numbers determined by plate counts. Dale (1974) observed a similar relationship for direct counts of bacteria on intertidal sediment grains. Over 80% of the variance in abundance was accounted for by difference in particle grain size.

While accessible space is an important feature for all sizes of benthic organisms, high permeability also permits a greater influx of water-containing dissolved nutrients, oxygen, and particulate matter. Drainage in littoral sediments and pumping due to wave action are dominant methods of introducing water into interstitial spaces (Steele *et al.*, 1970). Riedl and Machan (1972) used thermistor probes to provide the first direct measurements of hydrodynamic flow patterns within marine sediments. Their observed flow patterns on intertidal high-energy beaches varied in a predictable although complex way with tidal cycle. Interstitial currents were variable and reached high velocities (up to 2 mm/s) relative to the modal body length of the fauna (1·3 mm). Maximum water movement by wave action and tidal pumping through sediments would be expected on exposed beaches, with similar although less intense flow in deeper water (Riedl *et al.*, 1972).

Bulk water movement through sediment is of primary importance in determining oxygen supply to benthic organisms. Oxygen uptake by aerobic heterotrophic micro-organisms, which are numerous and have a high weight-specific metabolism, rapidly depletes dissolved oxygen within surface sediment layers. A high rate of supply is necessary to prevent anoxic conditions. Microalgae in surface layers of intertidal and shallow water sediments exposed to light also produce dissolved oxygen within the sediment pore water through photosynthesis. Subsurface maxima may be created to a depth of several millimeters in fine grain sediments (Revsbech *et al.*, 1980).

Despite the supply of oxygen by advective flow or photosynthesis, a reduced sulfide-containing zone underlies the oxidized layer in all sediments except those exposed to wave action. Partial oxidation of organic matter results in the accumulation of reduced compounds which diffuse upwards towards the surface. The boundary between oxidized and reduced conditions is operationally measurable as a discontinuity in vertical profiles of oxidation-reduction, sulfide ion potentials or dissolved oxygen using platinum or silver electrodes (Whitfield, 1969; Revsbech *et al.*, 1980). Since the sulfur cycle of bacterial conversion of sulfate to hydrogen sulfide often predominates anaerobic decomposition in marine sediment (Section 6.4.1.1), redox potentials (E_h) and sulfide activity may be correlated and used jointly to indicate the presence or absence of reducing conditions.

Numerous studies have shown that the absence of oxygen, or presence of sulfide, is a significant factor affecting the distribution of bottom organisms (Theede *et al.*, 1969; Fenchel and Riedl, 1970). The vertical distribution of meiofauna in sulfide-rich sediments can often be related to the distribution of oxygen. Fenchel (1969) reviewed previous studies and showed that both microfauna and microflora in sediment could be characterized by their vertical zonation relative to oxidation–reduction potentials and associated chemical conditions (Fig. 91). Powell *et al.* (1980) also observed that a sulfide detoxification system exists in the body wall of a variety of interstitial meiofauna. Detoxification through oxidation was dependent upon the availability of dissolved oxygen. Striking discontinuities in species distribution occur at the interface between oxidized and reduced sediment layers which could reflect the inability of some species to withstand the toxic effects of sulfide ions. The thickness of the oxygenated layer, or conversely the upward extent of sulfide in sediment, can be measured as the depth of the 'redox discontinuity'. Its depth is determined by a balance between processes which supply dissolved oxygen to surface sediments and those which remove it. Since the oxidative state of sediments reflects and results from the interaction of numerous physical, chemical, and biological (particularly by micro-organisms) effects, E_h profiles provide a useful integrative measure of ecological conditions within

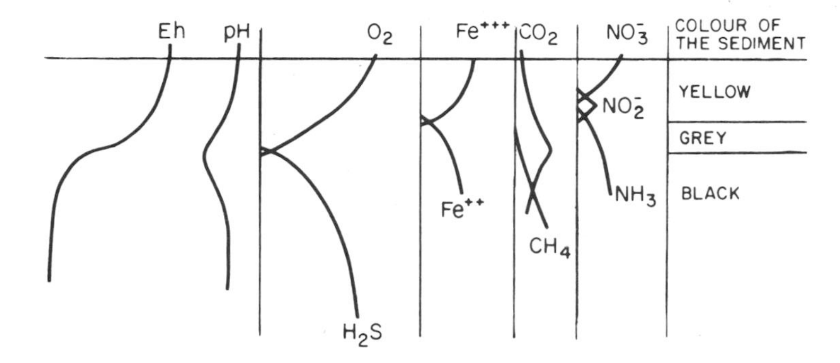

FIG. 91. Diagrammatic vertical profiles of chemical conditions within marine sediments. Redrawn from Fenchel (1969).

sediments, although information about the specific oxidation–reduction reactions which occur is not provided.

Sediment organic matter, either as a nutrient source or an attractive factor which induces settlement, has also been compared to the distribution of benthic organisms. Organic matter may enhance bacterial activity in localized areas of sediment to produce an accumulation of partially oxidized metabolites and reduced E_h potentials. Distinction must be made between non-living organic matter detritus (debris) in sediments and that associated with organisms, particularly bacteria. Organic matter in micro-organisms is recognized as a primary food source for invertebrates which ingest sediment (Section 6.4.2.2). However, much of the total organic content in sediments occurs as refractory compounds, some fraction of which may be digestible by deposit-feeding invertebrates (Foulds and Mann, 1978; Cammen, 1980a). Bader (1954) reported a linear relation between benthic pelecypod (mollusc) populations and the acid-soluble (non-refractory) fraction of sediment organic matter. There was no significant relation between population abundance and total sediment organic content.

Many studies have observed an absence of correlation between faunal abundance and total sediment organic content. Sanders and Hessler (1969) pointed out that variation in biomass between different areas is usually so much greater than variation in organic content that other factors must control distribution patterns. However, little information exists concerning small-scale variations of sedimentary organic matter. Studies on benthic organism microdistribution discussed previously have not been accompanied by measurements of organic content. Sediments in the deep sea contain less than 1% of organic carbon except in areas of upwelling (Bezrukov *et al.,* 1977). If deposition occurs as fine particulate matter, a rather homogeneous distribution pattern on a small scale should exist which would not conform to differences in faunal abundance in closely spaced samples (Jumars, 1975 a, b).

Despite the desirability of differentiating available and non-assimilable fractions of sediment organic matter for more meaningful comparisons with fauna distributions, several authors have observed linear correlations between biomass of certain invertebrates and total sediment organic carbon and nitrogen (Newell, 1970; Longbottom, 1970). The comparisons are usually confounded, however, by significant inverse correlations between sediment particle size and organic content. Thus, regression lines for biomass on particle diameter are almost identical to those for organic matter (Longbottom, 1970). Quantitative evaluation of the effects of organic content *per se* separate from associated changes in particle size on abundance of deposit-feeding benthos could be made since methods for differentiating lipids, proteins, and carbohydrates in sedimentary particles have been developed (Whitlach and Johnson, 1974). Separation of food (organic detritus) and non-food (inorganic) particle types available for deposit-feeding species in an intertidal area showed that both particulate and bulk sedimentary characteristics were related to species diversity (Whitlach, 1981).

Organic compounds in sediment pore water and associated with solid surfaces attract benthic organisms. Experiments to test for substrate selection by various benthic invertebrate species (Gray, 1966, 1967) show that organic films produced by bacteria attached to particle surfaces can render sediments attractive. Organic films of different origin can also be differentiated. Numerous studies of settling behaviour by planktonic larvae of benthic invertebrates (oysters, mussels, and barnacles) show correlations between settlement activity and the presence of acidic proteins or protein–carbohydrate complexes which may be species specific (Meadows and Campbell, 1972; Larmen and Gabbott, 1975).

Inter-relations between sediment grain size, organic content, and oxidation–reduction conditions as they affect total numbers (by direct counts) of bacteria in sediments in a shallow coastal inlet were quantified by Dale (1974). Water motion (waves and tides) was considered to be the dominant factor which determined accumulation of organic matter, grain size, and levels of oxygenation (E_h potentials) in sediment. Partial correlation coefficients were highest ($>0{\cdot}9$) between bacterial numbers, carbon, and nitrogen. Coefficients were reduced but still significant when grain size was controlled. Thus, not all of the covariance between bacterial numbers and organic content was attributable to the bacteria:grain size relationship alone. Micro-organisms are also not uniformly distributed over sediment grain surfaces. Small grains which may be pitted by weathering provide recessed microsites for bacterial and diatom growth where raised surfaces are abraded and free of micro-organisms (Nickels *et al.*, 1981). Topography of sediment particles is thus also important in permitting colonization of protected microsites.

Many studies have considered the effect of other abiotic factors which can affect the distribution of benthic organisms. Tolerance and preference experiments have shown that two broad classes of organisms (euryhaline and stenohaline) exist with respect to salinity response (Jansson, 1968). Variation in this factor is greatest in estuaries where brackish surface water may periodically reach organisms normally exposed to higher salinities. Gage and Tett (1973) considered that stress induced by reduced and unpredictable (fluctuating) salinity with distance along fjordlike inlets might affect pelagic larval stages of benthos. Since altered salinity conditions exist mainly near the surface, however, effects after settlement would be minimal. Changes in factors like salinity, pressure, and temperature would not be expected to have localized effects which would induce changes in small-scale patterns of distribution especially in deep water where these physical conditions are very stable.

Light is a factor of considerable importance to the vertical distribution of some benthic organisms. Sediments which lie within the photic zone (intertidal to 150 m in tropical oceans) usually contain both algal and photosynthetic bacterial populations which are vertically stratified. Patterns of vertical distribution result from a combination of factors. Although many types of algae and bacteria in sediment are capable of restricted movement (up to several millimeters) in response to light and tidal changes, living cells may occur to several centimeters depth (Fenchel, 1969). Pigmented cells in deep aphotic sediments could be displaced from the surface by sediment mixing through physical or biological reworking.

Fenchel and Straarup (1971) measured light penetration and the vertical distribution of various photosynthetic organisms in shallow-water marine sediments. Photosynthetic pigments in sediments have been measured but no previous investigation had compared absorption of different wave lengths of light and pigment distribution. The depth of the photic zone (light intensity $<1\%$ of surface values) in silt and very fine sand was about half that observed in coarse sand. The vertical distribution of living algal cells and chlorophyll *a* corresponded and indicated maximum light penetration in the upper 6 mm of the fine sand sediments. Fine-scale vertical profiles of dissolved oxygen before and after exposure to light have shown that photosynthesis by benthic microalgae is also confined to the upper several millimeters of sediment (Revsbech *et al.,* 1980).

While abiotic factors may affect organism distribution patterns on a small scale, the benthos itself alters physical and chemical features in sediments. These effects may induce heterogeneity on a scale comparable to habitat dimensions of the organisms themselves. Reworking, through digging and burrowing, and feeding activity (tube building, particle selection, fecal production) creates structures which locally modify sediments. Lee and Swartz (1980) describe numerous examples of the effects of biodeposition and bioturbation by macrofauna. Increased surface area, both external and internal, provide attachment sites for micro-organisms which can serve as a food source (Section 6.4.2). Irrigation and particle selection by deposit-feeding fauna enhances the growth of digestible micro-organisms (Hylleberg, 1975). Fecal pellets are usually enriched with organic matter and certain invertebrates may also stimulate bacterial growth by mucus production during feeding. The spatial extent of such enrichment would correspond to organism size and the amount of feeding activity.

Burrowing, crawling, and feeding activity of benthic fauna may lead to more or less consolidated sediments. Mills (1967) described the sediment sorting effect of feeding by a tube-dwelling, selective deposit-feeding amphipod *Ampelisca abdita.* Tube formation stabilized sediment and production of fecal pellets consolidated ingested fine material. However, feeding increased the proportion of fine particles at the sediment surface leading to increased sediment instability within aggregations. Polychaetes and bivalves also create topographic relief in sediments which, because of increased roughness in the boundary layer, can decrease the critical entrainment velocity (Nowell *et al.,* 1981). These authors observed that fecal mounds held by mucous adhesion were less easily resuspended than ambient sediment. Eckman *et al.* (1981) also found that the existence of stable beds of the tube-building polychaete *Owenia fusiformis* was attributable to mucous binding despite local scour caused by individual tubes.

Rhoads (1973, 1974) and others, on the other hand, have noted that biologically reworked sediments are usually more easily suspended than consolidated deposits not exposed to bioturbation. Rowe *et al.* (1974) attributed high rates of erosion at the upper end of Hudson canyon to excavation activity of large decapod crustaceans. Holothurian populations of only a few individuals per square meter increased water content, vertical sorting to 20 cm, and produced topographic relief (Rhoads and Young, 1971). In such locations, the activities of relatively large burrowing animals may dominate the physical structure of the sediment for smaller more abundant organisms. Rhoads and Young (1970) termed the interaction 'trophic amensalism' when organisms like deposit feeders restrict the presence of suspension feeders and attachment by sessile epifauna through creation of unstable sedimentary conditions.

The role of organism feeding activity and predation as a structural force in benthic communities is demonstrated by results of exclusion experiments and observations of the impact of keystone species in rocky intertidal areas (Section 6.1.2.1). Glasser (1979) has reviewed studies which illustrate the general idea that predators stabilize co-existing prey populations by feeding preferentially on those which are most abundant. There are many examples of the impact which predators can have in alteration of benthic population structure. Fager (1964) speculated that feeding by bat rays destroyed dense beds of tube-building polychaetes *(Owenia).* Dominance structure in a benthic macrofauna community may also be altered by fish predation. Levings (1973) observed that predation by the flatfish *Pseudopleuronectes* reduced absolute numbers and changed the relative abundances of macrofauna species following the onset of feeding in the spring.

Predator–prey interactions have also been thought to permit the co-existence of species in a state of dynamic equilibrium. Mann (1977) described such an interaction between lobsters, sea urchins, and kelp in a marine coastal area. Destruction of seaweed beds by sea urchin grazing, once a balanced predator–prey relationship between lobsters and sea urchin was upset by exploitation of lobster stocks, was hypothesized. Similar balanced relationships may exist in the deep sea. Dayton and

Hessler (1972) advanced the hypothesis that high faunal diversity on the sea floor is maintained by predictable disturbance through cropping. Disturbance is considered to allow the co-existence of species which share the same limited resources (thought to be food supply) reducing effects of competition. Jumars (1975a), however, disagreed with the view that predators in the deep sea operate in a homogeneous environment. He viewed predation as one of many organism-induced effects which create a highly variable environment. This may be a major source of disturbance in the deep sea where other factors vary little over time.

Exclusion experiments in rocky intertidal areas have shown that removal of predators from an area can lead to a rapid increase in abundance of organisms previously limited by competition with one or a few species becoming dominant (Paine, 1966). Predation is viewed as preventing competitive exclusion by maintenance of species diversity. Similar experiments in soft estuarine sediments, however, have shown that numbers and species richness increase after removal of predators with no tendency towards competitive exclusion by a few dominant species (Peterson, 1979). It is possible that three-dimensional space in soft sediments allows a refuge in the vertical axis which is not present on hard rocky substrates. Alternatively, competition for food resources rather than space may predominate in communities of infaunal organisms.

Caging (exclosure) experiments in soft sediments may also involve changes in more than the presence or absence of predators alone. Observations that removal of a sedentary polychaete resulted in an increase in a mobile deposit-feeding species (Woodin, 1974) can not be considered only to arise from competitive release. Sedimentary habitat changes created by cages, exclusive of the presence or absence of predators, can also significantly alter species composition of polychaetes in soft sediments (Hulberg and Oliver, 1980). Deep burrowing of small organisms may be less affected by the presence or absence of surface-feeding predators. Holland *et al.* (1980) observed that only infauna species living near the sediment–water interface increased in abundance in the absence of predators. The structure of soft sediment can thus create refuges for organisms (Woodin, 1978) which prevents the impact of predation visible in rocky intertidal areas.

Fauna which crop micro-organisms may also enhance bacterial turn-over rates and nutrient cycling in sediment. Hargrave (1976) observed that fecal pellets produced by a variety of benthic invertebrates are foci of increased oxygen uptake due to enhanced bacterial growth. Mineralization of detritus is also increased through protozoan grazing on bacteria and numbers of micro-organisms on detrital particles can be significantly increased by such grazing pressure (Fenchel and Jorgensen, 1978). Cropping could have a localized effect in selection for rapidly growing micro-organisms at sites of more intensive feeding. Effects may not always be positive, however. Grazing by the intertidal snail *Nassarius obsoletus* on benthic microalgae decreased algal biomass at natural levels of abundance (Pace *et al.,* 1979).

Thorson (1957) proposed that aggregations of many benthic species which produce pelagic larvae result from settlement in restricted areas of suitable substrate due to favourable hydrographic conditions. Many studies have directly linked localized abundance of some species of macrobenthos to settlement success. Curtis and Petersen (1978), for example, observed patches of the bivalve *Macoma* and the polychaete *Pectinaria* dominated by a few year classes. The patches were of a few square meters in size surrounded by other patches dominated by other year classes of the same species. Eagle (1975) also observed dramatic variation in community structure which appeared to reflect settlement success. Populations of non-selective deposit-feeders occurred in an unstable substrate which was unsuitable for spat settlement. Juveniles could co-exist with filter feeders and selective surface feeders which did not prevent larval settlement. Interactions of dominant species through adverse effects of feeding on larval and juvenile survival was thought to restrict dominant animals to single year classes. Such interactions, which may account for non-overlapping distributions and the formation of infaunal suspension-feeding and deposit-feeding assemblages, are consistent with Sanders' (1958, 1960) and Rhoads and Young's (1970) observations of specific animal–sediment relations. Woodin (1976) reviewed evidence of adult–juvenile macrobenthos interactions through competition for space or food. Interference through sediment destabilization was considered a major factor which separated assemblages of burrowing deposit-feeders, suspension-feeders, and tube-building infauna.

Spacing between individual benthic organisms, particularly sedentary species, has been related to intraspecific competition for space. Knight-Jones and Moyse (1961) reviewed early studies which indicate how spacing behaviour during larval settlement serves as an adaptation to prevent overcrowding. Studies of micro-organisms attached to particle surfaces have also shown that cells cover only a few percent of the total available space (Hargrave and Phillips, 1977; Marsh and Odum, 1979). Although abundance can be increased through nutrient enrichment, cell distribution tends to be over-dispersed. Holme (1950) found over-dispersion of the surface deposit-feeding mollusc *Tellina* which implied separation due to foraging activity in a territory around each individual. Crowded populations were more randomly distributed. Mobile deposit-feeders like *Nucula*, which may not benefit from territoriality, display random distribution patterns (Levinton, 1972).

Fenchel (1975) concluded that the small-scale distribution of local sub-populations of four species of mud snails (Hydrobiidae) in a complex estuarine environment resulted from interaction of habitat selection, dispersal and colonization rates, interspecific competition and extinction. The relation between body size and sediment particle size ingested was the same for three of the four species which were of similar size when they occurred separately. However, when stable populations co-existed, the size frequency distribution of the species differed and feeding occurred on sediment particles of different sizes. Character displacement was thought to reduce competition by permitting partitioning of a limited food resource. This type of dynamic interaction between marine benthos and their biotic and abiotic environment contrasts Thorson's (1957) concept of fixed and stable organism assemblages. It also confirms Mills' (1969) idea that a variety of associations may occur from closely linked species groups to loose aggregations due to co-occurrence. Small-scale patterns of distribution in co-existing species may reflect the nature and degree of association in such groupings.

6.1.3.3 Geographic variation. A century ago it was commonly believed that no life existed in any ocean below 550 m. The idea, termed an 'azoic hypothesis', originated from samples taken by dredge in the Mediterranean in 1841 by Edward Forbes (Merriman, 1965). It was described by Spratt and Forbes (1847) as follows: "As we descend deeper and deeper in this region, its inhabitants become more and more modified, and fewer, indicating our approach to an abyss where life is either extinguished, or exhibits but a few sparks to mark its lingering presence." Collections of dredge samples by Sars, Thompson, and Carpenter off Norway and Scotland, the recovery of animals attached to cables raised from the Mediterranean and bottom samples collected during the voyage of the *Challenger* (1872–76) proved that animals lived at least to 5500 m depth (Mills, 1973). The fauna, although collected inefficiently, was diverse, widely distributed, and generally of small size with many carnivorous species. The conclusion that these features were adaptations to a limited food supply in the deep sea (Honeyman, 1874; Murray, 1895) has been supported by Thiel (1975). Reduction in average size of individuals of macrobenthic infauna is thought to reflect the shortage of organic matter in these environments.

Wherever light sufficient for photosynthesis reaches sediments or solid substrates, attached and epibenthic algae and macrophytes occur. Kelps are found below low-tide level on almost all temperate rocky shore-lines while sea grasses occur in more sheltered areas of fine sediment. Mangrove swamps are also extensive in tropical latitudes throughout the world. Organic matter produced by these plants in both particulate and dissolved form enriches all coastal waters and provides a detritus supply that may extend to the deep sea (Wolff, 1979). The importance of light, substrate, and nutrient supply for production of benthic microalgae will be considered below (Section 6.4.1.1).

The effect of depth on the distribution of numbers and biomass of benthic organisms has been shown by quantitative core and grab samples for collections of macro-, meio, and micro-organisms. Surveys by photographs, submersibles, trawls, and dredges, when made quantitatively, have been used to determine megafaunal numbers and biomass on the bottom at various depths. Thiel (1975) reviewed many previous studies of biomass present in deep sea sediments and concluded that both biomass and average size of metazoan organisms decreased with increasing depth. He attributed this change in size structure to a differential decrease in macrofauna biomass over that of meiofauna with increasing depth.

There have been few studies which provide quantitative estimates of biomass of micro-organisms and microfauna in sediments from a broad range of depths. Burnett (1981) calculated that 2 to 5 g/m^2 was present in surface sediments from 1200 m in the San Diego Trough as micro-organisms (excluding bacteria) and microfauna (up to 50 μm in size). Higher values (up to 100 g/m^2, assuming 1 cm^3 = 1 g) occur in intertidal and shallow coastal sediments in this size interval (Schwinghamer, 1981). Surveys of changes in meiofauna biomass over depth have been more numerous. Small decreases, relative to those observed from macrofauna, appear to occur with increasing depth. Numbers of individuals of meiofauna decreased by an order of magnitude (900 to 74/10 cm^2) between 800 and 4000 m off North Carolina, for example, where biomass decreased from 1·6 g/m^2 between 250 and 750 m to 0·1 g/m^2 between 2800 and 3200 m (Thiel, 1975; Coull *et al.*, 1977).

Large decreases in macrofauna biomass occur with increasing depth. For example, macrobenthic infauna off the Pacific coast of the U.S.A. varied from 20 to 200 g/m^2 (fresh weight) between 400 and 1200 m, while average values of 1·4 g/m^2 occurred at 2000 m, 0·2 g/m^2 at 2000 m and 0.02 g/m^2 at 4000 m (Vinogradova, 1962). Values between 0·01 and 0·1 g/m^2 have been reported for abyssal and hadal depths (Thiel, 1975) with lowest values in central oceanic regions. Zenkevitch *et al.* (1971) summarized the distribution of macrofauna biomass to show that maximum values occur on continental shelves with lowest values in central ocean gyre regions (Fig. 92). The regional differences generally correspond to the distribution of phytoplankton production on a global scale (Section 1.3.2).

Sample collections at different depths and distance along transects from shore (Rowe, 1971a; Carey, 1981) have confirmed a general trend of decreasing biomass with depth and distance from land. The rate of decrease in different areas reflects varying levels of organic supply from phyto-plankton production and horizontal transport of organic material off continental shelves adjacent to abyssal plains. High productivity in overlying water may also reach hadal benthic communities in deep ocean trenches. Jumars and Hessler (1976) attributed unexpectedly high standing crops of macrofauna in the Aleutian Trench to enhanced organic-matter supply.

Biomass of benthic megafauna (invertebrates and fish on or near the sediment surface) determined from trawls on the continental slope, rise, and abyssal plain in the western Atlantic south of New England varied from a maximum of 5·8 g/m^2 at 1300 m to less than 0·1 g/m^2 between 3900 and 5000 m (Haedrich and Rowe, 1977; Haedrich *et al.*, 1980). These authors noted, however, that distinct faunal assemblages with characteristic catch rates, diversity, and dominant species occurred on the shelf, upper, middle, and lower continental slope and rise and within transition zones between the slope and rise and lower continental rise and abyssal plain. The three major taxa present in trawl samples (decapod crustaceans, echinoderms, and fish) were not represented in similar proportions within different assemblages and biomass did not decrease in a regular way with increasing depth (Fig. 93).

The question of whether benthic fauna changes, qualitatively or quantitatively, in a continuous or discontinuous manner with depth has served as a focus for a variety of zoogeographic studies (Rex, 1981). The greatest change in faunal composition occurs on the outer margins of continental shelves at depths between 200 and 300 m possibly reflecting the boundary between shallow and deep water

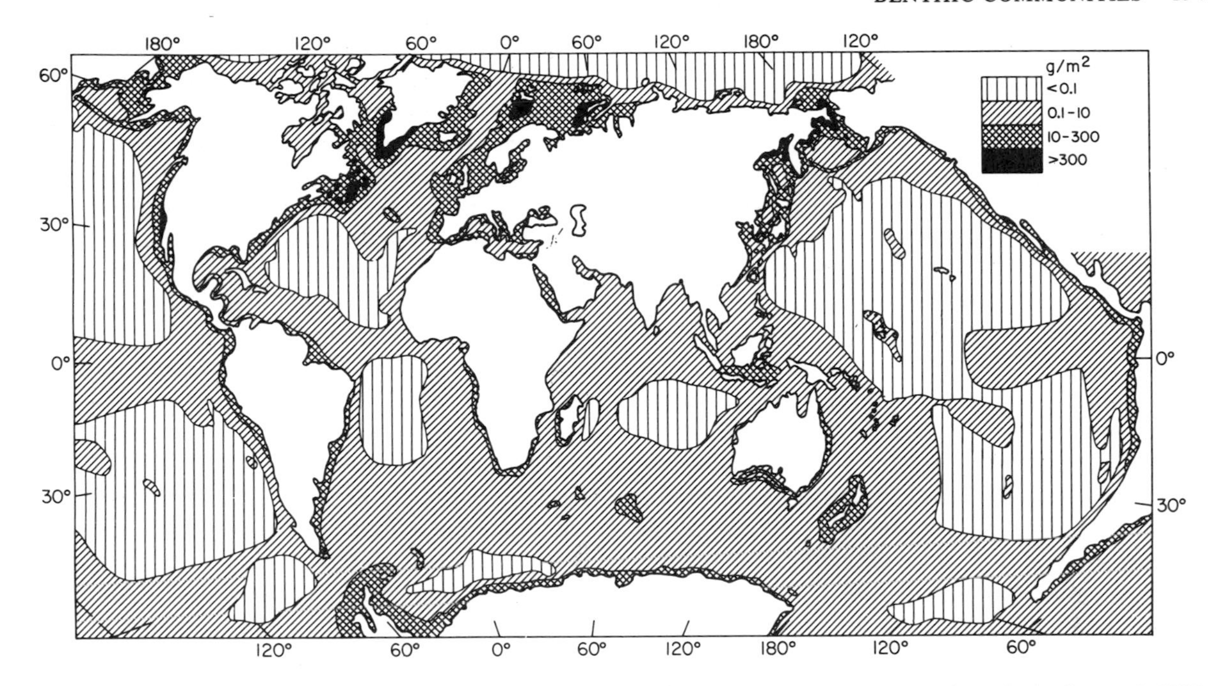

FIG. 92. Distribution of benthic macrofauna biomass (g fresh weight/m²) in the World Ocean (redrawn from Zenkevitch *et al.*, 1971).

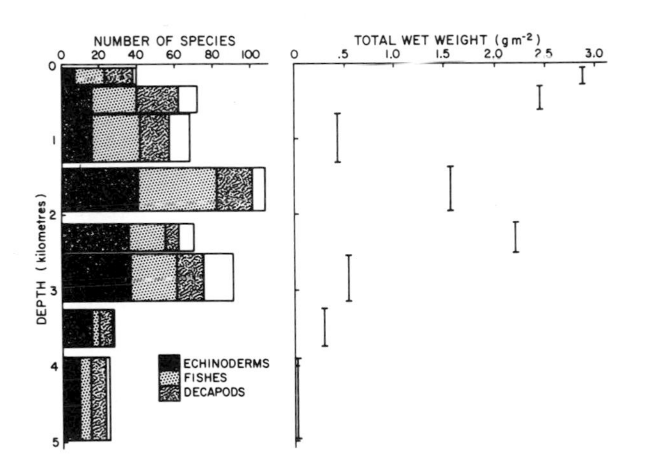

FIG. 93. Distribution of number of species and total biomass of benthic megafauna taken by bottom trawl samples at increasing depths from the continental shelf to the abyssal plain in the western Atlantic south of New England (redrawn from Haedrich *et al.*, 1980).

masses (Grassle *et al.*, 1979). Sediment texture, stability, and the relative amounts of assimilable organic matter in sediments and in suspension changes above and below the edge of a continental shelf and this should be reflected in changes in benthic community composition. At deep depths, the relative numbers of sessile and mobile species such as polychaetes which feed by filtration or burrowing could reflect variations in rates of organic matter supply in different areas (Fauchauld and Jumars, 1979). Sokolova (1972) differentiated eutrophic and oligotrophic oceanic regions on the basis of differences in macrobenthos feeding type. Collections of large benthic fauna allowed identification of eutrophic conditions in peripheral and equatorial regions. Conclusions that there is a dominance of deposit-feeding macrobenthos in sediments which contain excess assimilable organic matter are in part based on the large size of animals collected. Oligotrophic areas do occupy central areas of the three oceans where low concentrations of organic carbon <0·25% dry weight of more refractory sedimentary organic matter occur. A predominance of suspension-feeding species may not have appeared, however, if smaller-sized organisms (primarily deposit feeding species) had been collected.

Depth and the amount of food supply to benthic organisms are inversely related since decomposition removes utilizable energy during sinking of particles. Depth *per se,* or some factor related to it, seems to be the most critical overall factor limiting benthic biomass. Sanders and Hessler (1969) observed that changes in faunal composition over a vertical distance of 100 m were equivalent to those which occur horizontally over thousands of kilometers. Also, as mentioned above, abrupt faunal discontinuities tend to occur at shallow depths (200–300 m). These changes are usually not accompanied by large changes in total sediment organic content perhaps indicative that this is a poor measure of sediment nutritive value. Changes in fauna biomass below a few hundred meters depth, however, do generally correspond to the vertical distribution of other variables which decrease rapidly with increasing depth.

Several studies have shown that organic carbon and total particulate volume in the water column decrease rapidly in the upper 200 m but vertical distributions (plotted arithmetically) appeared constant with depth below 200 to 800 m (Menzel, 1974). Sheldon *et al.* (1972), however, observed that total particle concentration per unit volume decreased with depth when data were plotted on logarithmic axes. Wangersky (1976) and Gordon (1977) also compared numerous depth profiles of particulate carbon and nitrogen from different areas on this basis but the rate of decrease with depth was not quantified. Vinogradov (1968), Rowe (1971b), Rowe *et al.* (1974a), and Wishner (1980), on the other hand, used exponential and semi-logarithmic relations to linearize depth-dependent decreases in numbers and weight of macro-zooplankton and macrobenthos. Johnston (1962) also used a logarithmic curve to describe vertical distributions of fish and plankton biomass.

Vinogradov (1968) emphasized that generalized curves for vertical distributions of zooplankton biomass from the surface to maximum depths may transect several depth zones. Haedrich *et al.* (1980) observed rapid changes of megabenthic fauna across certain depth zones but there was not a continuous decrease with depth (Fig. 93). Despite these discontinuities, linearized profiles permit comparison of depth-related decreases. Hyperbolic, logarithmic, and exponential curves may describe depth profiles equally well. A power function, however, provides the best fit to depth profiles of different variables in many studies (Table 37). Even though only a few data points were used to calculate some curves, slope coefficients show that the rate of decrease of planktonic and benthic biomass is greater than that of variables related to suspended particulate matter. The sedimentation of fine particulate matter (as fecal pellets and detrital debris) (discussed in Section 6.4.1.2) is also a depth-dependent function. Data summarized by Suess (1980) show that 74% of the variation in vertical flux of organic carbon measured in various oceanic areas is attributable to variation in depth. When plotted as a power curve, the decrease in sedimentation with increasing depth occurs at a rate which is approximately double that for variables which measure standing stocks of non-living suspended particles (Table 37). The high values of slope coefficients for depth-related decreases in biomass of macrobenthos, and bathypelagic plankton imply that food sources for zooplankton and benthos in the deep sea may be similar (Rowe *et al.,* 1974a).

Sanders and Hessler (1969) observed that the abundance of macrobenthos decreased rapidly on the North Atlantic continental slope (the regime of greatest rate of depth change, Fig. 83), and thus depth rather than distance from land was inferred to determine food supply to benthic fauna. Differences in primary production in surface waters are thought to be small in this region and thus a lower food supply must account for reduced biomass with increasing depth. Rowe (1971a) also attributed higher average biomass of macrofauna in regions of higher phytoplankton production to a greater supply of organic matter. There may be a mid-slope region of enrichment to the benthos where frontal zones divide water over a continental shelf from offshore waters. Settling of products of primary production which are not consumed by pelagic herbivorous zooplankton (Iverson *et al.,* 1979) could enhance benthic secondary production in sediments directly below, or deposited material could be transported down slope to deeper water. Carey (1981) attributed higher biomass of benthic macrofauna in an abyssal plain area close to the

TABLE 37. INTERCEPT (a) AND SLOPE (b) COEFFICIENTS OF LINEAR REGRESSION CURVES COMPARING LOG_{10} DIFFERENT VARIABLES (Y) WITH LOG_{10} DEPTH (X)

Variable (Y)	Depth range (m)	No. of data points	a	b	Location and reference
Particle volume (ppm)	1–4500	47	0·63	−0·35	Various stations in North Atlantic and South Pacific (Sheldon *et al.*, 1971).
Total organic carbon (mg/l)	35–4000	6	9·33	−0·37	Central Pacific (Station Gollum) (Gordon, 1971);
Particulate organic carbon (μg/l)	25–3000	5[a]	105·4	−0·52	North Atlantic (Gordon, 1977);
	0–4000	6[a]	131·2	−0·51	Central Pacific (Gordon, 1971)
Particulate nitrogen (μg/l)	25–3000	5[a]	14·7	−0·56	North Atlantic (Gordon, 1977);
	0–4000	6[a]	48·2	−0·71	Central Pacific (Gordon, 1971);
	5–4000	5[a]	642·0	−0·69	South Atlantic, Pacific (Gordon, 1971)
	5–4500	5[a]	180·0	−0·48	(Wangersky, 1976).
DNA (μg/l)	25–3000	9	8·8	−0·55	North Atlantic (Gulf Stream) (Holm-Hansen *et al.*, 1968).
ATP (mμg/l)	10–1025	10	3810	−1·09	
	0–2400	19	10,177	−1·30	
'Living carbon' (ATP equivalent) as percent of total	10–1025	10	599	−0·84	California coast Peruvian coast (Holm-Hansen, 1970a). Various locations in Atlantic, Pacific and Baltic (Holm-Hansen *et al.*, 1966)
Organic carbon Sedimentation (g C $m^{-2}y^{-1}$)	13–5582	40	5889	−1·09	(Suess, 1980)
Net mesoplankton biomass (mg/m^3)	25–6000	8	5383	−1·21	Tropical Pacific (oligotrophic) (Vinogradov, 1968)
	25–7000	9	40,044	−1·39	Tropical Pacific (eutrophic); (Vinogradov, 1968)
	50–7300	9	62,332	−1·18	Southwest Pacific; (Vinogradov, 1968)
	25–6000	9	84,313	−1·16	Subarctic; (Vinogradov, 1968)
	25–3500	8	4150	−1·12	North Atlantic (Sargasso Sea) (Vinogradov, 1968)
Benthopelagic plankton biomass ($g/1000\ m^3$)	1000–4700	15	$1{\cdot}6 \times 10^6$	−1·91	Western Pacific, Northeast Atlantic (Vinogradov, 1968)
Macrobenthos biomass (g/m^2)	126–6225	9	13875	−1·12	Peru–Chile Trench (Frankenberg and Menzies, 1968)
	200–3900	8	36·7	−1·70	Gulf of Mexico, north Peruvian coast (Rowe, 1971a)
	200–6300	8	2547	−1·27	
	30–4950	22	62,098	−0·58	Northwest Atlantic shelf to abyssal plain
	550–2080	5	$1{\cdot}5 \times 10^7$	−1·24[b]	(Rowe *et al.*, 1974a)

[a]Means derived for depth intervals. [b]Continental slope stations only.

northwest Pacific coast to enhanced organic supply from the adjacent continental shelf.

There is often no significant correlation between depth and biomass or numbers of benthic fauna when sampling of benthic fauna is restricted to continental shelves (<200 m). Lie (1969) found that standing crop of macrobenthos off the coast of Washington was unrelated to depth, although mean biomass on the shelf was less than half that present in Puget Sound where phytoplankton production was considerably higher. Factors not necessarily related to depth affect biomass at shallow depths. Patterns of sediment distribution, hydrodynamic factors that influence sediment stability and interactions between tidal currents and the transport of suspended particles across the seafloor are not necessarily depth dependent over a shallow depth range on a continental shelf. However, these are critical variables which are correlated with the distribution and production of benthic organisms (Wildish, 1977; Warwick and Uncles, 1980).

The impingement of an oxygen minimum layer onto a continental slope, or reduced oxygen levels in deep basins with restricted circulation, may limit total biomass and species composition in benthic communities. The zonation of benthic meiofauna in the northern Baltic reflects the availability of dissolved oxygen in overlying water (Elmgren, 1975). Abundance and diversity decreased with depth with only nematodes tolerant of permanently reduced oxygen concentrations below 100 m depth. Sediments underlying upwelling areas off Peru are subject to oxygen stress and bacteria which utilize sulfides present in anoxic sediments are dominant organisms (Gallardo, 1977). High numbers of micro-organisms may occur in such areas due to their ability to use excess assimilable organic matter accumulated in anoxic sediments.

6.1.3.4 Temporal variation. Massé (1972) identified three time-related changes which affect the abundance and biomass of all benthic organisms:

(1) short-term changes correlated with altered hydrodynamic conditions or feeding activity of large organisms (physical disruption);
(2) seasonal changes related to reproduction and recruitment;
(3) long-term changes resulting from the successful recruitment of a previously non-abundant species.

Storm-induced water motion and excessive salinity or temperature changes can drastically alter community structure in near-shore areas (Glémarec, 1979). Such aperiodic disruptions, which maintain communities in an early successional state characterized by small organisms with a rapid growth rate, may be superimposed on seasonal and decadal patterns of population change associated with climatic variations. Regular seasonal changes in benthic species assemblages have been observed by Muus (1967) and Levings (1975) where the appearance of predators which utilized abundant prey species in the spring paralleled an increase in temperature. Bodiou and Chardy (1973) only observed seasonal changes in harpacticoid species groups at 10 to 15 m. No seasonal associations occurred at 20 m where abiotic factors had less variation.

Benthic communities may be physically or biologically controlled depending upon the degree to which abiotic factors vary. Temperature is considered to control the onset of breeding by invertebrates in marine boreal water (Thorson, 1966), but thereafter predation or interspecific competition for space and food supply may regulate species abundance and biomass. The importance of reproduction as a dispersal mechanism for benthic fauna is also demonstrated by the large number of species which have planktonic larval stages (over 80% in temperate and tropical areas with a slightly lower proportion in the Arctic and deep sea) (Thorson, 1950). Species with short pelagic lifetimes usually produce a few larvae from eggs rich in yolk (lecithotrophic) and recruitment does not vary greatly from year to year. This reproductive strategy is characteristic of cold-water species which demonstrate k-selection for maximum competitive advantage in a space or other resource-limited environment (Clarke, 1979). On the other hand, Vance (1973) has suggested that strategies of r-selection for maximum rate of population increase are desirable when food availability for pelagic larvae is unpredictable. Thus larvae with a long planktonic phase are usually produced in large numbers and survival depends on a variety of environmental factors.

Buchanan *et al.* (1974) classified species of dominant benthic macrofauna on the basis of fluctuations in abundance from year to year. They differentiated

(1) volatile species (alternate between high and low abundance),
(2) opportunistic species (high reproductive potential capable of rapid increase in numbers in response to elimination of previously abundant species), and
(3) conservative species (stable population numbers unaffected by fluctuations of volatile species).

Opportunistic species tend to be of small body size (<500 μm) which permits a rapid increase in numbers. Fenchel (1974) observed that the intrinsic rate of natural increase per day was inversely related to body weight for a variety of organisms. Thus, small organisms like bacteria and protozoa are capable of rapid increases in abundance when local conditions favour growth. Larger metazoans have much lower reproductive potentials.

Benthic fauna in the deep sea have been considered to reproduce throughout the year. However, several recent studies have provided evidence for synchronous annual reproductive cycles in deep sea mollusc, echinoderm, and fish species (Lightfoot *et al.*, 1979; Gordon, 1979; Gage *et al.,* 1980). Species in shallow-water temperate areas usually show a marked seasonality in reproductive activity. Bourget and LaCroix (1973), for example, observed that the settlement of benthic epifauna on hard substrates in a subarctic estuary occurred only during a short summer period after which peak abundance declined as a result of competition for space, sediment accumulation, and hydrodynamic factors. In contrast, constant conditions of temperature and salinity in the deep sea, deep-water polar regions and tropics results in benthic communities which are biologically accommodated. Seasonal variations in biomass are small compared to those occurring in temperate waters and changes which do occur may reflect predation effects which prevent monopolization of space by single species (Dayton, 1975).

Few studies have been of sufficient duration to quantify long-term changes which occur between years in benthic communities. Massé (1972) observed large differences in macrofauna abundance over 4 years at shallow depths in five areas of sand sediments in the Mediterranean. Differences were greatest for small-bodied short-lived species which appeared subject to more year-to-year variation in recruitment success. Communities dominated by large long-lived species demonstrated more constant biomass levels since reproductive success or failure in any one year has little effect on total biomass or numbers. Buchanan *et al.* (1974) documented changes in numbers of species, number of individuals and production of macrofauna in mud deposits off the Northumberland coast in the North Sea. While the total number of individuals more than doubled over 4 years, the total number of species and production was similar. Although proportions due to various species changed between years, biomass and production levels were maintained.

Continuation of these observations (Buchanan *et al.,* 1978) has shown that the broad features of the faunal associations at different stations have remained stable for more than a decade. There were changes in the pattern of relative species abundance which reflected the severity of cold winter sea temperatures. However, numbers of individuals of macrofauna demonstrated year to year coherence, increasing progressively towards an equilibrium set by a balance between density-dependent mortality and recruitment. Year to year changes were also observed in Puget Sound. Species composition of the fauna remain unchanged at four stations sampled 3 years apart, although biomass was reduced and the relative dominance of species varied (Lie and Evans, 1973). Rachor and Gerlach (1975) also concluded that despite frequent disruptions due to storms, species abundance of macrofauna has remained substantially the same at 25 m depth in the North Sea during the last 25 years.

Benthic species associations, biomass, and production may be drastically altered on a long-term basis by the feeding activity of a single species. Eagle (1975) noted that the presence of a single year-class in deposit-feeding species led to dramatic variations in abundance of other species. Predation by plaice on siphons of the mollusc *Tellina* can also reduce growth and interfere with reproduction (Trevallion *et al.,* 1970). The effect was greatest when mollusc abundance was high. A threshold in fish predation existed such that during years of low numbers, plaice consumed alternate food sources which permitted increased *Tellina* population growth. The functional relation between predator and prey thus induced a density-dependent recruitment. Mann (1977) has documented the long-term (decadal) decline in subtidal kelp *(Laminaria)* in a coastal bay due to extensive grazing by sea urchins *(Strongylocentrotus)*. Overfishing of lobsters, a natural predator of urchins, was thought to have allowed an increase in sea urchin populations. Grazing by large populations of urchins can completely destroy a kelp bed and even a small residual population may prevent recolonization by all but encrusting species. The longterm effect is to restrict kelp to exposed sites where water movement prevents sea urchin attachment while the urchins dominate large barren areas. These events may represent a successional phase in an unstable predator–prey oscillation or they may lead to a stable situation which only permits kelp growth in refuge areas.

It is useful to distinguish adjustment stability (some measure of how rapidly a community returns to its original state) and persistence stability (the constancy of community structure over time) (Margalef, 1969). For example, Peterson (1975) compared benthic macrofauna communities (primarily bivalve molluscs) in two lagoons over 37 months. The lagoon with the lowest species diversity was composed of species whose populations exhibited a high temporal variability (thus a low stability of component species) but the proportionate community composition of each lagoon varied to an equal degree through time (equivalence in stability of community composition). Stability at the species level and stability at the community level are thus not equivalent and the degree of difference is determined by the interdependence among species. The lagoon with the lowest species diversity was populated by species which had a high level of synchrony in their natural temporal fluctuations and thus changes in community composition over time were less than would be expected if all species fluctuated independently. Peterson suggested that community persistence stability was a more conservative community property than stability at the species population level.

This conclusion was supported by observations of macrobenthic infauna communities at 100 and 300 m off the Swedish west coast (Josefson, 1981). Communities at both depths showed high temporal stability over 5 years. Persistence stability was greater at the shallower depth where sediments were consolidated by large tube-dwelling polychaetes. Epifaunal species populations, particularly in shallow-water areas, may not be as stable as infaunal populations. Davis and Van Blaricom (1978) observed that significant changes in the temporal and spatial patterns of abundance occurred in a community of epifaunal invertebrates over 17 years on a shallow subtidal sand plain. The authors concluded that irregular recruitment was a major cause for long-term fluctuations in species abundance. They also emphasized that stability in a community of long-lived animals must be assessed over a period of time greater than the life spans of the dominant organisms.

6.2 CHEMICAL COMPOSITION

6.2.1 Sediments

Since benthic organisms affect and are themselves affected by the chemical composition of bottom deposits, it is useful to briefly consider the chemical characteristics of marine sediments as they relate to biological processes within benthic communities. In all but well-flushed coarse sediments, concentrations of biologically important nutrients (silicate, nitrate, ammonia, and phosphate) increase with depth to levels which are high relative to those in overlying water. The major transport process for exchange across the sediment surface is thought to be diffusion and models for diffusive flux have been proposed (Morse, 1974; Berner, 1980). Diffusion across a stagnant benthic boundary layer (up to 1 cm thick) may control flux into a layer where turbulent mixing occurs (Caldwell and Chriss, 1979).

There are several other processes which transport soluble nutrients and gases out of sediments in addition to diffusion. Waves may impinge on the bottom to cause bulk movement (pumping) of pore water (Riedl *et al.,* 1972). Density-driven displacement by water of higher salinity can also occur under certain hydrodynamic conditions. Thorstenson and MacKenzie (1974) and Smetacek *et al.* (1976) observed seasonal or episodic changes in dissolved ions in pore water or water trapped in a column over sediment which they inferred was due to flushing caused by gravity displacement. Aller

(1978, 1980) has also shown that animal burrows impart a three-dimensional gradient to concentration fields of dissolved ions in pore water. Since the shape of a vertical profile of dissolved nutrients in sediments arises from the combined effects of *in situ* reaction rates and bioturbation, diffusion rates in sediments where animal burrows are numerous must be calculated on the basis of horizontal as well as vertical gradients. Cylindrical bubble tubes, formed by rapid CH_4 production in organic-rich coastal sediments, may also lead to an abiogenic enhancement of vertical transport. Martens *et al.* (1980) observed that measured chemical fluxes of methane and ammonium were 3 times greater than those predicted on the basis of molecular diffusion in such deposits. They hypothesized that this was due to creation of additional exposed surface area and increased concentration gradients between overlying water trapped in tubes and deeper interstitial water enriched in dissolved CH_4 and NH_4^+.

Whatever the mechanism of release, where vertical transport is sufficient and regenerated nutrients enter the euphotic zone, they may be utilized for phytoplankton production. Zeitzschel (1980) concluded that in shallow-water areas up to 100% of the nutrient requirements for phytoplankton could be provided from this source. This contrasts Rittenberg *et al.*'s (1955) calculation that less than 1% of the annual inorganic nitrogen requirement of phytoplankton could be supplied by ammonia regeneration from sediments in a deep anoxic basin off southern California. They assumed that the difference in decrease between organic nitrogen and ammonia observed over depth in surface sediment with an estimated deposition rate was a measure of ammonia transferred from the sediment. This amounted to approximately 2 μM $NH_4^+/m^2/h$.

Rowe *et al.* (1975) and Smith *et al.* (1978, 1979) measured changes in dissolved nutrients directly in water trapped over sediments in a variety of shallow- and deep-water locations. Over 35% of nitrogen estimated to be required for phytoplankton production could have been supplied by flux out of the sediments on the continental shelf and off the Spanish Sahara, with a well-mixed water column. Nixon *et al.* (1980) measured a seasonal release of inorganic phosphorus from sediments in Narragansett Bay which was strongly correlated with temperature. Rates varied from near zero in winter to almost 60 μM/m^2/h in summer which on an annual basis was enough to support about 50% of the phosphorus required for phytoplankton production. Orders of magnitude differences can exist in nutrient flux in different areas, however, Smith *et al.* (1979) observed rates of ammonia release at two eastern North Pacific sites (10–20 μM/m^2/h) which were one to two orders of magnitude greater than rates measured in an oligotrophic area of the northwest Atlantic. Benthic nutrient regeneration may reflect the magnitude of input of products of phytoplankton production to sediments and the degree to which pelagic and benthic systems are coupled. The impact of benthic nutrient regeneration on primary production in surface waters depends on water-column depth and vertical mixing rates when levels of solar radiation are sufficient to permit phytoplankton growth.

The existence of high concentrations of dissolved inorganic ions in sediment pore water which is not rapidly exchanged with overlying water results from the metabolic activities of bacteria and other microorganisms. Baas Becking *et al.* (1960) identified pH and oxidation–reduction potential boundaries which are characteristic for various types of micro-organisms in natural marine sediments. pH generally decreases with depth. High values, in excess of 9, may occur in sediments in shallow water (Gnaiger *et al.*, 1978) due to photosynthesis of benthic microalgae while low values with depth reflect CO_2 production or sulfate reduction by bacteria. These processes may be highly stratified in sediments and distinct pH minima correspond to the redox discontinuity layer (Fig. 91). Depth related changes in pH vary with sediment type (ZoBell, 1946) but the range of values is restricted. Precipitation of sulfides prevents the occurrence of values below 6·9 (Ben-Yaakov, 1973) while the upper limit is controlled by buffering of the carbonate system.

Jørgensen (1977) has discussed the importance of H_2S formation by sediment sulfate-reducing bacteria (such as *Desulphovibrio*). The oxidation–

reduction potentials in sulfide-rich marine sediments may be controlled by the sulfide–sulfur half cell. Both Berner (1963) and Fenchel (1969) observed a linear relation between Eh and pS^{-} for a variety of natural H_2S-containing sediments and artificial sulfide systems. Jørgensen and Fenchel (1974) and Jørgensen (1977) found that 50% of the total mineralization of organic matter in sand microcosms enriched with *Zostera detritus* and in sediments of a Danish fjord was due to sulfate reduction. Initially most sulfide was precipitated as FeS but with time an increasing fraction remained in solution and diffused to the sediment surface where elemental and organic sulfur accumulated in bacterial plates. Both Hallberg (1968) and Jørgensen and Fenchel (1974) observed that insoluble sulfides could be formed at localized sites to produce reduce microniches within oxidized surface sediment. Thus, while oxidized and reduced sediment layers are generally separated vertically (Fig. 91), localized concentrations of organic matter in aerated surface layers create heterogeneous redox conditions which permit a close association of aerobic and anaerobic micro-organisms in oxidized sediment. Iron sulfides may be transformed by biological and inorganic processes to pyrite which is the major end product of sulfate reduction in reduced sediments (Howarth and Teal, 1979).

The presence of biological structures on the inner and outer surfaces of manganese nodules implies that accretion of manganese around basaltic fragments may be caused in part by micro-organisms. Some foraminifera species build tubes which incorporate manganese micronodules and at least one species may actively concentrate the element during test construction (Greenslate, 1974). Thus, whereas most planktonic foraminifera use calcium carbonate and radiolarian species use silica for test construction, some benthic species may use manganese.

Measures of dissolved gases in sediment pore water have been made with precautions against contamination and degassing during sampling. Martens (1974) confirmed earlier observations which showed that argon and nitrogen gases are present in surface sediments. Concentrations are near those of overlying water and they vary seasonally. Argon and nitrogen concentrations decreased with sediment depth below 25 cm while methane concentrations increased below this depth. The changes were thought to reflect mixing to 25 cm and the selective removal of nitrogen by stripping caused by release of bubbles of methane. Methane is usually only detectable in anoxic sediments when sulfate is not present. Martens and Berner (1974) observed that methane does not occur in anoxic sediments until about 90% of the sulfate is removed by sulfate-reducing bacteria. Anaerobic oxidation of methane at depths where sulfate is depleted may account for subsurface maximum in measures of sulfate reduction in anoxic sediments (Devol and Ahmed, 1981). Upward diffusion of methane and its production in sulfate free microenvironments could explain the co-existence of methane and sulfate in subsurface sediments.

Sediments within the photic zone are colonized by populations of benthic microalgae which photosynthesize and enrich interstitial water with dissolved oxygen. Heterotrophic respiration by organisms and chemical oxidation of reduced inorganic and organic compounds occurs in all sediments, however, and unless water movement maintains a high rate of supply, dissolved oxygen is only present within a thin surface layer (Revsbech *et al.,* 1980). Oxygen consumption and CO_2 production in sediments can affect concentrations of these gases in overlying water particularly when circulation of water over sediment is restricted. Low dissolved oxygen in stagnant marine basins reflects consumption at both the sediment surface and within the water column. The degree of oxygen depletion depends on basin morphometry (the total amount of oxygen present and the ratio of sediment surface-to-water volume), the rate of deposition of organic matter, and the amount of advective water exchange which can supply oxygenated water. Despite problems of interpretation, measures of oxygen uptake by undisturbed sediment cores are used as an index of metabolism by benthic communities. Relations between this measure and the total metabolic activity (both aerobic and anaerobic) in sediments will be considered in Section 6.4.5.

Organic matter exists in sediments as dissolved,

colloidal, and particulate material in concentrations per unit volume which exceed those in overlying water. As in sea water, organic compounds in sediments are complex mixtures which may be derived from terrestrial and aquatic sources and which may exist in either original or altered form. Studies of the organic origin, composition, and fate of sedimentary organic matter are of interest to chemists, geologists, and biologists and they form subject matter for the field of organic geochemistry.

Bordovsky (1965) reviewed biological and physical transformations which modify organic substances accumulated in marine deposits and Eglinton and Murphy (1969) edited a collection of papers which provides a review of the field. For example, Moore (1969) discussed changes which may occur through microbial decomposition of organic matter in sediments. Differences in the relative abundance of amino acids, carbohydrates and lipids reflect diagenic changes due to microbial utilization, transformation and solubilization. The identification of gross molecular composition, however, gives no information concerning rates of these changes. Some organic compounds are useful biogenic markers since their origin is known and they are preserved in deposits. Plant pigments (chlorophyll derivatives, carotenoids, and porphyrins), for example, can accumulate (Orr *et al.*, 1958) and these compounds must originate from deposited phytoplankton or macrophyte debris. Cyclohexanes and numerous aromatic hydrocarbons found in petroleum do not usually exist in organisms and their presence may indicate petroleum residues in surface sediments (Farrington and Meyer, 1975). Hedges and Mann (1979) have also applied methods to characterize plant tissues by their lignin oxidation products to identify gymnosperm wood and angiosperm tissues in sediments on the shelf and slope of the northwestern Pacific. These land plant remains were concentrated over a narrow depth range near the 100-m contour.

Waksman (1933) noted that humic substances present in many marine sediments were refractory to biochemical degradation. These compounds are high molecular-weight products condensed or polymerized from sugars, amino acids or polyphenolic and proteinaceous material. They may be derived from lignin of higher plants or synthesized by microbial activity. The organic polymers are converted to fulvic acids and precipitated to humic acids as hydrophilic groups are lost through the formation of complexes (Nissenbaum *et al.*, 1972). They are generally characterized by being acid- and alkali-insoluble and termed 'crude fibre'. These stable compounds can account for 30 to 50% of the total organic matter in sediments and the proportion may increase with depth indicative of sediment accumulation (Kato, 1956) or a terrestrial origin (Hedges and Mann, 1979).

A large proportion of humic-like compounds could account for the often observed lack of correlation between bulk organic content and benthic fauna biomass (Sanders and Hessler, 1969) since little of the accumulated organic matter in marine deposits may be available for metabolic utilization. Whitlach (1974), for example, used histological stains to determine the percentage of particles in salt marsh sediment which might be available to the deposit-feeding polychaete *Pectinaria*. Less than 0·4% of the particles stained with mercuric bromphenol blue indicative of protein-containing material. An average of 14% of particles were stained with periodic acid Schiff reagent specific for carbohydrate–protein complexes.

Particular organic matter in sediments is measured as the percentage weight loss on combustion (550 C) or by elemental analysis of carbon after removal of inorganic carbonates. Organic matter can constitute up to 30% of sediment weight in salt marshes or coastal areas where upwelling occurs and organic carbon can account for up to 50% of this amount. Open ocean sediments, however, usually contain less than 1% organic carbon (Bezrukov *et al.*, 1977). Organic matter in sediments under upwelling regions (off Peru) and in deep anoxic basins may exceed 10% in organic carbon (Reimers, 1982) indicative of preservation which occurs under reduced conditions. Nitrogen concentrations are an order of magnitude less than those of organic carbon since most marine sediments have C:N ratios in excess of 10.

An inverse relation is often observed between

sediment particle size and organic content which reflects the high surface area for organic adsorption on fine-grained deposits (Dale, 1974). Fine-grained sediments usually accumulate in depositional areas to form sediments rich in organic matter. Seasonal cycles of organic matter may occur in sediments, particularly in shallow water where vascular plant and algal production results in periods of increased organic input. Volkmann and Oppenheimer (1962) found organic matter in sediments of a shallow marine lagoon to be minimal (0·1%) and more refractory during winter. Maximum concentrations (3·7%) occurred in coarse sediments nearshore during summer possibly reflecting production by photosynthetic algae. Organic matter may be more or less concentrated with depth indicative of accumulation of refractory material, burial through animal reworking or depositional events. Both temporal and vertical differences may be less in offshore sediments due to the considerable decomposition which has occurred during deposition.

The relative proportions of dissolved, colloidal, and particulate organic matter present in sediments have not been extensively studied. The few observations of dissolved organic material which exist show that pore water may contain concentrations one to two orders of magnitude above those in sea water. High values (25–150 mg/l) have been reported in reduced sediments of an anoxic fjord (Nissenbaum *et al.*, 1972) and at the landward end of an estuary (Martens and Goldhaber, 1978). High concentrations immediately above the bottom and decreases with depth imply active transport and/or consumption across the sediment surface. Stevens (1967) considered the uptake of dissolved organic matter of low molecular weight as a nutrient source for various marine invertebrates. Glucose and twelve neutral and acidic amino acids occurred in interstitial water of a mud flat from trace amounts to concentrations of $2{\cdot}5 \times 10^{-5}$ M/l. Experiments with labelled compounds showed that even at these low concentrations the compounds were accumulated by several soft-bodied invertebrates. The relative metabolic importance of dissolved organic compounds for organisms in sea water and sediments remains unknown, however. Bacterial uptake and physical adsorption by surface-active organic and inorganic material should maintain low concentrations of these simple compounds. However, low molecular weight organic acids have been extracted from organically rich sediments (Miller *et al.*, 1979). Filters (usually 0·45 μm or 0·8 μm) are usually used to separate dissolved and particulate fractions. While operationally necessary, separation by dialysis membranes or molecular sieve filtration would be more informative for characterizing the degree to which compounds are actually in true solution.

6.2.2 Benthos

Studies of the chemical composition of macroscopic attached benthic algae, macrophytes, and detritus derived from these sources indicate that, in contrast to phytoplankton (Section 2.2), carbohydrates predominate other constituents. Over 50% of dry weight of *Spartina* exists as carbohydrates and while protein content of detritus derived from dead *Spartina* may reach 20%, living plants contain less than 10% protein on a dry-weight basis (Odum and de la Cruz, 1967; Hall *et al.*, 1970). Carbohydrates may also accumulate seasonally. Harrison and Mann (1975) observed that *Zostera* leaves lost organic carbon but not nitrogen during the fall. Himmelman and Carefoot (1975) also observed distinct late summer–winter maxima (up to 3·5 kcal/g dry weight) in calorific values of three seaweed species; however, the composition of major metabolites was not determined.

The relative proportions of fats, carbohydrates, and proteins and calorific content in detritus and benthic algae are of obvious importance to benthic consumer organisms. Paine and Vadas (1969) concluded that food preference reflected more the availability than the calorific value of benthic algae for a variety of invertebrate herbivores. However, nutritional quality does affect the rate at which calories can be obtained and feeding avoidance may occur when protein-deficient food sources are present (de la Cruz and Poe, 1975). Seasonal and geographic differences in chemical composition are also of importance for commercially harvested species of benthic algae. Polysaccharides like carrageenan may form up to 60% of the dry weight of species such as *Chondrus* with concentrations varying seasonally and in response to nutrient conditions. Buggeln and Craigie (1973) have reviewed aspects of the biochemical composition of this commercially important species.

Fatty acids and hydrocarbons, particularly n-alkanes with odd-carbon-number chains, are synthesized by both plants and animals. The specificity of these compounds for a particular class of phytoplankton has been mentioned (Section 2.2). Jeffries (1972) also observed that saltmarsh grasses have a terrestrial pattern of fatty acids rich in sixteen to eighteen carbon chains while fish (*Fundulus*) and a detritus-feeding shrimp (*Palaemonetes*) have a marine pattern dominated by long-chain poly-unsaturates. The relative distribution of various fatty acids in detritus and digestive tracts of these organisms was used to infer the proportion of detritus and animal tissue in their diet. Ackman (1965) noted the occurrence of an unusually high proportion of odd-numbered fatty acids in the detritus-feeding mullet and suggested a dietary source from algae. However, differences in lipid and fatty-acid composition characteristic for two species of oysters could not be substantially altered by feeding with phytoplankton containing different fatty acids (Watanabe and Ackman, 1974). Thus, although some hydrocarbons may not be modified during passage through a food web (Farrington and Meyer, 1975), some species rapidly convert unusual fatty acids to a species-specific composition.

The proximate composition of benthic crustaceans generally corresponds to that of zooplankton (Section 2.3). Water usually accounts for 70 to 80% of wet tissue weight which is slightly lower than values for planktonic animals (Table 11). Lipid content reflects age, reproductive, and feeding conditions but values seldom exceed 15% of dry weight (1 to 3% wet weight). Ansell (1975) found that carbohydrate (mostly glycogen) and lipid content in the bivalve mollusc *Astarte* decreased during winter following spawning. Values for tissue carbohydrate (15%) and lipid (5·9%) were maximum prior to spawning during late summer when calorific content was also maximum (4·85 kcal/g). Moore (1976), on the other hand, observed high lipid content (20 to 25% of dry weight) in five species of benthic crustaceans and neither this nor ash content (20%) varied seasonally. Optimal feeding conditions may have caused the accumulation of lipids in these species since pronounced seasonal changes in lipid content (from 15% of dry weight to a maximum of 26% in spring coinciding with reproduction) occurred in a species of Arctic amphipod (Percy, 1979). Ackman and Cormier (1967) observed that lipid reserves decreased from 4·8 to 1·2% (wet weight) during 2 weeks of starvation of the periwinkle *Littorina*. Ansell and Sivadas (1973) also noted that starvation resulted in rapid losses of carbohydrate, lipid, and protein in the bivalve *Donax*. These changes in tissue biochemical composition imply that the relative proportions of the three organic fractions indicate the nutritional state of organisms.

Concentrations of phytoplankton pigments and derivatives in benthic fauna have been determined to measure the utilization of algae by bottom-feeding invertebrates (Fox *et al.,* 1948; Ansell, 1975). Absorption and fluorescent spectroscopy were used to compare spectral patterns in tissue extracts and phytoplankton thought to serve as a food source. Caution is necessary in interpreting spectra, however, since material synthesized by invertebrates themselves may interfere. An alternative approach to identifying energy sources for benthic invertebrates has been to measure the potential for enzymatic digestion. Hylleberg (1972) compared the carbohydrase activity in twenty-two species of marine benthic invertebrates. Hydrolysis of laminarin, glycogen, and amylose occurred in all species but enzyme activity was greatest in crustaceans. Little hydrolysis of oligosaccharides or structural polysaccharides occurred. Differences in the relative activity of carbohydrases between three co-existing species of deposit-feeding molluscs, however, could indicate a quantitative enzymatic response to divergent food sources (Hylleberg, 1976). Polysaccharides such as cellulose may by hydrolyzed by cellulase enzymes present in some benthic species. Wildish and Poole (1970) identified the carboxymethyl–cellulase component of cellulase in the hepatopancrease of the amphipod *Orchestia*. At least part of the cellulase activity was attributed to the presence of symbiotic bacteria. Foulds and Mann (1978) concluded that *Mysis* which assimilated ^{14}C-labelled cellulose contained micro-organisms in its gut which permitted digestion of this otherwise refractory compound.

Separation of proteins from tissue homogenates using gel electrophoresis serves as a tool for quantifying the amount of genetic variation in natural populations. Selective enzyme assays may be used to identify activity associated with particular protein bands on electrophoretic gel. Banding patterns are characteristic for each species and they can be used to test for genetic differences between populations. Tracey *et al.* (1975) used these methods to confirm that lobsters (*Homarus*) from various areas off the eastern coast of North America were genetically similar but subdivided into isolated inshore and offshore populations.

Differences in particular protein (or enzyme) banding patterns may also occur within a species, however, and the variation can be used as a measure of genetic diversity (polymorphism). Doyle (1972) observed the frequency of a polymorphic esterase isoenzyme in the brittle-star *Ophiomusium* in the deep sea using these techniques. Variability of the esterase isoenzyme was high contrary to the idea that selectively significant genetic variation should decrease with depth when physical conditions become progressively more stable. Banding patterns, assumed to represent gene frequencies, were similar over horizontal distances up to 200 km along isobaths but varied considerably over 8 km in a direction normal to this. Differences in the frequency of occurrence of the isoenzymes across isobaths showed that polymorphism of this enzyme system is depth related. Similar investigations of genetic polymorphism at different enzyme loci in eight species of *Macoma* and the mussel *Mytilus* showed that environmental heterogeneity rather than temporal environmental variability may select for polymorphism (Levinton, 1975; Gartner-Kepkay *et al.,* 1980). Genetic variability thus exists in organisms from the deep sea as well as those from shallower depths although the selective value of variation in esterase isoenzymes is unknown.

Concentrations of nucleic acids (RNA, DNA) have been used to measure growth in cell suspensions made from a variety of organisms. Sutcliffe (1969) observed a direct relation between growth rate and RNA concentrations in a variety of invertebrates. The relationship implied that just as chlorophyl concentration might be used as a measure of relative photosynthetic potential in phytoplankton, the ratio and absolute concentration of RNA and DNA might measure the capacity of cells for protein synthesis. Pease (1976) reviewed evidence that nucleic acid concentrations and growth rates of organisms are related. He compared RNA/DNA and protein synthesis in oysters during one year. The RNA/DNA ratio was directly related to growth rate for a specific oyster year class and to the rate of protein synthesis during the preceding month. Uniformity in the ratio for adductor muscle from all sizes of oyster was thought to reflect the constant composition and function of this tissue. Differences in ratios derived from whole-body analyses could have been due to the presence of gonads at certain times in specific size classes.

ATP, which does not exist outside of living organisms, has been used to measure biomass of benthic organisms in sediments. While the ATP content of bacteria and algae depends on nutritional state and metabolic activity, average content in a variety of aquatic micro-organisms is approximately 0·4% of cell carbon (Hamilton and Holm-Hansen, 1967). Comparison of ATP and total sediment carbon in surface salt-marsh sediments showed that microbial biomass would account for less than 1% of organic carbon and values decreased with depth (Christian *et al.*, 1975). The pool of non-living detrital carbon was very large relative to that present in organisms in these deposits. In contrast, Yingst (1978) observed that large numbers of newly settled bivalves and juvenile polychaetes in the topmost centimeter of sediment in Long Island Sound during summer and fall accounted for the total ATP present in bulk sediment extractions. Standing stocks of micro-organisms and meiofauna in subsurface layers could be determined by ATP measurements, but most of the living biomass at the surface was not in these small organisms. Yingst reviewed estimates of

ATP concentrations in a variety of aquatic sediments. Concentrations of 0·1 to 4 μg ATP/ml wet surface sediment have been measured in shallow-water areas while samples from 200 to 6000 m contain an order of magnitude less.

Measurements of elemental concentration in marine organisms often show that concentrations reflect those present in the seawater or sediment from which the organisms were obtained (Section 2.2). Stable isotopic ratios, such as $^{13}C/^{12}C$, have been thought to be useful as an indicator of the origin of carbon derived from plants which differ in pathways of carbon assimilation. The slightly different weights of these carbon isotopes lead to a specific fractionation during photosynthesis which is dependent upon pathway of carbon fixation. Phytoplankton possess a C_3 (phosphoglyceric acid) pathway which results in a $\delta^{13}C$* value between −18 and −24‰ whereas in *Spartina*, a plant with a C_4 carbon fixation pathway, values are much lower (<15‰) (Fig. 94). Since animals do not appear to alter the $^{13}C/^{12}C$ ratio of material they ingest and assimilate, the ratio in their tissues may be assumed to reflect their primary food source. Haines and Montague (1979) have compared values of ^{13}C in various organisms from a Georgia salt marsh and attempted to differentiate food pathways for consumer organisms derived from *Spartina* and phytoplankton.

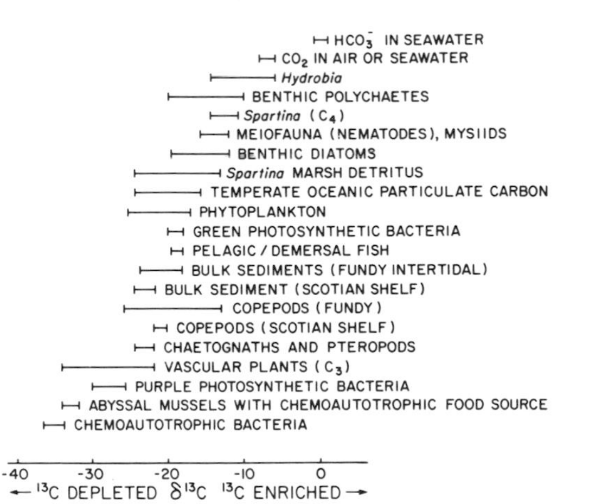

FIG. 94. $^{13}C/^{12}C$ ratios, expressed as $\delta^{13}C$ values, in microbial, plant, and animal tissues from various intertidal, subtidal benthic and pelagic areas. Data summarized from Schwinghamer et al. (1983); Peterson *et al.* (1980); Mills *et al.* (1982).

While there is a clear difference in ^{13}C ratios in *Spartina* and plants possessing the C_3 pathway of carbon fixation, there is considerable overlap of values in various organisms (Fig. 94). Peterson *et al.* (1980) have pointed out that ratios will be reduced in environments where bacterial chemosynthetic fixation of CO_2 occurs since carbon dioxide in interstitial water arising from decomposition of organic matter is depleted in ^{13}C. Rau and Hedges (1979) observed such a depletion in tissues of mussels collected from deep sea volcanic thermal vents where chemosynthetic bacteria are likely to be a food source for filter-feeding invertebrates. Values typical of salt-marsh detritus and sediment (15–20‰), which are intermediate between values for *Spartina* and plants with C_3 carbon pathways, could reflect composite values due to mixtures of bacterial and *Spartina* carbon. It is likely that benthic microalgae, phytoplankton, macrophytes, and terrestrial vegetation combine with above and below ground products of *Spartina* production to supply organic matter to salt-marsh ecosystems. All of this material is subject to bacterial utilization. When this occurs through chemosynthesis, ^{13}C ratios will be altered from those in source material and food sources for consumer organisms can not be differentiated.

6.3 BENTHIC AUTOTROPHIC PROCESSES

6.3.1 Photosynthesis and Production by Benthic Macro- and Microflora

Macroscopic and microscopic algae and photosynthetic bacteria develop extensive populations in and on sediments and solid substrates which receive light. Highest biomass and production are observed in salt marshes, intertidal sediments, and in littoral areas up to 20 to 30 m deep where clear water permits adequate light penetration. Although production by benthic macro- and microalgae may

* $$\delta^{13}C = \left(\frac{^{13}C/^{12}C \text{ sample}}{^{13}C/^{12}C \text{ carbonate standard}}\right) - 1 \times 1000.$$

occur in benthic communities to the compensation depth for phytoplankton (approximately 1% of surface radiation), the significance and magnitude of this source of production relative to that of phytoplankton decreases as depth increases.

The potential for production by benthic microalgae may be inferred by high cell counts and pigment concentrations which occur in some intertidal sediments. Plant pigments in sediments in deep water are usually present as accumulated degradation products (pheophytins, pheophorbides) arising from settled phytoplankton or macrophyte debris. Shallow-water sediments, however, may also contain a variety of undegraded pigments (chlorophyll *a* and *c*, diatoxanthin, diadinoxanthin, fucoxanthin, carotene, phycocyanin, and bacterial chlorophylls from green and purple sulfurbacteria, Fenchel and Straarup, 1971). Absorption spectra of intertidal sediments extracted with organic solvents or water (Fig. 95) indicate differences in peak absorption for cells of different taxonomic groups and they demonstrate the difficulty of distinguishing between algal and bacteriochlorophylls. Absorption spectra cannot be used to quantify pheophytin pigments because of spectral similarities to undegraded chlorophyll. Sedimentary chlorophyll concentrations are thus usually calculatd by measuring light extinction at 665 nm and assuming a specific absorption coefficient or by relating extinction to arbitrary units. Chromatographic separation permits the identification of specific pigments and spectrofluorimetric techniques have been used to quantify chlorophyll *a* and pheophytin in sediments. High-pressure liquid chromatography may also be used to quantitatively identify undegraded pigments from breakdown products which are common in sedimentary material (Brown *et al.*, 1981).

The vertical distribution of benthic algae and bacterial photosynthetic activity corresponds to the marked attenuation of light which occurs in the upper few millimeters of sediment. Perkins (1963) observed that moist mud 1·0 and 2·0 mm thick reduced light intensity to 2% and 0·4% of incident illumination. One per cent of red and infrared light penetrates to 4 and 4·8 mm respectively in rinsed quartz sand while penetration only occurs to 3 mm in undisturbed estuarine sediment (Fenchel and Straarup, 1971). These authors also found bacteriochlorophyll *a* (present in purple sulfur-bacteria) down to 6 mm depth with maximum concentrations between 2 and 3 mm. Decreases in

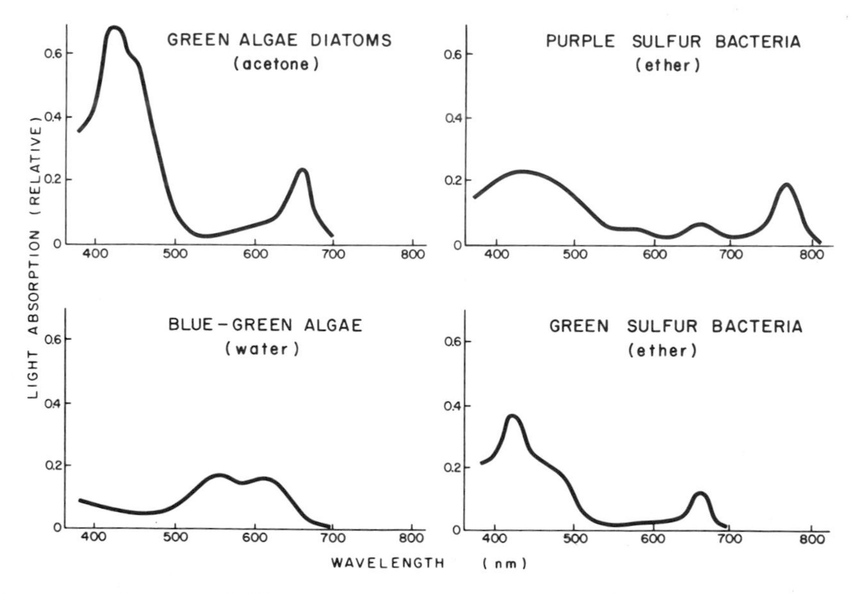

FIG. 95. Absorption spectra of extracts of different groups of sediment microalgae (redrawn from Fenchel and Straarup, 1971).

sulfide concentration due to these photosynthetic bacteria were restricted to the upper 3 mm of sediment cores (Blackburn *et al.*, 1975). Changes in fine-scale vertical profiles of dissolved oxygen also occur when sediments which contain microalgae are darkened (Revsbech *et al.*, 1980). The changes were restricted to the upper 6–8 mm at the sediment surface showing that photosynthetic activity was restricted to this depth.

Attenuation coefficients for light in sediments depend on grain size and wavelength. Long-waved light penetrates deeper than short-waved radiation and fine-grained deposits attenuate light more than coarse sediments. Taylor (1964) and Fenchel and Straarup (1971) observed attenuation values between 0·52/mm and 1·39/mm. Unlike the water column where light extinction is primarily a function of phytoplankton and other suspended matter, non-biological absorption (reflection, refraction, scattering, and light absorption) by sediment particles accounts for the rapid attenuation of light with sediment depth. Fenchel and Straarup (1971) estimated that a maximum of 20% of infrared, red, and blue light absorption could be attributed to plant pigments in intertidal mudflat sediments. The actual efficiency of light absorption depends on the abundance and vertical stratification of photosynthetic organisms within surface sediment layers.

Values up to 100 mg chlorophyll a/m^2 are common in intertidal sediments but the amount usually decreases with water depth and increasing amounts of wave action (exposure). Moss (1968) reviewed measures of chlorophyll *a* content in algae attached to rock ('epilithic'), sand grain ('epipsammic'), and larger plant ('epiphytic') surfaces. Although concentrations up to 2000 mg/m^2 may occur under conditions of high nutrient supply and on stable substrates, values from 10 to 100 mg/m^2 are typical for a variety of surfaces. Fenchel and Straarup (1971) noted the similarity of these values to those observed for standing crops of phytoplankton when integrated per square meter for the photic zone. These values are below theoretical maximum chlorophyll *a* concentrations in a water column of 400 to 800 mg/m^2 estimated to attenuate light to 1% of surface values (Steeman Nielsen, 1962). This is not surprising since sediment particles absorb and scatter most of the incident radiation. When physically stable conditions exist, however, photosynthetic organisms grow in thin layers, concentrated so as to maximize the efficiency of light absorption. In these cases, the microorganisms do not compete with their substrate for light and much higher chlorophyll concentrations may be achieved.

Photochemical processes during photosynthesis depend on light absorption by photosynthetic pigments while enzymatic conversions of organic compounds produced depend on enzyme concentrations and temperature. Photosynthetic organisms adapt to varying conditions of light and temperature by altering the relative proportions of photosynthetic enzymes and pigments. These effects of light and temperature on photosynthesis have led to the development of two different methodologies to estimate production by benthic microalgae. *In situ* light and dark exposure is used with undisturbed sediments to estimate gross and net production (Pamatmat, 1968; Joint, 1978) while incubation of sediment samples removed from specific depth layers and stirred under artificial light provides a measure of potential production when corrected for chemosynthetic uptake of $H^{14}CO_3^-$ (Gargas, 1970). *In situ* light measurements are used to extrapolate these measurements to rates expected under field conditions. ^{14}C-techniques have also been used to estimate production of macrophytes (UNESCO, 1973). These methods offer sensitivity when production is low, and when samples are mixed during incubation, production by attached and free-living algae can be separated. The light response and vertical distribution of microalgal production can also be directly measured by these methods. Measures of oxygen exchange across undisturbed sediment surfaces in the light and dark, on the other hand, quantify phototrophic and heterotrophic activity under *in situ* conditions. Hunding and Hargrave (1973) compared estimates of benthic primary production on a sandy beach measured under *in situ* conditions with laboratory ^{14}C methods. Both techniques gave similar measures of the magnitude of production.

Production by marine macrophytes (seaweeds,

sea grasses) which grow between high water and a depth of approximately 30 m has been assessed by harvesting (cropping) to estimate seasonal increments in biomass (Bellamy *et al.*, 1968). Tissue growth of the kelp *Laminaria* can be measured directly by punching holes in blades and measuring their relative position with time (Mann, 1972b). Rapid tissue growth at the base of blades compensated for erosion from the tip and thus measures of standing crop alone would have underestimated production which had occurred. Comparison of estimates of annual primary production by macrophytes and benthic microalgae (Table 38) indicates that carbon fixed per unit area in these communities may equal or exceed phytoplankton production. Phytoplankton production predominates as a source of synthesis of new organic matter in the open ocean but macrophytes and benthic microalgae make significant contributions in coastal regions. Only a fraction of macrophyte production may be grazed directly. Almost all of it is thought to enter the water

TABLE 38. PRIMARY PRODUCTION BY BENTHIC MICROALGAE AND MACROPHYTES[a]

Location/Source	Technique	$g\ C/m^2/yr^a$	Author
Benthic microalgae:			
Georgia salt marsh	O_2, CO_2	200	Pomeroy (1959)
Delaware salt marsh	O_2	38–99	Gallagher and Daiber (1973)
California salt marsh	O_2	217–400	Zedler (1980)
Massachusetts salt marsh	^{14}C (shaded)	106	Van Raalte *et al.* (1976)
	(unshaded)	165	
Intertidal sandflat	O_2	143–226	Pamatmat (1968)
Intertidal sandflat	O_2	0–325	Riznyk and Phinney (1972)
Intertidal sandflat	^{14}C	4–9	Steele and Baird (1968)
Intertidal mudflat	^{14}C	31	Leach (1970)
Estuarine subtidal	^{14}C	116	Grontved (1960)
	^{14}C	90	Marshall (1970)
	^{14}C	180	Joint (1978)
Wadden Sea sand flat	^{14}C	58–177	Cadée (1980)
Sea and marsh grasses:			
Thalassia beds	O_2	520–640	Westlake (1963)
Spartina (Georgia)	O_2	257–897	Teal (1962)
(North Carolina to Nova Scotia)	cropping	130–256	Mann (1972b)
(Massachusetts)	cropping	1100–2300[b]	Valiela *et al.* (1976)
Mangrove swamps:			
Florida (net prod.)	O_2 (+ litter)	400	Mann (1972b)
Kelps:			
Laminaria (Nova Scotia)	cropping	1900	Westlake (1963)
(England)	cropping	1225	Bellamy *et al.* (1968)
(Nova Scotia)	blade renewal	1750	Mann (1972b)
Macrocystis	cropping	400–820	Clendenning (1972)
Littoral seaweeds:			
Fucus	O_2	<3000[c]	Kanwisher (1966)

[a]Only studies which provide annual estimates are included. Mann (1972b) and Van Raalte *et al.* (1976) provide extensive comparisons of macrophyte and microalgal production rates respectively in different areas.

[b]Includes above- and below-ground biomass and assumes 35% by weight as carbon.

[c]Calculated from hourly rates as net oxygen production by a dense stand of fronds covered by filamentous brown and red seaweed epiphytes.

column as dissolved and particulate organic matter and some fraction may be incorporated in detritus-based food webs within sediments. Vascular plant detritus in salt marshes can be accumulated and decomposed locally and/or transported offshore. The relative importance of these pathways for detritus utilization may vary in different areas (Haines, 1977; Peterson *et al.*, 1980).

High levels of productivity by emergent and submerged macrophytes and microalgae in intertidal and shallow neritic areas, which per unit area are among the most productive plant communities in the world, are thought to be sustained by a high rate of nutrient supply (Mann, 1973). Wind, estuarine, and tidal mixing ensure a supply of nutrients by upwelling of deep water and by vertical transport of nutrients regenerated from sediments. The importance of nutrients for production by benthic plants was demonstrated when inorganic salts added to stimulate phytoplankton production were found to enrich sediments and cause increased growth of benthic microalgae (Marshall and Orr, 1948). The addition of sewage sludge, urea, and other nitrogen fertilizers enhanced production of salt-marsh grasses 2- to 3-fold (Valiela and Teal, 1974). These and other studies (Valiela *et al.*, 1976) have established that organic nitrogen, rather than nitrate or phosphorus, is responsible for increased production. Phosphate and nitrate may be present in sediment pore water in concentrations sufficiently high so as not to be limiting. Concentrations may exceed those required for optimal growth, however. Admiraal (1977) observed that ammonia concentrations higher than 0·5 mg-at N/l were inhibitory for photosynthesis by benthic diatoms.

Nutrient supply is not the only determinant of benthic microalgal production. Enrichment of salt-marsh sediments did not increase chlorophyll concentration per unit area (Estrada *et al.*, 1974) and production by benthic microalgae was only increased by fertilization during early spring before a canopy of grasses (*Spartina, Distichlis*) had formed (Van Raalte *et al.*, 1976). Light availability and desiccation were thought to be more important than nutrient supply for production of salt-marsh algal mats. Pamatmat (1968) and Hartwig (1978) also concluded that incident radiation was of primary importance in determining seasonal patterns of production by microalgae on an intertidal sandflat and subtidal sediments although other factors (tidal exposure, resuspension and temperature) also significantly affected production rates.

Measurements to quantify the effect of light intensity on benthic primary production have established two characteristic responses: benthic phototrophic organisms demonstrate maximum photosynthesis over a wide range of light intensities, thus light saturation may occur at very low light intensities (curve type 1, Fig. 36) and inhibition of photosynthesis seldom occurs even at high light levels. Burkholder *et al.* (1965) observed that planktonic algae generally show lower I_k values (discussed in Section 3.1.3) than benthic algae, but seasonal changes due to adaptation and differences in species composition make comparisons difficult. Gargas (1971) observed that I_k values for benthic microalgae decreased from 21 klx during summer to 4 klx in winter which followed changes in incident light. Also, I_k values for algae at various depths in sediment cores were equivalent to those for surface populations and they similarly declined with seasonal reductions in illumination. This implies that benthic algal cells may not remain stratified at one depth or that maximum photosynthesis may be maintained by prevailing light intensities even at low light levels.

Littoral seaweeds and benthic algae reach photosynthetic saturation at light intensities between 10% and 50% of incident levels (Brinkhuis *et al.*, 1976; Dawes *et al.*, 1978). Taylor (1964) observed maximum carbon fixation by benthic diatoms at 12 g cal/cm^2/hr (14% of incident radiation). Other studies have shown that linear correlations exist between *in situ* production by benthic microalgae and light intensity with no indication of saturation of photosynthesis (Van Raalte *et al.*, 1976). The absence of photoinhibition, even under full sunlight, and the concentration of microalgal cells at the sediment surface implies that, as in seaweeds, growth of benthic microalgae may proceed until reduced by self-shading. Adaptation for maximum utilization of light is demonstrated by estimates of photosynthetic efficiency. Values may reach 3% of energy in incident light photosynthetically fixed when light intensity is less than 0·15 g cal/cm^2/min but lower efficiencies (0·1% to 1%) characteristic of phytoplankton occur at high illumination (Pamatmat, 1968). Pamatmat also noted diurnal changes in photosynthetic efficiency with higher values tending to occur at low light levels.

Marked vertical stratification and zonation of benthic algae and bacteria can in part be related to changes in the vertical distribution of light quality as well as quantity. *Chondrus* exhibited maximum photosynthesis under red light (Mathieson and Prince, 1973). Photosynthetic sulfide oxidation was also stimulated by long wavelengths and red light had a greater effect than infrared light (Blackburn *et al.*, 1975). This would be expected if green sulfurbacteria (peak absorption near 650 nm) were more important in oxidizing sulfide than purple sulfurbacteria (peak absorption near 780 nm) (Fig. 95). Maximum concentrations of bacteriochlorophyll *a* also occurred at 2 to 3 mm depth in estuarine sediment which corresponded to the depth at which red and infrared light decreased to 1% of surface values (Fenchel and Straarup, 1971). The observations show that some photosynthetic organisms may concentrate at depths which maximize the efficiency of light absorption or at least correspond to the depth of penetration of light characteristic for specific photosynthetic pigments. Benthic diatoms are also motile and capable of vertical migrations to concentrate in layers only a few cells thick at the sediment surface (Brown *et al.*, 1972). Such movements expose cells to maximum light intensities. Expected time-related changes in profiles of chlorophyll concentration in intertidal sandflat sediment due to vertical migration of cells during the day have not been observed (Pamatmat, 1968).

Calculation of net photosynthetic production requires that respiration be known. While this is usually assumed to be light-independent, there is increasing evidence for photorespiration during photosynthesis (Tolbert, 1974). Respiration appears to be replaced by a different process of CO_2 production during photosynthesis in some species of marine benthic algae (Brown and Tregunna, 1967). Net photosynthesis cannot be calculated by assuming a light-independent respiration rate if respiration is inhibited during photosynthesis. Respiration (as dark O_2 uptake) is often assumed to be between 10% and 30% of gross photosynthesis. The compensation point at which photosynthesis and respiration are equal for diatoms is about 0·3% of midday surface radiation (Taylor, 1964). Thus, cells at 4 mm depth in fine sand sediment are near their compensation point while cells at 2 mm depth photosynthesize at more than 90% of their maximum capacity. The compensation point for the seaweed *Chondrus* occurs at less than 0·5% of full sunshine and this permits photosynthesis to occur under water of 20 m depth on a sunny day (Mathieson and Prince, 1973).

Calculations of annual primary production by ^{14}C or O_2 methods are usually based on experiments of a few hours' duration on different days throughout the year. Diurnal changes are thus estimated by extrapolation from hourly rates. Burkholder *et al.* (1965) used a relationship between light and photosynthesis to estimate daily production by benthic algae. Pamatmat (1968) calculated multiple linear regressions and substituted average daily values for variables which accounted for a significant part of the variation in measures of oxygen to make annual estimates. Pamatmat's measurements demonstrated an endogenous tidal rhythm in oxygen uptake and photosynthesis in settled intertidal sand samples held in the laboratory in continuous darkness or in alternating periods of light and darkness. Rates of photosynthesis and respiration were in phase and depressed during times which corresponded to low and high water in the field. Increased rates occurred during flood and ebb tidal periods. The maximum variation in photosynthesis (10 to 55 ml O_2/m^2/hr) and respiration (3 to 21 ml O_2/m^2/hr) corresponded to approximately 50% of the variation in values observed seasonally when bell jars were used to cover undisturbed sediments during tidal exposure. Different time-related changes have been observed in other studies. Beyers (1963) repeatedly found an evening burst in oxygen uptake in laboratory sediment–water microcosms even though they were exposed to continuous light. Gallagher and Daiber (1973) did not observe any diel variation in respiration of sediment cores from a *Spartina* salt marsh but an endogenous photosynthetic rhythm with a 10-fold amplitude and maximum oxygen production at midday existed under constant light.

Variation of oxygen production by benthic algae incubated under *in situ* conditions over 24 hr occurs because light is necessary for photosynthesis. However, increased oxygen uptake by undisturbed benthic communities at night, increased respiration induced by exposure to light and changing relations between photosynthesis and illumination during different times of day, indicate that complex relationships may exist between heterotrophic and photosynthetic activity. Hunding (1973) postulated that these responses are enhanced where inorganic nutrient supply is low since benthic algal growth during the day may deplete nutrients. Organic substances could be released in increasing amounts during the afternoon to progressively stimulate microbial respiration. Alternatively, algal and bacterial cells could adapt to changing light conditions during the day or alter their position in the sediment thereby affecting the ratio of production:respiration. Measurements which demonstrate the appearance of extracellular photosynthetic products in sediments during the day and independent estimates of algal and bacterial respiration are needed to clarify these relationships. Saks and Khan (1979) have shown that diatoms can compete successfully with bacteria in salt-marsh sediments to assimilate low molecular weight organic substrates at low concentrations (1–10 μM), although the metabolic importance of such heterotrophic growth for natural populations of benthic microalgae has not been established (Darley *et al.*, 1979).

Measures of specific production (assimilation numbers) have been determined for benthic microalgae on sandy beaches and intertidal sandflats (Steele and Baird, 1968; Admiraal and Peletier, 1980). Values varied from 0·1 to 1·4 mg C/mg Chl *a*/hr for cells growing on undisturbed sediment surfaces with higher numbers (3·4 to 13) for cells stirred in suspension under natural light levels. Although these measures of specific production may not represent levels of maximum photosynthesis, they are similar to values observed with phytoplankton grown in cultures when nutrient supplies are not limiting (Table 17).

6.4 BENTHIC PRODUCTION

6.4.1 Mechanisms of Organic Supply

6.4.1.1 Photosynthetic, heterotrophic and chemosynthetic production. Light, particulate, and dissolved organic matter and reduced inorganic compounds supply energy for production by benthic organisms. The processes of synthesis through oxidative and reductive pathways which these energy sources drive are interrelated and their relative importance depends on the availability of light, oxygen, and various organic and inorganic hydrogen acceptors (Fig. 96). The stratification of these processes, which may exist in a water column between aerobic and anaerobic water layers (Section 5.2), forms a major structural feature in almost all sediments since light and oxygen are only supplied across the sediment surface.

The cycling of elements within detritus, sediments, and across the sediment–water interface are due largely to the metabolic activities of autotrophic and heterotrophic bacteria (Fenchel and Jørgensen, 1977). Heterotrophic bacteria decompose organic matter to CO_2 with oxygen as the terminal hydrogen acceptor in aerated sediments as in the water column, where turbulence and diffusion maintain an adequate oxygen supply. However, microorganisms use both organic and inorganic ($SO_4^{=}$, NO_3^-, CO_2) compounds as hydrogen acceptors in subsurface anaerobic sediments. Fatty acids, alcohols, and other simple organic compounds are produced as products of fermentation if organic substrates are used. Utilization of inorganic hydrogen acceptors produces reduced inorganic compounds (CH_4, NH_3, H_2S). These compounds accumulate in depth zones where metabolic activity occurs (Fig. 91) and upward diffusion provides substrates for chemoautotrophic bacteria at the oxidized interface. The energy released by oxidation

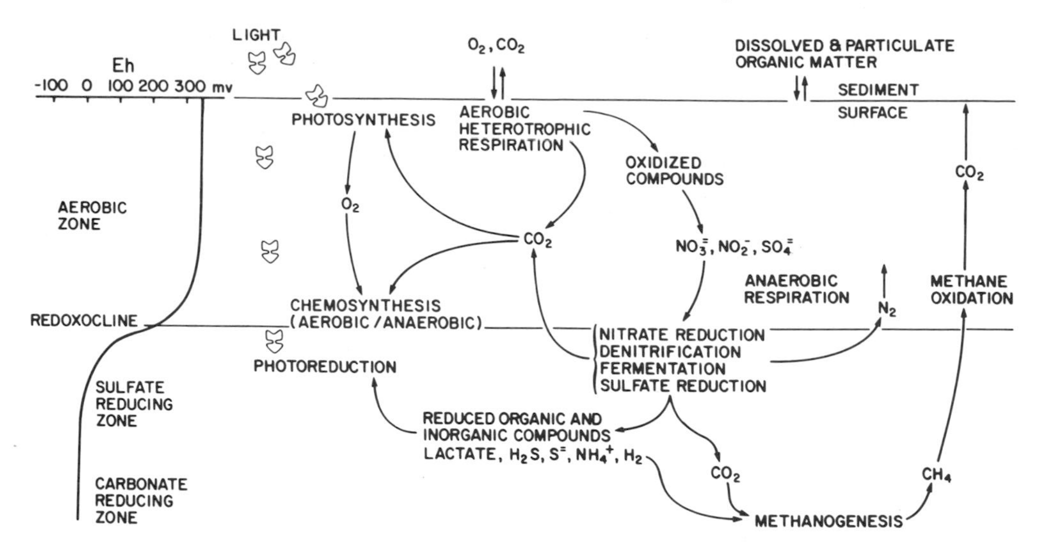

FIG. 96. Inter-relationships between photosynthetic, heterotrophic, and chemosynthetic processes in sediments. Photosynthesis and photoreduction only occur in the presence of light. *Aerobic metabolic processes:* heterotrophic respiration (oxidation of simple reduced organic compounds with possible reduction of CO_2); photosynthesis (reduction of CO_2 to carbohydrates using H_2O and light); aerobic respiration (reduction of oxygen to water with organic compounds as electron donors); aerobic chemosynthesis (oxidation of CH_4, H_2S, NH_3, Fe^{++}, H_2 to form organic carbon compounds by fixation of CO_2). *Anaerobic metabolic processes:* anaerobic respiration (oxidized inorganic end products of aerobic decomposition used as hydrogen acceptors for the oxidation of organic matter); fermentation (organic compounds used as hydrogen acceptors to produce CO_2, H_2O and reduced organic compounds such as lactate, glycollic acid, H_2S, NH_3); photoreduction (reduced compounds used to reduce CO_2 to carbohydrates in the presence of light with H_2S, SO_3, S, H_2 or reduced organic compounds serving as hydrogen donors); anaerobic chemosynthesis (oxidize inorganic compounds H_2, H_2S, Fe^{++} $NO_2^=$ and use energy to reduce CO_2 to carbohydrates) (redrawn from Fenchel, 1969, with modifications).

of these compounds is utilized to reduce CO_2 to carbohydrate in interstitial water. When reduced sediments are exposed to light (as in shallow water where organic enrichment leads to highly reduced conditions) photosynthetic bacteria utilize reduced inorganic and organic compounds to photosynthetically reduce CO_2. These processes are vertically stratified in sediments (Fig. 97) such that metabolic end products produced within a given depth layer diffuse upwards to provide substrates for microbial metabolism at a shallower depth. Aller and Yingst (1980) postulated that substrate limitation controls the depth distribution of microbial activity in sediments from Long Island Sound since rates of $SO_4^=$ reduction and NH_4^+ production decreased exponentially with depth as did bacterial numbers and ATP content.

When algae are not present, bacterial chemosynthesis has been measured as $^{14}CO_2$ uptake in the dark (Sorokin, 1965). Aerobic heterotrophic bacteria generally fix only a few percent of carbon as CO_2 in the dark, but chemosynthetic bacteria which oxidize simple reduced organic compounds may obtain 30% to 90% of their carbon supply from this source. Chemoautotrophic bacteria, like *Desulfovibrio* and *Thiobacillus*, utilize only CO_2 as a carbon source. Kepkay and Novitsky (1980) attributed enrichment of organic carbon in subsurface layers of a sulfide-rich sediment to the metabolic activity of sulfur oxidizing bacteria. Assimilation of CO_2 does not measure total chemosynthetic production but, if corrected for heterotrophic uptake, it does provide a comparative measure of metabolic activity for certain groups of micro-organisms. Dark CO_2 assimilation relative to photosynthesis by attached and free-living microalgae within sediments may change seasonally and increase with sediment depth just as ^{14}C dark fixation expressed as a percentage of total carbon fixation (in light) increases with water depth as

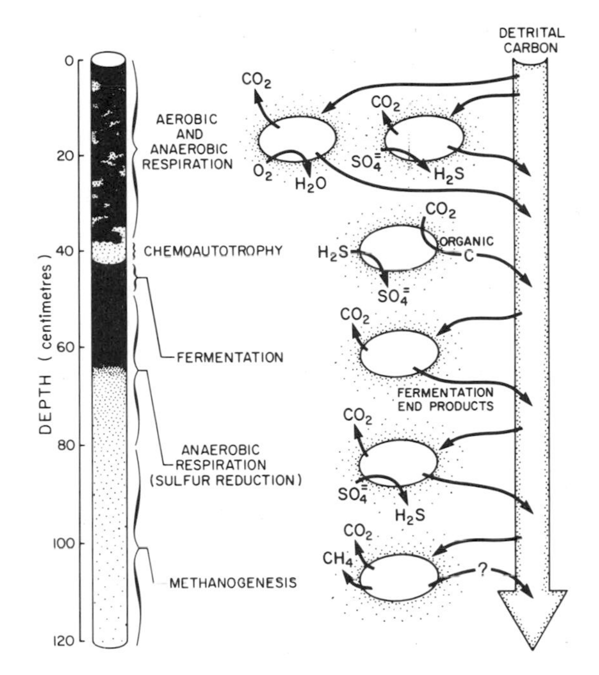

FIG. 97. Vertical stratification of aerobic and anaerobic metabolic processes carried out by bacteria in sediments (redrawn from Kepkay and Novitsky, 1980).

photosynthesis decreases (Gargas, 1970). Absolute rates of chemosynthesis, however, may not increase with depth because of lower numbers of metabolically active cells. Maximum chemosynthesis occurs when Eh values reach −20 mV (Sorokin, 1965). This would occur just below the aerobic/anaerobic interface (Fig. 96).

Micro-organisms which undergo aerobic heterotrophic growth, anaerobic fermentation or photoreduction all metabolize organic compounds. Thus. ^{14}C-labelled organic substrates added to sediments are incorporated and respired as $^{14}CO_2$ if they are fully oxidized. Dissolved organic compounds may also be physically absorbed to sediment particles, however, and uptake kinetics are usually not simple first-order reactions often observed for algae and bacteria in water samples. Addition of labelled substrates to open ocean water, shallow and deep sea sediments have not always produced the expected saturation-curve response (Section 3.2.2). Either no significant respiration or an irregular uptake of added substrate at increasing concentrations may be observed. Induction of a kinetic-like response can also be achieved by pre-incubation of samples with a particular substrate. Williams and Gray (1970) observed that the amount of induced activity was proportional to increases in substrate concentration. Either inducible enzyme systems could be present or selective growth through enrichment could occur particularly in prolonged incubations. The observations show that natural microbial populations have an ability to respond to increased substrate concentrations by increased metabolic activity. The measures do not quantify *in situ* metabolism.

Hall *et al.* (1972) found that temperature, nutrient availability, and the nature of the organic substrate affected the diffusion rate and uptake of dissolved organic compounds in undisturbed lake sediments. Up to 80% of ^{14}C-glycine added to stirred surface sediment was respired within 2 hr but acetate and glucose were mineralized to a lesser extent (10–25%). Also, when undisturbed sediment cores were used and a range of concentrations of substrate added uptake mechanisms were not saturated, even at concentrations as high as 1 mg/l (values in pore water equalled 50 μg/l). Wood and Chua (1973) and Novitsky and Kepkay (1981) similarly observed that up to 80% of various carbohydrates added to stirred sediments was respired in 2 hr with no relation between uptake rate and substrate concentration. Jones and Simon (1975), however, achieved saturation at concentrations >100 μg/l in undisturbed cores and stirred sediment with uptake equivalent to 400 to 600 $\mu g/m^2/hr$. Uptake of glucose by stirred autoclaved mud amounted to 10% of rates in fresh samples. Turnover times, while dependent on substrate type, are usually higher (minutes) in sediments than in water (hours) as would be expected from the greater microbial biomass per unit volume of sediment. Rates are also greatly affected by stirring. Uptake and mineralization can be increased by one to three orders of magnitude by mixing and dilution and thus measurements with stirred samples are not representative of those which occur in undisturbed sediment.

The assessment of microbial growth or metabolic

transformation of energy sources in natural populations is a central problem in aquatic microbiology. Metabolic regulation, the interdependence of different groups of micro-organisms for essential nutrients conditions and selective conditions for growth imposed by incubation prevent the extrapolation of data obtained from cultures to natural mixed populations. *In situ* measures of growth are required. Brock and Brock (1968) described microscopic autoradiographic techniques which permit a quantitative assay of uptake of labelled substrate by micro-organisms. Exposure of a culture of filamentous marine bacteria (*Leucothrix mucor*) to tritiated thymidine showed that dividing cells synthesizing DNA accumulated the label in a linear fashion and that 1% of the cells were labelled in 0·002 generations. Generation times of 660 to 685 min were observed in natural populations while values of 94 min occurred in pure cultures growing attached to solid surfaces. Stull *et al.* (1973) used similar methods to quantify species specific incorporation of $^{14}CO_2$ during phytoplankton photosynthesis. Generation times varied from over 1000 hr for some diatom species to less than 2 hr for several species of green algae. These techniques have not been widely applied to studies of microbial growth in sediments. The conversion of grain counts to radioactivity requires many assumptions and quantitative results may be difficult to obtain. Evidence that metabolic rate (oxygen uptake) and the degree of attachment of bacterial cells to particles are related to dissolved nutrient concentrations and the nature of the substrate (Jannasch and Pritchard, 1972) could be assessed by these methods.

Pütter (1909) first hypothesized that dissolved organic matter in water might serve as a nutrient source when directly absorbed by metazoan organisms. Experimental evidence for heterotrophic utilization of organic substances thought to be truly dissolved was summarized by Stephens (1967) who demonstrated the uptake of various labelled organic compounds by numerous soft-bodied marine and estuarine invertebrates. Net influx of amino acids into veliger larvae of oysters occurs at ambient substrate concentration as low as 9 μM (Rice *et al.*, 1980). Pogonophora species, which possess no internal digestive system, accumulate small organic molecules (glucose, amino acids, fatty acids) which could only occur through external body surfaces (Southward and Southward, 1974). These authors also calculated that some benthic species could accumulate sufficient quantities to meet respiratory requirements at concentrations which exist in sea water and sediments (10^{-9}–10^{-5} M/l). Weight specific uptake of amino acids varied inversely with body weight in the polychaete *Nereis* and logarithmic relations between substrate concentration and uptake rate existed for nereids and pogonophores (Southward and Southward, 1972).

Methodological problems persist in interpreting measures of uptake of dissolved organic compounds, however. Physical adsorption or uptake by micro-organisms on external surfaces must be controlled through the use of sterile and non-living organisms. Also, the addition of labelled compounds to filtered water provides no assurance that particulate matter is not present. Debris may be formed during filtration and even the smallest pore size of membrane allows particles to pass (Riley, 1970). The incorporation of a labelled organic compound in body tissue from dilute solutions and the evolution of $^{14}CO_2$ indicate active accumulation and metabolism of dissolved substances. The relative importance of heterotrophic uptake for nutrition remains in doubt, however. Excretion of various dissolved organic compounds by aquatic invertebrates can be comparable to respiratory energy loss (Section 6.4.3) and Johannes and Webb (1970) observed a net release of amino acids by the polychaete *Clymenella*. Southward and Southward (1974) reviewed previous studies and concluded that the nutritive value of epidermal uptake by animals capable of particulate feeding was unknown. Pogonophores, however, do apparently exist only on absorbed soluble organic matter.

Dissolved organic carbon in sediment pore water decreases in concentration with depth in the upper 10 to 15 cm of abyssal sediments (Karl *et al.*, 1976). Maximum bacterial numbers and metabolic activity also usually occur in surface layers of sediment where the turnover of these soluble organic compounds is probably maximum. Lewin and Lewin (1960) and Andrews and Williams (1971) have measured heterotrophic uptake of low molecular-weight organic compounds by numerous species of planktonic and attached microalgae. The ability of many benthic algal species to remain viable despite prolonged burial could be due to heterotrophic utilization of dissolved organic compounds in interstitial water.

Carbon:nitrogen ratios in sediments and suspended organic detritus are usually higher than values typical of living organisms (5 to 6) which implies that nitrogen is often in short supply relative to carbon in these substrates. Various micro-organisms which fix and transform nitrogen provide an additional source of supply of this essential element for aquatic production. Nitrogen fixation (the synthesis of cellular nitrogenous compounds from elemental nitrogen) is carried out by both aerobic and anaerobic micro-organisms which are either free-living or in symbiotic associations.

Increases in Kjeldahl-N or NH_4^+–N have been used as a measure of N_2 fixation but ^{15}N incorporation provides a more sensitive and reliable method. Recent techniques depend on the ability of N-fixing micro-organisms to reduce acetylene to ethylene which is measured by gas chromatography. Jones (1974) used this method to measure annual rates of nitrogen fixation in various zones of a salt marsh. Values varied from 0·4 g N/m² on bare mud to 46 g N/m² in pools which contained blue–green algae (*Nostoc*). ^{15}N-labelled products of nitrogen fixation by micro-organisms were transferred to the roots and leaves of higher plants in the salt marsh. Nitrogen fixation by micro-organisms (primarily *Desulfovibrio*) in estuarine sediments reached maximum rates (1·07 ng N/g dry sediment/hr) under aerobic conditions during summer (Herbert, 1975). Incubations with anaerobic cores resulted in higher maximum rates of fixation (1·84 ng N/g/hr) and the addition of glucose to cores increased both aerobic (×3) and anaerobic (×2·5) rates.

Many micro-organisms capable of nitrogen fixation possess respiratory systems which use elements other than oxygen as a terminal electron acceptor. For example, sulfate in anoxic sediments may be reduced to sulfide by anaerobic chemautotrophic bacteria such as *Desulfovibrio* which oxidizes H_2 by reducing $SO_4^{=}$. Similarly, *Thiobacillus denitrificans* oxidizes H_2S by reducing NO_3^-. Energy obtained by these processes is used to reduce CO_2 to carbohydrates (Fig. 96) and no other organic compounds are required for growth. The oxidation of organic matter by these anaerobic pathways, particularly in organically rich sediments, is often dominated by the sulfur cycle (Fig. 98). Sulfate reduction accounted for over 50% of the total mineralization of marsh grass detritus added to experimental aquaria and in natural *Spartina* marshes (Jørgensen and Fenchel, 1974; Howarth and Teal, 1979). H_2S and FeS which accumulate during mineralization may be stored within the sediment or oxidized abiotically. Chemo- and photoautotrophic bacteria which developed as a sulfuretum at the sediment surface also oxidize organic and inorganic sulfur compounds.

Microbially controlled transformations of organic and inorganic substances are stratified not only by the presence of oxygen and Eh gradients, but by the presence of substrates produced and consumed during bacterial metabolism. Thus,

	Micro-organisms capable of nitrogen fixation in sediments	
Type	Aerobic	Anaerobic
Heterotrophic	*Azotobacter* *Pseudomonas*	*Clostridium, Aerobacter* *Pseudomonas, Bacilluspolymyxa*
Autotrophic	–	*Methanobacterium*
Photosynthetic	*Nostoc, Calothrix* *Anabaena*	*Chromatium, Achromobacter* *Chlorobium*
Symbiotic	*Puccinella*	

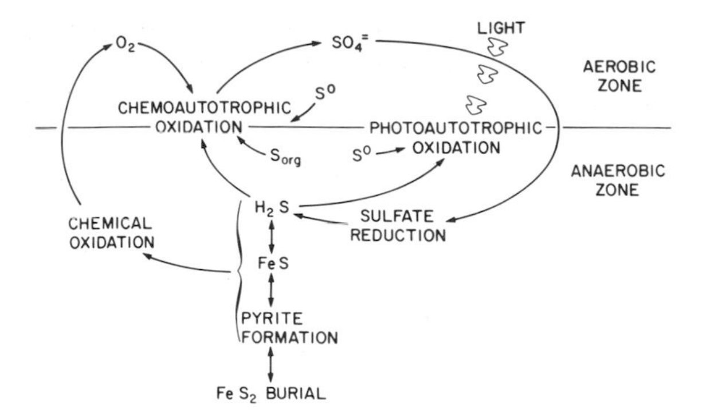

FIG. 98. Processes in the cycling of sulfur in marine sediments (redrawn from Blackburn *et al.*, 1975).

although cycles of carbon, nitrogen, and sulfur in sediments are intimately related, the organisms and metabolic transformations are vertically separated in a sediment column (Fig. 97). Wiebe (1979) has described how nitrification (the oxidation of ammonia to nitrate and nitrite) is an obligately aerobic process. Although micro-organisms which produce nitrate are present below the depth of oxygen penetration, active metabolism is probably restricted to oxygenated surface sediments (Henriksen *et al.*, 1981). Denitrifying bacteria, on the other hand, convert nitrate to nitrite and elemental nitrogen only under anaerobic conditions in deeper sediment layers. Thus the production and dissimilation of nitrate are vertically separated (Fig. 99).

Jørgensen and Fenchel (1974) described stratification in micro-organisms metabolizing sulfur in an artificial sulfuretum. Chemoautotrophic white sulfur

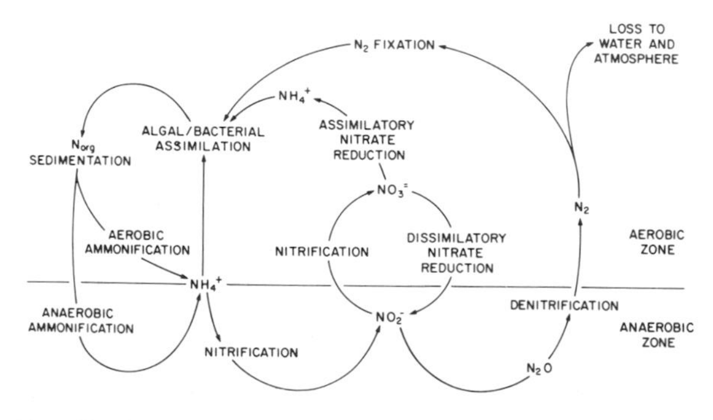

FIG. 99. Processes in the cycling of nitrogen through marine sediments (redrawn from Wiebe, 1979).

bacteria (such as *Beggiatoa* and *Achromatium*) catalyzed the biological oxidation of sulfide and formed a dense plaque on the sediment surface when sulfide was present only below the sediment photic zone. As sulfide concentrations increased within the photic layer, photoautotrophic purple (*Chromatium, Thiopedia*) and green (*Chlorobium*) sulfurbacteria developed in sequential layers overlying deposits of FeS. Layers of purple and green sulfurbacteria became sufficiently dense to restrict the upward diffusion of sulfide. Reduced sulfide supply caused oxidation of stored elemental sulfur to sulfate and destruction of bacterial plates followed in reverse order to their formation.

There are many other examples which show that microbially mediated reactions are linked and vertically separated within sediments. Martens and Berner (1974), for example, observed that methane production and sulfate reduction are mutually exclusive processes in anoxic sediments. Methane production, which does not occur until dissolved sulfate is totally consumed, occurs below the zone of sulfate reduction (Fig. 97). The interdependence of these processes ensures that if oxidation of organic matter is incomplete, a series of end products are produced which in turn serve as substrates for other metabolic pathways. The diversity of metabolic processes allows continued bacterial transformation of sedimentary organic material regardless of oxidation–reduction conditions.

6.4.1.2 Sedimentation, resuspension, and horizontal transport of particulate matter. Energy for benthic or pelagic organisms not exposed to light may be obtained from dissolved and particulate organic matter and reduced inorganic material. The importance of chemosynthetic pathways of energy flow has been demonstrated by discoveries of specialized communities associated with deep-sea hydrothermal vents. Areas affected by the flow of hot (300–400°C) water from volcanic rocks at oceanic spreading centers are small but high biomass of suspension-feeding macrofauna clustered near vents is thought to utilize chemosynthetic bacteria abundant in the vent water (Lonsdale, 1977; Rau and Hedges, 1979), Dissolved and particulate organic matter which could serve as a food source may also be enriched immediately around vents. The relative importance of heterotrophic uptake of dissolved organic matter for metabolism of these as for other benthic communities is unknown.

Particulate sedimentation as a 'rain of detritus' has been thought to be the primary mechanism of organic and inorganic material transport to deep sea sediments. Numerous studies which described the dynamics of natural phytoplankton populations and suspended particulate material have calculated that a relatively small daily loss occurs through sedimentation (Riley, 1970). Nakajima and Nishizawa (1972) estimated that a daily elimination of 2 to 4% of particulate carbon would be required to maintain profiles of decreased concentration in the upper 50 to 90 m of the Bering Sea described by eqn. (13). Taguchi and Hargrave (1978) calculated daily rates of organic carbon deposition at various depths in a marine embayment as a percentage of the concentration suspended in the water column above the depth of sediment-trap exposure. Daily loss rates varied from 1% to 5% with highest values during times of lowest stratification. The amount and quality of material removed from the water column by sedimentation varies seasonally, as discussed below, and loss rates may be enhanced at certain times. Smetacek *et al.* (1978) observed increased deposition of material following the spring bloom in Kiel Bight which provided a large supply of organic matter to the benthos in a short period of time.

Direct measurements of particle settling rate in laboratory and field experiments have provided values which, when combined with vertical profiles of mass, number or volume of suspended particles, can be used to estimate vertical flux by applying Stoke's Law. Riley (1970), for example, expressed the net transport of particles (F) in a water column where concentrations exist at steady-state as the difference between two vertical flux rates where

$$F = SC - A\,\frac{\Delta C}{\Delta Z}. \tag{143}$$

The products of setting rate (S) and concentration (C) corrected for flux due to eddy diffusion (calculated as the eddy diffusivity coefficient × the mean vertical concentration gradient (ΔC) over a depth interval (ΔZ)) gives the net flux. Riley observed that small particles settled in test cylinders in the laboratory at rates between 0·25 and 1·0 m/d. Much higher rates (10–1000 m/d) have been observed for fecal pellets and organic aggregates (Komar *et al.*, 1981; Bruland and Silver, 1981).

Sinking rates determined in still water may not be similar to those which occur in natural waters. Burns and Rosa (1980) have shown that *in situ* settling velocities of suspended particles are independent of size. Flagellated phytoplankton actively migrated downwards at sunset while species of diatoms, blue-green, and green algae altered their buoyancy in relation to changing light conditions. Turbulence may also reduce loss and permit dense particles such as diatoms to remain in suspension. Suspended particulate matter may not truly sink but only have an apparent downward movement due to advection. Sedimentation probably reflects conditions of turbulence more than settling of particles in water where horizontal and vertical turbulence exist (Murray, 1970).

Kranck (1975) observed that particles in sea water flocculate into aggregates of characteristic stable size distributions dependent on the grain size of inorganic particles. She hypothesized that grains flocculate until all particles have approximately equal dynamic transport speed and thus further collisions are minimized. In a subsequent study (Kranck, 1980) changes in concentrations and the shape of particle size spectra over short time periods were thought to arise by the transformation of organic matter between dissolved and particulate states. Aggregates appeared to form and break up in response to variations in levels of turbulence. Aggregation will significantly affect sedimentation since particle mass and settling velocity are increased. The effect may be most significant for small particles. Size fractionation has shown that only 2% of total organic carbon in open ocean water is in particles larger than 0·8 μm, 25% is colloidal and the remainder is 'dissolved' (Sharp, 1973). Fifty-eight per cent of particles larger than 1 μm in surface water are between 1 and 10 μm and 30% are between 10 and 95 μm (Mullin, 1965). Light scattering and electronic particle counters have been used to quantify the overwhelming abundance of particles <2 μm (Lerman *et al.*, 1977). The cumulative number of particles (N) larger than a given diameter (d) can be described by a negative exponential equation (hyperbolic function) of the form

$$N = ke^{-md} \quad (144)$$

or a power function as

$$N = kd^{-m} \quad (145)$$

where k and m are constants (McCave, 1975). The relations both resemble a hyperbola on linear scales but the fit of the equations to actual data often only applies over a small diameter range as an approximation and different curves may be required to describe the relationship over broad depth ranges.

McCave (1975) calculated negative exponential (hyperbolic) distributions of suspended particle number and volume from data provided by previous studies and found slope values between −2·4 and −3·6 for suspended material below 200 m. Density was assumed (with empirical evidence) to decrease with increasing particle size and histograms of the frequency of volume distribution were converted to mass distributions. Stokes velocities, calculated for each particle size class, showed that most of the particle flux (settling velocity × mass) occurs as large particles. Particles between 90 and 362 μm diameter may settle with velocities between 10 and 100 m/day. Suspended material <32 μm appears to constitute a 'background' of particles which does not contribute substantially to downward flux due to its low settling velocity (0·01–1 m/day). This conclusion is supported by calculations of an analytical model to determine the concentration and vertical flux of pellets produced by different stages of the copepod *Paracalanus* (Hofman *et al.*, 1981). Nauplii were estimated to produce an average of 50% of the mass of pellets released each day while adults produced 13%. However, the adult pellets, because of their mass, accounted for 63% of the depositional flux while those from nauplii only contributed 4%. These authors concluded that most of the pellets produced by nauplii would be consumed in the water column because of their low sinking rate.

Fecal pellets produced by planktonic organisms near the sea surface have been identified as vehicles for the rapid transport of material to the seafloor. McCave (1975) discussed observations of the localization of biogenic compounds under areas of high surface production, the presence of short half-lived radionuclides in deep-sea invertebrates and similarities between atmospheric dust and seabed sediments which imply transport of particles from the surface on a time scale of weeks to months. Biogenically produced aggregates contribute to a close coupling of phytoplankton production at the surface of the Sargasso Sea and sedimentation at 3200 m. Deuser *et al.* (1981) estimated a transit time of less than 60 days to this depth for material produced at the surface. The equivalent sinking rate (53 m/day) is similar to that observed *in situ* for macroscopic aggregates (68 m/day) in settling chambers in Monteray Bay (Shanks and Trent, 1980). Much higher velocities (10^2–10^3 m/day) of euphausiid and zooplankton fecal pellets (Section 1.3.5) and their resistance to physical breakage show that grazing activity could provide a supply of rapidly sinking material. Fecal pellets can be abundant in material deposited in marine bays. Hargrave *et al.* (1976) observed 10^4 to 10^6 pellets/m²/day deposited in traps suspended in an enriched coastal embayment although numbers settled in oceanic areas are two to three orders of magnitude lower (Honjo, 1980). Biodegradation is rapid in warm surface waters and this could lead to a reduced supply of rapidly sinking pellets in deep water. Once pellets sink into colder water, however, decomposition rates are lower and the integrity of particles not consumed by bathypelagic zooplankton may be preserved until they reach the bottom.

Particulate material suspended above the sediment and throughout the water column may not only be derived from sinking products of phytoplankton and zooplankton. Resuspension of bottom sediments and lateral transport in nephaloid layers or concentration in turbidity maxima may occur in estuaries, coastal areas, and the deep ocean. Postma (1967) discussed processes of sediment transport which prevent loss of suspended matter from nearshore areas at rates expected from removal of coastal water. Settling and scour lags can combine with tidal movement to cause residual transport towards a coast. Density differences due to river input may concentrate materials in turbidity maxima where sedimentation of fine-grained particles also increases due to flocculation. Relations between erosion, transportation, deposition, and inorganic sediment grain size cause deposition to occur at all particle sizes unless current velocity exceeds 15 cm/s. Clay and silt size particles at the sediment surface, however, are usually incorporated into an organic matrix which may result from sediment reworking by deposit-feeding invertebrates (Johnson, 1974). These aggregates have different sedimentological properties from individual mineral grains of comparable size.

Increased concentrations of suspended particulate matter within a near-bottom nephaloid layer may arise due to resuspension through tidal or other currents. Walsh *et al.* (1981) have hypothesized that horizontal transport of recently settled material constitutes an important pathway for removal of organic matter from continental shelf and slope regions to deeper water. Particulate matter may move along the bottom (as bedload transport) or suspended materials may remain concentrated (a nephaloid layer) above the sediment and become detached where currents flow across steep bottom gradients. Turbidity layers separated from the bottom and extending at the same depth into the slope water column have been observed to be associated with thermal gradients as small as 0·1°C (Drake, 1971). Turbidity currents, gravity currents with an excess of specific weight due to suspended sediment, can also arise from submarine slumps and the influx of water across a sill (Sholkovitz and Soutar, 1975). The thickness of the turbid layer and its horizontal extent will depend upon the mean current flow, turbulence, and the settling rate of suspended material. The presence of sand layers typical of continental shelf deposits buried within uniform hemipelagic clay sediments in distal abyssal plains shows that turbidity flows can occur over hundreds of kilometers and transport material to the bottom across ocean basins.

The direct measurement of sedimentation of particulate matter is made by mooring collectors (cylinders or funnels open at the upper end) on fixed or floating moorings. Deposition is proportional to mouth-opening area in calm water when there is no resuspension. Collector design can determine both the quantity and quality of settled material collected (Gardner, 1980 a, b) and traps must be constructed to reduce internal turbulence. The placement of traps relative to the bottom is also a critical factor which affects the amount and nature of settled material. Studies in coastal waters with traps placed at various heights above the sediment generally have shown that increasing amounts of material are deposited in collectors placed closest to the bottom (Davies, 1975; Smetacek *et al.*, 1978). Chlorophyll degradation products, high amounts of inorganic matter and organic material with high carbon:nitrogen ratios in material sedimented near to the bottom are indicative of resuspension caused by tidal and wave action. Correction for resuspension requires that the origin of sedimenting material at a specific depth be known or a linear profile of sedimentation with depth must be assumed (Davies, 1975). Non-vertical supply may cause anomalous sedimentation patterns (high or low values between successive sampling depths) which indicate that material trapped at different depths is of different origin.

Collectors are usually exposed for periods of time sufficient to accumulate material for microscopic examination and chemical analysis. In deep oceanic waters this may require several weeks dependent upon the size of trap used. Bacterial growth can alter organic carbon and nitrogen content of freshly settled material and plant pigments can degrade over collection periods of even a few days' duration. Iturriaga (1979), for example, observed that settled debris derived from phytoplankton and zooplankton in the Baltic Sea was mineralized by 35% and 18% per day respectively at 20°C. However, mineralization was reduced by approximately an order of magnitude at 5°C. On the other hand, Seki *et al.* (1968) used a bacterial growth bioassay to show that only a few percent of the carbon content in sedimented material was utilized in 3 weeks. The difference between these observations may reflect rapid losses which occur immediately after sedimentation when labile organic matter is degraded. Trapped material collected after a few days' or weeks' exposure may best be considered as the supply of relatively stable organic-matter residual after bacterial attack if decomposition of settled material is not measured.

Studies of sedimentation have shown a variety of seasonal patterns. A bimodal distribution in sedimentation was observed by Stephens *et al.* (1967). Phytogenous material was deposited in Departure Bay during May–July, two months after the spring bloom, and terrigenous material sedimented during October–December, the time of peak runoff. Maximum carbon deposition occurred during winter in St. Margaret's Bay, possibly due to detritus supply from seaweeds (Webster *et al.*, 1975). Studies in other coastal waters show seasonal patterns which indicate that resuspension occurs at certain depths during the year (Davies, 1975; Hargrave, 1980). Only one study has been conducted to demonstrate seasonality in rates of sedimentation in an oceanic environment. Deuser *et al.* (1981) showed that the flux of fine particulate matter at 3200 m in the Sargasso Sea reflected the annual cycle of phytoplankton production at the surface with highest rates between January and May.

High rates of sedimentation during and following a spring bloom (Smetacek *et al.*, 1978; Hargrave, 1980) show that at certain times intact phytoplankton cells and products of primary production may reach the bottom as organic debris in the early stages of decomposition. This is particularly likely to occur in shallow water where sinking material rapidly reaches the sediment surface. Stratification measured as the depth of the mixed layer is important for determining the direct supply of phytoplankton cells to benthic organisms (Fig. 100). Stratified conditions which persist for at least several weeks allow pelagic populations of grazing organisms to become established. Sedimented material may then be dominated by zooplankton fecal pellets and debris arising from grazing activity. However, pelagic grazer populations may not always utilize phytoplankton populations efficiently. In temperate oceanic waters this occurs during periods of minimum stratification in winter

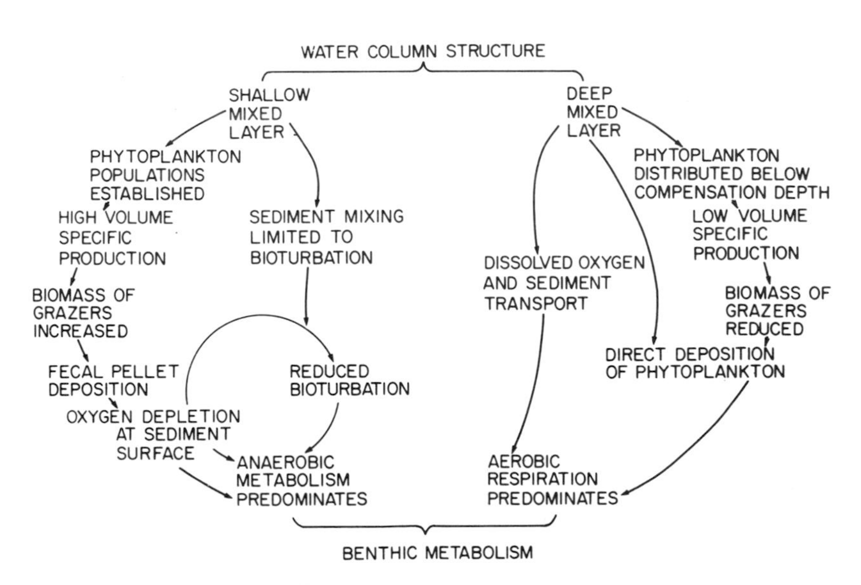

FIG. 100. Factors which determine pathways and rates of supply of particulate organic matter produced by planktonic organisms for metabolism by the benthos (from Hargrave, 1980).

and early spring. Blooms of phytoplankton may exhaust reservoirs of dissolved nutrients and then senescent cells settle rapidly. Walsh (1980) concluded that products of the spring bloom in continental shelf areas are not utilized in the water column but settle directly to the bottom. Attempts to compare the relative rates of energy flow through pelagic and benthic compartments of marine food webs are discussed in Section 6.4.4.

Bacterial degradation and consumption by suspension- and deposit-feeding invertebrates may occur either at the site of deposition or in adjacent areas where material is transported by lateral advection. Wildish and Kristmanson (1979) have postulated that the supply of seston utilized by filter-feeding invertebrates in the benthos may be described as the product of horizontal transport through tidal currents and the concentration gradient existing across the boundary layer above the bottom. They have used the ATP content in suspended material as a measure of assimilable organic matter available to the benthos. Supply is then dependent upon the rate of horizontal transport and the vertical concentration gradient near the sediment.

Other attempts to quantify the supply of particulate matter to the benthos through sedimentation have emphasized vertical transport of material through gravitational settling. This view has arisen from studies which show that deposition of organic matter in a deep-water column should be inversely related to depth because material is consumed by organisms and bacterial decomposition during sedimentation. Ohle (1956) suggested, for example, that if production and decomposition rates in surface waters are closely linked, then only a small fraction of the products of phytoplankton growth will actually be deposited. Many studies have shown that 70% to 90% of the organic matter produced in the ocean is mineralized within the upper layers of the water column. Riley (1970) calculated that 75% to 80% of the carbon fixed by phytoplankton in the Sargasso Sea was consumed by zooplankton and bacteria between the surface and 900 m. The remaining particulate carbon was consumed between this depth and the bottom (4000 m). Thus only a few percent (1–2 g $C/m^2/yr$) of the carbon fixed by phytoplankton could reach the bottom. Deuser *et al.* (1981) have confirmed that sedimentation rates of this magnitude do occur in the Sargasso Sea. Other calculations for various regions in the Pacific Ocean and Black Sea show that between 5% and 10% of organic matter produced at the surface would reach depths below 2000 m to 3000 m (Deuser, 1971). As a contrast, Riley (1956) calculated that between 30% and 40% (60–80 g $C/m^2/yr$) of primary production would reach the sediments in Long Island Sound — an amount consistent with observations in other coastal waters.

High organic input to coastal sediments relative to that in the deep sea would be expected on the basis of higher levels of primary production and shallow water depth. This general relationship was confirmed by Suess (1980) who reviewed published observations of organic carbon sedimentation measured by exposure of sediment traps in different marine locations. When these data are used to compare measures of sedimentation (C_f, g $C/m^2/yr$) and total water depth (Z, m) the equation

$$C_f = 5889Z^{-1 \cdot 09} \qquad (146)$$

can be derived. Depth is inversely related to sedimentation and accounts for 74% of the variance in the data. Suess expressed flux below the euphotic zone (50 m) as a function of annual phytoplankton production (C_p, g $C/m^2/yr$) and water depth by the equation

$$C_f = \frac{C_p}{0 \cdot 024 - 0 \cdot 21Z} \qquad (147)$$

which accounted for 79% of the variance in tabulated data. Organic carbon sedimentation would be expected to show some proportionality to phytoplankton production and be inversely related to depth because of consumption of material during sedimentation. Data reviewed by Suess can be replotted after logarithmic transformation to express sedimentation as a function of the ratio of primary production and depth (Fig. 101). Calculations of annual sedimentation as a percentage of phytoplankton production plotted against depth

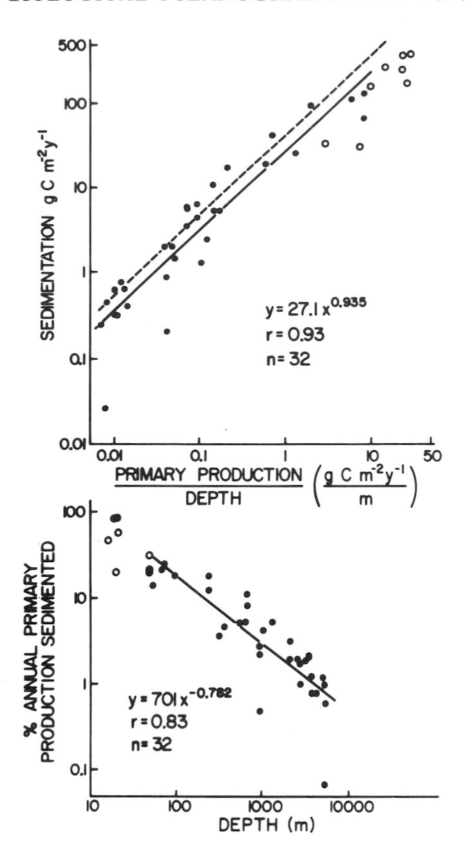

FIG. 101. Comparisons of estimates of annual sedimentation of particulate organic carbon and the ratio of phytoplankton production:water column depth (upper panel) and annual sedimentation expressed as a percentage of annual carbon production plotted against depth (lower panel) from data tabulated in Suess (1980). Measurements in areas where depth was <50 m (open circles) are not included in the regression calculation (solid line). Dotted line is described by eqn. (147) as derived by Suess (1980).

confirm that flux below 1000 m amounts to a few percent of the surface supply in these oceanic locations.

6.4.1.3 Macro-detritus, carcass deposition, and vertical migration. Studies with sediment traps only provide measures of fine particle deposition. Organic and inorganic material also reaches both shallow- and deep-ocean sediments in the form of macroscopic debris. The relative importance of fragments of macrophyte debris, wood, leaves, and *Sargassum* input to that supplied as zooplankton feces and finely dispersed detritus has not been evaluated in the deep sea. However, these plant remains represent a food supply for several animal groups when they are present (Wolff, 1979). Macroscopic debris may also be a major source of organic supply to sediments in shallow water, near-shore locations. Daily input of macroalgal fragments and worm tubes to nearshore sediments exposed to wave surge on the coast of California was two orders of magnitude less than fallout of finely dispersed debris (Hartwig, 1976). However, 90% of the organic carbon deposited was resuspended and thus large particles of organic debris which were not removed after deposition were a significant source of carbon input.

The importance of large particles for the supply of material to the deep sea has been inferred from the attraction of organisms to bait. Turner (1973) found that wood blocks placed on the seabed were attacked by boring pelecypods within 3 months. The implication of the observation is that wood must be supplied in sufficient quantity to allow populations to persist. Micro-organisms also rapidly colonize organic nutrient substrates placed at the sediment surface (Jannasch and Wirsen, 1973). The rapid formation of aggregations of mobile scavenging invertebrates (amphipods) and fish attracted to baited cameras (Isaacs and Schwartzlose, 1975) show that populations of organisms live in the deep sea with efficient sensory mechanisms capable of locating carcasses. Dispersed scavengers are rapidly concentrated and appear in photographs as long as food is available (Dayton and Hessler, 1972). Evolution of behavioural and physiological traits which permit the location of carcasses (Dahl, 1979) implies that this food supply exists in the deep sea although the quantitative importance has not been assessed. Many species of deep-sea fish have mouths adapted to ingest only large pieces of food and gut-content analyses show that certain predators do feed on carcasses (Clarke and Merrett, 1972).

A final pathway for the transfer of material through the water column to the benthos involves the vertical migration of organisms. Andersen and Zahuranec (1977) edited a collection of papers concerning the ecology of sound-scattering organisms. The directed and synchronous active movement of a variety of species shows that food ingested at one depth could be egested at another depth to move material in both vertical and horizontal directions. Riley (1951) and Vinogradov (1955) proposed that predator–prey links between migrating zooplankton and nekton would result in a rapid step-wise downward transport of material by interlocking cycles of vertical migration (Fig. 102). Organisms present in oceanic sonic scattering layers, primarily fish and crustaceans, generally occur between 450 m and 750 m during the day. Some species rise into the epipelagic zone (upper 200 m) at night while others remain at depth. There is a diurnal feeding rhythm in some species such as the small non-migrating fish *Valenciennellus tripunctulus* (Baird and Hopkins, 1981) with maximum feeding after midnight. Feeding by other species near the surface, where prey organisms are more abundant, could produce rapidly settling fecal matter. In addition, when these migratory species descend at dawn, they carry material ingested at the surface within their guts. Organisms themselves provide a source of organic supply for benthic predators where migration concentrates animals immediately above the bottom. This occurs on continental shelves and in coastal areas where light reaches the bottom. The rate of movement of nekton participating in vertical migrations of several meters/minute (Section 1.3.4) shows that the rate of transport of biomass through the water column may be very rapid.

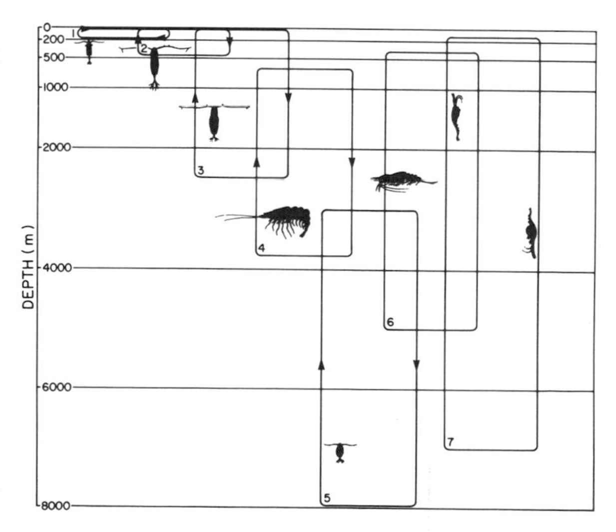

FIG. 102. Conceptual scheme of vertical migration by oceanic plankton and nekton for the active transport of organisms between the surface and the deep sea (1 — diurnal migration in epipelagic zone; 2, 3, and 4 — diurnal, seasonal and ontogenetic migrations within and between the epipelagic, mesopelagic and bathypelagic zones; 5, 6 and 7 — ontogenetic and irregular feeding migrations between the benthic and bathypelagic zones) (redrawn from Vinogradov, 1968).

6.4.2 Feeding Processes

6.4.2.1 Mechanisms. Invertebrate and vertebrate feeding habits may be classified on the basis of whether nutrients are obtained as a liquid or as small or large particles. This operationally useful separation of food sources avoids the necessity of assigning animals to specific trophic levels — a particularly difficult choice for benthic animals which often have catholic diets. Extensive reviews that differentiate liquid and particulate feeding (Pandian, 1975; Conover, 1978) summarize feeding mechanisms and nutritional physiology in various marine organisms. Examples of both modes of nutrition exist in the benthos.

Direct utilization of dissolved organic matter occurs in benthic plants, micro-organisms, and internal parasites of vertebrates and invertebrates. Absorption occurs across external surfaces and compounds are rapidly incorporated into cellular material. The direct uptake of dissolved organic compounds by soft-bodied marine invertebrates, however, is still a topic of controversy since excretion of similar substances also occurs (Section 6.4.1.1). Net uptake of dissolved material across external body surfaces has not been demonstrated under field conditions and since excretion is a little studied variable in energy budget calculations, the physiological significance of the phenomenon for metazoans remains unknown. The reciprocal transfer of soluble organic compounds between symbiotic algae and host organisms (various corals, anemones, turbellarian flat worms, molluscs), however, does illustrate the direct utilization of dissolved organic material by co-existing organisms. Conover (1978) has reviewed studies which have measured the rapid movement of dissolved organic compounds into and between symbiotic organisms. It is seldom clear how essential soluble materials are for nutrition. Corals, for example, may obtain sufficient energy by particle feeding when prey numbers exceeded minimal levels (Coles, 1969). Yet, intact coral heads appeared to meet only a small part of their respiratory requirements by capturing zooplankton (Johannes *et al.*, 1970). Perhaps pathways of nutrition in these organisms are regulated depending on the relative availability of dissolved and particulate organic matter.

Benthic animals which ingest small particles do so by one of three mechanisms which may be selective or non-selective:

1. filter feeding,
2. browsing (rasping),
3. deposit feeding.

To some extent, macrofauna which feed by different means are vertically stratified within benthic communities. Filter feeders usually remove suspended material from water over the sediment, browsing organisms scrape material from solid surfaces, and deposit feeders ingest sediment particles directly. The feeding methods are not mutually exclusive. Taghon *et al.* (1981) showed that the feeding behaviour of three species of spionid polychaetes varied from deposit feeding at low current speed to suspension feeding at higher velocities. *Scrobicularia,* a deposit-feeding mollusc, obtains particles through filtration by holding its inhalant siphon above the sediment surface when not directly ingesting sediment particles (Hughes, 1969). Also, it was earlier thought that these methods of feeding were generally non-selective. Filter feeders like bivalve molluscs, barnacles, and sponges were thought to take what came to them. These organisms use cilia and setae to create water currents which cause fine particles to collect on feeding surfaces where mucus bands carry food to the digestive tract. Annelids, polychaetes, and echinoderms which lack piercing and sucking mouth parts and ingest sediment were also considered to feed indiscriminately. Recent studies of invertebrate feeding have shown that food-particle gathering is seldom automatic or indiscriminate.

Suspension-feeding molluscs use gill surfaces for both gas exchange and food-particle collection. Particles above 4 μm are usually retained with 100% efficiency with more efficient retention of smaller particles in species which possess latero-frontal ciliary tracts (Møhlenberg and Riisgård, 1978). Feeding and respiration are also linked, as demonstrated by the effect of current speed and body size on pumping rate. Walne (1972) observed a power relation between water flow and filtration in five species of bivalves which demonstrates the importance of water movement past sedentary filter feeders.

Pumping, as a metabolic process, is also related to body weight by a power function but exponents of regressions of body weight on filtration rate are highly variable (−0·3 to −0·8). Mohlenberg and Riisgard (1979) observed values between 0·62 and 0·75 for five species of bivalves when experiments were performed above critical levels of water flow. If pumping is directly related to respiration, a standard metabolic weight exponent of −0·2 to −0·35 would be expected. Regressions permit calculation of weight-specific pumping rates although comparison must be made with caution when different methods are used to obtain data. Comparisons of various bivalve species on this basis show that mobile, rapidly growing species such as scallops have weight-specific filtration rates an order of magnitude greater than more sedentary species (Winter, 1978). Ali (1970) similarly observed lower pumping per unit weight by oysters and a rock-boring mollusc *(Hiatella)* in comparison to other species. In addition, Hughes (1969) found that the square root of gill area was linearly related to shell length in four species of molluscs, thus pumping rates per unit gill area could be used as a standardized measure of filtration rate. When animals of the same shell length were compared, *Scrobicularia,* primarily a deposit feeder, had the lowest pumping rate per square millimeter of gill surface. The three species with the highest gill area-specific pumping rate *(Venus, Mytilus, Cardium)* filter water from above the sediment surface. *Scrobicularia* and *Mya,* which inhale water and suspended matter at the sediment surface, had lower rates on this basis.

Invertebrates which feed by scraping substrates (plant surfaces, rocks, and sand grains) may ingest particulate matter directly or absorb fluids and soft tissue by piercing and sucking. Various meiofauna such as Tardigrades have tube-like mouths with stylets to pierce plant cell walls and a muscular pharynx to suck out cell contents. Snails and limpets scrape surfaces through rotation of a radula which may permit the removal of micro-organisms which adhere to particles (Lopez and Kofoed, 1980). This feeding action in intertidal areas often keeps rock surfaces free of epilithic algae (Castenholz, 1961). Moore (1938) observed that the size of the feeding area (y as cm^2) and volume of the limpet *Patella* (x as cm^3) were linearly related by the equation:

$$y = 0{\cdot}0125x \qquad (148)$$

which can be calculated from her data. Castenholz (1961) did not measure feeding area, but he did demonstrate that volumes of littorine snails greater than 0·2 cm^3/dm^2 and limpet volumes over 0·8 cm^3/dm^2 kept rocks free of diatoms during summer. Since natural snail abundance in tide pools exceeded these levels, feeding areas must have overlapped with rock surfaces being continually rescraped depending upon duration of immersion. Browsing by sea urchins also prevents colonization of rocks by macroalgae (Bernstein *et al.,* 1981).

Deposit-feeding organisms may ingest particles from on or under the sediment surface by a variety of methods. Some holothurian species, for example, appear to consume plant and animal debris from rock, sand, and mud surfaces (Yingst, 1976) while other species feed on subsurface sediment and apparently accumulate interstitial meiofauna selectively (Walter, 1973). Holothurian species in deep-water areas of the Bay of Biscay, on the other hand, do not ingest living meiofauna but have gut contents enriched with small particles (Khripounoff and Sibuet, 1980). Selective feeding has also been clearly demonstrated for various polychaete and mollusc species. *Macoma* and *Scrobicularia* (bivalves) rotate their inhalant siphon to suck in the top millimeter of sediment from a circular area around the siphon-tube opening (Hylleberg and Gallucci, 1975). Initial intake appears indiscriminate, but separation occurs on the gills and palps where coarse organic and inorganic particles and small particles low in organic matter are ejected as pseudofeces. Hughes (1975) examined the mechanisms of sorting by the bivalve *Abra* and found that the upper size limit of particles entering the mantle cavity was determined by the diameter of the inhalant siphon. Material ingested was usually smaller than that taken into the mantle cavity implying that sorting occurred by pallial organs.

Whitlach (1980) postulated that food resources are allocated among deposit feeders in two ways — with respect to particle size and according to vertical location in the sediment. Behavioural, morphological, and physiological differences thus permit the co-existence of many deposit-feeding species. When related *Hydrobia* species (surface deposit-feeding gastropods) occurred together, Fenchel (1975) observed that a narrower range of particle sizes was ingested than when each species occurred separately. Character displacement apparently caused a size separation which led to differences in the range of particle sizes ingested. *Hydrobia* often co-exists with a burrowing amphipod *(Corophium)* and whereas *Hydrobia* generally consumes particles between 60 μm and 300 μm diameter, *Corophium* selects smaller particles and ingests a greater proportion of bacteria (Fenchel *et al.,* 1975). Similarly, as was described for holothurians above, burrowing lugworms selectively ingest organically rich fine-grained sediment (Hylleberg, 1975) and polychaetes *(Pectinaria)* select particles in direct proportion to their body size (Whitlach, 1974).

Mechanisms of selective feeding by deposit-feeding species are varied and selection is not always for the smallest particle. *Macoma* and *Scrobicularia*, for example, separate particles on the gills and palps and coarse organic and inorganic particles and small particles low in organic matter are produced as pseudofeces (Hylleberg and Gallucci, 1975). Selective ingestion by surface of deposit-feeding ampharetid polychaetes is dependent upon worm size and varies with the specific gravity and surface texture of particles (Self and Jumars, 1978). The observation that particles of higher specific gravity have a shorter gut-residence time implies that gut-content analysis alone cannot provide evidence for selective ingestion. The presence of an organic coating on particles also leads to preferential ingestion by various species (Taghon, 1982).

Burrowing deposit-feeding species also show selective feeding although different methods are involved. Studies by Whitlach and Hylleberg mentioned above show that *Abarenicola* uses its proboscis to excavate a water-filled pocket. Coarse sand accumulates at the bottom and fine particles are kept in suspension. *Pectinaria* uses palps to dig at the base of a vertically facing tube and ciliated grooved tentacles independently select particles near the mouth. Electivity indices for such organisms are often difficult to calculate since there may be a vertical gradation in particle size and individuals of different size may feed at different depths. A similar problem exists in assessing feeding activity in numerous fish and crustaceans which feed on and within sediments. While they are predators, inorganic and organic debris often accounts for the bulk of their gut contents. Odum (1970) found that detritus-feeding fish *(Mugil)* either browse epiphytic algae or consume macroscopic debris and select for fine particles, depending on their relative availability.

All of these studies demonstrate discrimination in feeding which permits some degree of food-resource partitioning based on particle-size selection. In contrast, benthic micro- and meiofauna which feed selectively are much more highly specialized with different kinds and sizes of micro-organisms (bacteria and algae) serving as specific food sources for particular species. The high diversity of microfauna in sediments allows this specificity of food type. A lack of food specialization in some detritus- and deposit-feeding invertebrates, however, is balanced by the development of alternative feeding methods. For example, *Ampelisca,* a tube-building amphipod, is a predator, but when food is unavailable antennae are used to scrape up surface sediment (Kanneworf, 1965). Newell (1970) described feeding by many benthic invertebrates like *Nereis,* a polychaete, which generally filters water through a secreted mucus bag, but inorganic sediment particles and other organisms may also be ingested. Similarly, *Hydrobia,* which usually ingests individual mineral grains, may browse flat surfaces and perhaps trap particles in mucus in the surface film of water (Fenchel *et al.,* 1975). Since switching of feeding methods appears to depend on food availability, experimental measures of food intake may not represent rates which occur in nature.

Benthic animals which ingest large particles are generally carnivores. While sedentary predators do exist in the benthos (attached coelenterates and echinoderms), most are mobile organisms which use visual and chemical senses to perceive prey. Epifauna are generally more vulnerable to predators than infauna, but any prey may be swallowed whole, crushed or drilled and partially consumed depending on the feeding method of the predator. Limits to ingested prey size are often dependent on predator size. For example, the carnivorous mollusc *Navanax* can only consume prey up to a certain maximum size and during growth the ability to manipulate and swallow small prey is lost (Paine, 1965). Many invertebrate predators (starfish, shell-drilling molluscs) only ingest a portion of their prey and demersal fish often browse on extended siphons of buried molluscs. While many of these predators are morphologically and behaviourally adapted to feed on selected prey species, food selection is often inferred from identification of prey in gut contents. Sympatric species-pairs of skates, for example, utilized similar species as prey organisms but in different proportions and with differences in dominance of epifauna and infauna (McEachren *et al.,* 1976). The differences largely reflected the availability of different prey in the benthic communities in which the species-pairs occurred. Levings (1973) also concluded that flatfish predation in a shallow-water benthic community was opportunistic with large vulnerable macro-fauna being the primary food supply.

6.4.2.2 Food availability. Energy uptake by organisms, either as light, soluble nutrients or particles, has often been related to the abundance of the energy source. Light-photosynthesis curves (Section 3.1.2), substrate and nutrient uptake by micro-organisms (Sections 3.1.6 and 3.2.2), particle retention by suspension feeders (Section 4.1.1) and predation by both invertebrates and vertebrates, for example, have been represented by rectilinear or curvilinear equations which have threshold values where energy uptake becomes saturated. Many of these response curves have a shape similar to a rectangular hyperbola and such a formulation has been used to describe light–photosynthesis relations, the kinetics of nutrient uptake and population growth of micro-organisms (Section 3.2.2). Although photosynthesis–light responses in laboratory cultures of phytoplankton may be described by such a formulation Jassby and Platt (1976) found that a hyperbolic tangent function was a better representation for light response in natural phytoplankton populations. Uptake of substrates by mixed populations of micro-organisms and metazoans in sediments is not always of the form expected from Michaelis–Menten kinetics (Section 6.4.1.1). Interactions and conditions exist in planktonic and benthic communities which prevent the occurrence of idealized response curves observed in cultures of single species.

There have been frequent attempts to use a hyperbolic equation or related hyperbolic functions to describe food supply and feeding in predator–prey systems. While these may suitably represent predation based on a two-phased feeding habit of searching and handling individual prey items before ingestion, they are generally not applicable to the process of filter feeding where food is gathered and processed simultaneously. Lehman (1976) proposed a more realistic model of selective feeding by suspension feeders based on the relative abundance, size, and digestibility of particles. The assumption of the model was that feeding behaviour maximizes the net energy gain. Taghon *et al.* (1978) applied the same model to deposit feeders by assuming that food supply is proportional to surface area rather than volume of particles. This reflects the empirical evidence that deposit feeders digest the microbial epigrowth on the surface of particles. The general equation to predict how a deposit feeder would adjust its feeding to maximize its net energy gain was:

$$Q = E_{\alpha} - E_{c} - E_{r} \tag{149}$$

where Q (calories/time), the net energy gain, is calculated from the energy gained by assimilation of microbial biomass on particle surfaces corrected for the energy used to collect (E_c) and reject (E_α) sediment particles. Feeding is assumed to maximize Q by striking an optimal balance between E_α, E_c, and E_r. Doyle (1979) pointed out that the predictions that ingestion rates will vary with particle size and that they will rise as food concentration rises to level off or decrease at higher concentrations are also general properties of mechanistic feeding models which do not assume energy maximization. He used mixtures of natural sediments and glass beads to show that the parameters of a hyperbolic model to describe ingestion as a function of particle size for an individual deposit-feeding amphipod *(Corophium volutator)* were dependent upon the nutritional value of material ingested — an assumption of the feeding model described by Taghon *et al.* (1978). Behavioural responses for selective ingestion of particles with organic coatings has been confirmed for several species (Taghon, 1982).

The response of suspension-feeding invertebrates to changes in particle number and type is variable, but in general filtration declines at high particle concentration and ceases entirely when clogging impedes ciliary or appendage movement. Winter (1969) proposed a seven-step behavioural response to changes in particle concentration in molluscs. Food intake by mussels and oysters, for example, reached a plateau above a certain concentration while other species either demonstrated no threshold or decreased ingestion at high particle number. Lam and Frost (1976) similarly identified three phases in the feeding response of adult *Calanus* to differences in food concentration. Functional relations between filtering rate and food concentration for each phase of feeding behaviour permitted calculation of net energy gain expected dependent on food abundance.

Many studies of suspension feeding have shown

that food-particle quality as well as its size and abundance influences ingestion. Conover (1978) cited numerous examples of adaptation of feeding rates which appear to maximize useful ration. Thus, when suspension feeders consume inert or less nutritious particles, greater ingestion rates may ensue. Emlen (1973) proposed that such adaptation demonstrated a foraging strategy for selection of food on the basis of its energy content. Carefoot (1967), however, found no relation between consumption rate and calorific content of macrophytic seaweeds ingested by the mollusc *Aplysia.* Many predators seem less dependent on total energy content of prey than on their accessibility. Carnivorous starfish *(Lepasterias)* provide an exception to this generalization since when moderately energy-rich prey disappeared during winter, rarer species with a high energy content were selectively consumed (Menge, 1972).

Detritus- and deposit-feeding species may also alter feeding rates in response to changes in food quality. Since Newell's (1965) observations on the feeding of the molluscs *Hydrobia* and *Macoma* on sediment particles, it has been assumed that micro-organism attached to particle surfaces, rather than non-living organic debris, provides the main food source for deposit-feeding invertebrates. Many studies have presented evidence for high ingestion rates by deposit-feeding species with little or no assimilation of non-living organic detritus. While micro-organisms are assimilated, often with a high degree of efficiency, observations by Cammen (1980a) showed that non-living detritus may also be digested by some deposit-feeding organisms. A direct assessment of the role of non-living organic debris was provided by exposing mysiids to labelled cellulose under sterile conditions (Foulds and Mann, 1978). Chemical treatment of sterile material, however, provides substrates unlike those present in natural sediment and the extent of digestion of these materials in laboratory preparations may not reflect the availability of natural substrates.

Tenore (1981) has shown that the nutritional value of detritus derived from a variety of plant species for the deposit-feeding polychaete *Capitella capitata* varies with the source of material. Vascular plants contain a high percentage of material low in nitrogen and energy which must be converted to microbial biomass for utilization. Detritus derived from seaweeds, on the other hand, is richer in nitrogen and higher biomass levels of *Capitella* were supported with food prepared from this source. Egested material from deposit feeders, which contains indigestible organic matter, may be colonized by micro-organisms and again become suitable food. Increases in nitrogen content have been used to document the colonization process (Levinton, 1979). Selective ingestion may concentrate organically rich material in feces and even after assimilation of the digestible fraction, organic content may exceed that present in bulk sediment (Khripounoff and Sibuet, 1980). Hylleberg (1975) described the effect of selective ingestion and the production of organically enriched fecal material by the lugworm *Abarenicola.* The concept of 'gardening' in deposit-feeding invertebrates is implied by the effects which some benthic-deposit feeders have in reducing detrital particle size, thereby increasing surface area for microbial growth and altering metabolism of microflora in sediment (Levinton, 1979).

Studies of food intake by a variety of benthic organisms have provided quantitative comparisons of feeding rates and food supply. Starvation and reduction in food quality can cause increased feeding and assimilation in some species (Conover, 1978) and thus serve as adaptations to an altered food supply. Models of optimal diet discussed above have assumed that size and abundance are important factors which control a prey's availability and energetic value to a predator. Deposit-feeding species do not pursue prey and maximum energy gain is realized by sorting and selectively feeding on energy-rich particles or by maximizing the amount of food consumed. Besides the widespread occurrence of selective feeding, a general tendency towards the latter strategy is implied by the dependence of sediment turnover on body size in various deposit-feeding invertebrates (Hargrave, 1972b). While there was not a single curve common to all species, the data conformed to a power function ($b = 0{\cdot}68$) between body size and ingestion in terrestrial arthropods. The rate of food passage may be controlled towards an upper limit determined by body size (or gut volume). This is to be expected if energy intake is maximized.

Cammen (1980b) summarized published values for ingestion by nineteen species of deposit-feeding benthic invertebrates (excluding molluscs) to show that where estimates of organic content of food ingested at 15°C and body size were given, feeding rate (I, mg dry material/day) by individual animals was related to body weight (W, dry weight) by the equation

$$I = 0{\cdot}295W^{1{\cdot}12}. \quad (150)$$

Ingestion rate of dry material also varied inversely with organic content of the food and organic-matter ingestion (C, mg organic material/day) was described by the equation

$$C = 0{\cdot}381W^{0{\cdot}74}. \quad (151)$$

Cammen postulated that deposit-feeding species could maintain a given rate of intake of organic matter by actively adjusting ingestion in response to changes in organic content, an idea implicit in the hypothesis that food intake occurs at rates which maximize energy accumulation.

6.4.3. Energy transformation and elemental budgets

Measures of energy flow and elemental cycling through individual organisms, populations, and communities have been made with the hope that patterns might emerge which would provide a common basis for comparison of different ecological systems. Lindeman's trophodynamic model has provided a paradigm for many of these studies but observations of mixed diets and the apparently ubiquitous role of detritus (non-living debris and attached fauna and flora) in both pelagic and benthic food webs has made evaluation of pathways and transfer efficiencies under natural conditions difficult to assess (Pomeroy, 1974). Laboratory measurements necessitate assumptions for extrapolation to nature and often balanced budgets are obtained by calculating unmeasured terms by difference. Despite these problems, budgets provide an organized way of describing how an organism partitions ingested energy or materials. The comparisons may indicate mechanisms of control of population size and community structure insofar as such studies permit growth and reproduction to be predicted.

IBP handbooks edited by Ricker (1971) and Holme and McIntyre (1971) summarize the separate processes involved in the passage of energy and materials into and out of all heterotrophic organisms by an expanded version of eqn. (92),

$$R = P + G + T + E + U. \quad (152)$$

This equation equates R, total food material or energy intake, to the sum of P, that incorporated into biomass (growth), G, gonads, T, the amount used in metabolism, E, that released as undigested feces, and U, excreta (secretion as mucus, urine, moults). Differentiation of gonad production from growth may be important seasonally particularly in organisms which store energy for gonad maturation at a later time. Also, although few studies have separated soluble excretory products from non-digestible feces, this permits evaluation of A_b, the energy or elements absorbed or assimilated, where

$$A_b = R - E = P + G + T + U. \quad (153)$$

This is not equivalent to A_r, the retention of energy or elements by an organism, since soluble and particulate excretory losses occur. Thus,

$$A_r = R - (E + U) = P + G + T. \quad (154)$$

The difference between measures of absorbed and retained energy or elements has been stressed by Johannes and Satomi (1967) and several studies with benthic invertebrates summarized below have demonstrated that the release of soluble excretory products occurs at rates which equal or exceed loss through respiration. Kleiber (1961) further differentiated 'net energy gain' as retained 'metabolizable energy' corrected for the energy lost in food conversion (specific dynamic action) and 'surplus energy' ($P+G$) which remained after metabolic costs of maintenance and activity. These distinctions are necessary if patterns of energy allocation in animals are to be quantified.

Attempts to construct short-term energy or elemental budgets for individual organisms have generally showed that metabolic losses (as respiration and excretion) account for a large portion of assimilation (Conover, 1978). Little measurable residual energy or material may be stored for growth over a short time but exceptions can occur in young animals or reproductively active adults when ingestion and growth per unit weight are most rapid. Conditions under which feeding and respiration are measured often affect rates which are observed and the usual assumption that unmeasured excretory losses are small or a constant proportion of assimilation does not permit realistic assessment of rates of energy or material transfer. Assimilation may be difficult to accurately determine particularly when small amounts of ingested material are absorbed as by many deposit-feeding species (Cammen, 1980a). Changes in weight or chemical content permit direct calculation of A_b from eqn. (153) when the amount of food ingested is known and feces are quantitatively collected. This method is widely used to quantify feeding by fish (Ivlev, 1961) but it requires holding animals under laboratory conditions and the provision of an unnatural food supply. The measurements are often difficult for deposit-feeding invertebrates when egested material cannot be quantitatively recovered. When comparisons have been made, however, individual dietary components or specific elements are assimilated with a greater efficiency than total organic matter (Conover, 1978). Proteins and nitrogen are generally assimilated more efficiently than carbon, carbohydrates, or lipids, but digestibility depends greatly on the nature of material offered as food (Tenore, 1981).

Radioisotopic methods can provide sensitive and specific measures of transport of elements from food into organisms. However, unless the specific activity in food and feces are directly compared over a time period sufficiently short to permit uptake and release processes to be differentiated, instantaneous rates cannot be properly calculated (Conover and Francis, 1973). Determination of release and accumulation rates of specific elements is possible when homogeneous labelled food is used and recycling quantified. The use of ^{14}C-labelled food also permits quantitative measures of $^{14}CO_2$ respired and the importance of soluble excretion can be evaluated. Soluble excretion of carbon by the benthic snail *Hydrobia* estimated in this way amounted to more than 30% of that assimilated (Kofoed, 1975). Conover and Francis (1973) emphasized these and other errors which can occur when isotopic methods are improperly used to estimate feeding and assimilation rates.

Radioisotopic methods have also been used to estimate food uptake and elimination under natural conditions. Kevern (1966) calculated the amount of ^{137}Cs which would have to be ingested by young carp living in a pond contaminated with the isotope if an equilibrium body burden was to be maintained. The isotope was assumed to be in equilibrium in the fish and the pond environment and uptake by the carp was thought to be primarily by ingested food. Thus, with a steady state,

$$RA = Q_e k \tag{155}$$

where R, the rate of ingestion of ^{137}Cs/day, and A, the fraction consumed assimilated, was balanced by Q_e, the body burden corrected for k, the fraction eliminated per day. When the concentration of ^{137}Cs in the food, D, was known, the rate of ingestion of isotope could be converted to the rate of food ingestion, R', since

$$R' = R/D \tag{156}$$

Thus the rate of ingestion of a specific dietary component required to maintain an equilibrium body burden was

$$R' = \frac{Q_e k}{AD}. \tag{157}$$

Elimination of an isotope (by physical decay and biological removal from different metabolic pools) can also be measured to provide an estimate of metabolism in animals exposed to natural conditions. If labelling has occurred to steady-state and exponential loss is assumed to occur from all pools, Kevern (1966) calculated the amount of isotope remaining at any time, A_t, as

$$A_t = A_0\, e^{-kt} \text{ or } A_t/A_0 = e^{-kt} \quad (158)$$

where A_0 was the initial amount assimilated from a single feeding and t time. The length of time for 50% elimination (biological half-life), T_b, was calculated as

$$A_t/2A_t = e^{-kT_b} \text{ or } \ln(0{\cdot}5) = -kT_b \quad (159)$$

and

$$k = 0{\cdot}692/T_b. \quad (160)$$

Since elimination occurs from metabolic pools with different turnover times, k values can be estimated by plotting A_t/A_0 against t on semi-logarithmic axes. The elimination of ^{65}Zn from snails and plaice under field conditions was followed in this manner (Mishima and Odum, 1963; Edwards, 1967). Once non-equilibrated label was eliminated, the long-term metabolic loss for free-living animals was inversely related to body size and temperature and paralleled oxygen uptake in snails. Both studies demonstrated metabolic losses at least twice as high as those measured in laboratory studies. Edwards *et al.* (1969) found that only plaice feeding at high ration levels had comparable metabolic rates. The observations support the view that active metabolism may be double that measured as 'standard' rates in the laboratory (Conover, 1978).

A third group of methods used to measure assimilatory uptake of material by aquatic organisms involves comparing the ratio of inert compounds to nutrients in food and feces. Conover's ash-ratio method [eqn. (94)] was developed on this basis. The technique does not require the quantitative recovery of fecal material for estimating assimilation, but selective feeding or changes in ash content during passage through the gut prevent its use, although corrections for the latter problem are possible (Conover, 1978).

In a more general inert indicator method, a non-digestible additive, such as Cr_2O_3, is mixed with a food supply. Percent assimilation is calculated from

$$A_b = (1 - N{:}R/N{:}E) \cdot 100 \quad (161)$$

where $N{:}R$ and $N{:}E$ are the ratios of marker, N, in food, R, and feces, E. Calow and Fletcher (1972) used a double labelling method with the inert tracer ^{51}Cr (as $^{51}CrCl_3$) which is not absorbed and ^{14}C to label bacterial and algal cells offered as food to surface browsing snails. Attempts were made to ensure that both labels were homogeneously distributed in diatoms offered as food and both tracers were assumed to move uniformly through the gut and maintain a stable ratio in feces. Corrections for slight absorption of ^{51}Cr were applied and no leaching from feces was observed. Thus, the assimilation efficiency was calculated as

$$A_b = (1 - (^{51}Cr{:}^{14}C \text{ food})/(C^{51}Cr{:}^{14}C \text{ feces})) \cdot 100. \quad (162)$$

Assimilation efficiencies, in excess of 80%, were comparable to those estimated from the difference between ^{14}C lost from food discs and ^{14}C appearing in feces. The technique has the advantage that total fecal production need not be observed and fecal material derived from food can be distinguished from that derived as intestinal secretions.

Cammen (1980c) described an *in situ* method for measuring ingestion rates of infaunal deposit feeders without collecting feces which consisted of labelling sediment with fluorescent particles. These were ingested during feeding and they could be used to trace the movement of ingested material along the gut of the polychaete *Nereis succinea.* Ingestion, expressed per unit body weight, increased with temperature and decreased with body size. The technique was used *in situ* to show that more than 80% of the population of worms in an estuarine *Spartina* marsh ingested fluorescent particles at temperatures above 14°C while the percentage dropped to 50% at 10·5°C and 35% at 7°C.

Conover (1978) reviewed many studies which

TABLE 39. BUDGETS FOR ENERGY AND ELEMENTS AS A PERCENTAGE OF TOTAL AMOUNTS ASSIMILATED FOR VARIOUS BENTHIC ORGANISMS.
Letters refer to terms defined in eqn. (152).

Organism	Basis	P	G	T	U	Reference
Modiolus demissus (mollusc)	calories	25	5	70		Kuenzler (1961a)
	P (annual)	6	3		90	Keunzler (1961b)
Aplysia punctata (mollusc)	calories (annual)	21	15	51	13[a]	Carefoot (1967)
Scrobicularia plana (mollusc)	calories (annual)	13	12	90		Hughes (1970)
Asellopsis intermedia (harpacticoid copepod)	C (annual)	5	11	82	2[b]	Lasker *et al.* (1970)
Idothea baltica	calories (daily)	4	—	94	2[b]	Tsikhon-Lukanina and Lukasheva (1970)
Hyalella azteca (amphipod)	calories (annual)	15		49	36[b]	Hargrave (1971)
Tellina tenuis (mollusc)	C (annual 3 yr)	14	1	84		Trevallion (1971)
		5	8	87		
		12	16	71		
Nerita (3 spp.) (mollusc)	calories (annual)	12	2	86		Hughes (1971a)
Fissurella barbadensis (mollusc)	calories (annual)	24	3	73		Hughes (1971b)
Tegula funebralis (mollusc)	calories (annual)	13	1	77	10[a]	Paine (1971)
Hippoglossoides platessoides (teleost)	calories (annual)	31	20	72		MacKinnon (1973)
Stronglyocentrotus droebachiensis (echinoderm)	calories (annual)	6	1	25	68[b]	Miller and Mann (1973)
Tigriopus brevicornis (harpacticoid copepod)	N (annual)	4	23	73	0·4[a]	Harris (1973)
Neanthes virens (polychaete)	calories (monthly totals)	71		26	3	Kay and Brafield (1973)
Hydrobia ventrosa (mollusc)	C (daily)	29		39	26	Kofoed (1975)
Parechinus angulosus (echinoderm)	calories (annual)	7	1	33	59	Greenwood (1980)
Chlamys islandica (mollusc)	calories (annual)	11	2	23	63	Vahl (1981)
Ilyanassa obsoleta (mollusc)	calories, C N and P (annual)	1	2	16	74–81	Edwards and Welsh (1982)

[a]Assumed value for mucus production, calculated by difference or assumed as a constant proportion of assimilation.
[b]Loss through moulted exoskeletons.
[c]Calculated from measured dissolved organic carbon released.
— Indicates that estimates were not made.

have attempted to quantify some or all budget terms of eqn. (152) for various aquatic organisms (Table 39). Calories, carbon, nitrogen, phosphorus, and sulfur have been used as a basis for calculation but few studies have estimated all terms as a check on the accuracy of measurements. Also, only one study (Edwards and Welsh, 1982) has calculated an energy and elemental budget simultaneously. The mud snail *Ilyanassa obsoleta* assimilated calories, carbon, nitrogen, and phosphorus with different efficiencies (62%, 63%, 58%, and 91% respectively). Mucus production was estimated to account for up to 80% of assimilated calories, carbon, and nitrogen while somatic tissue growth and reproduction accounted for 2 to 3% of assimilation.

Although carbon is most often measured, growth, as protein synthesis, may be better assessed by measurement of nitrogen. For example, Tenore (1981) observed that caloric content was most significantly correlated with growth when the polychaete *Capitella* was cultured on detritus prepared from various plant species with high nitrogen content. Growth was most significantly related to nitrogen content when diets low in nitrogen were offered. Also, while carbon and calories are lost directly as CO_2 and heat during respiration, other elements may be metabolized and excreted in soluble form. It is thus difficult to systematically compare the fate of assimilated material measured in different forms. In addition, although growth of individual organisms of small body or cell size can be measured in short-term experiments, studies of population growth and estimates of production of larger organisms measured over long periods must usually be extrapolated to a reduced time scale. Conversely, short-term measurements of respiration and excretion must be extended to annual rates to be consistent with production estimates.

Some general conclusions may be drawn although these problems limit the accuracy of many budget calculations. Growth and assimilation vary greatly with food supply and rates of production (tissue and gonad growth) are often more dependent on food supply than are rates of respiration (Schiemer *et al.*, 1980). Also, the amount of assimilated material stored for growth varies seasonally and with age and it is clearly dependent on reproductive state. The proportion of assimilated energy and material utilized for metabolic activity (respiration and excretion) increases as organisms grow in size. These pathways of energy loss may account for almost all of that which is assimilated in mature slow-growing individuals. Estimates of assimilation calculated by difference after summation of other terms in a budget calculation can be in error if loss of soluble excretory material is not measured.

The inter-related nature of respiratory and excretory processes was demonstrated by Satomi and Pomeroy (1965) who observed a linear relation

$$y = -3{\cdot}1 + 0{\cdot}015x \qquad (163)$$

between inorganic (molybdate reactive) phosphate release (y as μg-at PO_4/hr) and respiration (x as μg-at O/hr) per gram dry weight in net zooplankton, oysters, and sea anemones. Soluble organic phosphorus was released in a constant proportion to phosphate (P–PO_4:total P = 0·6). Webb and Johannes (1967) also observed that planktonic and benthic invertebrates released ammonia and dissolved free amino acids in a ratio of approximately 4:1. Complete oxidation of plankton with a C:N:P ratio of 106:16:1 would require 276 atoms of oxygen per atom of phosphorus and 16 atoms of oxygen per atom of nitrogen (Redfield, 1958). The O:P ratio of 267 observed in oysters and values of 54 to 72 in zooplankton (Satomi and Pomeroy, 1965) and O:N ratios as low as 1 in some invertebrates during winter (Conover, 1978) imply variable and incomplete oxidation of food. Complete protein oxidation should yield an O:N ratio of 8 and much lower values could reflect processes such as gluconeogenesis which would alter expected relations between respiration and excretion during periods of low food supply (Mayzaud, 1973).

Respiratory and excretory processes respond similarly to changes in temperature and body size. Miller and Mann (1973) calculated an average Q_{10} value of 2·05 for respiration by fourteen species of macrofauna. Price and Warwick (1980) derived similar values (1·2 to 2·2) for three meiofauna species. Increases in size-corrected respiration rate (R_o) with increasing temperature (T) were adequately described in all studies, by an equation of the form

$$\log_{10} R_o = a + bT \qquad (164)$$

where a and b are constants which may differ for different species. Similarly, respiration may be standardized with respect to organism size as described for planktonic animals in Fig. 62. Exponents of the size–metabolism relationship may vary slightly during the year and with experimental temperature (Newell and Roy, 1973). Changes in nutritional state would also occur throughout the year and measures of respiration should reflect this after standardization for size and temperature effects. Bayne (1976) used measurements of respiration after they were standardized for these variables as a physiological indicator for the nutritional status of populations of the mollusc *Mytilus.* Harding (1977) reviewed previous studies with aquatic organisms and quantitatively tested the idea that metabolic rate per unit weight decreases with increasing size of organisms but is constant per unit surface area. He measured the external surface area of the euphausiid *Thysanöessa* and found this to be proportional to wet weight$^{0·84}$ while oxygen consumption was proportional to weight$^{0·82}$. Respiration was thus proportional to body surface area in this species.

There have been few *in situ* measures of respiration with marine organisms. Usually incubations are made in the laboratory under conditions which may differ significantly from those in nature. Comparisons may be made, however, when experimental conditions are standardized. Torres *et al.* (1979), for example, measured the respiration of bathypelagic fish caught from the surface to a depth of 1000 m and incubated in respirometers on-board ship. A decrease in respiration (T, μl O_2/mg wet weight/hr) with respect to the minimum depth of occurrence (D, m) was described by the equation

$$T = 0·29D^{-0·49}. \qquad (165)$$

Respiration decreased from values near 0·1 μl O_2/mg wet weight/hr for species caught in the epipelagic zone near the surface to an order of magnitude less for those collected below a few hundred meters depth. A depth-dependent decrease in respiration of pelagic fishes would be expected if food supply and metabolic activity decreases with depth as discussed in Section 6.4.1.2. Smith and Laver (1981) also observed that night-time levels of respiration of the gonostomatid fish *Cyclothone* measured by *in situ* incubation at 1300 m off southern California were 3 to 5 times higher than rates measured during the day. Elevated rates of respiration associated with digestive activity (specific dynamic activity) has been observed in shallow-water species (Ivlev, 1961) and nocturnal feeding of this bathypelagic species could explain a diel rhythm. Enhancement of metabolism associated with feeding should occur in deep water just as at shallower depths.

Excretion of inorganic and organic substances by planktonic and benthic organisms, like respiration, is positively correlated with temperature and inversely related to body size. Exponential or logarithmic functions usually provide the best empirical description of the relationships. Peters and Rigler (1973), for example, summarized earlier studies which compared weight specific release of dissolved phosphorus (E, μg P/mg/hr) and body dry weight (W, mg) when temperature, food concentration, and phosphorus content in food were fixed. The form of the equation derived for *Daphnia*

$$E = 0·16W^{-0·49} \qquad (166)$$

was similar to those derived in previous studies with a variety of benthic and planktonic organisms although exponents varied from −0·33 to −0·89. Methodological differences in various studies may account for some of the variance since both radiochemical and colorimetric methods have been used to follow the time course of phosphorus excretion. Johannes (1964a) attempted to evaluate the relative importance of dissolved phosphorus excretion by different sizes of marine invertebrates by calculating the time required for an animal to excrete an amount equal to its total phosphorus content (TT_B, the body-equivalent excretion time). A power relation existed between this measure and body dry weight (W) with the equation

$$TT_B = 1584W^{0.67} \quad (167)$$

derived for animals greater than 1 mg in weight. The equation

$$TT_B = 398W^{0.33} \quad (168)$$

described the relation for animals between 10^{-7} μg and 1 mg dry weight. O:P ratios were observed to decrease markedly with decreasing animal size when Johannes compared weight-specific phosphorus excretion rates calculated from these equations with estimates of oxygen uptake based on body size. Thus, oxygen consumption was much less affected than apparent phosphorus release with decreasing size. Problems arise in interpreting the data, however, since excretion by animals weighing less than 1 mg was determined with ^{32}P while colorimetric measurements were used for larger animals. The phosphorus content of food ingested by different sizes of animal was also probably different. In addition, the fraction of total soluble phosphorus released as organic phosphorus could have been different in different sizes of animals. Analyses did not permit separation of the two fractions.

Egestion of particulate matter as feces, leakage across the body wall, and true excretion are all involved in the release of dissolved organic compounds by aquatic organisms. There may also be an uptake, at least by some soft-bodied species, which may equal but seldom exceed loss. Few budget studies have quantified net flux due to both processes. Rice *et al.* (1980) monitored the influx of labelled amino acids into larval and juvenile oysters and determined net flux by following the disappearance of primary amines from solutions. Rates of influx and net flux were similar for juvenile *Ostrea* at ambient amino acid concentrations (2 μM). Net influx occurred into larval stages at all concentrations down to 9 μM.

Miller and Mann (1973) suggested that herbivorous sea urchins, like many animals which feed on a largely carbohydrate diet, may secrete excess soluble organic carbon to retain sufficient nitrogen. The identification of nitrogen-fixing bacteria and measurement of nitrogen fixation in the gut contents of the sea urchin *Strongylocentrotus* supports this idea and provides an additional pathway for nitrogen supply (Guerinot *et al.,* 1977). In addition, Fong and Mann (1980) demonstrated that bacteria present in the gut of this species have the ability to digest cellulose. These micro-organisms synthesize essential amino acids and provide urchins with a source of protein which is not available in the natural diet of kelp. These observations show that an organism's food supply may not only consist of ingested material. The presence of symbiotic micro-organisms in the gut content of many species may provide essential nutrients not present in the food.

An aspect of biological transformation of energy and materials of importance in benthic communities concerns the biogeochemical effects of feeding, egestion, and secretion of organic matter. Deposit-feeding invertebrates which burrow, selectively feed, and have a high egestion rate can physically and chemically alter the sedimentary environment. These secondary effects may equal or exceed those directly associated with their own metabolism. For example, the conversion of fine-grained deposits into fecal pellets affects sediment stability and nutrient recycling (Aller, 1980). Fecal material and mucous secretions may also provide sites for enhanced microbial activity (Hargrave, 1976). From a community point of view, these effects might be greater than the metabolic activities of organisms themselves. In addition, benthic suspension feeders are an important link between suspended and sedimented organic particles since the production of biodeposits may equal or exceed gravitational sedimentation. Only about 11% of the calories theoretically available were retained in an annual gross energy budget of a mature Pacific oyster and 60% of these were deposited as feces (Bernard, 1974) (Fig. 103). The high organic content of such material (28·5% in Bernard's study) provides an enriched substrate for deposit-feeding species.

Sediment reworking by deposit-feeding in-

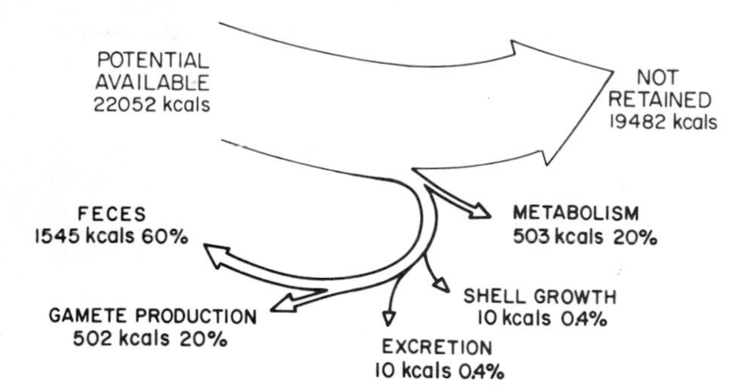

FIG. 103. Percentage distribution of annual calories available and utilized by a 1-m oyster *(Crassostrea gigas)* reef (redrawn from Bernard, 1974).

vertebrates, with feces deposited either within burrows or at the surface, provides a constant renewal of surfaces exposed for microbial growth. Some pellets are resistant to bacterial and mechanical breakdown although fecal pellets of many benthic species are rapidly colonized by bacteria (Fenchel and Jorgensen, 1977). Observations of a deposit-feeding polychaete *Heteromastus* in intertidal sediments of the Wadden Sea, for example, showed that pellets were resistant to breakdown (Cadée, 1979). An average population of 500 individuals/m^2 could transfer a 4-cm-thick sediment layer from 10 to 20 cm, the depth of feeding, to the surface annually. The maximum population observed (9000/m^2) could rework a 20-cm sediment column in 4 months. The influence of such high rates of bioturbation on microbial activity and chemical processes within surface sediments cannot be underestimated.

6.4.4 Population Growth and Production

A direct measure of growth rates in natural populations is possible if individual organisms can be marked and recaptured at a later time (Ricker, 1971). The method is widely used to study fish populations and it provides an estimate of total population size if homogeneous mixing of marked and unmarked animals is assumed and if marking does not affect growth and mortality rates. The method may be applied to molluscs with hard shells and Mann (1972a) used a variation of the technique to measure length increments of seaweed blades. Large crustaceans can also be tagged such that moulting frequency and growth are determined simultaneously (Newmann and Pollock, 1974).

The direct observation of increments in size permits calculation of an individual's specific or instantaneous growth rate

$$G_i = \frac{1}{t} \ln \frac{W_t}{W_0} \qquad (169)$$

where W_0 and W_t are weight (or length) before and after time period t. Values of G usually decrease as organisms age. The data may also be used to consider weight changes with time as a function of maximum attainable weight. This reflects the sigmoid growth pattern of organisms which Beverton and Holt (1957) quantified, following Bertalanffy, by the equation

$$W_t = W_\infty \left[1 - e^{-K(t-t_0)}\right]^3 \qquad (170)$$

where W_∞ represents weight at maximum size, W_t is weight at time t, with t_0 the time of no weight, and K a growth constant. K can be determined by plotting $e\sqrt{W_t}$ after the time interval $t + 1$ against t and if a constant time interval is used, a line with the slope K results which intersects the origin at $W_\infty^{1/3}$. Length–age curves, transformed to straight lines of slope e^{-k} by plotting L_{n+1} against L_n, are termed 'Ford–Walford plots' (Crisp, 1971). The method has been used primarily to study fish growth but it has also been applied to compare growth rates of various bivalve molluscs (Warwick and Price, 1975). Length–frequency analysis of individuals from populations can only provide estimates of specific growth rates of different sizes if all ages are sampled with similar efficiency. Mean length as well as standard deviations in length for various age classes provide parameters useful for estimates of growth at different ages (Schnute and Fournier, 1980).

Total population production (P) over time (t) is given by

$$P = \sum_{t=0}^{t=1} \sum_{0}^{N} G_i W_i \Delta t \qquad (171)$$

if the instantaneous growth rate (G_i) of an individual of weight (W_i) is known. This calculation is based on survivors which are enumerated and it ignores the biomass eliminated by predators. A measure of elimination, either through natural processes or that due to harvesting, is necessary when yield from a population is of interest. Thus, if B_0 is the initial population and B_t that remaining after time t

$$P = (B_t - B_0) + E = \Delta B + E \qquad (172)$$

where E is loss through mortality. E, the instantaneous mortality rate of an individual, may be calculated by determining the slope of the survivorship curve of a single age class of organisms. Plots of abundance over time may be fitted by empirical equations for various portions of the curve. Usually a curvilinear plot best describes changes in abundance, particularly during early life-history stages; however, a linear decrease may occur during certain seasons or in larger weight classes.

Mann (1969) differentiated between these two measures of production by reference to early studies of Boysen Jensen (1919) who measured numbers and weights of macrofauna species annually in the Limfjord, Denmark. Numbers of specific age classes of molluscs and polychaetes decreased with time, but survivors increased in their individual weight. Therefore E ('elimination' as production consumed by predators or lost by natural death if no immigration or emigration occurred) was calculated as

$$E = (N_1 - N_2) \cdot 1/2(W_1 + W_2) \qquad (173)$$

and production by the remaining animals which had increased in weight was determined as

$$P = 1/2(N_1 + N_2) \cdot (W_2 - W_1). \qquad (174)$$

The overall biomass change, ΔB, between two sampling times is

$$\Delta B = N_1 W_1 - N_2 W_2 \qquad (175)$$

and from eqn. (172)

$$\Delta B = P - E. \qquad (176)$$

Wildish and Peer (1981) used these and other calculations to estimate total production when cohorts of similarly aged amphipods in a population were distinguished. Weight–survivorship curves were obtained for a single recruitment over a time interval equal to the lifetime of one generation. Waters (1979) emphasized that production estimates by most methods depend on recognition of discrete cohorts. Instantaneous growth or mortality rates as a function of size must be determined and summed throughout the year for total numbers in each size class when age classes cannot be recognized. Menzie (1980) described a modification of a size-frequency (Hynes) method based on this approach. Estimates of numbers of individuals that develop into each size class are used to calculate the losses in numbers between size categories. Production is assumed to be the sum of biomass lost between successive size categories.

A third type of approach to measurements of production of biomass is based on direct measures of growth. Groups of organisms can be isolated in cores or enclosures which permit feeding. Changes in biomass are measured over time. Comparison of numbers and biomass of isolated individuals with those in the natural population may be used to estimate mortality through natural sources assuming immigration or emigration has not occurred. Rates of growth and mortality should be equivalent in populations where biomass is constant from year to year ($\Delta B = 0$).

Winberg (1968) proposed a 'physiological approach' to calculate growth efficiency (E_g), the total energy of production of body tissue and gonads $\Delta B = P + G$ as a fraction of the energy ingested (R). Thus,

$$E_g = \frac{\Delta B}{R} = \frac{(P + G)}{R}. \qquad (177)$$

This ratio should be calculated on an energy rather than weight basis since fats and carbohydrates have different weight-specific energy content and their proportion as storage products may change seasonally or with animal size. ΔB and R are expressed as weight and E_g becomes a 'conversion efficiency' if caloric content of tissues is unknown. R is seldom known for natural populations and it is usually assumed to be approximated by the sum of $P+G+T$, with U neglected. While this is not justified (Section 6.4.3), many workers have followed the convention which then permits production to be calculated from estimates of respiration and growth efficiency by the equation

$$E_g = \frac{\Delta B}{\Delta B + T} \tag{178}$$

and

$$\Delta B = E_g\ (\Delta B + T). \tag{179}$$

Estimates of respiration and growth efficiency at different temperatures must be independently established. This introduces additional errors due to artificial experimental conditions. The method has the advantage that ΔB may be estimated from size-frequency histograms and no age-class separation is necessary.

The physiological approach provides no estimate of elimination from a population. A conservative estimate of this removal may be derived, however, when organisms consumed as food by specific predators can be measured. Since predation rates must be known or predicted for comparison with loss from a spatially defined prey population, the technique may be most useful when experimental enclosures are used to isolate populations. An example of the approach was used by Trevallion *et al.* (1970) for plaice feeding on siphons of the mollusc *Tellina*. Only extended siphon tips were cropped and regeneration was possible. *Tellina* and plaice were held in experimental tanks in varying numbers and energy requirements by plaice were compared to calculate energy requirements for siphon regeneration. Energy utilized to regenerate siphons (elimination) was equivalent to estimated energy requirements of the plaice populations.

Generation time (also termed 'doubling time' or 'turnover rate') has been used to estimate production from biomass determinations. The figure can be multiplied by standing crop values of natural populations to provide a measure of production potential when life cycle or laboratory culture studies permit estimates of generation time. Gerlach (1978) and Schwinghamer (1981) used the approach to estimate annual production which might be achieved by benthic bacteria, meiofauna, and macrofauna. The technique is most applicable to small organisms with a rapid generation time but growth and development rates in the laboratory are often more rapid than would occur in nature.

The concept of reproductive or replacement rate is implicit in this approach to measures of population growth. Smith (1954) suggested that this could be related to generation time (t) by the equation

$$R_0 = e^{r_m t} \tag{180}$$

where R_0 is the sum of the offspring of each female per unit time (reproductive rate), and r_m is the growth rate in the absence of any limiting conditions (intrinsic rate of natural increase). A population replaces itself each generation if $R_0 = 1$. Although factors such as food availability and temperature greatly affect measures of t, particularly in small organisms in culture, Bonner (1965) and Sheldon *et al.* (1973) observed that doubling time was related to body size by a power function. The equation

$$\log_{10} t = 0{\cdot}64 + 0{\cdot}80 \log_{10} L \tag{181}$$

can be derived from their data which compare doubling time (t) in days with body length (L) in millimeters. Fenchel (1974) also found that values of r_m are related to average body weight (W) in various organisms by the equation

$$\log_{10} r_m = a - 0{\cdot}275 \log_{10} W \tag{182}$$

where the constant a differed for unicellular (−1·94), heterotherm metazoan (−1·64), and homiotherm metazoan (−1·4) organisms. Blueweiss (1978) used a larger data base to derive a similar equation with a slope of −0·26. The relationship is similar to that which describes metabolic rate per unit body weight and body size with an exponent of −0·249. Fenchel concluded that r_m is a measure of the maximum productivity (production potential) of an exponentially growing population which must be correlated with metabolic rate.

These equations may be useful for predicting the relative importance of production by different sizes of organisms in a community. They may also prove useful for observing regularities in size composition of some ecosystems. Kerr (1974), for example, by assuming that predator and prey sizes were simply related and that growth and metabolism were both power functions of body size, was able to calculate that the standing stock of prey organisms should typically be in the order of 1·2 times that of their predators. Sheldon *et al.* (1972) provided empirical evidence for this in pelagic marine ecosystems. Concentrations of particles (1 to 10^6 μm diameter) suspended in surface waters of various oceanic areas were relatively uniform, with a slight decrease when data were grouped in logarithmic size intervals. Schwinghamer (1981) used a similar approach to calculate the magnitude of production expected in various size classes of benthic organisms. His calculations showed, as did those of Gerlach (1978), that bacterial production was much greater than that of any other size group of organisms. This result arises from the larger biomass of micro-organisms present in most fine-grain sediments. However, Stevenson (1978) has pointed out that, as in soil, most aquatic bacteria may exist in a dormant state. Calculations of production based on microbial biomass estimates and assumed turnover rates will be erroneous if this is the case.

Since growth and respiration rates are both correlated with body size, population production and respiration should also be related. Respiration is a variable which can be directly measured over short time intervals, and this provides an attractive alternative to measures of growth. Humphreys (1979) analyzed published energy budgets for terrestrial and aquatic animal populations to reassess earlier observations which compared annual caloric values (kcal/m²/yr) of production (P) and respiration (T). The equation

$$P = aT^b \tag{183}$$

described the relationship derived in previous studies. Data for fourteen taxonomic groups yielded significant relationships but the slopes were not similar. Exclusion of data for insectivores and molluscs yielded parallel lines with a common slope which did not differ significantly from 1·0. A conservative estimate of assimilation (A_r) as the sum of P and T is provided by these studies. Then production efficiency is equivalent to

$$P_{eff} = \frac{P}{A_r}. \tag{184}$$

Humphreys found that invertebrate groups were separable into trophic categories on this basis with herbivores having the lowest and carnivores the highest production efficiency. There was no large difference in these values for poikilothermic and homiothermic animals as was implied in earlier studies.

Another approach to estimate production makes use of the concept of 'turnover time', the reciprocal to the finite death rate. A population with a daily instantaneous death rate of 0·69 has a finite death rate of ln(0·69) = 0·5 and its turnover time (reciprocal) is 2 days. At average biomass $\bar{B}$, production is $0{\cdot}5\bar{B}$ per day. The finite death rate is also equivalent to the ratio $P/\bar{B}$ (the 'turnover ratio' TR) which, as weight-specific production, allows production by populations of different biomass to be compared on a common basis. Attempts to make general statements about turnover ratios in aquatic organisms are confounded by effects of temperature, organism size, and the duration of production and biomass estimates relative to the total life cycle. Regularities have appeared, however. Waters (1969, 1979) calculated that within the life span of an age class (i.e. per generation), populations with different mortality and growth patterns have turnover ratios between 2·5 and 6, with a modal value of 3·5.

Annual turnover ratios are equivalent to this figure multiplied by the number of generations per year.

Sufficient data have accumulated to allow a test of Waters' idea of an approximately constant turnover ratio per generation. Studies* ($n = 13$) in which annual production and average annual biomass have been estimated for various macrobenthic species ($n = 55$) were compared on the basis of generation time and the equation

$$\log_{10} \text{TR} = 0{\cdot}69 - 0{\cdot}14t \qquad (185)$$

was derived where t, mean lifespan in years, accounted for 60% of the variance in TR. Substitution of $t = 1$ into the equation yields a turnover rate of 3·55, precisely the modal value suggested by Waters. The semilogarithmic nature of the relation, also observed by Zaika (1973), may arise from the lack of inclusion of data from organisms with short (less than 1 year) generation times. Robertson (1979) made a similar comparison ($n = 49$) for marine macrobenthos with a lifespan >1 year. He derived the power function

$$\log_{10} \text{TR} = 0{\cdot}66 - 0{\cdot}73 \log_{10} t \qquad (186)$$

which accounted for 70% of the variance in the regression.

It is often difficult to relate published $P/\bar{B}$ estimates to specific cohort generation times for short-lived organisms and thus the conservative nature of lifetime turnover ratios cannot be assessed when $t < 1$. Gerlach (1971) estimated life-cycle duration in two species of nematodes and calculated cohort turnover ratios between 2 and 3. Meiofauna populations with 3 to 12 generations per year could have annual turnover ratios between 10 and 36, values comparable to those for zooplankton populations. Allen (1971) pointed out that the relation between P and $\bar{B}$ over time depends on the shape of growth and mortality curves. Mean age and mean lifespan both equal the reciprocal of the $P/\bar{B}$ with constant exponential mortality. For other mortality functions, if growth in weight is linear, the $P/\bar{B}$ ratios equals the reciprocal of mean age. There is no simple relation if growth follows a non-linear pattern.

High annual turnover ratios for small-bodied organisms are predictable from the relationship between body size and generation time [eqn. (181)]. Banse and Mosher (1980) compared specific production rates per unit biomass ($P/\bar{B}$) for thirty-three invertebrate species of different body mass (M, kcal). An inverse power curve

$$P/\bar{B} = 0{\cdot}65M^{-0{\cdot}37} \qquad (187)$$

described the relationship for terrestrial and aquatic invertebrates as had been identified in earlier studies. Much of the variability of measured values of $P/\bar{B}$ against those predicted from this equation was attributable to variability in the ratio of annual production/respiration. Small metazoan species tended to have lower $P/\bar{B}$ values than indicated by the relationship for larger invertebrates. Turnover ratios for small organisms may be more variable than those for larger organisms since their rapid generation time provides more opportunity for environmental factors (food availability, temperature) to affect growth and development rates. High energy flux through populations of small organisms such as bacteria and meiofauna is also usually accompanied by high spatial and temporal variability. In contrast, fish and macrobenthos, with their large body size and longer generation time, are less affected by short-term environmental variability once critical early growth periods are past. Banse and Mosher (1980) pointed out that in stable populations the specific mortality rate equals $P/\bar{B}$ and hence this also declines by $0{\cdot}65M^{-0{\cdot}37}$. Presumably mortality rates are also more variable for small organisms than for larger ones.

The importance of temperature on turnover rates, largely independent of the effects of body size and generation time, was demonstrated by Johnson and Brinkhurst (1971a). Estimates of production by major taxonomic groups of freshwater macrobenthos (all with 1- to 2-year life cycles) in four different areas were derived from calculations of instantaneous growth. Although turnover ratios

---

*Boysen Jensen, 1919; Sanders, 1956; Richards and Riley, 1967; Peer, 1970; Zaika, 1973; Hughes, 1970, Miller and Mann, 1973; Burke and Mann, 1974; Buchanan and Warwick, 1974; Warwick and Price, 1975; Chambers and Milne, 1975; Klein *et al.*, 1975; Nichols, 1975.

differed between groups, values were related to annual mean temperature by the equation

$$TR = T^2/10 \quad (188)$$

where TR is the annual turnover ratio and T is the annual mean temperature (°C). Annual production was described by the equation

$$P = \bar{B}\,[T/10]. \quad (189)$$

The relation implies that production may be calculated from measures of standing stock if annual mean temperature is known. The authors suggested that biomass of organisms at a particular site may be adjusted to utilize energy supplied consistent with a turnover ratio set by temperature. The importance of temperature in shortening generation time of pelagic organisms of different size was emphasized by Sheldon *et al.* (1973) and empirical relationships between temperature and development time for zooplankton were discussed in Section 4.2.1. Palmer and Coull (1980) determined development rates for egg, naupliar, and copepodite stages of a meiobenthic copepod held over a range of temperatures. Belehradek's function and a curvilinear model adequately described relationships between growth and temperature.

Estimates of production by benthic macrofauna populations have been made by applying different methods of calculation in a single study. Kuenzler (1961a) compared estimates of instantaneous growth and mortality to calculate annual production by a population of the mussel *Modiolus* but poor agreement was achieved. Production by several freshwater macrobenthic species calculated by the growth efficiency–respiration technique was similar to that estimated from instantaneous growth rate although some data were common to both sets of calculations (Johnson, 1974). Production estimates for the amphipod *Pontoporeia femorata* calculated by summation losses of cohorts, growth increment survivorship curve methods and the size-frequency method yielded values between 11·1 and 13.6 g weight with annual turnover ratios between 3·6 and 4·8 (Wildish and Peer, 1981). P and $\bar{B}$ values estimated by the size-frequency method were higher than for the cohort-based methods due in part to corrections for subsampling.

Calculations of energy flow through an entire food chain, community, or ecosystem have been made in order to describe how production is distributed amongst co-existing organisms (Section 4.2.3). Ecological transfer efficiencies for particular predator–prey associations cannot be calculated since most organisms derive food from several sources. Also, the precision of production estimates for various groups of organisms is not equivalent with many values calculated by difference or through application of assumed $P/\bar{B}$ values. However, sufficient data have been collected in some locations to encourage comparison of measured primary production with production by major groups of organisms (herbivorous copepods, carnivorous zooplankton, benthos, and fish). Production by various groups can be compared to calculate ecological transfer efficiencies without specifying actual food supply.

Steele (1974) and Anderson and Ursin (1977) used this approach to summarize existing information on production by various organisms in the North Sea. A similar description has been made for the western Baltic (Arntz, 1978), the Scotian Shelf off eastern Canada (Mills and Fournier, 1979), and the Beaufort Sea shelf (Walsh and McRoy, 1979). There are some common features in all of these attempts to connect consumer organisms with their food supply on the basis of production. The role of bacteria, which convert dissolved and particulate matter into biomass available to consumer organisms in the water column and sediment, was not quantified but such transfers were identified as being critical in determining the amount of deposited organic matter available for benthic production. In addition, invertebrate carnivores, both planktonic and benthic, consume unmeasured amounts of energy which reduces the amount available to larger predators. The yield of commercially exploited pelagic and demersal fish in all areas varied between 4 and 10 kcal/m^2/yr. Yet, given reasonable assumptions concerning food-web structure, ecological efficiencies and using measured values of phytoplankton production, each study showed that there were difficulties in providing enough energy to support the observed levels of fisheries yield.

Pomeroy (1979) developed a compartmental model of energy flow through a hypothetical continental shelf ecosystem which linked energy flow from phytoplankton, through detritus, micro-organisms, and dissolved material to terminal consumers. Although no study has provided biomass estimates for all of these compartments, data from the North Sea and the Scotian Shelf can be used in Pomeroy's conceptual food web to assess the adequacy of energy transfer to support estimated rates of production by higher trophic levels (Fig. 104). The absence of data for energy flow through detritus, dissolved organic material, and micro-organisms necessitates a condensation of the web of interactions proposed by Pomeroy, but the configuration provides sufficient energy flux to support both pelagic and benthic consumer organisms even assuming that current estimates of phytoplankton production are not greatly in error. Gross growth efficiencies of 50% for bacteria, 30% for grazers, mucus-net feeders, benthic invertebrates and carnivorous zooplankton, and 10% for fishes and other carnivores permitted Pomeroy to increase production expected by terminal consumers to levels equal to or above those rates estimated from commercial yields. Steele's (1974) observations of demersal and pelagic fisheries yield could also be reconciled with observed or calculated production by benthos and plankton by assuming growth efficiencies of 20% or greater for these organisms.

Mills (1975) discussed the idea that benthic communities with an abundance of epibenthic invertebrates might support lower biomass of demersal fish because of competition for a limited common food source between fish and invertebrates. He suggested that benthic communities dominated by epibenthos are functionally distinct from those where infauna (bivalves, polychaetes) predominate. While the effects of such structural differences on pathways and efficiencies of energy flow within and out of the benthos are unknown, observations of predation effects on macrobenthos show that predator–prey associations in food webs

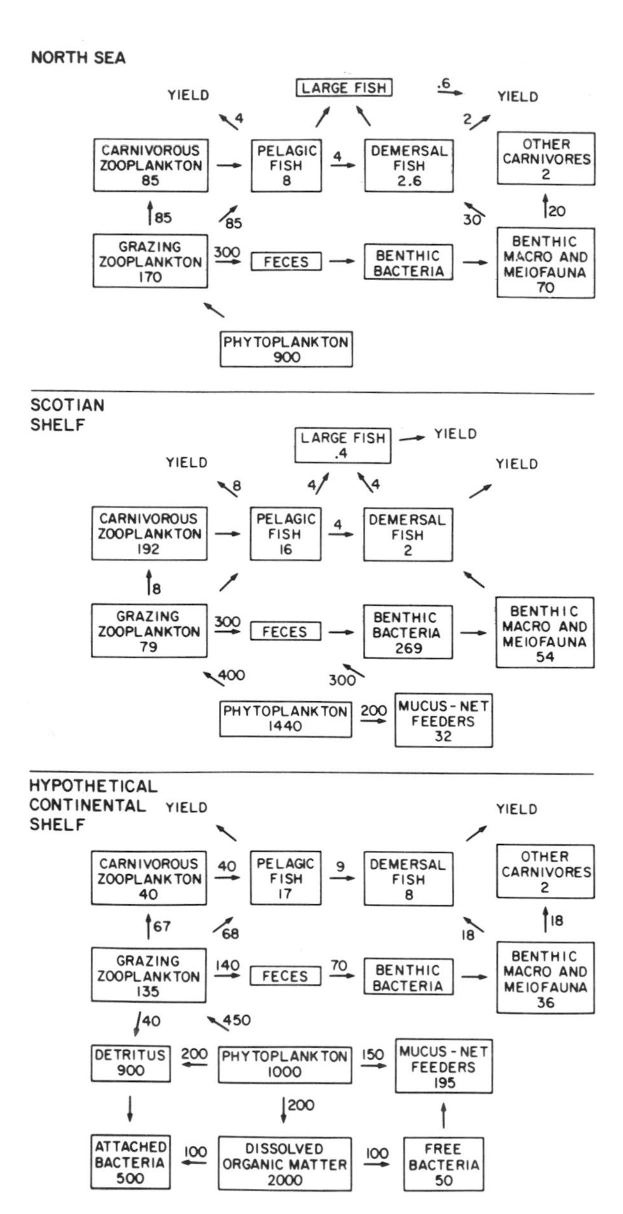

FIG. 104. A diagrammatic food web of major groups of organisms and their estimated annual production (kcal/m^2) in continental-shelf ecosystems. The network of interactions was used by Pomeroy (1979) to develop conceptual models for a shelf ecosystem. Data from the North Sea and the Scotian Shelf were taken from summaries presented in Steele (1974) and Mills and Fournier (1979).

may be tightly linked. Feeding by plaice eliminated approximately half of the annual macrofauna production and also dramatically altered community structure of benthic fauna in a coastal marine bay (Levings, 1975; MacKinnon, 1973). Arntz (1978) estimated that demersal fish, primarily cod and dab, consumed 4 g $C/m^2/yr$ in the western Baltic. An annual turnover ratio of 0·4–1·0 for the standing stock of total macro-benthos would yield a net production between 5 and 14 g $C/m^2/yr$. Cropping could thus remove a substantial amount of annual production and be of major importance for limiting the development of biomass of epifauna — those species at the sediment surface most vulnerable to predation by fish.

6.4.5 Community Metabolism

Energy exchange in biological systems can be measured as either synthetic (anabolic) or degradative (catabolic) processes. Organisms use diverse methods to accumulate energy and thus a measurement of production by all species in a community is difficult to obtain. Energy release, however, as aerobic (oxidative) or anaerobic (sulfate and nitrate reduction, denitrification, and fermentation) metabolism involves exothermic biochemical reactions. Thus measures of heat release provide a common basis for comparing metabolic activity in all organisms.

Calorimetry has been widely used for measuring metabolic heat production in homoiotherms but measurements with small organisms like aquatic invertebrates are difficult because of their low rates of metabolism. Also, endothermic reactions involved in chemical precipitation, dissolution, and ionization in water and sediments can result in heat absorption which exceeds low rates of heat release by micro-organisms and invertebrates present in samples (Pamatmat, 1975). The use of dense cultures of micro-organisms or double-chambered calorimeters with organisms confined to one chamber may avoid these problems (Brown, 1969).

Pamatmat and Bhagwat (1973) were able to measure metabolic rates of 3 to 17 mcal/g/hr when sediment from Lake Washington was placed in a single chamber calorimeter. Long periods (over 12 hr) were required for temperature stabilization and thermal equilibrium was not established. Also, thermochemical processes resulted in net heat absorption when intact sediment cores were incubated which prevented detection of metabolic heat production (Pamatmat, 1975). More recent studies (Pamatmat *et al.,* 1981) with beach and subtidal sand showed that heat production was not consistently correlated with adenosine triphosphate (ATP) concentration (assumed to measure metabolically active biomass) or electron transport activity (ETS) (activity of dehydrogenase enzymes). Heat production/ATP ratios in batch cultures of an obligate anaerobic fermenting bacteria *(Bacteroides)* were relatively stable during active growth but values decreased during stationary and senescent phases. A variable ratio of heat production/ATP would be expected if only part of the biomass of micro-organisms in sediments is actively growing and the remainder exists in a dormant state (Stevenson, 1978). The direct measure of anaerobic metabolism as heat production by bivalves held so as to prevent aerobic respiration (Pamatmat, 1979) showed that anaerobic activity in large organisms can be measured as heat production.

Chemical methods which measure metabolism by indirect means have been used more widely than the determination of heat production. Chemical measures of metabolic activity are based on the fact that all metabolism involves the transfer of H^+ through successive steps of dehydrogenation. High-energy ATP is formed when electrons are passed from dehydrogenase enzymes to various electron acceptors. The final transfer is to oxygen in aerobic respiration and thus oxygen uptake (or CO_2 release) is used as an indirect measure of hydrogen production. However, electron acceptors (H_2S, CH_4^+, and Fe^{++}, NO_3^- or reduced organic compounds in fermentation) other than oxygen are used in anaerobic metabolism. Thus, whereas oxygen consumption only reflects aerobic metabolic activity in organisms, total (aerobic + anaerobic) metabolism may be measured as the rate of hydrogen or electron transfer by the activity of dehydrogenase enzymes which participate in the transfer.

Pamatmat (1975) reviewed past studies which have used artificial electron acceptors and carriers to measure rates of electron transfer. Reduction of TTC (2,3,5-tri-phenyltetrazolium chloride) to formazan has been used as a technique for measuring total dehydrogenase activity (DHA) in plankton (Packard, 1971) and sediments (Pamatmat and Bhagwat, 1973). Refinement of the technique (Wieser and Zech, 1976) avoids prolonged incubation of samples by extraction and solubilization of enzymes and determination of initial and final activity by standardized kinetic analysis with added substrate. Incubation of extracted enzyme with high concentrations of substrates may represent the metabolic potential of organisms rather than *in situ* metabolic rates. Electron transport activity in anoxic Lake Washington sediment was positively related to metabolic heat release (Pamatmat and Bhagwat, 1973) and to ATP content in anaerobic Norwegian fjord sediments (Pamatmat and Skjoldal, 1974). Wieser and Zech (1976) estimated that 80% to 90% of total electron transport activity in carbonate beach sand was associated with particle surfaces. Thus, interstitial organisms were assumed to be of little importance relative to attached microfauna and fauna for total energy release by the benthic community.

ATP content both in absolute concentrations and relative to other phosphate compounds containing lower amounts of energy have been used to estimate metabolic activity in sediments. Decreased ATP content and increased DHA with increased sediment depth (Pamatmat and Skjoldal, 1974) could reflect changes in species composition, metabolic type and activity or concentrations of reduced substances. Christensen and Devol (1980) investigated losses which occur during extraction of ATP from sediments. Modification of extraction methods can overcome this limitation. These authors estimated adenosine diphosphate (ADP) and adenosine monophosphate (AMP) by enzymatic conversion to ATP. The three adenylate concentrations were used to determine the energy charge (EC) ratio defined as

$$EC = (ATP + 0{\cdot}5\ ATP)/(ATP + ADP + AMP). \qquad (190)$$

The ratio approaches 1·0 in bacterial cultures when ATP forms almost all of the total adenylate concentration under conditions favourable for growth and active metabolism. It falls to lower values (below 0·8) when growth is depressed. Profiles of total adenylates in sediments from Puget Sound and the Washington continental shelf were similar (4–7 nmoles/cm^3 in surface sediment) but EC values in Puget Sound samples were higher (0·62) than those from the continental shelf (0·49). This implies that microbial populations in Puget Sound sediments were in a higher state of metabolic activity but neither location supported bacterial populations in a state of rapid growth. The conclusion was supported by higher rates of benthic oxygen uptake in Puget Sound compared with those on the continental shelf.

Difficulties exist in interpreting measures of DHA activity and ATP concentrations in metabolic terms, however, since neither TTC reduction nor ATP content can readily be converted to a measure of metabolic rate. Studies with freshwater sediments showed that chemical reduction of TTC to formazan occurred at similar rates before and after poisoning with formaldehyde (Pamatmat, 1975). The positive correlation between TTC reduction and ATP content observed by Pamatmat and Skjodal (1974) would be expected if dehydrogenase activity and ATP content are present in microorganisms in a constant ratio in proportion to metabolic activity.

Measurements of metabolism calculated by conversion of oxygen uptake or carbon dioxide release to energy equivalents are widely used techniques for determining metabolic rates in organisms. However, estimates provided by these methods can be in error for two reasons. Reduced substances are formed which are not immediately oxidized if anaerobic processes occur. Secondly, actual heat production is less than values calculated by conversion of oxygen uptake by application of oxy-calorific equivalents if energy released is not converted into heat but stored as high-energy compounds. Crisp (1971) discussed these and other problems associated with interpretation of laboratory measurements of respiration. However, in view of the many other errors involved, the assumption that 1 ml of oxygen consumed at n.t.p. is equivalent to 4·8 calories (1 mg O_2 = 3·34 cal) as a mean value for oxidation of different substrates introduces a small error.

A more serious problem for conversion and interpretation may exist when community oxygen uptake is measured with undisturbed sediment cores or stirred samples. Oxygen may be consumed by non-biological chemical oxidation (measured as residual oxygen uptake after poisoning of biological respiration with buffered formalin or $HgCl_2$) if reduced inorganic (HS^-, Fe^{+++}) or organic compounds are present (Dale, 1979). Respiratory quotients ($\Delta CO_2/\Delta O_2$) are not comparable to those observed with individual organisms since carbon dioxide is not evolved in these oxidation reactions. Also, chemoautotrophic bacteria may use free carbon dioxide as a carbon source or it may be produced by fermentation without oxygen uptake where aerobic and anaerobic heterotrophic and autotrophic organisms co-exist, as in sediments.

Oxygen and carbon dioxide exchange have been widely used as measures of biological activity in studies of metabolism in soil, water, and sediments despite these problems. Oxygen depletion from bottom layers of stratified lakes was used as a basis for lake classification (Hutchinson, 1938) and Ohle (1956) used carbon dioxide accumulation in a similar way as a bioactivity index for whole lake metabolism. The enclosure of water samples in dark bottles, with and without concentration of particulate matter, has been used to measure respiration in mixtures of planktonic organisms concentrated by reverse filtration (Pomeroy and Johannes, 1968). Oxygen consumption was predominantly due to microflagellates (2 to 10 μm) and bacteria associated with flocculated organic aggregates. Maximum respiration rates (up to 14 mg-at O/m^3/day) occurred in surface waters above the continental slope in the western North Atlantic and off the coast of Peru. Respiration below 500 m was estimated to be several orders of magnitude less than rates in surface water. Comparison of respiration rates with estimates of primary production determined with ^{14}C in other studies indicated that large spatial and temporal differences may exist between production and consumption of organic matter. The imbalance could arise from advective transport of particulate matter to or from other areas and loss through sinking.

Input of energy or elements to any ecosystem under steady-state conditions will be balanced by loss and no accumulation. Seki *et al.* (1968) used this idea to calculate organic-matter turnover (decomposition) rates in nearshore sediments from the ratio of supply/steady-state concentrations of carbon, nitrogen, and phosphorus. Hargrave and Phillips (1981) made similar calculations to compare the annual release of CO_2 and the concentration of organic carbon in subtidal sand sediments. Carbon present in the upper 2 cm of sediment was sufficient to supply the measured annual flux. In ecosystems where no allochthonous supply of organic matter occurs, the ratio of photosynthesis:respiration (P/R) characterizes the balance between energy input and release (Odum, 1971). Physical removal (burial or transport) or accumulation of incompletely oxidized compounds resulting from anaerobic metabolism can cause an underestimate of metabolic loss. RQ values in a salt marsh fell between those expected for aerobic oxidation of fats and carbohydrates (0·7–1) when community metabolism was determined by following changes in dissolved oxygen and carbon dioxide over 24 hr ('diel curve method') (Nixon *et al.,* 1976). Simultaneous measurements of communitity photosynthesis yielded PQ values between 1·1 and 1·2. Thus a balance existed in the marsh where material produced during the day was respired at night. Production and respiration rates in laboratory microcosms also showed a tendency to converge (P/R = 1) (Beyers, 1963) which implies a tight coupling between metabolism and production in spatially limited ecosystems. Pomeroy (1970) considered this to be characteristic of systems where nutrients available for production are largely determined by recycling mechanisms. This would explain Copeland's (1965) observation that gross photosynthesis and respiration in a laboratory microcosm returned to original levels when light intensity was changed despite completely altered species composition. Total nutrient content remained unchanged and photosynthesis was re-established at a rate dependent on nutrient regeneration.

Teal and Kanwischer (1961) proposed that total oxygen uptake by undisturbed sediments be considered to be an integrated measure of aerobic and anaerobic metabolism by an entire benthic community. Pamatmat (1975) questioned this idea since reduced metabolic by-products of anaerobic metabolism accumulate in many sediments and chemical oxidation of these compounds may comprise a variable proportion of total oxygen uptake. Dale (1979) also emphasized that chemical oxidation cannot be assumed to be proportional to anaerobic metabolism. Pamatmat and Bhagwat (1973) calculated that rates of metabolism measured by electron transport activity were up to 39 times greater than rates estimated from chemical oxidation. On the other hand, Jørgensen and Fenchel (1974) observed that H_2S and F_eS accumulation in detritus-enriched sand in microcosms approximated the rate of sulfate reduction. Part of the sulfide that diffused to the sediment surface was oxidized by CO_2 produced by photoautotrophic sulfurbacteria which would not be detected by measures of chemical oxidation. Total net oxygen uptake (corrected for oxygen produced by algal photosynthesis) over 6 months (470 μM O_2/cm^2) could have oxidized 1·4 mg of carbohydrate per square centimeter to CO_2 which was approximately equal to the observed loss in organic matter. Thus, although total oxygen consumption may underestimate benthic community metabolism, the measure still appears to be useful as an indicator of the amount of organic matter oxidized in sediments.

Gradients in dissolved oxygen which form across sediment and particle surfaces show that uptake may be limited by rates of diffusion. Several studies have utilized the basic equation

$$\frac{dD}{dt} = D \cdot \frac{d^2C}{dz^2} \qquad (191)$$

where C is the concentration of diffusing material, t is time, D is the diffusion coefficient and z is distance, to describe changes in dissolved oxygen concentration with depth in sediment and soil. Hutchinson (1957) discussed ideas developed by Grote which expressed the depth of oxygen penetration into sediment as a function of oxygen concentration at the surface, molecular diffusion, and metabolic consumption. Thus, the depth at which oxygen disappears with a linear concentration gradient in the sediment is given by

$$Z_0 = k(O_2)/R \qquad (192)$$

where k is the molecular diffusion coefficient (cm^2/s) for oxygen in the sediment, (O_2) is the oxygen concentration at the sediment surface (mg/l), and R is the rate of oxygen consumption per unit volume of sediment with unlimited oxygen supply.

Murray and Grundmanis (1980) observed that dissolved oxygen in interstitial waters of pelagic red clay and carbonate ooze sediments from the central Pacific decreased with depth to level off at non-zero values even 20–40 cm below the surface. They hypothesized that a low supply of oxidizable organic carbon mixed into surface sediments by bioturbation limited respiration of heterotrophic organisms at depth. They fitted data to an exponential equation of the form

$$O_2(Z) = O_2{}^1 + \Delta O_2 \exp(-BZ) \qquad (193)$$

where ΔO_2 was the total decrease in oxygen concentration from the interface to the depth where a constant value ($O_2{}^1$) was achieved and z was depth (cm). B (a constant to reflect the first-order consumption of oxygen in proportion to the concentration of particulate organic carbon within the zone of bioturbation) with units of cm^{-1} was calculated as

$$B = -\frac{w-(w^2 + 4kK)^{1/2}}{2K}. \qquad (194)$$

Values for w (the sedimentation rate), k (the first-order rate constant for organic carbon respiration), and K (the bioturbation mixing coefficient) were taken from published estimates. The shapes of the oxygen profiles showed that diffusive flux should occur from overlying water into the sediments. Average estimates of this flux using Fick's first law [eqn. (191)] amounted to $8{\cdot}8 \times 10^{-14}$ mole/cm^2/s (0·08 ml/m^2/h), a value comparable to *in situ* measurements in the North Atlantic and Pacific at depths between 4000 and 5000 m with chambers inserted into sediments. Smith (1978) observed no measurable oxygen uptake in these experiments when water enclosed over sediment-contained formalin. If formalin addition halts biological respiration, the observation shows that no chemical oxygen demand is present in these deep-sea sediments. However, oxygen gradients observed by Murray and Grundmavic should still exist after formalin poisoning and diffusive flux into the sediment should still have occurred.

Mass transport (bulk water movement) caused by respiratory and burrowing activity of benthic macrofauna may serve to move oxygen into sediments in addition to diffusive flux. McCaffrey *et al.* (1980) concluded that the activity of organisms was of equal importance to molecular diffusion for transport of dissolved substances between sediment and water in a shallow coastal bay. Diffusive fluxes were estimated from assumed diffusion coefficients and measured concentration gradients. Advective fluxes were calculated in laboratory tracer experiments using ^{22}Na uptake by the undisturbed sediment. Calculated (advective plus diffusive) fluxes were found to be similar to observed rates measured *in situ* with chambers inserted into the sediment.

Despite the role which bulk water movement may play in transport across the interface, various diffusion models have been used to relate oxygen uptake to the formation of an oxidized microzone in aerated sediment cores held in the laboratory. Bertru (1971) observed that the formation of an oxidized surface layer (to a maximum thickness of 4 cm in 10 days) was proportional to the square root of time. Oxygen demand was considered to represent a steady-state biological consumption and a variable uptake dependent on oxygen concentration was due to oxidation of diffusable material. A biological demand of 46 μg O_2/cm^2/day (50% of the total) was estimated — a value similar to that calculated from various diffusion models with and without anoxic sediment layers discussed by Bouldin (1968).

The effect of limited oxygen diffusion into sediments was demonstrated by Baity (1938) who noted that oxygen uptake from water over sludge was not proportional to deposit depth but was described by the equation

$$R' = 2700D^{0{\cdot}49} \qquad (195)$$

where R' is oxygen consumption (mg/m^2/hr) and D is depth (cm). Depth had no significant effect on measured rates when cores were deeper than 4 cm which implied that most oxygen consumption occurs in near-surface layers. Boyton *et al.* (1981) discussed observations which show that increased oxygen uptake occurred in sandy-silt deposits exposed to variable stirring speeds. Water movement as turbulence or advection across the sediment surface might be expected to increase uptake, particularly if an oxygen-depleted boundary layer exists. Berner (1976) and Suess (1976) discussed physical and chemical gradients which occur a few meters above deep-water marine sediments. Little is known, however, concerning conditions of water motion immediately above the bottom on a scale (cm to mm) which would affect the exchange of dissolved nutrients and gases. Caldwell and Chriss (1979) have used thermistor probes to show that a non-turbulent viscous sublayer of this scale can exist above the seabed. The role of roughness features on natural sediment surfaces which may affect the formation of such layers is unknown.

The distribution of dissolved oxygen in shallow-water marine sediments has been measured directly by use of membrane-covered platinum electrodes (Revsbech *et al.,* 1980). The small size of the electrodes (2–8 μm tip diameter) allowed measurements at mm intervals from the surface downwards and profiles in a fine sand sediment showed that oxygen existed only in the upper 3 to 5 mm in the absence of light. The presence of photosynthetic microalgae caused oxygen to be produced on exposure to light, but even then oxygen was not present below 10 mm depth. The observations confirm calculations which show that oxygen penetration into fine-grained sediments by diffusion only occurs to very shallow depths. Greenwood (1968) demonstrated a similar effect by calculating the volume of a sphere where anaerobic processes occur. If particles are spherical with a radius r, a volume V_2, with an internal diffusion coefficient k, an oxygen-free volume at their centre of V_1, and a rate of oxygen uptake per unit volume of R, then the volume of the anaerobic sphere is given by

$$6k(O_2) = Rr^2 \,[1 + 2V_1/V_2 - 3(V_1/V_2)^{2/3}. \quad (196)$$

Substitution of experimentally derived values for R (10^{-5} ml/ml/s) and k (10^{-5} cm^2/s) for moist soil aggregates with values for air-saturated water (6 ml O_2/1) and no anaerobic zone ($V_1 = 0$) gives an r value of 2 mm. Zones farther than this distance from the gas/water interface would become anaerobic. Withers (1978) made a similar calculation to compare the metabolic requirements of animals of different body mass in burrows of cylindrical and spherical shape. Burrowing invertebrates consume oxygen at rates which exceed those which could be met by diffusion alone if they are greater than 1 mm in diameter.

The importance of diffusion-related processes for oxygen uptake by aquatic sediments has also been inferred from significant correlations between measured rates and dissolved oxygen concentration in overlying water and temperature. Baity (1938) observed that shallow layers (0·5 cm) of sludge consumed oxygen at rates which were independent of oxygen concentration between 2 and 5 mg/l. However, with deeper layers of sediment, Edwards and Rolley (1965) and Hargrave (1969) found that the equation

$$R' = a(O_2)^b \qquad (197)$$

described oxygen uptake R' (ml O_2/m^2/hr) where a and b were constants with an oxygen concentration in overlying water between 2 and 8 mg/l. A relation of this form implies that diffusion and active consumption occur simultaneously. Values of b calculated in the separate studies were 0·45 and 0·66 and the relationship was not adequate to describe data obtained from sediments which contained high numbers of macrofauna. In contrast to these experimentally based studies, Pamatmat and Banse (1969) reported that seasonal changes in sediment oxygen uptake in Puget Sound were unrelated to changes in dissolved oxygen.

Seasonal variation in benthic oxygen consumption has also been related to temperature differences. Hargrave (1969) compared measures of oxygen uptake (R') (ml O_2/m^2/hr) by sediments from different locations with temperature (T) (°C) and derived the equation

$$R' = 0{\cdot}27T^{1{\cdot}74} \qquad (198)$$

Different slope coefficients can be derived from other studies with a minimal value (0·36) calculated by Smith (1973) for a sublittoral benthic community off Georgia. Since seasonal fluctuations in factors other than temperature also affect oxygen uptake (for example, changes in the size of populations of respiring organisms, relative rates of aerobic and anaerobic metabolism), Q_{10} values cannot be interpreted on a physiological basis. However, highest values (4–11) occur in studies in northern latitudes (45–40°N) (Smith, 1973). This has been interpreted as an indication of a lack of temperature compensation and metabolic regulation at higher latitudes (Duff and Teal, 1965) but it could also reflect differences in seasonal patterns of organic input to sediments. Davies (1975) observed the effect of temperature alone ($Q_{10} = 1{\cdot}9$) by incubating individual sediment cores from a Scottish sea loch at different temperatures.

Separate effects of season and temperature on sediment oxygen uptake were experimentally evaluated by Edwards and Rolley (1965) by incubating cores collected at different times and three different temperatures. A temperature coefficient

$$t = \frac{1}{T_2 - T_1} \log_e \frac{R_2'}{R_1'} \qquad (199)$$

was calculated where R_1' and R_2' were oxygen consumption rates (g O_2/m^2/hr) measured at temperatures T_1 and T_2 (°C). Average temperature coefficients of 0·065 and 0·077 (per degree centigrade) between 10°C and 20°C were similar for sediments from different locations. Although periods of acclimatization were insufficient for complete temperature adaptation, minimum and maximum oxygen consumption occurred during winter and spring at each experimental temperature. Increased uptake during spring was attributed to growth of benthic microalgae and other benthic populations rather than only to an effect of increased temperature. Pamatmat (1971) also observed seasonal changes in total and chemical oxygen consumption which were independent of temperature at two sites in Puget Sound. The degree of seasonal effect was greatest at a location where biological respiration accounted for 48% of total consumption. There was a reduced seasonal difference after correction for temperature at a site where chemical oxidation accounted for most of the total annual consumption. Pamatmat concluded that differences between locations (a maximum value during summer of 45 ml $O_2/m^2/hr$ in shallow water to a minimum value of 9 ml $O_2/m^2/hr$ in deep water) reflected a lower organic supply or the presence of less oxidizable organic matter in deep-water sediment.

Comparison of benthic oxygen consumption with ^{14}C estimates of phytoplankton primary production in Puget Sound showed that total benthic oxygen consumption at two stations was equivalent to 17% and 25% of phytoplankton production (Pamatmat and Banse, 1969). A seasonal cycle in benthic chemical oxidation and respiration corresponded to a summer maximum in phytoplankton production. Hargrave (1973) compared annual estimates of total benthic oxygen uptake and primary production in different freshwater and marine locations. Other sources of organic matter were not thought to be important in the sites which were compared and estimates of primary production provided a measure of organic carbon supply. The expression

$$C_0 = 55\,[C_s/Z_m]^{0\cdot39} \qquad (200)$$

was empirically derived from multiple linear regression analysis which showed that benthic oxygen consumption (C_0) |(1 $O_2/m_2/yr$) was inversely related to mixed-layer depth (Z_m) [depth of the thermocline (*m*) during stratification], and directly correlated with annual primary production (C_s) (g $C/m^2/yr$).

Rates of benthic oxygen uptake measured *in situ* in the deep sea are one to two orders of magnitude less than rates in coastal areas (Smith, 1978). These values are less than those predicted from eqn. (200) using values for C_s and Z_m typical of the open ocean. This may reflect the lack of observations from deep-sea locations in the earlier comparison of data from different areas. It may also arise from the fact that only a small proportion of organic matter produced at the surface actually reaches the sediment at depths greater than 1000 m (Section 6.4.1.2). No clear relationship between production at the surface and decomposition in very deep water need exist if almost all of the consumption and decomposition of organic matter occurs within the upper few hundred meters. In contrast, studies of particulate matter deposited in sediment traps in shallow coastal waters show that sedimentation of freshly produced material can occur from the euphotic zone. Products of a spring bloom may arrive in a fresh state for utilization by benthic micro- and macro-consumers when the water column is shallow or mixing is deep enough to transport material from the euphotic zone to the sediment (Fig. 100).

Hargrave (1973) assumed that sediment oxygen uptake could be converted to carbon release by applying an RQ value of 0·85. Comparison of calculated values of benthic carbon respiration with primary production indicated that increased rates of phytoplankton production resulted in higher absolute rates of organic input to the benthos at a given depth but a proportionately smaller fraction was respired at the sediment surface. Metabolic loss to planktonic organisms could increase at higher level of primary production, or rates of burial of unoxidized organic matter could be higher. Pamatmat (1975) suggested that the latter seems most probable since with increasing deposition anerobic metabolism may be expected to become more important than aerobic respiration. These and other data imply, despite the underestimte of total metabolism, that measures of annual oxygen uptake by sediments may be regarded as a fair estimate of the fraction of annual organic supply which is aerobically respired during a given year.

A significant effect of total water-column depth on benthic oxygen consumption, in comparison to other factors, has been observed over an extended depth range (Smith, 1978). Smith compared total oxygen uptake (R, mg O_2/hr) with depth (D, m), temperature (T, °C), organic carbon and nitrogen in sediment (C and N, mg/g), annual phytoplankton production (P, g C/m^2/yr), benthic biomass (B, mg wet weight/m^2), and dissolved oxygen (O_2, ml/l) to derive the equation

$$\ln R_n = 0{\cdot}59 - 0{\cdot}09\,\frac{D}{2} - 1{\cdot}61T + 0{\cdot}18(C/N) + \frac{0{\cdot}12}{N} + 0{\cdot}01P + \frac{98{\cdot}98}{B} \qquad (201)$$

The regression accounted for 98·9% of the variation in the data where depth varied from 40 to 5200 m. Depth alone accounted for 97·1% of the variation in total oxygen uptake and temperature and carbon: nitrogen in sediment only accounted for an additional 1% of the variance. Respiration, the fraction of oxygen uptake removed after addition of formalin to chambers, was represented by the equation

$$\ln R = 3{\cdot}82 - 0{\cdot}09\,\frac{D}{2} + 0{\cdot}16(C/N) - 0{\cdot}16(O_2) + \frac{0{\cdot}13}{N} - \frac{142{\cdot}7}{P} + \frac{92{\cdot}61}{B}. \qquad (202)$$

Depth accounted for 97·4% of the variation in the regression with carbon:nitrogen in sediment and dissolved oxygen responsible for an additional 0·7% of the variance. Annual benthic respiration at stations between 2200 and 3650 m expressed as CO_2 released (1 to 2 g C/m^2/yr) corresponded to 1 to 2% of estimated phytoplankton production — a value consistent with observations discussed in Section 6.4.1.2.

Comparison of carbon released by benthic respiration and annual organic carbon sedimentation estimated for oceanic stations sampled by Smith (1978) indicated that only 15 to 29% of organic matter deposition would be respired by the benthic community. Smith noted that other organisms (macro-epibenthic and benthopelagic species) could also consume organic carbon deposited on the seabed. Hinga *et al.* (1979) made similar comparisons of particulate organic carbon sedimentation and benthic respiration to show that flux at 3500 and 4000 m in the North Atlantic (0·3 to 1·2 g C/m^2/yr) exceeded respiration by 3 to 4 times. At shallower depths, however, respiration equalled or exceeded estimates of sedimentation. The transport of organic matter by vertically migrating organisms and horizontal transport of particulate matter on or near the seabed could supply organic matter at depths on the continental shelf (60 to 1300 m) where this imbalance occurred.

The observations that carbon sedimentation and benthic oxygen consumption are both related to the ratio of carbon supply to either mixed-layer or total water column depth [eqns. (146) and (200)], show that the flux of organic matter (particulate sedimentation) to benthic communities may be quantitatively related to benthic oxygen uptake. Smith (1974) observed that oxygen consumption by sediments under enriched surface water in an upwelling area off California was significantly higher, equivalent to 3·5% of estimated annual primary production, than uptake at comparable depths in more oligotrophic areas. However, comparison of a larger data set (Smith, 1978) did not show either total sediment oxygen uptake or benthic respiration to be significantly related to phytoplankton production once differences in depth between sites had been accounted for.

Linkage between oxidation of material deposited at the sediment surface and its supply from the water column may depend upon the origin and age of settled material. For example, material derived from terrestrial debris is mineralized slowly and may be buried without extensive degradation (Seki *et al.*, 1968). Johnson and Brinkhurst (1971b) clearly demonstrated the effect by comparing sedimentation and benthic oxygen consumption at four locations where terrestrial debris contributed varying amounts to sediments. The energy equivalent of annual sediment oxygen uptake was

23% of that sedimented and macroinvertebrates assimilated 2% when terrestrially derived material was important. At other locations, where sedimented material was derived mostly from phytoplankton, 90% of the organic matter deposited was respired and macrofauna assimilated 30% of the supply. The annual carbon equivalents of benthic oxygen uptake were similar to estimates of organic carbon sedimented in semi-closed marine bays where phytoplankton was the main source of organic matter (Davies, 1975; Hargrave, 1980). However, Hargrave could not demonstrate a significant correlation between total sediment oxygen uptake or benthic respiration and organic carbon sedimentation in a marine bay sampled on a monthly basis throughout one year. A significant correlation did exist with nitrogen sedimentation, but respiration was only linearly related to deposition between March and July. Seasonal changes in the organic composition of settled material, not only the total organic content, were judged to be of importance for regulation of benthic respiration.

Other studies have also shown that supply and oxidation of organic matter in sediment may be linked to various degrees. Comparisons of annual benthic microalgal production and community respiration have shown that consumption measured as oxygen uptake amounted to only half of that produced in salt-marsh sediments (Pomeroy, 1959). Pyrite (FeS) and other reduced materials are major end products of sulfate reduction in salt-marsh ecosystems and oxygen uptake alone cannot be used to estimate respiration of organic matter in such sediments. In more oxygenated sediments, where reduced compounds accumulate to a smaller degree, aerobic respiration may be more closely related to the supply of oxidizable material. Hartwig (1976) described a sublittoral marine area where carbon deposited or produced by benthic microalgae was either utilized immediately or resuspended. Marshall (1970) and McIntyre *et al.* (1970) also observed apparent rates of organic supply which were insufficient to meet organic carbon removed calculated from oxygen-demand measurements in subtidal and artificial sand columns. Imbalances in such budget calculations may arise from methodological problems (such as some unmeasured fraction of dissolved or particulate matter). The calculations show that in sediments with low organic content but high rates of oxygen consumption, organic supply must be consumed rapidly to prevent accumulation.

This view of a rapid flux and microbial utilization of organic matter in many benthic communities is consistent with the lack of strong correlations between benthic oxygen consumption, standing stock of macrofauna and organic content (Smith, 1978). Usually less than 20% of total oxygen uptake can be attributed to macrofauna respiration (Banse *et al.,* 1971). High numbers of benthic macroinvertebrates can increase rates of sediment oxygen consumption but not necessarily in proportion to increased abundance. Interactions between organisms and their food supply when this is primarily micro-organisms modify any simple relation which might be expected. This supports Marshall's (1970) idea that it is the flux of oxidizable organic matter and not that remaining as refractory substrates which fuels benthic metabolism. Waksman and Hotchkiss (1938) reached similar conclusions by observing oxygen uptake in stirred sediment samples. Hargrave (1972a) emphasized the importance of particle surface area in such measurements as a controlling factor for oxygen consumption. Positive correlations between oxygen uptake, bacterial biomass, and organic content also imply that microbial aerobic respiration may be limited by factors other than only available organic matter.

In situ measurements of microbial respiration in sediments have been attempted by using antibiotics as respiratory inhibitors (Smith, 1974). Differences in rates of total oxygen uptake in cores with and without antibiotic treatment were assumed to reflect microbial respiration. Yetka and Wiebe (1974), however, examined the effectiveness of different antibiotics for respiratory inhibition in six strains of marine bacteria. Percent inhibition varied with growth phase, antibiotic, and bacterial type and it was concluded that these antibiotics could not be used to halt bacterial respiration in mixed microbial communities. The method may also have little quantitative value as a measure of bacterial respiration if most populations in nature exist in a dormant state (Stevenson, 1978). Antibiotics which interfere with cell division can have little inhibitory effect if most cells in a population are in a dormant condition.

General quantitative relations may exist between benthic community oxygen uptake and production by benthic fauna. Sokolova (1972) correlated the biomass of meiobenthos (0·5 mm–0·5 cm) in eutrophic and oligotrophic regions of the Pacific on a semilogarithmic basis with heterotrophic activity of microflora and oxygen consumption by stirred samples of surface sediment. The relationships were thought to reflect the supply of organic matter to sediments in different areas and its transformation by micro-organisms to produce levels of food supply which would support different amounts of meiobenthic biomass. Quantitative differences in abundance of macrobenthos between areas were attributed to the level of oxidizable organic matter in surface deposits.

Observations by Johnson and Brinkhurst (1971b) also linked the production of benthic macrofauna in a Lake Ontario bay to the supply of oxidizable organic matter. Macrofauna production (*P*) was described by the equation

$$P = c(a \cdot b \cdot I_m) \quad (203)$$

where *c* was the proportion of growth to respired and exported energy by macrofauna (i.e. growth efficiency), *a* was the proportion of imported energy (I_m) (sedimentation) used by the total benthic community, and *b* was the proportion of this used by macroinvertebrates. All three coefficients must remain constant if production is proportional to sedimentation. However, Johnson and Brinkhurst observed that the proportion of sedimented energy used by macrofauna decreased with increasing input. This could have been due to the importance of terrestrial debris at the site of highest deposition which was utilized by the macrofauna with a low efficiency. It could also reflect added respiratory costs associated with conditions of high sedimentation. In either case, the value of *a* (which approaches 1 when the total supply is fully oxidized) may be expected to vary directly with the proportion of material derived from autochthonous production. *b* may also decrease as the importance of refractory organic matter deposited increases. The lower its value, the greater is the fraction of total community respiration attributable to non-macrofauna consumption. The relationship implies that all three proportions may decrease as I_m increases, particularly if this is associated with the input of material which is largely refractory to utilization by either micro-organisms or macro-invertebrates. With a non-refractory organic supply, macroinvertebrate production should increase with increasing organic sedimentation up to a maximum level where the product $a \cdot b$ (the proportion of sedimentation used by macro-invertebrates) is maximized. If these relations have general application, it seems possible that macro-fauna production could be predicted from measures of organic input and oxygen uptake by benthic communities.

Tenore and Rice (1980) examined factors which affect the utilization of detritus by macro-invertebrate consumers. They reviewed observations which show that the source of detritus, amount and state (age) of decomposition affect nutritional quality. They proposed that the ratio of the net production of macroconsumers (*P*) to the oxidation of the detritus (and associated micro- and macro-organisms) (*O*) be used to represent the relative amounts of detrital energy that is conserved (*P*) and expended (*O*) in food-chain transfer from detritus to macroconsumers. Experimental studies with laboratory microcosms showed that microbial activity increased the nutritional value of detritus and resulted in an increase in ratio value. The source of detritus, its age, and the presence of ciliates affected the ratio. The lowest value (0·01) was observed for meiofauna and the polychaete *Nephthys* feeding on *Zostera* detritus with the highest value (0·75) for polychaetes *Aspidisca* and *Nereis* feeding on detritus derived from the seaweed *Gracilaria*. The ratio provides a quantitative measure of the efficiency of incorporation of energy into a detritus-based food web which could be used for mixed natural populations.

CHAPTER 7

SOME PRACTICAL PROBLEMS IN BIOLOGICAL OCEANOGRAPHY

Before the turn of the last century the earliest publications in biological oceanography were primarily concerned with the taxonomy and distribution of marine planktonic organisms. With the introduction of extensive nutrient analyses in the 1930s, biological oceanography entered a more quantitative era and some efforts were soon made to use biological oceanographic information for the solution of practical problems. Early attempts to relate nutrient distributions to fisheries were not successful but the exercise served to demonstrate where more knowledge was required. Riley's (1946) work on modelling the plankton community led the way to a much greater understanding of integrated processes. The impediment to further development of models then lay in the inability to collect sufficient data over an area and time scale applicable to the models. This problem is being solved with the use of automated equipment such as autoanalysers coupled to shipboard computers (e.g. Walsh and Dugdale, 1971). Throughout this brief history of development, however, it has been continually necessary to summarize relationships between plankton and the environment in a variety of empirical expressions, many of which have been discussed in earlier sections of this text. These expressions (and future modifications) are the building blocks on which practical solutions to biological oceanographic problems must be attempted. While it is recognized that more knowledge would be helpful, solutions to many of today's problems are required today, and where answers are required they should be given to the best of the state of present-day knowledge.

In the following section some attempt has been made to identify a number of problems for which solutions may be attempted from a knowledge of biological oceanographic processes. The actual application of biological oceanographic information to the solution of a problem will depend heavily on the local environment and for this reason the following subjects are only considered by example, or from a point of view of general relationships.

7.1 SOME GENERAL CONSIDERATIONS REGARDING MARINE POLLUTION

Any form of pollution should be considered from the point of view of whether it is acute or chronic and further whether the results are lethal or sub-lethal. An example of acute pollution would be an oil spill, while a chronic form of pollution in the marine habitat would be the gradual accumulation of heavy metals in organisms surrounding a sewage outfall. Lethal effects of pollutants are best diagnosed through the laboratory approach of LC_{50} responses (i.e. diagnosis of the lethal concentration, LC, which will kill 50% of a population in some specified time period — e.g. 48 or 96 hours, etc.). This lethal test of pollutant toxicity is also a test of acute toxicity response. The more difficult problem facing the biological oceanographer is the sub-lethal response of organisms which is generally a chronic problem. Thus the neoplastic response of mussel tissue to low levels of petroleum hydro-carbons, or the carcinogenic effect of some heavy metals on local populations of benthic fish, are examples of pollutant effects which are both sub-lethal and chronic. It is apparent, however, that more subtle long-term chronic effects may occur in the ecology of the sea due to man's activities, all of which in the broadest sense can be considered as 'pollutant' if they affect the natural ecology of the sea. Thus the removal of 60 million tons of top predator by the fishing industry must be considered as an ecological disturbance, the consequence of which has received the barest consideration in the literature. Other long-term, sub-lethal effects may or may not be attributable to man. For example, the increase in the amount of mercury in marine environments can be shown to be associated with some sewage outfalls, but much larger quantities are encountered in association with undersea volcanisms.

The transfer of pollutants, such as heavy metals and chlorinated hydrocarbons, in the marine biota involves a study of the food web of organisms in benthic and pelagic environments and as such it is an area of practical interest to the biological oceanographer. Several general theories exist on the way in which pollutants are transferred between different trophic levels. These include a bio-amplification theory which may operate in a case where a small amount of pollutant is selectively absorbed by certain tissues (e.g. DDT in fatty tissues); this then becomes more and more concentrated since the pollutant is not readily destroyed but the amount of tissue it is dissolved in becomes a smaller and smaller fraction of the total biomass as it moves towards the top of the food pyramid. This theory depends on the presence of a structured food chain (cf. Isaacs, 1973) in which there is a direct flow of food from the lowest trophic levels to the highest (e.g. the Bermuda petrel, Wurster and Wingate, 1968). However, other animals are very diverse feeders and as predators of many different types of food (e.g. sea gulls) they have unstructured food chains and are less subject to the accumulative effect of a pollutant. This has given rise to another theory on the concentration of pollutants which indicates that it may be more associated with the surface area of organisms and the adsorptive properties of these surfaces. For example, metals appear to be adsorbed by the plankton community and much larger amounts are adsorbed by the phytoplankton than the zooplankton (Mayzaud and Martin, 1975). On the other hand, Kerr and Vass (1973) conclude that the uptake of another group of pollutants (chlorinated hydrocarbons) is a function of metabolic rate and diffusion across respiratory surfaces. It is probable that the biological transfer of pollutants is in fact a function of any one or more of the three theories put forth above, and many in addition involve other processes as yet undefined.

Substances which have been shown to have caused some form of pollution in the marine hydrosphere include three major groups of elements or compounds. These are heavy metals, hydrocarbons (including both petroleum and chlorinated), and radionuclides. Many papers and reviews of these subjects have been written and it is not our intention to give any exhaustive treatment to the subject of contaminants in the hydrosphere. However, we hope to show, by example, where biological oceanographers might be involved in some practical aspects of pollution research.

Certain geographical areas should be considered of greater importance with respect to pollution, and this is particularly true of estuaries. Estuarine areas are of general interest to biological oceanographers, both from the point of view of their unique biological activity and because they tend to be centres of human activity which includes the development of ports. The river above the estuary may be considered as a corridor for fish migrations, ship transport, and disposal of natural and man-made products which are carried from the land to the sea. In this situation a conflict often exists between the natural ecology of the estuary and its modification due to man's activities which have generally resulted in pollution. Obvious forms of pollution, such as organic and inorganic industrial poisons, cause direct and catastrophic effects on the natural estuarine environment and remedial action is generally best handled through chemical evidence and legal procedures. Slightly more subtle, however, are the effects of non-toxic substances which may decrease light penetration in the water column, take up oxygen, or cause toxic substances to accumulate through precipitation in the sediment near the river mouth. Substances which decrease the penetration of light into the water (e.g. silt, coal dust, pulp mill effluent) can cause a decrease in the primary productivity of the water column; this in turn reduces the amount of oxygen in the water and decreases the supply of phytoplankton at the primary trophic level. Since the effect of light-attenuating substances can be expressed in terms of changes in the extinction coefficient (k) of the water, a diagnosis of the overall effect of this form of pollution might be modelled through some of the expressions discussed in this text. Oxygen depletion over a period of time and distance can also be modelled using experimental data on the Biochemical Oxygen Demand (BOD) of samples of the polluting water (for BOD methodology see Amer. Pub. Health Assoc., 1965).

7.1.1 Monitoring Programs

The problem with all pollution-monitoring programs is to be able to establish the baseline natural conditions in order to evaluate the extent of change caused by a pollutant. Monitoring may involve changes in the physiology or biochemistry of single species, behavioural or genetic changes within populations, or ecological changes in the whole food web of the sea. As one ascends from observations of individual species, to populations and finally to ecological communities, it becomes more and more difficult to establish the baseline of natural events. For example, the heavy-metal content of a group of organisms from one area can be compared statistically with that from another area; however, to determine if the food web of the sea is undergoing a change which will produce more jellyfish and less commercial fish (e.g. Greve and Parsons, 1977) is a much larger problem which requires a great deal of understanding of biological oceanographic processes. For example, in a recent report (Myers *et al.*, 1980) it was found that marked declines in four species of littoral organisms which were noted following a tanker explosion in January 1979 could be attributed to natural events over a longer time scale which happened to coincide with the accident.

The choice of a single organism which has a wide global distribution has been adopted as one way in which to monitor ocean pollution. The mussel, *Mytilus edulis*, has been adopted in one program (Goldberg, 1975) and comparisons of its contamination with various pollutants have been made throughout the coastal regions of the hydrosphere. Unfortunately, in spite of having genetically similar animals, some differences in environmental factors make the results of the comparisons less easy to understand than was at first anticipated. Thus differences in the ratio of metals present in a polluted environment can affect the absolute concentration of any one metal, as is the case with copper and zinc (for discussion see Phillips, 1977).

Changes in the distribution of species have been advocated as a useful means of demonstrating pollution effects at the community level. This can be done by the simple use of diversity indices or by plotting the distribution of individuals among different species. Among communities which have reached an equilibrium state, the latter distribution is log-normal, but under conditions of pollution some of the abundant species may become more abundant resulting in a change in the log-normal distribution (Phillips, 1977; Gray and Pearson, 1982).

In both the above examples of monitoring, reliance is placed on demonstrating a statistically valid change before and after pollution. However, to really understand the processes involved one should undertake a more diagnostic, determinate approach. This is particularly important in being able to gain an understanding of a pollution change which may be taking place *before* it becomes statistically significant, as in the two examples above. A diagnostic approach to a pollution problem requires the measurement of a number of parameters so that the sequence of interconnecting events can be established. This can be seen by example in Fig. 105 from Pearson (1980) where changes in four parameters are plotted with distance from the source of a pulp-mill outfall. Although the graph shows distance along the abscissa, it would be equally valid to monitor at a point over time in order to diagnose any small change in one parameter (E_h) which could lead to

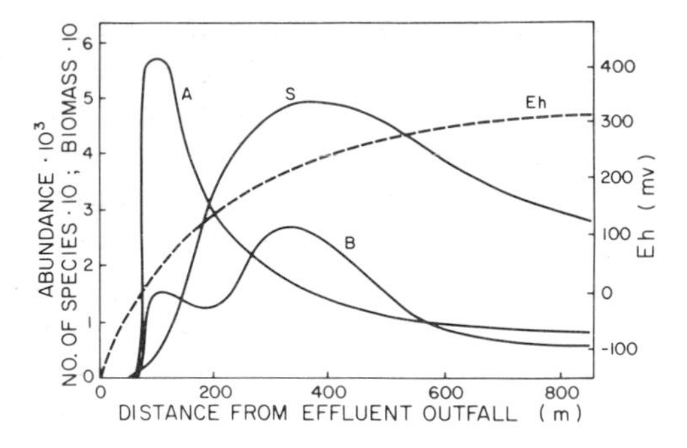

FIG. 105. Diagrammatic representation of changes in benthic macrofauna in terms of (A) number of animals, (S) number of species, and (B) the total biomass (from Pearson, 1980).

changes in other parameters, such as has become established over distance in the example shown.

The general problem of monitoring pollution effects can be followed in the simple diagram shown in Fig. 106. Here it is assumed that environmental stress can be detected by means of environmental monitoring; that the short-term accommodation to this stress can be measured through metabolic changes in organisms; and that finally, population analysis will demonstrate changes in the ecology. The monitoring of the environment and population analysis can be carried out by a variety of techniques and the interpretation of results forms much of the background information in this text. The need for metabolic studies of marine organisms requires some explanation which is given by example in the next section.

7.1.2 The Metabolic Response of Organisms to Pollutants

The ability of organisms to detoxify certain pollutants, particularly under the impact of chronic low-level pollution, is well recognized. Thus an expected impact of a pollutant on some parts of an ecosystem may not occur if the species in the system have some defense mechanism against the level of pollution encountered. The subject has been reviewed by Bryan (1980) who shows that heavy-metal concentrations in organisms from two appreciably polluted estuaries in England resulted in concentrations of some metals of 20 to 100 times values which would be considered 'normal'. Yet it appears that over a 30-year period the number of species has remained relatively constant. The form of the metabolic response to heavy-metal pollution has frequently been shown to involve metallothionein production. This is the production of a glycoprotein within certain organs (particularly the liver) which binds and detoxifies many metals, such as Cd^{+2} and Hg^{+2}. The metallothionein protein is

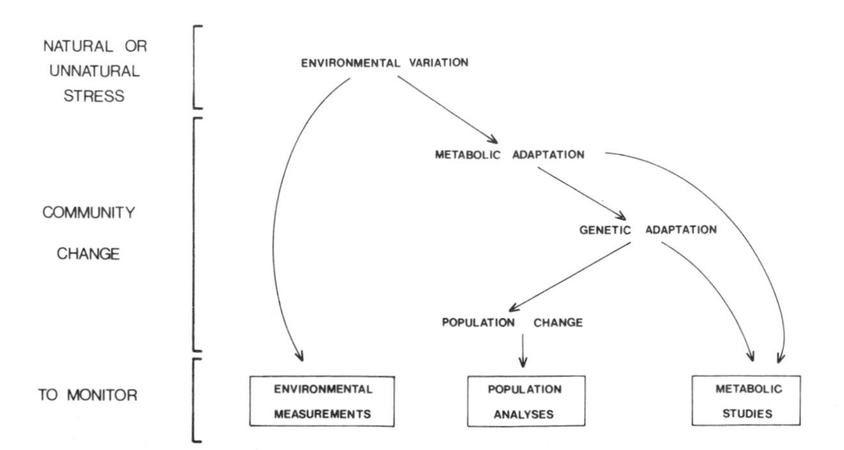

FIG. 106. A simplified scheme showing the general stages of biological change following environmental stress (modified from Pearson, 1980).

produced in response to the presence of heavy metals. However, there is some question as to the rate at which it can be induced in order to protect the organism; in this respect Brown and Parsons (1978) suggested that if insufficient metallothionein was produced by an organism, then heavy metals could 'spill over' from the protective SH-groups of the metallothionein onto the SH-groups of enzymes, resulting in the toxic syndrome and lethal effects of acute heavy metal poisoning.

Another example of metabolic response to marine pollutants is the induction of mixed function oxidases (MFOs). The subject has been reviewed by Payne (1977) and Moore *et al.* (1980). In the latter reference the authors have pointed out that the mechanism of detoxification of organic contaminants (e.g. petroleum hydrocarbons) may result in the formation of oxidized compounds which are more toxic than the original substances. However, the general detoxification process via MFO enzymes is to convert substances which are difficult to excrete as hydrophobic compounds, into polar metabolites which may be easier to handle as an excretory product. MFO analysis as a method for detecting hydrocarbon pollution has been established (e.g. Payne, 1976), although in many cases the role of the detoxification process is not clearly understood.

7.1.3 Experimental Approaches to the Ecological Effects of Pollutants

During the past decade considerable advance has been made in the use of experimental ecosystems for the study of the effects of pollutants on the ecology of the sea. Unlike the LC_{50} type response of single organisms to a pollutant, the problem of an ecological response of a community is much more difficult to diagnose. This problem should be investigated *before* a pollutant has had time to impact on a community. Therefore a system was required which allowed 'field testing' of the effect of potential pollutants on whole communities. The problem is solved in part by the use of large controlled experimental ecosystems (CEEs). A variety of these have been described (e.g. Parsons *et al.*, 1973) and many results of experiments using these ecosystems have been reported by Grice and Reeve (1982) and Boyd (1981). In the latter reference the author concludes by contrasting the experimental approach of CEEs with studies made at sea. It is unfortunate that such a contrast is made because in fact in experimental approaches, as compared with surveys discussed in Section 7.1.1, there are recognized limitations to *both* avenues of scientific investigation. The correct conclusion to draw from alternate approaches is that neither one should be allowed to dominate the science. (An unfortunate example of the dominance of the stock:recruitment paradigm in understanding fisheries problems is given in Section 7.2.)

Essentially the CEE is some facility which allows one to capture and control a sufficiently large body of water to contain several trophic levels, starting with natural radiation and the *in situ* nutrients and ending with the production of higher trophic levels such as fish. Perturbations in the form of pollutant stress can then be imposed on these systems. The results of a whole variety of pollution experiments using CEEs have been reported on with respect to the effect of heavy metals (e.g. Thomas and Seibert, 1977), petroleum hydrocarbons (e.g. Lacaze, 1974), eutrophication (e.g. Harrison and Davies, 1979), and studies on higher trophic levels including ctenophores and fish (e.g. Mullin and Evans, 1974; Koeller and Parsons, 1977). In some cases the equipment can be used to measure ecological change following natural perturbations such as the effect of mixing (e.g. Eppley *et al.*, 1978) on the species composition of phytoplankton. The general conclusion to these experiments has been that small amounts of pollutants do not have drastic, lethal effects on all forms of life in the ocean. Rather, the ecology of the ocean may be changed in some direction which may result in a greatly increased production of some species and a depression in the production of others. This kind of scientific finding is in contrast with speculations which have appeared (e.g. Thor Heyerdahl, *Saturday Review*, 29 Nov. 1975) on 'How to kill an ocean'. It appears, in contrast to these speculations, that the ocean ecosystem is a very resilient system which has survived, as the cradle of life, for many millions of years before terrestrial life appeared. It may in fact survive future catastrophes more readily than life forms living on the land.

7.1.4 A Summary of Specific Examples of Marine Pollution

7.1.4.1 Sewage pollution and eutrophication. Sewage in coastal areas is often disposed of at sea. This may involve untreated sewage or sewage which has undergone some level of treatment either to remove organic material, reduce the coliform count and, very occasionally, to remove nutrients and heavy metals. In general, however, one may assume that the forms of pollution from sewage entering the marine environment will be (1) pathogenic organisms, (2) eutrophication from nutrients, (3) biological oxygen demand to decompose the organic matter, (4) heavy-metal contamination from industrial outfalls being combined with domestic outfalls, and (5) if chlorine is used as part of the treatment, chlorinated organic compounds may occur which are potentially carcinogenic.

Probably the most common form of coastal pollution from sewage is the contamination of beaches and adjacent shellfish fisheries with pathogenic organisms. Jones (1971) has estimated that $4{\cdot}4 \times 10^7$ lb of feces are released per day in the United States and that an appreciable proportion of this is disposed of in the near-shore environment. While the distribution of this material on entering the sea is a problem in physical oceanography, the survival of pathogenic bacteria in sea water is a biological problem. The subject has recently been reviewed by Mitchell (1968) and Jones (1971). Of the many bacteria entering the sea, the presence of *Escherichia coli* is usually taken as an index of sewage pollution. The organism is not itself a pathogen but its presence in large numbers may indicate the presence of pathogenic bacteria and viruses. The actual 'acceptable' level of coliform contamination may vary depending on national standards and use of surrounding beach areas; generally a level of less than 100 coliforms/100 ml in 90% of the samples averaged over a 30-day period is considered an acceptable limit while a level of 1000 coliforms/100 ml would be considered contaminated.

Several mechanisms are considered to cause the destruction of terrestrial bacteria in the sea. The production of antibacterial substances by marine plankton (e.g. Duff *et al.*, 1966; Burkholder *et al.*, 1960) has been shown to include substances active against *Staphylococcus* although only about 40% of the algal species tested by Burkholder *et al.* (1960) displayed activity against this pathogen. A specific antibacterial activity of the phytoplankter, *Phaeocystis*, was identified by Sieburth (1960) as acrylic acid. In addition to the antibacterial activity of marine phytoplankton it has also been suggested that sunlight, temperature, and specific bacteriophages may cause mortality of terrestrial bacteria in seawater. However, most of these mechanisms might be considered to be equally effective against the indigenous species of marine bacteria and not specific agents in causing the high mortality of terrestrial species. A mechanism which is more specifically related to a difference in properties between freshwater and seawater is the salt content of the latter; salt (NaCl) in itself, however, is not considered to be the lethal agent since several authors have shown that some terrestrial bacteria can tolerate (but not necessarily grow in) NaCl concentrations higher than are found in the sea (Korinek, 1927; Burke and Baird, 1931). On the other hand, Jones (1963) found that *E. coli* was killed by natural seawater; a process which could be prevented by the addition of a chelating agent. Jones (1963, 1964 and 1967) concluded that it was the heavy metal content of seawater together with the low concentration of natural organic chelators which was lethal to *E. coli*, and other terrestrial bacteria. Under these circumstances it is apparent that near-shore environments which contain a large amount of organic material (from pollutants or natural sources) will allow for a greater *survival* of fecal bacteria in the marine environment. In addition to heavy metal toxicity, however, it is also apparent that substrate concentrations are generally much lower in seawater than in terrigenous environments; thus the *growth* of terrigenous bacteria would be inhibited in seawater compared with the natural marine bacteria which have adapted to very low substrate concentrations.

In summary it appears that a function of the salt content of seawater (i.e. heavy metals) together with the presence or absence of chelating agents and organic substrates are the most likely factors determining the growth and survival of terrestrial bacteria in the sea. On the other hand, the division of estuarine bacteria in terms of their tolerance to sodium chloride may be a useful ecological classification with which to differentiate between autochthonous and allochthonous populations. Larsen (1962) classified micro-organisms as halophobic or halophilic and described the former as organisms which grow best on a medium containing less than 2% NaCl. From an estuarine study, Seki *et al.* (1969) determined a regression line between the ratio of freshwater:saltwater colonies (y) and salinity (x). From this equation

$$y = -2{\cdot}2x + 5{\cdot}2 \qquad (204)$$

it is apparent that bacteria in the estuarine environment studied by Seki *et al.* (1969) grew best in a freshwater medium ($y > 1$) when the salinity was less than 1·9% or approximately 19‰.

The assessment of microbial activity in polluted waters has been reviewed by Jannasch (1972). The author emphasizes the need to make a number of routine microbiological measurements in order to determine the biochemical activity of polluted waters, rather than simply obtaining an index of pollution, such as a coliform count. Five kinds of determinations are suggested; these are biomass determinations, respiratory activity, special metabolic activities (such as nitrification), growth-rate determinations, and specificity of substrate utilization.

The eutrophication of a near-shore environment from sewage may be enhanced by other nutrients coming into the same coastal systems from agricultural and horticultural run off. Since estuarine environments are also enriched by entrainment of deep-water nutrients from the sea, the degree to which nutrients are contributed by the land versus the sea is sometimes difficult to diagnose. Classical examples of eutrophied estuarine areas can be found among local rivers (e.g. Jeffries, 1962; Barlow *et al.*, 1963) and off large rivers where natural upwelling may contribute to the enrichment process (e.g. the New York Bight, Ketchum, 1967; the Columbia River estuary, Haertel *et al.*, 1969). In addition, in some limited areas a nutrient 'trap' may develop where the organisms grown at the surface, sediment to a layer where they are carried back into the inlet by the counter-current of the surface flow. This situation was first described by Redfield (1955) and examples of self-enriching processes can generally be found in fjords with shallow outer sills; in some cases the water in the inner basins becomes anoxic due to the accumulation of organic materials. In other cases, such as the Nile river before the construction of the Aswan dam, the river itself supplied nutrients to the nutrient impoverished Mediterranean Sea (Halim, 1960).

The identification of nutrient-polluted estuaries has been discussed by Ketchum (1967). In the simplest case it is apparent that if the nitrate or phosphate content of estuarine waters is higher than the maximum nutrient content of surrounding deep saline water, then a terrestrial source of nutrients may be expected. In the case of only two water types (i.e. freshwater and saltwater) the fraction (F) can be obtained from the expression (eqn. 31),

$$F = 1 - \frac{S}{S_0}$$

where S is the salinity of the sample and S_0 is the salinity of the source seawater. Thus the concentration of a nutrient (or some other potential pollutant) at any point in the estuary, can be determined from the salinity of a sample and a knowledge of the nutrient content of the source waters. In a system in which there are three water types Ketchum (1967) assumed that *total* phosphorus could be used as a conservative property (since loss would only be by a small amount of sedimentation of particulate phosphorus) and together with the salinity, the fraction of any three water types could be found from the salinity and total phosphorus content of a sample. Thus in

Ketchum's example three water types were identified in the New York Bight; these were

A. Brackish river water:
30‰ S_a and total P_a 2·9 μg at/l
B. Surface coastal water:
30·95‰ S_b and total P_b 0·5 μg at/l
C. Deep ocean water:
34‰ S_c and total P_c 1·25 μg at/l

Then the equations used to determine the fraction of the three water types are

$$S_x = AS_a + BS_b + CS_c, \quad (205)$$

$$P_x = AP_a + BP_b + CP_c, \quad (206)$$

$$A + B + C = 1. \quad (207)$$

Where A, B, and C are the volume fractions of each water type and S and P are the salinities and total phosphorus contents of the three water types and the unknown sample, x.

This rather simplified diagnosis of water types can be used in conjunction with other environmental data to analyse changes in the estuarine environment. For example the maximum level of primary productivity is often found at some distance from a river mouth and is not necessarily associated with the maximum availability of nutrients. This is caused by a number of factors including the increased availability of light due to sedimentation of silt, decreased mixing processes with distance, and a time factor which allows the seed population of phytoplankton to increase exponentially as the waters move away from the river mouth; this situation has been documented as occurring under natural conditions in the Fraser river plume (Parsons *et al.*, 1967) and is predictable from a simulated model of a sewage outfall entering the sea (Walsh, 1972) and from actual observations (Caperon *et al.*, 1971). In the latter reference the authors also use a model (similar to Walsh's — 1972) to predict the course of future and past eutrophication of a tropical estuary. The model employs a four-component food chain in which each component is related to the rest by a hyperbolic function (Fig. 37); feedback is also introduced through nutrient regeneration. The model shows that in an oligotrophic environment, small changes in the input of nutrients are taken up rapidly by the population acting over the maximum response region of the hyperbolic function. However, the effect of a large perturbation, in terms of a massive injection of nutrients (e.g. sewage), could not be dampened out once the rate-compensating capacity of any trophic level had been exceeded. This resulted in a build-up of nutrients and the establishment of a large biomass of a new population at the primary trophic level; in other words, the original ecosystem had become destabilized.

7.1.4.2 Petroleum hydrocarbons. Hydrocarbons are natural constituents of seawater since they are produced by both phytoplankton and zooplankton (cf. Chapter 2). The background level of non-volatile hydrocarbons dissolved in ocean waters is probably in the range 1 to 10 ppb (Brown and Huffman, 1976), but may be much higher in areas of tanker traffic, natural oil seeps or oil spills. In addition the persistence of long-chain petroleum hydrocarbons (>C30) in the marine environment gives rise to tar blobs which float on the ocean surface and may persist for many years. According to Wong *et al.* (1976), the concentration of these tar blobs is greatest along oil tanker routes (e.g. 2·1 mg/m^2 in the Kuroshio current) and least in South Pacific waters (e.g. 0·0003 mg/m^2). However, their presence in relation to the marine biota may be relatively insignificant compared with the local effects of a large oil spill. Many tar blobs in fact become colonized with seaweeds and barnacles as they float around the ocean.

Petroleum oils entering the oceans probably amount to about 6 × 10^6 tons per year of which a large fraction comes from uncontrollable sources such as river runoff, atmosphere, and natural seeps. The value of 10^6 tons represents about 0·2% of the world's hydrocarbon production and this should also be compared to *ca.* 10^{12} tons of dissolved organic matter occurring naturally in the sea (Gunkel and Gassman, 1980). Less than half of the petroleum oils entering the oceans come from the transportation and exploration of oil by the industry. However, due to the high concentration of petroleum hydrocarbons carried by a tanker, or occurring as a blowout, it is the industrial source that has received the most attention. During the last 30 years over 400 oil spills (>200 tons) from tankers have occurred. The largest of these were the *Amoco Cadiz* (22,000 tons) and *Torrey Canyon* (117,000 tons) while many others were considerably smaller (i.e. <10,000 tons). In addition, large blowouts have occurred such as in the Gulf of Mexico in 1979 when 430,000 tons of crude oil entered the marine environment. As a result of these accidents many thousands of research papers have been written on the effect of hydrocarbons on the marine biota. A few of these effects are discussed below as being of interest to the biological oceanographer. However, in general it can be said that there are few, if any long-lasting (>10 years) effects of petroleum hydrocarbons on the marine environment which have been demonstrated to date.

The reason petroleum hydrocarbons do not pose a long-term effect on the marine biota is that, being natural sources of energy, they are readily decomposed by bacteria and animals, either independently or in association with bacteria in their guts. The distribution of hydrocarbon-utilizing bacteria can be taken as an indication of the presence of hydrocarbons since they increase by several orders of magnitude following an oil spill (Seki *et al.*, 1974; Stewart and Marks, 1978). In addition certain polychaetes (*Arenicola marina*) have been shown to tolerate and decompose petroleum hydrocarbons (e.g. Gordon *et al.*, 1978).

The general sequence for the disappearance of oil following a spill is that volatiles are lost in a matter of hours or days, while biological degradation of <C26 hydrocarbons proceeds rapidly during the first year. After 4 to 6 years, depending on environmental conditions, it is usually difficult to visually find oil from a near-shore spill, and after 10 years the effects have generally disappeared (Creteney *et al.*, 1978; Southward and Southward, 1977; Mann and Clark, 1978). Arguments against this long-term prognosis for recovery can be advanced if one considers that rare organisms which may occur at the limit of their geographical range may not recolonize following an oil spill, or that alternatively, the sensitivity of chemical hydrocarbon analysis can demonstrate the occurrence of petroleum hydrocarbons many years after a spill, in, however, minute traces.

The general effects of an oil spill on the marine biota vary with location, weather conditions, the amount spilled, and the type of oil. Generally the more volatile, low molecular weight oils are the more toxic in terms of narcotic and carcinogenic activity towards higher organisms. However, these are also the compounds which are most readily evaporated and decomposed by bacteria. Oil spills occurring on the high seas have resulted in very few demonstrated effects on ocean ecosystems and this is probably a result more of the vast size of the oceans relative to the size of an oil spill, rather than to the known experimental effects of oil on marine organisms. which at high enough concentrations is definitely harmful. Thus the attenuation of oil is one factor which limits its harmful effects following an open ocean oil spill. However, in coastal areas the presence of oil on littoral organisms can result in the total annihilation of most populations of fauna and flora. Here again, however, the area covered is generally <200 km of coastline, even for large spills. Thus on a world scale this is a small event, although locally it may be catastrophic. The most severely decimated populations of marine organisms which also have a relatively slow rate of recovery are the marine birds. Small amounts of petroleum hydrocarbons on the feathers can cause penetration of water to the bird's body and eventually death from exposure (Levy, 1980); larger quantities of oil prevent feeding and flying. In the *Amoco Cadiz* spill a conservative estimate of bird mortality was >3000 animals (Jones *et al.*, 1978). The elimination of intertidal fauna and flora by smothering with oil is followed by a period of recolonization, usually giving rise to bizarre growths of seaweeds before the benthic grazers arrive, and subsequent normalization of near-shore communities occurs. This process takes from 1 to 10 years, depending on surf zone activity which aids in weathering (e.g. Jacobs, 1980; Creteney *et al.*, 1978).

Early attempts to protect the biota by using dispersants to remove the oil are believed to have resulted in greater damage from the dispersants than from the oil (Southward and Southward, 1978). More recent dispersants have been found to be far less biologically harmful and in some cases these can aid in bacterial degradation (e.g. Creteney *et al.*, 1981).

The effects of petroleum hydrocarbons on lower trophic levels have been studied extensively. In some cases hydrocarbons may enhance the growth of certain species of phytoplankton but in many cases the effect is inhibitory (e.g. Dunstan *et al.*, 1975). LC_{50} tests on zooplankton have shown that petroleum hydrocarbons are toxic to zooplankton in concentrations of *ca.* 10 mg/l which is generally much higher than concentrations encountered *in situ* (e.g. Wells and Sprague, 1976). Fish eggs and larvae under experimental conditions may also be killed or undergo abnormal development (cf. Duval *et al.*, 1981). Pelagic bacteria have been reported to be enhanced by low concentrations of hydrocarbons (*ca.* 10 μg/l) but inhibited at concentrations >300 μg/l (Hodson *et al.*, 1977). Considering the rapid reproductive rate of the plankton community and the relatively small size of the populations that could be affected by an oil spill, it is probable that the lower trophic levels are least affected by hydrocarbon development. A good review on the effect of hydrocarbons on plankton is given by Corner (1978).

7.1.4.3 Chlorinated hydrocarbons. Chlorinated hydrocarbons are used both in agriculture (e.g. dichlorodiphenyltrichloroethane, DDT) and in industry (e.g. polychlorinated biphenyls, PCBs). These substances are generally toxic and in some cases lethal at quite low concentrations (i.e. *ca.* 30 to 100 ppm, Woodwell *et al.*, 1967). Unlike petroleum hydrocarbons which are a part of the natural world, the halogenated hydrocarbons have been prepared specifically by man. As such they are new to nature and therefore potentially nature may be less able to cope with this form of pollution. In addition the use of DDT-type compounds for spraying gives rise to appreciable quantities in the atmosphere which can then be transported globally and eventually deposited very widely in the oceans. For example, Tanabe *et al.* (1982) found concentrations of hexachlorocyclohexane (HCH) of 3 to 7 ng/l in southeast Asian surface waters and *ca.* 0.2 to 0.9 ng/l in Antarctic waters, both values being largely attributed to atmospheric transport.

The question of bioamplification of chlorinated hydrocarbons as opposed to simple bioaccumulation or bioconcentration of pollutants must also be examined, particularly in reference to halogenated hydrocarbons. Bioaccumulation (or bioconcentration) simply refers to the increased concentration of a substance by transfer across the gill surface or by absorption from the gut of an animal. Bioamplification implies that more of the substance is retained by higher organisms. Thus in the case of lipid-soluble pollutants such as DDT, one might suppose that as lipid is metabolized by successive animals in the food chain, and DDT is not (while still being part of the lipid pool), so the concentration of DDT in lipids must increase from the lowest to the highest steps in the food chain. A clear example of what is meant by bioamplification has been given in relation to metals for the cesium/potassium ratio which is found to increase with increasing trophic position (Young *et al.*, 1981). Some evidence of DDT bioamplification is indicated by earlier references (e.g. Woodwell *et al.*, 1967). However, Harvey *et al.* (1973) examined PCBs and DDT in plankton, planktivorous fish, and sharks from waters 66°N to 35°S. Mixed plankton populations gave PCB concentrations of *ca.* 200 μg/kg, flying fish 4–50 μg/kg and sharks >1200 μg/kg. Thus in this food chain, from plankton to flying fish and their predators, there does not seem to be any bioamplification and the authors concluded that some fish can rid themselves of PCB accumulations, while others (e.g. sharks) may not have this ability.

The persistent reports of chlorinated hydrocarbons at great distances from their source include their presence in Antarctica (George and Frear, 1966) and at the top of the food chain among marine birds (Riseborough *et al.*, 1967). Also the concentration of PCBs in marine organisms is strongly related to the industrial source. Harvey *et al.* (1973) showed that PCBs in the Georges Bank biota, close to the industrial east coast of the United States, were 10 times higher than fish from

Denmark Strait, which is over 2000 km downwind of the nearest American sources.

Wurster (1968) and Menzel *et al.* (1970) demonstrated photosynthetic inhibition of phytoplankton by chlorinated hydrocarbons. However, while there is no disagreement that qualitatively chlorinated hydrocarbons are toxic to the marine environment, the quantitative aspects of the problem indicate that present DDT levels are generally well below any dangerous concentrations on a global basis, with possible exceptions to this to be found in some local near-shore areas and in particular among some marine birds; in the latter example it now appears well documented that egg-shell thinning and reduced breeding success was associated with the presence of DDT residues in these animals (e.g. Hickey and Anderson, 1968; Wurster and Wingate, 1968).

7.1.4.4 Heavy metals. It has been shown that the concentrations of heavy metals have increased in some marine environments, especially in coastal areas. Since heavy metals such as lead, mercury, copper, and cadmium can be shown to be toxic to biological organisms, their presence is generally considered undesirable. Examples of intensive local pollution in near-shore areas are well known (e.g. the Minamata disease). However, it is quite another question whether or not traces of these elements in the oceanic environment of the sea are having any large-scale effect on the pelagic ecosystem. At present the lack of catastrophic changes in the marine biota of oceanic areas can generally be considered as indicating that the presence of a pollutant at very low concentrations (i.e. parts per billion) may not be harmful. However, the more subtle effects of such low concentrations really have not been investigated and the following brief review of current findings with respect to mercury serve as an illustration of where more research is needed by biologists.

The distribution of dissolved mercury in the eastern neritic zone of the Irish Sea shows values in excess of 200 ng/l (Gardner and Riley, 1972). These particularly high values have been in part attributed to the presence of sewage and sludge. Fish caught in the same area had an average mercury content of 0·53 ppm dry weight, or more than double the value for fish caught in the rest of the sea (average 0.21 ppm). Mercury may also enter the surface waters of much larger marine areas by being airborne. For example, Gardner (1975) found that the mean mercury content of North Atlantic waters under the northern jet stream was 26·5 ng/l compared with an average southern hemisphere value of 15·9 ng/l. However, the source of this mercury is not clearly established and may be natural phenomenon caused either by degassing of the earth's crust in the northern land mass or by volcanic activity, such as in the region of Iceland (Siegel *et al.*, 1973). Alternatively the high surface mercury values may be due to atmospheric transport of mercury from the urban industrial complex of North America. However, analysis of fish samples from the northeast Atlantic (Leatherland *et al.*, 1973) failed to show any specific mercury contamination. In fact the highest mercury values in fish have been reported for the Black Marlin (7·3 ppm wet wt.; Mackay *et al.*, 1975) caught off the coast of northeast Australia where man is unlikely to have caused extensive pollution of the ocean habitat. Thus from chemical analyses there are a number of unexplained differences in the distribution of mercury in the marine environment. Since the transfer of mercury into the food chain of the sea is in part a biological phenomenon, involving the formation of organic mercury compounds by bacteria (Jensen and Jernelöv, 1967; Wood *et al.*, 1968; Fagerström and Jernelöv, 1972), it appears that biological oceanographers should be involved with the chemists in this type of problem.

The rate at which heavy metals are eliminated from animals depends on the species of animal and the element. Biological half-times range from days to years (Kullenberg, 1982); the path of elimination also varies considerably with the element as indicated in Table 40 for the efflux of various elements accumulated by a euphausiid.

A review of heavy-metal accumulations in the sea, particularly with reference to coastal environments is given by Bryan (1980). Heavy-metal accumulations in many organisms in estuarine environments are in excess of 100-fold; while this

TABLE 40. RELATIVE DISTRIBUTION OF ELEMENT EFFLUX[a] THROUGH THE EUPHAUSIID *Meganyctiphanes norvegica* BY MOLTING, DEFECATION, AND SOLUBLE EXCRETION (values accumulated by Kullenberg, 1982)

Element	Molting (%)	Defecation (%)	Soluble excretion (%)
Zn	1·1	92·6	6·3
Cd	3·3	84·5	12·2
Sc	2·4	54·4	43·2
Hg (inorg.)	2·5	29·1	68·4
^{239,240}Pu	0·8	98·6	0·6

[a]Calculations based on euphausiids grazing under sufficient food conditions during the non-egg-laying period.

makes these organisms unsafe for human consumption, it may not result in their elimination due to biochemical protection within the organisms when pollution has been introduced over a relatively long time period (see Section 7.1.2); the same levels of metals introduced as an acute dosage may have very different effects on organisms and their predators.

7.1.4.5 Radioactive materials. The distribution of radioactive pollutants in the sea is a special case of marine pollution associated with nuclear power stations, dumping, and earlier atmospheric testing of nuclear bombs. The distribution of isotopes, as with heavy metals and other pollutants, is a function of both biological and geochemical processes. The destructive effects of highly active isotopes on biological materials are well known and need not be discussed here further. However, it is of considerable interest that small traces of active isotopes, both artificial and natural, can be used to follow the transport of materials in the food web of the sea. For example, Williams *et al.* (1970 have shown that ^{14}C from weapons testing in 1961–2 has in some cases been rapidly transported into the tissue of bathypelagic animals, while other animals in the same habitat contained very little ^{14}C. Since ^{14}C has labelled the surface bicarbonate and has been taken up by phytoplankton in the euphotic zone it appears that in some cases there is a rapid vertical transport of this material to bathypelagic organisms (e.g. via overlapping vertical migrations, Vinogradov, 1955; or through fecal pellet sedimentation, Sasaki and Nichizawa, 1981), while in other cases, bathypelagic organisms must rely for food on an organic carbon source which is 'old' relative to the age of surface bicarbonate.

A review of problems associated with the deep-sea dumping of radioactive wastes has been presented by Gerlach (1981). In this review it is pointed out that some of the waste products have radioactive half-lives which are extremely long, relative to the generations of man and indeed most of the biosphere (e.g. plutonium-239, 24,000 yr; neptunium-237, $>2\times 10^6$ years). This eliminates the simple dumping of materials in the deep sea since they are likely to re-emerge in <1000 years at the very latest. Deep-sea burial into the sediment to depths of up to 5000 m below the sea bottom has been suggested but this presents considerable problems in deep sea drilling.

The accumulation of radioisotopes varies considerably with species as indicated in Fig. 107 for

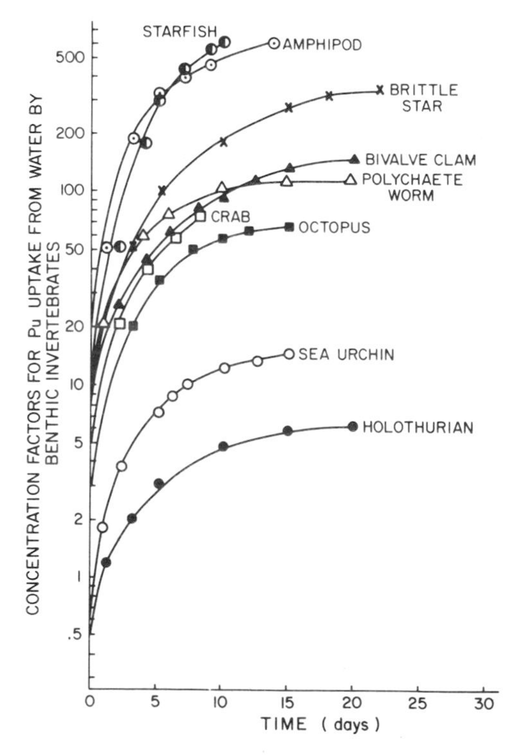

FIG. 107. Concentration factors for plutonium (^{238,239}Pu) — with permission from S. W. Fowler, International Laboratory of Marine Radioactivity, Monaco.

plutonium. Present evidence indicates that both food sources (e.g. Pentreath, 1976) and surface or gill transfer may be operative to a different extent in different animals. In addition the ability of organisms to purge themselves of isotopes varies with the metal species, as much as with the animal species (cf. Table 40).

The radioactivity of ocean waters has a level of around 320 pCi/l while sediments are much higher (*ca.* 20,000 pCi/kg). Molluscs and fish may range from *ca.* 1000 to 3000 pCi/kg, but near a recycling plant, values in the flora and fauna may be 100 times background levels (Gerlach, 1981). In view of the increasing need to dispose of radioactive wastes, the problem of how to do this in the ocean habitat, or if indeed this is the best place to use as a disposal site, is one which needs thorough research.

The chronic pollution by nuclear processing plants or power stations is really a separate subject to the problem of radioisotope disposal. As an example, the contamination of the seabed–seawater interface in the vicinity of the nuclear reprocessing plant at Windscale on the northwest coast of the United Kingdom is shown in Fig. 108. According to Woodhead (1981), although the levels of radioactivity are readily mapped in association with this treatment facility, the actual biological effects are insignificantly small. This is surmised from the fact that the radioactive dose for an individual fish in the area would be an order of magnitude below that at which minor physiological disturbance would occur (i.e. *ca.* 40 mrem h^{-1}) and at least two orders of magnitude below that at which reproduction of the population would be affected.

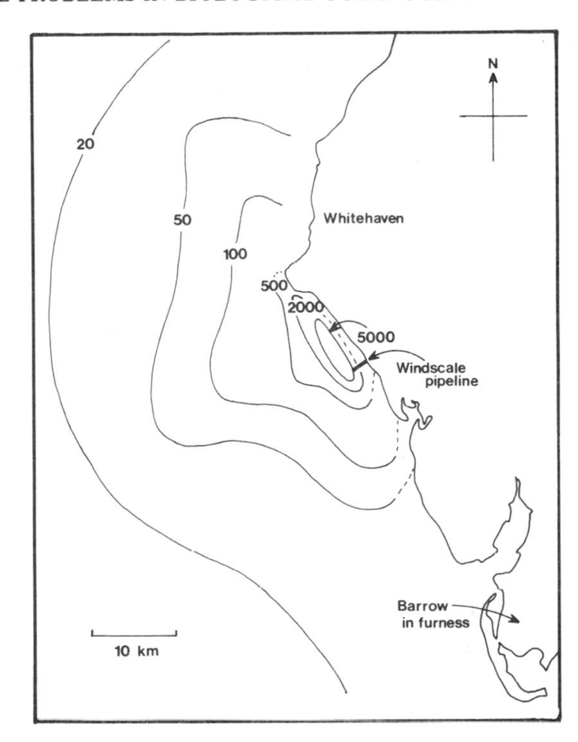

FIG. 108. Dose-rate contours on the sea bed in the vicinity of Windscale, United Kingdom. Dose rate in μrem h^{-1} (from Woodhead, 1980).

7.2 OCEANOGRAPHY AND FISHERIES

The management of a fishery is a complex problem involving not only theories on the production of fish in the sea, but economics, traditions, unions, fishing gear selection, and a host of other considerations. The solution to all these problems, however, would be made simpler if man understood why some species of fish were so successful in the sea that they reach an abundance that attracts the attention of man (and other predators). This is a question in biological oceanography and there has been little effort expended on finding an answer. Where major effort has been expended is in answering another question: how much fish can man take from the sea? In attempting to answer this question, fisheries scientists have studied the fish stocks of the sea, but seldom the environment which produces the stock. Further, the question of how much fish can be harvested is tied to economic considerations and as such it falls outside the scope of the more basic, purely biological question, which is concerned with what makes some species of fish superabundant in the oceans. The connection between the purely biological question and fisheries management lies in the common need to understand: (1) recruitment of young fish to adult stock (or why the natural survival of larval and juvenile fish varies) and (2) changes in the growth of fish. Both survival and growth are spread out over the entire life cycle of any fish species. For all fish species survival is very low for the early life stages but increases with age, while growth can be represented as an S-shaped curve. Both these curves are illustrated in Fig. 109 which, with appropriate adjustments to the time, length, and fecundity scales, might represent a large variety of fish from different habitats such as tuna, herring or plaice.

For the biological oceanographer, the problem in the two curves, illustrated in Fig. 109, is that both mortality and growth may vary considerably under natural conditions. For the fisheries manager the problem is even greater because the effect of fishing on a stock may also cause further variation in both mortality and growth.

In examining the history of fisheries management science, one is forced to ask why it has not been more successful in an age which has seen great advancements in other fields of science, such as space research, or among the biological sciences, such as agriculture and medicine. There are some obvious answers to these questions which relate to the amount of money put into research in other

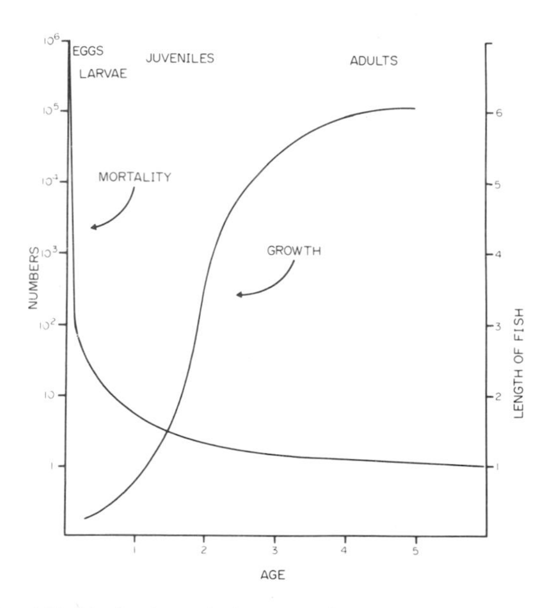

FIG. 109. Idealized population mortality and individual growth curves for a species of fish during its life cycle from egg to adult (arbitrary units of age and length depending on species).

fields compared with fisheries, as well as to more tangible aspects related to the ease of studying terrestrial physics and biology compared with having to probe data from the hidden world of the oceans. However, man's ingenuity has generally been such as to overcome many obstacles to the establishment of scientific understanding. In the case of fisheries science, however, it appears that perhaps the paradigm of fisheries management technique was also seriously at fault and that this fact has only recently been recognized.

The problem in fisheries management can really be divided into two parts. The first is the belief in a stock/recruitment curve and the second is the use of that curve to determine the 'maximum sustainable yield' (MSY). The latter is defined by Ricker (1968) as "the largest catch which can continuously be taken from a stock under current environmental conditions [allowing that] for species with fluctuating recruitment the maximum might be obtained by taking fewer fish in some years than in others". Thus, stated in the simplest terms, it was assumed that the parent stocks of fish gave rise to certain numbers of new recruits (i.e. new fish entering the fishery at a size close to that of the mature adult stock) and that the numbers of these new recruits were generally in excess of a 1:1 replacement line required for the maintenance of the stock. The excess number of recruits could then be harvested and in fact there was a maximum theoretical harvest (the MSY) which could be taken each year. This is illustrated in Fig. 110 together with actual data points on which the particular example of a stock/recruitment curve was based. many examples of such curves can be found in the literature and their theoretical treatment goes far beyond the scope of this text. Two features should be noted, however, with respect to this rather general relationship. The first is that if it exists, then it indicates that the MSY is a fixed region against a variable background (the data points). Thus any lowering of recruitment, or the genetic elimination of certain fish (or of particular fish stocks) would tend to destabilize the stock/recruitment curve which could eventually eliminate the fishery for a fixed MSY (Larkin, 1977). Ricker's (1968) caveat of a flexible MSY which should respond to changes in

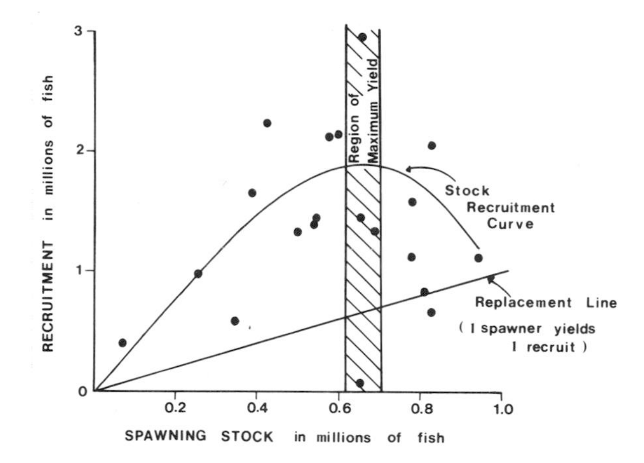

FIG. 110. Stock/recruitment curve for Babine Lake sockeye salmon modified from Healey (1982) to give replacement line, stock/recruitment curve and region of MSY.

recruitment does not work in practice since changes in recruitment are not known until it is too late. The actual effect of fishing pressure on a stock/recruitment curve is illustrated in Fig. 111(A) where fishing mortality (which is a non-linear function) is shown to intercept the stock/recruitment curve at two points. The first is a stable point where adequate stock remains; however, the second is an unstable point where the taking of stock will drive the fish population to extinction. Thus fishing mortality may have dramatically different effects, depending on the nature of the stock/recruitment curve; a curve which is not known with any reliability!

The second observation on Fig. 110 which must follow from the first is that in fact the stock/recruitment curve itself may be a hypothetical creation for which there is just enough evidence for a mathematical relationship to be formulated but which in fact is a misleading relationship — a paradigm in which many people have come not to believe. It is almost tautological to say that even if a statistical relationship between stock and recruitment can be demonstrated, this does not explain the basis for that relationship. It could be fortuitous, or it must in fact depend on some secondary factor, such as predation of juveniles or a disease vector, the function of which may be biologically understandable. Neither of these alternatives justify the use of the stock/recruitment curve as a management strategy. Perhaps a better representation of the stock/recruitment curve is given in Fig. 111(B) (Anon., 1980). Here a broad region (II) is depicted in which environmental factors determine the relationship between stock and recruitment (density independent) while in a narrower region (I) stock and recruitment are related (density dependent) such that the taking of stock (fishing), or the replacing of stock (transplants), can move the curve down or up, to extinction or into an area of density independence, respectively. In this one small region, therefore, at low stock density, the number of recruits will be proportional to the size of the parent stock. In the broad region of the curve (II) it is assumed that fish are primarily r-strategists and not K-strategists as is implied by the original presentation (Fig. 110). Thus the laying of large numbers of eggs from which only a few adults survive is a typical r-strategy, the success of which will be governed by the environment and not the numbers of parent stock. At some lower level of

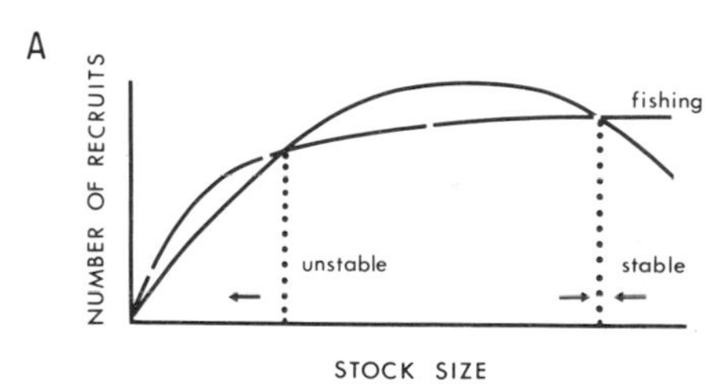

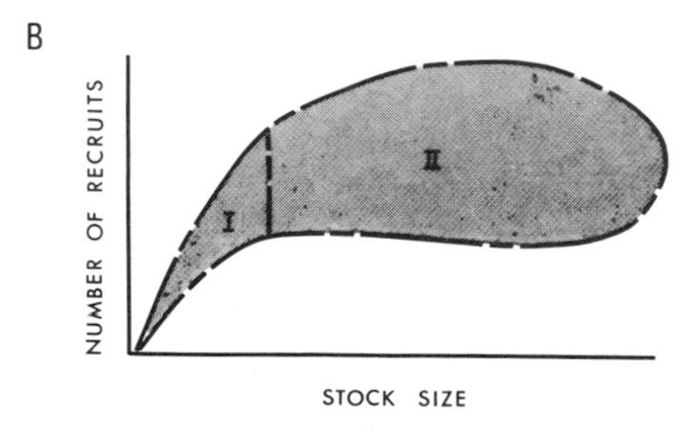

FIG. 111. (A) Fishing mortality imposed on a stock/recruitment curve (Anon., 1980). (B) General relationship between stock and recruits. I. Density dependent region. II. Density independent region.

survival (e.g. see larval fish survival, Section 7.2.1) density dependence in terms of the amount of food per organism may play a determinate role. However, the production of food organisms relative to fish feeding is still a density independent process.

7.2.1 Larval Fish Survival

The next question to pursue is to ask what kind of hypothesis can be formulated that would replace the stock/recruitment curve with a more reliable management strategy (i.e. in the broad region II, Fig. 111(B), where there is density independence). During their life cycle fish belong to several different populations, usually as plankton when they are eggs or larvae, later as juveniles and finally as adults, they are part of the nekton community. From Fig. 109 it is apparent that the egg and larval phases of a fish's life are subject to high mortality and that this decreases towards adulthood. For example, Cushing (1975) summarizes the mortality for plaice as 80% per month for larvae up to the time of first feeding (*ca.* 4 months), 10% per month during the first winter (i.e. up to 1 yr old), and 10% per year during the life of adult fish under natural conditions. The important aspect of such a high mortality curve, particularly at the beginning of the animal's life (i.e. the planktonic stage), is not in the mortality of many, but in the variable survival of a few. For example, Gulland (1965) estimated that from several million eggs laid by one female cod, the early life mortality must be about 99·9999%; the statistical difference between this figure and 99·9998% is not detectable, but the survival increase of 100% is the important event to consider. At this stage in the fish's life it is impossible to conceive of a predator devouring the entire larval year class — the energetics of predation generally assume a threshold prey level [cf. eqn (101)] at which it is no longer economic for a predator to consume any more food items when they are very sparsely located. Therefore a predation mechanism (or even a disease vector) which affected 90% of the larvae would still leave large numbers surviving. The more important mechanism to consider is food available to the larvae. Here we have a hypothesis first formulated by Hjort (1914) who described larval survival as a 'critical phase' in the life cycle of fish. More recently Ivlev (1944) and Cushing (1975) have discussed what appears to be the kind of 'make or break' explanation for the sudden appearances of large year classes, or the virtual absence of fish due to poor survival, as seen in many fish stocks. Simply stated this hypothesis assumes that the fish larvae must either feed or die at the end of the yolk-sac period. While the lack of food could be due to a variety of physical factors, it is the concept of a match or mismatch in the timing of the arrival of hungry larvae with the appearance of suitable prey items which is important. The hypothesis is illustrated in Fig. 112 from Cushing (1975). In this illustration three possible time sequences in the production of copepod nauplii are shown (a, b, and c), together with one pulse of fish larvae. The larvae that coincided in one year with nauplii cycle 'a' would have very few food items per larva and consequently the whole year class of larvae would die. In years 'b' and 'c' a greater number of nauplii/larva are present and an increased survival is assumed. This sequence can be better understood

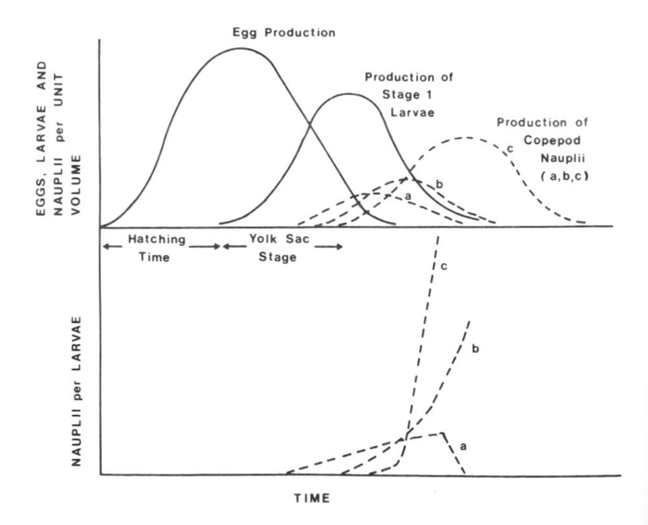

FIG. 112. The match or mismatch of larval production to that of their larval food. The numbers of nauplii/larvae represent the degree of feeding success. The three curves represent three conditions of copepod nauplii production and hence three conditions of feeding success, $a < b < c$ (from Cushing, 1975).

if one considers also Fig. 113 from Ivlev (1944). Here it is seen that in terms of prey concentration there is some critical number of nauplii needed, below which the time spent in hunting by the larva goes to infinity — at this point there is zero survival of any larvae since in the absence of their yolk sacs they are unable to live more than a few days. The match and mismatch hypothesis of larval fish survival is a plausible explanation which allows for a possible increase in larval *survival* (e.g. by several hundred percent) — a factor that does not significantly alter the mortality coefficient, as decribed above.

Experimental and field evidence for larval survival during this critical phase of the animal's life is difficult to obtain. From detailed observations on the survival of plaice larvae in the North Sea, Shelbourne (1957) has described larvae in a deteriorating condition during the month of January when suitable planktonic food items were scarce. Deterioration of larvae started as the yolk reserves became exhausted; similar observations conducted during March showed the presence of healthy larvae which were feeding on plant and

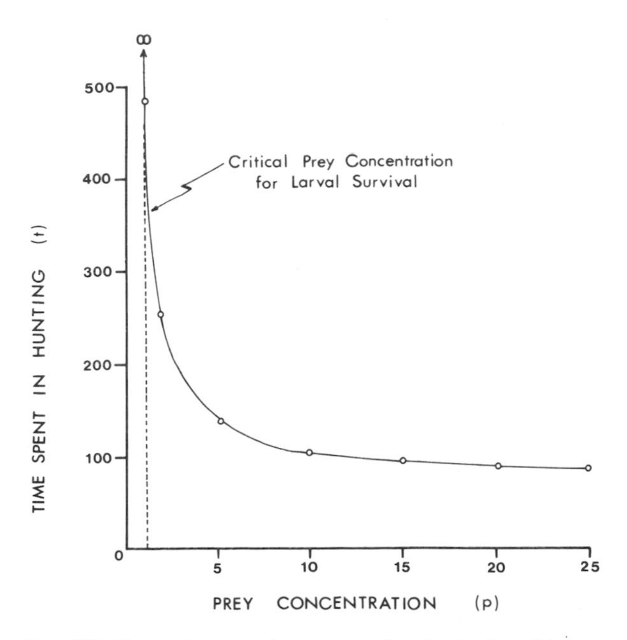

FIG. 113. Dependence on time spent in hunting by larval fish on the concentration of food organisms in arbitrary units (redrawn from Ivlev, 1944).

animal plankton. The importance of specific food items in the diet of plaice larvae was also observed; during the earliest feeding stage, the larvae fed off some large diatoms (e.g. *Biddulphia* and *Coscinodiscus*) but an abrupt change to a zooplankton diet (primarily *Oikopleura*) occurred soon after feeding started.

The feeding and survival of herring larvae has been studied extensively both in the field (e.g. Blaxter, 1963; Lisivnenko, 1961) and under laboratory conditions (e.g. Rosenthal and Hempel, 1970). Blaxter's (1963) detailed observations have shown that there is some obvious selectivity for prey items depending on size, while visual sighting and attacks on individual prey items depended to some extent on the amount of light available. Nutritional differences between prey items consumed were also demonstrated to have an effect on the growth and survival of the larvae. From field studies conducted by a number of authors, Blaxter (1963) concluded that larvae were most abundant when food items were present at a concentration of *ca.* 30 organisms per litre. This figure is similar to the concentration of prey items determined experimentally by Rosenthal and Hempel (1970), if it is assumed that only a certain fraction of the prey items seen by a larva are effectively captured. Lisivnenko (1961) studied the abundance of herring larvae and the concentration of food organisms over a period of 5 years in the Gulf of Riga; her results show a strong correlation between an approximate 5-fold increase in larval abundance and an increase in food items from *ca.* 5 to 20 organisms/litre.

Sysoeva and Degtereva (1965) showed that the main food item of the Arctic–Norwegian cod larvae and fry was the copepod, *Calanus finmarchicus*. Both larvae and fry fed off *C. finmarchicus* during different stages of the copepod's life cycle and this is illustrated in Fig. 114 together with the secondary importance of *Oithona* nauplii and some other food organisms. The absolute concentration of food organisms required by cod fry for successful feeding was expressed as the number of organisms per m^2 in a 50-m water column. Concentrations of >18,000 *C. finmarchicus*/m^2 provided a sufficient food supply while the intensity of feeding decreased over the range from 18,000 down to 5000 organisms/m^2.

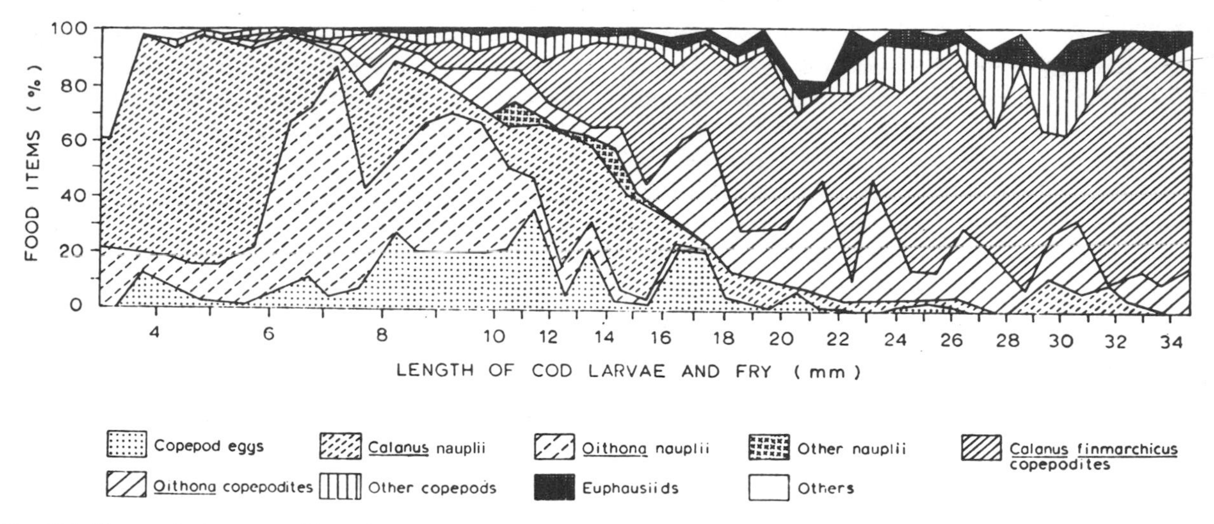

FIG. 114. Changes in the composition of the food of larvae and pelagic fry of cod with increasing length. (Redrawn from Sysoeva and Degtereva, 1965.)

Lasker (1975) in a unique series of experiments has combined both laboratory and field techniques to derive a specific relationship supporting Hjort's original 'critical phase' concept. In Lasker's (1975) experiments it was shown that first-feeding anchovy larvae require phytoplankton aggregations of >20 cells/ml within 2½ days after the larvae are ready to feed and that individual cells had to be about 40 μm diameter. The authors found that such feeding conditions were quite transient but usually associated with subsurface chlorophyll maxima.

In summary it appears that there may be one or more 'critical phases' in the life of larval fish and that these in fact form a critical period in which survival could be severely reduced. The 'first feeding' phase suggested by Cushing (1975) may be one of the critical occasions in the early life of a fish, but mortality could also be caused by the absence of a specific prey type, such as is suggested by Shelbourne's (1957) observations on plaice larvae when they switched from herbivorous to carnivorous feeding. The 'make or break' mechanism of population control referred to earlier may not be exactly the same in any one year. However, the importance of adequate food items in any such mechanism can not be over-stressed, whether it is to prevent starvation of a larval population at a particular phase, or even if it is to assure adequate health of the larvae under the stress of some secondary factor such as disease or a pollutant.

7.2.2 Juvenile Fish Survival

The second question regarding the early life stages of fish is to consider what might be the principal factors controlling juvenile fish survival once the fish have passed the larval stage. In the case of fish larvae there is no doubt that large numbers of larvae are eaten by predators and that this at the same time accounts for high mortality but not for high survival, as described above. When the fish become juveniles, they are much larger prey items and there are comparatively few of them, when measured against *ca.* 10^6 eggs compared with 10^2 juveniles, surviving from the original spawning (arbitrary relative survival from Fig. 109). Mortality of this group is much lower but now it becomes significant in controlling the population. As juveniles left alone without predators, the population can survive the lack of plankton, or it can seek out new patches of plankton on which to feed. However, the more plankton that juveniles eat, the faster they grow and vice versa. In the presence of predators (i.e. under natural conditions) juvenile fish which grow slowly will never outgrow attacks from predators who are also growing, but generally at a much slower rate. An example of this kind of juvenile survival hypothesis is given by Parker (1971). In the case of young pink salmon, Parker (1971) was able to show that the true growth rate was greater than that of the pink salmon's predator (2-year-old coho salmon) which grew at 0·7% per day.

Thus the survival of the juvenile pink salmon was assured as they 'outgrew' being suitable prey items for coho salmon. This type of mechanism is illustrated in Fig. 115 where it is seen that depending on the amount of zooplankton present young pink salmon may grow at different rates. Those that grow faster than the growth rate of the predator will eventually escape attacks as suitable prey items. Thus in this juvenile fish survival hypothesis, the quantity of planktonic prey items is again important. There is a difference in the mechanism for larvae and juveniles, however, in that for larvae it is a matter of survival while for juveniles it is a matter of growth rate.

Imposed on the adequacy of food items available to the juvenile fish are a host of other variables which can also affect survival. These would include different predators and variation in their abundance, disease including viruses and parasites, and condition factors of the juvenile fish as they emerge from their earlier life stages. The latter would include their genetic origin (e.g. hatchery fish vs. wild stock), their physiological state and the extent

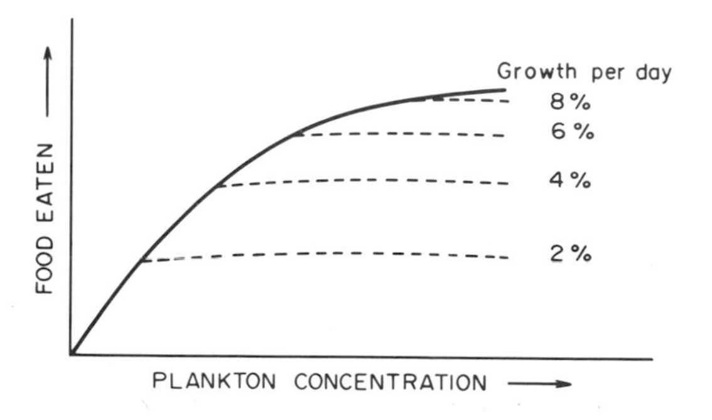

FIG. 115. Schematic relationship between plankton concentration, food eaten and the daily growth rate of juvenile fish.

to which they may have been subject to near-shore pollution during their early life stages. Thus any weakness in the juvenile population must also be considered as a determinate factor in survival and growth.

7.2.3 The Growth of Fish

The growth of fish has been the subject of extensive reviews by other authors and a good precise summary is given by Jones (1976). If one neglects the very earliest stages in a fish's life, then the growth of a fish can be represented as an asymptotic equation which represents most of the curve in Fig. 109 (Bertalanffy, 1938).

$$L_t = L_\infty (1 - e^{-K(t-t_0)}) \quad (208)$$

where L_t is the length at age t, L_∞ is the asymptotic length at $t = \infty$, and K represents the rate of growth of an individual between the time interval, $t-t_0$. The equation can be converted to a weight relationship on the assumption that weight is a cubic function of length for many species. Other growth equations have been used but for the purposes described here, the parameters employed in eqn. (208) are sufficient.

In general there is an inverse relationship between growth (K) and maximum length (L_∞); temperature tends to increase the rate of growth (K) up to some maximum above which the animal can not survive; maximum length (L_∞), on the other hand, generally shows a decrease with an increase in temperature. There tends to be a positive relationship between t_{max} (the time to reach maximum size) and L_∞, while within species, certain stocks may show differences in fecundity, egg size, and spawning cycle which can be related to the mean size of adults in different fish stocks (e.g. herring — see Cushing, 1975).

For the biological oceanographer the most important aspect of a fish's growth is the availability of sufficient prey items of the right type for the animals to feed on. The division of food energy will generally follow the same pattern as discussed for small crustaceans (e.g. Fig. 60). The most critical aspect of this division of energy can be summarized in the use of the growth-efficiency term ($K_1 = G/R$).

From extensive studies of fish-feeding data reported by various authors Paloheimo and Dickie (1966b) concluded that growth efficiency (K_1) was related to the amount of food eaten by an animal (R) such that,

$$\ln K_1 = -a - bR. \quad (209)$$

When $\ln K_1$ is plotted against R for different rations, different values of a and b are obtained. The relationship may not apply to hand-fed laboratory fish cultures but there is considerable justification for its use under field conditions. The importance of eqn. (209) from the point of view of the production of food items is that constants a and b must be related in some way to the metabolic cost of behaviour patterns associated with particular types of prey. As an example of this effect the authors suggested that the size of prey items might be one factor in determining differences between growth efficiency and ration (i.e. in the differences in a and b). Differences in the ability of young salmonids to graze off different-sized prey has been demonstrated in short-term experiments, such as those reported by Parsons and LeBrasseur (1970). In the latter experiments the same biomass of three different sized zooplankton (two species of copepods and a euphausiid) were fed to young salmon over a concentration range from 0·5 to 100 g/m^3 wet weight of zooplankton. The results showed that the salmon were able to eat more of the medium-size copepods per unit time than of the large euphausiids or the very small copepods. From this observation it may be concluded that not only is the absolute abundance of prey items important to the growth of juvenile fish, but also there are marked differences in the growth efficiency of fish depending on the type of prey as indicated by eqn. (209). With some fish species the methods of capturing prey change and this will also alter the rate of feeding at similar prey densities. This is illustrated in Fig. 116 from Leong and O'Connell (1969). The experimental data show that the Northern Anchovy obtained small *Artemia* nauplii by filter feeding at a slower rate than they obtained *Artemia* adults by raptorial feeding. Furthermore, the shapes of the feeding curves are different and show that the greatest difference occurs during the shortest feeding interval. In a later experiment O'Connell (1972) was able to show that raptorial feeding was the exclusive form of feeding down to an adult *Artemia* concentration of 7% of the total biomass of mixed adult and nauplii *Artemia* prey. Only when the large prey items were reduced to less than 2% of the total biomass did the form of feeding become greater than 50% in favour of filtering. In similar experiments using the much larger Pacific mackerel (*Scomber japonicus*), O'Connell and Zweifel (1972) showed that *Artemia* nauplii were ignored as food items but that *Artemia* adults were fed on by raptorial (biting) feeding at low prey densities (1 or 2/litre) and by filter feeding at high prey densities (22 to 112/litre).

In summary, it is apparent that the growth of fishes is a complex function of their energetics involving temperature, swimming, metabolism,

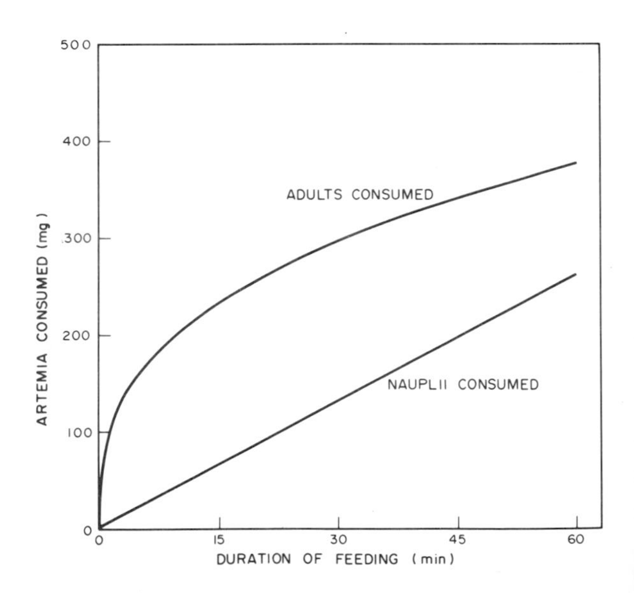

FIG. 116. Comparison of particulate and filter feeding rates for a 4-g anchovy. Curve showing 'adults consumed' was determined for *Artemia* concentrations of 2- to 50-mg adults/litre; curve showing 'nauplii consumed' was determined for *Artemia* concentrations of 4·4 mg nauplii/litre. (Figure redrawn from Leong and O'Connell, 1969).

food type, and reproduction. A good review of this subject is given by Brett (1970).

7.2.4 Some Aspects of Fisheries Management

The traditional management of a fishery has usually involved the establishment of catch quotas based either on a long-term average (i.e. independent of the population in one year) or on a year-to-year basis (i.e. in response to the availability of fish). Theoretically, both strategies applied to a randomly fluctuating population and its growth rate can lead to the decimation of the stock (e.g. Beddington and May, 1977). Actual examples of such effects may be found among a large number of very different stocks (e.g. Antarctic whales, Peruvian anchovy, or North Sea herring). Thus by far the greatest effect of man on the ecology of the oceans has been the effect of over-fishing; no form of marine pollution is in any way comparable to the ecological impact which occurs with the removal of *ca.* 70 million tons per year of predatory species from the ocean ecosystem. Fortunately for the fishing industry, the effect of decimating traditional fisheries has often initiated the establishment of new fisheries which result from an abundance of lower trophic level stocks after the removal of top predators. This has been postulated, for example, to explain the increase in stocks of sand-eel, Norway pout, and sprat in the North Sea where herring and mackerel have been over-exploited. Similar effects have been found in the northwest Atlantic pelagic ecosystem, and data are shown in Fig. 117 (Hempel, 1978; Sherman *et al.*, 1981).

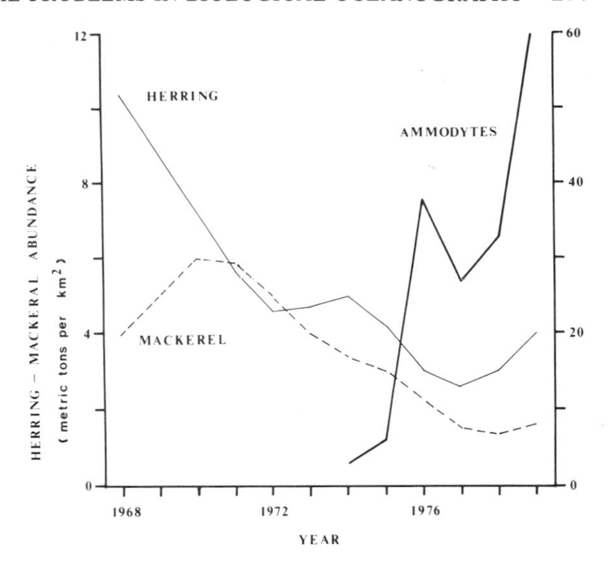

FIG. 117. Decline in herring and mackerel biomass in metric tons per km^2 and successive increase in larval *Ammodytes* in millions of larvae/km^2 for the northwest Atlantic from Cape Hatteras to the Gulf of Maine, 1974–1979 (redrawn from Sherman *et al.*, 1981).

Thus the problem of fisheries management today is first to understand the natural ecosystem in order to answer the question — how much fish can be taken from the sea? Secondly, it is necessary to understand the effect of removing large quantities of fish on the same natural ecosystem. However, this is still only part of the problem of commercial fisheries since the economic and social problems of the fishing industry should also be taken into consideration. In fact, the management of a fishery today is no longer to determine the 'maximum sustainable yield', but to make the 'best use' of the resource, where the latter term is defined as "the sum of net social benefits (personal income, occupational opportunity, consumer satisfaction, stock preservation, etc.) derived from the fisheries and the industries linked to them". Mitchell (1979) gives a complex model for embracing all these factors of which the ocean ecosystem (including the fish) are only a small part. However, it is the ecosystem which makes the rest possible and it is this aspect that can be discussed further relative to fisheries needs.

7.2.4.1 Ecosystem approaches to the management of marine resources. The amount of fish taken out of any ecosystem must be proportional to the amount of production within that ecosystem. Since nitrogen is generally regarded as the rate-limiting element in many (but not all) ocean ecosystems, it is appropriate to consider first how much new nitrogen enters an ecosystem each year. The alternate source of this element is recycled nitrogen. However, recycled nitrogen is largely retained within the system so that removal of nitrogen from such a closed system in the form of fish results in an ultimate decimation of the nitrogen reserves of the ecosystem. This problem has been discussed by Eppley and Peterson (1979) who compared the amount of new primary production in different ocean areas as a function of the total primary production for the same areas. In the Peruvian upwelling, the new production (i.e. from upwelled nitrate) was estimated to be 50% of the total production but for oligotrophic tropical waters it was less than 10%. On a global basis, using an estimate of 20 to 25×10^9 tons C/yr as the total annual primary production of the oceans, the authors estimated that less than 20% of this was due to 'new' production (i.e. entrainment of deepwater nitrate, with lesser quantities of nitrogen from N_2 fixation or allochthonous sources). If one assumes a transfer efficiency to a potential world fishery of 0·1%, then the total yield of fish that could be taken out of the system would be *ca.* 100×10^6 tons/yr (wet weight, assuming wet weight:carbon ratio of 20). As a global estimate, this is close to the present world catch of *ca.* 70×10^6 tons. However, obviously such an approximation is highly dependent on the conversion factors used. The main purpose of the exercise is to show that there is an upper limit to the amount of fish protein that can be harvested from the sea and that this will be very small in the case of oligotrophic waters (e.g. in coral reef communities) but much larger in strong upwelling areas (e.g. the Peru current).

Another ecosystem approach to estimating how much fish can be produced in a particular area is given in a series of papers by Sheldon *et al.* (1972, 1977, and 1982). In this approach it was observed that there are roughly similar amounts of material (biomass, not number of organisms), in logarithmically equal size intervals based on a diameter scale of 1 to 10^6 μm. This scale includes plankton, fish, and whales. Thus if the biomass of phytoplankton in an area is known, the biomass of fish can be estimated; if the growth rate of the fish is known, production and the potential yield to the fishery can be calculated. The advantage of this method is that it avoids making an estimate of the most difficult parameter which is the fish standing stock; phytoplankton standing stock is much more easily estimated. An example is given in Table 41 and Fig. 118, and as follows. From a primary productivity value for the Gulf of St. Lawrence of 200 g C/m^2/yr, Sheldon *et al.* (1982) estimate that this must come from a standing stock of 27 g/m^2, or 2·7 g/m^2 (wet wt.) of phytoplankton per size grade. This value should represent the amount of fish in a single size grade. It was then estimated that of the two principal fish, herring occupied two size grades and cod four. Thus the standing stock of fish is estimated as shown in Fig. 118 and from an approximate knowledge of the growth rate (doubling time) of the fish, the production of each size cohort is estimated. The method actually says nothing about the species of fish which can be harvested, only the amount to be produced in a particular size class. Further, the technique depends on two properties of the food chain: that large animals grow slower than small ones and that in the sea, predators are usually larger than their prey by a fairly constant amount. These two points are discussed in Sheldon *et al.* (1972 and 1977). The conclusion reached by the authors from a variety of fisheries was that in many cases, on a regional basis, the maximum amount of fish was already being harvested. However, this does not exclude higher or lower production occurring due to environmental change or by altering the size spectrum harvested. A theoretical treatment of this subject is given by Platt and Denman (1978).

The most popular form of ecosystem approach to fisheries production is the use of some form of energy budget in which standing stocks, production, and transfer coefficients are determined and a model built around the various interactions (e.g. Steele, 1979). One of the most extensive developments in this field is given by Odum (1967) who uses the analogy of electrical circuitry to convey energy flow and storage in natural ecosystems. An energy budget of an ecosystem is a useful way of comparing production of different ecosystems and illustrations from Cohen *et al.* (1982) are given in Figs. 119 and 120. From such a model it is possible to gain some insight into the correctness of assumptions on stock estimates as well as the consequences of changing the quantity of material harvested at any particular trophic level. In the examples shown in Fig. 119 it is

ND COD IN THE GULF OF ST. LAWRENCE AS ESTIMATED BY
IELDON *et al.* (1982)

Size grades	Standing stock (g/m^2)	Doubling time	Production ($g/m^2/yr$)
10	27	2 d	3400
2	5·4	4·7 yr	0·81
4	10·8	1·7 yr	4·41

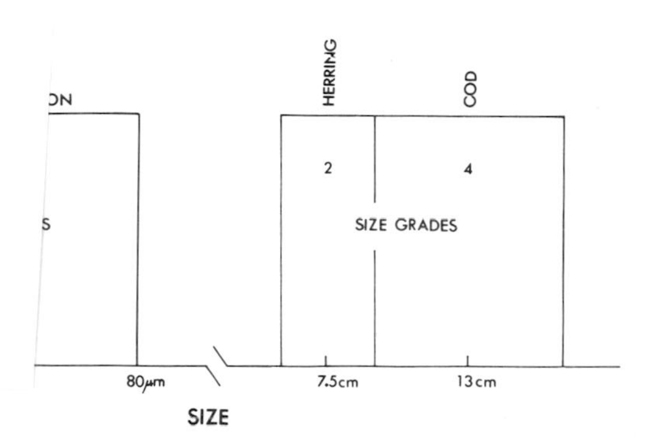

FIG. 118. Relationship between phytoplankton and fisheries in the Gulf of St. Lawrence (from Sheldon *et al.*, 1982).

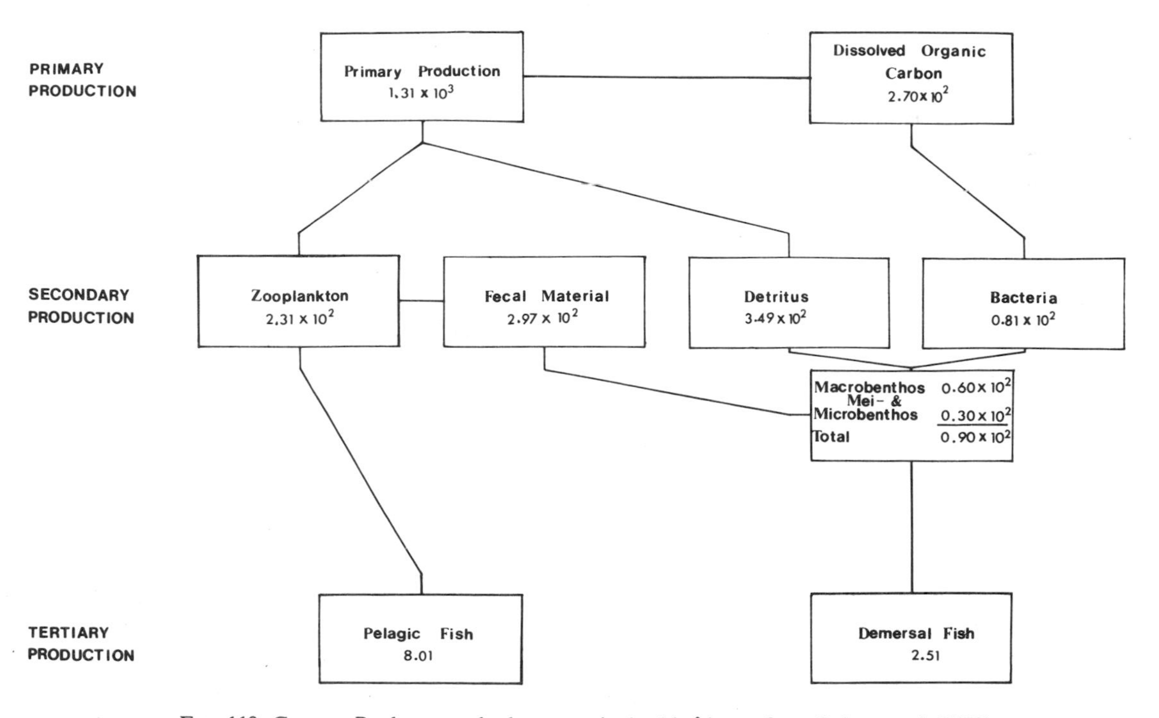

FIG. 119. Georges Bank energy budget — units $kcal/m^2/yr$ — from Cohen *et al.* (1982).

apparent that Georges Bank is very much more productive per m^2 at the primary level compared with the North Sea, Fig. 120. This is probably attributable to the former area being a frontal zone (see Section 1.3.5). However, much of this primary production must be exported off the Bank since for a *ca.* four-fold higher primary production, the fish production is only about 20% higher *in toto* per m^2. The effect of higher primary production on Georges Bank is in fact mainly seen in the benthic community.

Implicit in the ecosystem approaches to the understanding of fish production, as given above, is the fact that the driving force for biological production in the sea is the physical environment. Few fisheries managers consider this point and yet its effect has always been apparent in many papers scattered through the literature over the past 50 years. One summary of the physical effects on fisheries has been given by Wyatt (1980) and is illustrated in Fig. 121. In this figure it is assumed that the introduction of new nutrients by upwelling, either through the winter overturn of stratified water or on a more persistent time scale of >250 days (e.g. the Peru current) will produce more fish than an area which is either too stratified or too well mixed. The mixing process itself causes different phytoplankton species to grow (cf. Semina, 1972; Parsons, 1979 for discussion). The resultant food chain will give its highest fish production with moderate upwelling of long duration (Box 5) and its lowest production with low (or negative, i.e. a convergent gyre) upwelling of short duration (Box 1). An alternative low production area would occur with too much upwelling (Box 3) when waters are so turbulent that little primary productivity can occur. Wyatt (1980) also points out that systems can shift from left to right and vice versa over time, such as occurs with alternate seasons of high and low production (e.g. summer vs. winter). The persistence of upwelling regions in rather local environments is also as important to fisheries as the more global considerations given by Wyatt (1980). An example of local upwelling and fisheries is given by Iles and Sinclair (1982). In this report the authors draw attention to the fact that herring stocks spawn in

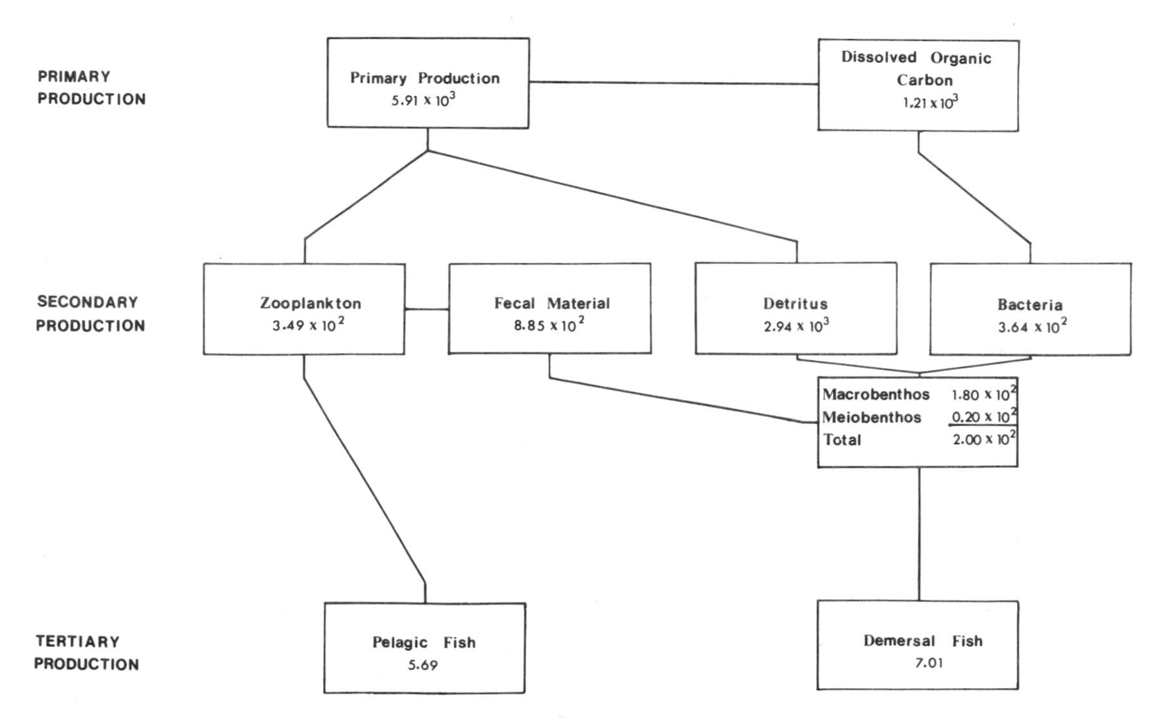

FIG. 120. North Sea energy budget — units kcal/m^2/yr — from Cohen *et al.* (1982).

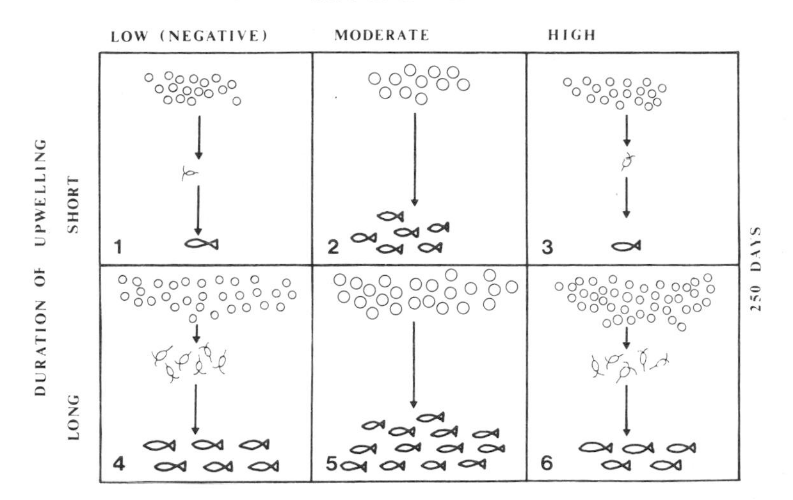

FIG. 121. Classification schemes for upwelling systems in terms of their duration and upwelling rates. The duration determines the absolute amount of production, and the upwelling rate the mean algal cell size (i.e. small and large circles in the upper part of each box). The latter controls the number of links in the food chain, and hence in conjunction with the absolute production, the amount of fish which can be sustained (from Wyatt, 1980).

relation to frontal zones in the North Atlantic and that these zones form discrete stocks. As a corollary of their description, the authors advocate fisheries policies which allow fishing only on the spawning grounds and not as mixed stocks. This strategy would give better control of the absolute size of the separate stocks while a mixed-stock fishery could result in the decimation of any one stock.

Changes in the structure of ecosystems due to changes in the presence or absence of particular fish species have been discussed by Cushing (1975) with particular reference to the *ca.* 40-year 'Russel' cycle in the English Channel. In this example it is believed that changes in the ecosystem were set in motion by a decline in the herring fishery during the 1920–30s. The herring were replaced in about three generations by pilchards, "but consequential changes affected the whole ecosystem, the quantities of macroplankton, the numbers of planktonic predators, the elasmobranchs, the number of spring and summer spawning fishes, and also the maximum in winter phosphorus itself" (Cushing, 1975). The return of herring stocks in the 1970s followed a reverse trend in community structure. Another example of changes in marine ecosystems due to the removal of top predators comes from the Antarctic. Hinga (1982) has estimated that since whaling began in the 1920s, approximately 96×10^6 tons of whale food per year is now available to other parts of the food chain due to the decimation of whales. Part of this has been identified as having resulted in large increases in the penguin and crabeater seal populations.

Large-scale oceanographic phenomena affecting fish populations are not confined to upwelling as discussed earlier. Favorite and Ingraham (1976) showed, for example, that there were cyclical changes in fish stocks and that some of these could be attributed to 11-year (sunspot) and an unidentified 6-year cycle in physical events in the northeast subarctic Pacific Ocean. The scales on which fish stocks may vary with climatic change can be both short term and long term; these may encompass frequencies that range from 1 to 100 years. Dickson and Lee (1972) showed variations in hydro-meteorological events affecting herring and cod stocks; they differentiated between short-term fluctuations of several years and longer-term trends lasting a number of decades. Until recently the exact nature of the forcing function (i.e. other than correlations with temperature and salinity) was not known. However, Mysak *et al.* (1982) have suggested that the magnitude of certain physical wave motions in the sea (e.g. Rossby, Kelvin waves) would be sufficient to alter the environment for fish. In many cases these environmental changes are out of phase with the physical event since they depend on both the life cycle of the fish and the periodicity of large-scale waves in the ocean. In coastal areas the fish ecosystem will be greatly affected by cyclical changes in large-scale run off from major rivers. An illustration of this is given by Sutcliffe (1973) who found correlations between seasonal river discharge and local landings of American lobster and Atlantic halibut in the Gulf of St. Lawrence.

Manipulation of some part of the food chain or life strategy of any marine organism (e.g. seaweed, oysters or salmon) can be included under the general heading of aquaculture if the process increases the yield. In many countries, particularly in tropical and subtropical waters, 'fish farming' or aquaculture has been practised economically for many centuries. A review of some of the methods used in estuarine habitats is given by Hickling (1970). In temperate latitudes aquaculture has generally not developed to the same degree, either because of the slower growth rate of organisms or because the labour-intensive practices of aquaculture have been generally uneconomical. However, with some species such as shellfish and salmon which have coastal habitats for all or part of their life cycle, some progress has been made in the enhancement of stock by various artificial means. For oysters, this includes the location of rafts in areas of high primary productivity, spat culture, and predator control. In the case of salmon, stream improvement for spawning, migration channels, and hatcheries have all played a part in producing more fish. In some cases, however, this may have occurred at the expense of selecting a genetically weak stock where a few adult fish may have been used to produce a large number of offspring in a hatchery.

One method of Pacific salmon enhancement which has been dramatically successful and is of interest to biological oceanographers because of the technique involved is lake fertilization. Low nutrient additions were made to a lake 34 km long and 1·5 km wide on the Pacific coast of Canada. The purpose of these nitrate and phosphate additions was to increase the supply of plankton for young salmon before they went to sea. It was anticipated that the better-fed salmon would be larger and survive better resulting in returns, 3 years later, which would be greater than under conditions of no nutrient additions (Parsons *et al.*, 1972; LeBrasseur and Kennedy, 1972; Barraclough and Robinson, 1972). The addition of nutrients was set up to meet the requirements of hyperbolic growth responses of phytoplankton (e.g. see discussion by Caperon *et al.*, 1971). As a result, it was found to be possible to increase the primary productivity of the lake without causing instability (i.e. undesirable eutrophication). Further, there is good evidence that the nutrient additions resulted in increased juvenile salmon production *within* the lake (cf. Barraclough and Kennery, 1972) as well as increased salmon returns as indicated in Table 42 from LeBrasseur *et al.* (1979 and references cited).

Finally, an examination of an ecosystem approach to fisheries management would not be complete without some mention of ecosystem modelling. Walsh (1976) has provided a good summary of marine ecosystem models; his

TABLE 42. COMPARATIVE SIZES OF ANNUAL GREAT CENTRAL LAKE SOCKEYE SALMON MIGRATIONS ASSOCIATED WITH UNFERTILIZED AND FERTILIZED YEARS

Régime	Mean number of fish migrating	Standard error
Unfertilized (1958–1972)	52,340	±12,000
Fertilized (1973–1976)	373,058	±89,000
Ratio of means (fertilized:unfertilized)	*ca.* 7:1	

conclusion is that "no model is a perfect representation of the real world". Nevertheless it is apparent that areas which are sufficiently restricted in space and time, such as upwelling ecosystems, will lend themselves to some forms of modelling. As defined by Walsh (1976) mathematical models may be of many different kinds, viz: steady state, time dependent, simulation, statistical and involve different mathematical methods for their solution. Verification of a model is difficult to achieve and if achieved, it does not mean that some other kind of model may not be equally versatile in fitting the facts. However, as Radford *et al.* (1981) have pointed out, the business of modelling is in itself a system which generates new ideas, new problems, experiments, and hypotheses. Another review on ecosystem models is given by Silvert (1981) while Laevastu and Favorite (1981) provide an actual example of a fisheries ecosystem model.

In summary to this section it is apparent that close scrutiny of life cycles, environmental data, and the scales of biological events has resulted in many examples of where fish production can be understood, predicted or even enhanced. These early results of ecosystem approaches to fisheries problems should be encouragement to others, especially those who have found the stock/recruitment paradigm of little lasting value outside of its limited function in portraying the general extent of overfishing.

7.2.4.2 Future fisheries. In the discussion above it is assumed that ecosystem approaches are needed for existing fisheries. However, the changing nature of fisheries is such that many of the traditional single-species fisheries are in the process of changing towards multispecies fisheries. In addition there exist new opportunities for fisheries at lower trophic levels as well as on stocks of more conventional fish which have not been exploited. This change again emphasizes our need to understand the ecosystems of new fisheries before the stocks are over-exploited, as has been the traditional approach in many cases. For example, the anchovy fishery of Peru was a recently developed fishery for which a traditional 'maximum sustainable yield' was determined. The fishery started in the late 1950s and the MSY was set at about $9{\cdot}5 \times 10^6$ tons/year (the largest single-species fishery in the world). By 1970 some catches had exceeded 12×10^6 tons but by 1973 the catch had collapsed to 2×10^6 tons. A more recent potential fishery to receive attention has been the Antarctic krill. It is not known how much krill can be harvested but some estimates have run as high as 100×10^6 tons/year. However, in the case of the Antarctic krill an ecosystem approach to the understanding of the biology and the environment of the species has been organized by the Scientific Committee on Oceanic Research (SCOR) and the Scientific Committee on Antarctic Research (SCAR). At present it appears encouraging that there has been little attempt to hurry the exploitation of this species until its ecology is understood.

Other potentially new fisheries in the world are squid, lantern fish, and grenadier. Squid fisheries have already developed extensively in some countries, particularly Japan, where about one-sixth of their total catch is now composed of this group of molluscs. Lantern fish are very abundant in the deep scattering layers of many oceans, while grenadier (a distant relative of the cod living off the edge of the continental shelf of the world's oceans) has been found in quantities indicating harvests of up to 10×10^6 tons/year. Taking into account these new fisheries, Gulland (1970) estimated that a reasonable extension of the world fish catch might be up to *ca.* 300×10^6 tons/year against a current catch of *ca.* 70×10^6 tons/year. However, it is clear that new technology for catching the organisms mentioned above will have to be developed. The question is whether the development of the catching technology will occur faster than the science of understanding their ecosystems?

REFERENCES

Abdullah, M. I., L. G. Royle and A. W. Morris 1972. Heavy metal concentration in coastal waters. *Nature,* **235,** 158–160.

Abele, L. G. and K. Walters. 1979. The stability-time hypothesis: reevaluation of the data. *Am. Nat.* **114,** 559–568.

Ackman, R. G. 1965. Occurrence of odd-numbered fatty acids in the mullet *Mugil cephalus. Nature,* **208,** 1213–1214.

Ackman, R. G. and M. G. Cormier. 1967. α-Tocopherol in some Atlantic fish and shellfish with particular reference to live-holding without food. *J. Fish. Res. Bd. Canada,* **24,** 357–373.

Ackman, R. G., C. S. Tocher and J. McLachlan. 1968. Marine phytoplankter fatty acids. *J. Fish. Res. Bd. Canada.* **25,** 1603–1620.

Ackman, R. G., C. A. Eaton, J. C. Sipos, S. N. Hooper and J. D. Castell. 1970. Lipids and fatty acids of two species of North Atlantic krill *Meganyctiphanes norvegica* and *Thysanoessa inermis* and their role in the aquatic food web. *J. Fish. Res. Bd. Canada,* **27,** 513–533.

Adams, J. A. and J. H. Steele. 1966. Shipboard experiments on the feeding of *Calanus finmarchicus* (Gunnerus). In *Some Contemporary Studies in Marine Science,* Ed. H. Barnes, Allen & Unwin Ltd., London, pp. 19–35.

Admiraal, W. 1977. Tolerance of estuarine benthic diatoms to high concentrations of ammonia, nitrite ion, nitrate ion, and orthophosphate. *Mar. Biol.* **43,** 307–315.

Admiraal, W. and H. Peltier. 1980. Influence of seasonal variations of temperature and light on the growth rate of cultures and natural populations of intertidal diatoms. *Mar. Ecol. Prog. Ser.* **2,** 35–43.

Ahmed, S. I., R. A. Kenner and T. T. Packard. 1977. A comparison study of glutamate dehydrogenase activity in several species of marine phytoplankton. *Mar. Biol.* **39,** 93–101.

Alam, M., T. B. Sansing, E. L. Busby, D. R. Martiniz and S. M. Ray. 1979. Dinoflagellate sterols. I. Sterol composition of the dinoflagellates of *Gonyaulax* species. *Steroids* **33,** 197–203.

Ali, R. M. 1970. The influence of suspension density and temperature on filtration rate of *Hiatella arctica. Mar. Biol.* **6,** 291–302.

Allan, G. G., J. Lewin and P. G. Johnson. 1972. Marine polymers. IV. Diatom polysaccharides. *Botanica Marina,* **15,** 102–108.

Alldredge, A. L. 1979. The chemical composition of macroscopic aggregates in two neretic seas. *Limnol. Oceanogr.* **24,** 855–866.

Alldredge, A. L. 1981. The impact of appendicularian grazing on natural food concentrations *in situ. Limnol. Oceanogr.* **26,** 247–257.

Allen, K. R. 1971. Relation between production and biomass. *J. Fish. Res. Bd. Canada,* **28,** 1573–1589.

Allen, M. B. 1956. Excretion of organic compounds by *Chlamydomonas. Arch. Mikrobiol,* **24,** 163–168.

Allen, M. B. 1963. Nitrogen fixing organisms in the sea. In *Symposium on Marine Microbiology,* Ed. C. H. Oppenheimer, C. C. Thomas, Springfield, Illinois, pp. 85–105.

Allen, M. B., T. W. Goodwin and S. Phagpolngarm. 1960. Carotenoid distribution in certain naturally occurring algae and in some artificially induced mutants of *Chlorella pyrenoidosa. J. Gen. Microbiol.* **23,** 93–103.

Allen, M. B., L. Fries, T. W. Goodwin and D. M. Thomas. 1964. The carotenoids of algae. Pigments from some cryptomonads, a heterokont and some Rhodophyceae. *J. Gen. Microbiol.* **34,** 259–267.

Aller, R. C. 1978. Experimental studies of changes produced by deposit feeders on pore water, sediment and overlying water chemistry. *Amer. J. Sci.* **278,** 1185–1234.

Aller, R. C. 1980. Diagenetic processes near the sediment–water interface of Long Island Sound. I. Decomposition and nutrient element geochemistry (S,N,P). *Adv. Geophys.* **22,** 237–250.

Aller, R. C. and J. Y. Yingst. 1980. Relationships between microbial distributions and the anaerobic decomposition of organic matter in surface sediments of Long Island Sound, U.S.A. *Mar. Biol.* **56,** 29–42.

Alverson, D. L., A. R. Longhurst and J. A. Gulland. 1970. How much food from the sea? *Science,* **168,** 503–505.

Amer. Publ. Health Assn., Amer. Water Works Assn., Water Pollution Contr. Fed. 1965. *Standard Methods for the Examination of Water and Waste Water including Bottom Sediments and Sludges,* 12th ed. Amer. Publ. Health Assn. Inc., New York. 769 pp.

Andersen, N. R. and B. J. Zahuranec (Eds.). 1977. *Oceanic Sound Scattering Prediction.* Plenum Press, N.Y. 859 pp.

Anderson, G. C. 1964. The seasonal and geographic distribution of primary productivity off the Washington and Oregon Coasts. *Limnol. Oceanogr.* **9,** 284–302.

Anderson, G. C. 1965. Fractionation of phytoplankton communities off the Washington and Oregon Coasts. *Limnol. Oceanogr.* **10,** 477–480.

Anderson, G. C. 1969. Subsurface chlorophyll maximum in the northeast Pacific Ocean. *Limnol. Oceanogr.* **14,** 386–391.

Anderson, G. C. and R. P. Zeutschel. 1970. Release of dissolved organic matter by marine phytoplankton in coastal and offshore areas of the northeast Pacific Ocean. *Limnol. Oceanogr.* **15,** 402–407.

Anderson, J. W., J. M. Neff, B. A. Cox, H. E. Tatem and G. M. Hightower. 1974. Characteristics of dispersions and water soluble extracts of crude and refined oils and their toxicity to estuarine crustaceans and fish. *Mar. Biol.* **27**, 75–88.

Anderson, K. P. and E. Ursin. 1977. A multispecies extension to Beverton and Holt theory of fishing with accounts of phosphorus circulation and primary production. *Meddr. Danm. Fisk-og-Havunders,* N.S. **7**, 319–435.

Andrews, P. and P. J. Le B. Williams. 1971. Heterotrophic utilisation of dissolved organic compounds in the sea. III. Measurement of the oxidation rates and concentrations of glucose and amino acids in sea water. *J. Mar. Biol. Ass. U.K.* **51**, 111–125.

Angel, H. H. and M. V. Angel. 1967. Distribution pattern analysis in a marine benthic community. *Helgol. wiss. Meeresunters.* **15**, 445–454.

Angel, M. V. and M. J. R. Fasham. 1973. SOND Cruise 1965: factor and cluster analysis of the plankton results, a general summary. *J. Mar. Biol. Ass. U.K.* **53**, 185–231.

Ankar, S. and R. Elmgerm. 1976. The benthic macro and meiofauna of the Asko-Landsort area (northern Baltic proper). A stratified random sampling survey. *Contrib. Asko Laboratory,* No. 11, Univ. of Stockholm. 115 pp.

Anon. 1980. Ocean Sciences Board, National Academy of Sciences. "Fisheries Ecology—Some Constraints that Impede Advances in our Understanding." pp. 16.

Anraku, M. and M. Omori. 1963. Preliminary survey of the relationship between the feeding habit and the structure of the mouthparts of marine copepods. *Limnol. Oceanogr.* **8**, 116–126.

Ansell, A. D. 1975. Seasonal changes in biochemical composition of the bi-valve *Astarte montagu* in the Clyde Sea area. *Mar. Biol.* **29**, 235–243.

Ansell, A. D. and P. Sivadas. 1973. Some effects of temperature and starvation on the bivalve *Donax vittatus* (da Costa) in experimental laboratory populations. *J. Exp. Mar. Biol. Ecol.* **13**, 229–262.

Antia, N. J., C. D. McAllister, T. R. Parsons, K. Stephens and J. D. H. Strickland. 1963. Further measurements of primary production using a large-volume plastic sphere. *Limnol. Oceanogr.* **8**, 166–183.

Antia, N. J. and E. Bilinski. 1967. A bacterial toxin type of phospholipase (Lecethinase C) in a marine phytoplanktonic chrysomonad. *J. Fish. Res. Bd. Canada,* **24**, 201–204.

Antia, N. J., J. Y. Cheng and F. J. R. Taylor, 1969. The heterotrophic growth of a marine photosynthetic cryptomonad *(Chroomonas salina). Proc. Intl. Seaweed Symp.* **6**, 17–29.

Antia, N. J., B. R. Berland, D. J. Bonin and S. Y. Maestrini. 1975. Comparative evaluation of certain organic and inorganic sources of nitrogen for phototrophic growth of marine microalgae. *J. Mar. Biol. Ass. U.K.* **55**, 519–539.

Armstrong, F. A. J. and W. R. G. Atkins. 1950. The suspended matter of sea water. *J. Mar. Biol. Ass. U.K.* **29**, 139–143.

Armstrong, F. A. J. and E. C. LaFond. 1966. Chemical nutrient concentrations and their relationship to internal waves and turbidity off southern California. *Limnol. Oceanogr.* **11**, 538-547.

Armstrong, F. A. J., C. R. Stearns and J. D. H. Strickland. 1967. The measurement of upwelling and subsequent biological processes by means of the Technicon Autoanalyser® and associated equipment. *Deep-Sea Res.* **14**, 381–389.

Arntz, W. E. 1978. The "upper part" of the benthic food web: the role of macrobenthos in the western Baltic. *Rapp. P.-v- Réun. Cons. int. Explor. Mer.* **173**, 85–100.

Aruga, Y. 1965a. Ecological studies of photosynthesis and matter production of phytoplankton. I. Seasonal changes in photosynthesis of natural phytoplankton. *Bot. Mag., Tokyo,* **78**, 280–288.

Aruga, Y. 1965a. Ecological studies of photosynthesis and matter production of phytoplankton. II. Photosynthesis of algae in relation to light intensity and temperature. *Bot. Mag., Tokyo,* **78**, 360–365.

Aruga, Y. 1966. Ecological studies of photosynthesis and matter production of phytoplankton. III. Relationship between chlorophyll amount in water and primary productivity. *Bot. Mag., Tokyo,* **79**, 20–27.

Aruga, Y. and S. Ichimura. 1968. Characteristics of photosynthesis of phytoplankton and primary production in the Kuroshio. *Bull. Misaki Marine Biol. Inst. Kyoto Univ.,* No. 12 (*Proceedings of the U.S.–Japan Seminar on Marine Microbiology*, August 1966 in Tokyo), 3–20.

Aruga, Y., Y. Yokohama and M. Nakanishi. 1968. Primary productivity studies in February–March in the Northwestern Pacific off Japan. *J. Oceanogr. Soc. Japan,* **24**, 275–280.

Baas Becking, L. G. M. and E. J. F. Wood. 1955. Biological processes in the estuarine environment. I and II. Ecology of the sulfur cycle. *Koninkl. Ned. Akad. Wetenschap. Proc.,* **B58**, 160–181.

Baas Becking, L. G. M., I. R. Kaplan and D. Moore. 1960. Limits of the natural environment in terms of pH and oxidation-reduction potentials. *J. Geol.* **68**, 243–284.

Baasham, J. A. and M. Kirk. 1962. The effect of oxygen on reduction of CO_2 to glycolic acid and other products during photosynthesis by *Chlorella. Biochem. Biophys. Res. Commun.* **9**, 376–380.

Bader, R. G. 1954. The role of organic matter in determining the distribution of pelecypods in marine sediments. *J. Mar. Res.* **13**, 32–47.

Bainbridge, R. 1957. The size, shape and density of marine phytoplankton concentrations. *Biol. Rev.* **32**, 91–115.

Baird, R. C. and T. L. Hopkins. 1981. Trophodynamics of the fish *Valenciennellus tripunctulatus*. III. Energetics, resources and feeding strategy. *Mar. Ecol. Prog. Ser.* **5**, 21–28.

Baity, H. G. 1938. Some factors affecting the aerobic decomposition of sewage sludge deposits. *Sewage Wks. J.* **10**, 539–568.

Bannister, T. T. 1974. Production equations in terms of chlorophyll concentration, quantum yield, and upper limit to production. *Limnol. Oceanogr.* **19**, 1–12.

Banse, K. 1964. On the vertical distribution of zooplankton in the sea. *Prog. Oceanogr.* **2**, 56–125.

Banse, K. 1974a. On the interpretation of data for the carbon-to-nitrogen ratio of phytoplankton. *Limnol. Oceanogr.* **19**, 695–699.

Banse, K. 1974b. The nitrogen to phosphorus ratio in the photic zone of the sea and the elemental composition of the plankton. *Deep-Sea Res.* **21**, 767–771.

Banse, K. 1975. Pleuston and Neuston: On the categories of organisms in the uppermost pelagial. *Int. Rev. ges. Hydrobiol.* **60**, 439–447.

Banse, K. 1977. Determining the carbon to chlorophyll ratio of natural phytoplankton. *Mar. Biol.* **41**, 199–212.

Banse, K. 1979. On weight dependence of net growth efficiency and specific respiration rates among field populations of invertebrates. *Oecologia,* **38,** 111–126.

Banse, K. and G. C. Anderson. 1967. Computations of chlorophyll concentrations from spectrophotometric readings. *Limnol Oceanogr.* **12,** 696–697.

Banse, K., F. H. Nichols and D. R. May. 1971. Oxygen consumption by the seabed. III. On the role of the macrofauna at three stations. *Vie Milieu*, Suppl. **22,** 31–52.

Banse, K. and S. Mosher. 1980. Adult body mass and annual production/biomass relationships of field populations. *Ecol. Monographs,* **50,** 355–379.

Barber, R. T. 1966. Interaction of bubbles and bacteria in the formation of organic aggregates in sea water. *Nature,* **211,** 257–258.

Barber, R. T., R. C. Dugdale, J. J. MacIsaac and R. L. Smith. 1971. Variations in phytoplankton growth associated with the source and conditioning of upwelling water. *Inv. Pesq.* **35,** 171–193.

Barlow, J. P., C. J. Lorenzen and R. T. Myren. 1963. Eutrophication of a tidal estuary. *Limnol. Oceanogr.* **8,** 251–262.

Barnes, H. 1949. On the volume measurement of water filtered by a plankton pump, with an observation on the distribution of plankton animals. *J. Mar. Biol. Ass. U.K.* **28,** 651–662.

Barnes, H. 1952. The use of transformations in marine biological statistics. *J. Cons. Int. Explor. Mer,* **18,** 61–71.

Barnes, H. 1956. *Balanus balanoides* (L.) in the Firth of Clyde: the development and annual variation of the larval population, and the causative factors. *J. Anim. Ecol.* **25,** 72–84.

Barnes, H. and S. M. Marshall.1951. On the variability of replicate plankton samples and some application of 'contagious' series to the statistical distribution of catches over restricted periods. *J. Mar. Biol. Ass. U.K.* **30,** 233–263.

Barraclough, W. E., R. J. LeBrasseur and O. D. Kennedy. 1969. Shallow scattering layer in the subarctic Pacific Ocean: detection by high-frequency echo sounder. *Science,* **166,** 611–613.

Barraclough, W. E. and D. Robinson. 1972. The fertilization of Great Central Lake. III. Effect on juvenile sockeye salmon. *Fish. Bull.* **70,** 37–48.

Bary, B. McK. 1959. Species of zooplankton as a means of identifying different surface waters and demonstrating their movements and mixing. *Pac. Sci.* **13,** 14–34.

Bary, B. McK. 1963. Distributions of Atlantic pelagic organisms in relation to surface water bodies. In *Marine Distributions,* Ed. M. J. Dunbar, Royal Soc. Canada, Sp. Publ., No. 5: 51–67.

Bary, B. McK. 1967. Diel vertical migrations of underwater scattering, mostly in Saanich Inlet, British Columbia. *Deep-sea Res.* **14,** 35–50.

Baylor, E. R. and W. H. Sutcliffe. 1963. Dissolved organic matter in sea water as a source of particulate food. *Limnol. Oceanogr.* **8,** 369–371.

Bayne, B. L. 1976. *Marine Mussels: Their Ecology and Physiology,* Cambridge Univ. Press, Cambridge. 494 pp.

Beattie, A., E. L. Hirst and E. Percival. 1961. Studies on the metabolism of the Chrysophyceae. Comparative structural investigations on leucosin (Chrysolaminarin) separated from diatoms and laminarin from the brown algae. *Biochem. J.* **79,** 531–537.

Beddington, J. R. and R. M. May. 1977. Harvesting natural populations in a randomly fluctuating environment. *Science,* **197,** 463–465.

Beers, J. R. 1966. Studies on the chemical composition of the major zooplankton groups in the Sargasso Sea off Bermuda. *Limnol. Oceanogr.* **11,** 520–528.

Beers, J. R., M. R. Stevenson, R. W. Eppley and E. R. Brooks. 1971. Plankton populations and upwelling off the coast of Peru, June 1969. *Fish. Bull.* **69,** 859–876.

Bell, R. K. and F. J. Ward. 1970. Incorporation of organic carbon by *Daphnia pulex. Limnol. Oceanogr.* **15,** 713–726.

Bell, W. and R. Mitchell. 1972. Chemotactic and growth responses of marine bacteria to algal extracellular products. *Biol. Bull.* **143,** 265–277.

Bellamy, D. J., D. M. John and A. Whittick. 1968. The 'kelp forest ecosystem' as a 'phytometer' in the study of pollution of the inshore environment. *Underwater Ass. Rep.,* 1968, 79–82.

Benson, A. A. and R. F. Lee. 1975. The role of wax in oceanic food chains. *Sci. Amer.* **232,** 76–83.

Ben-Yaakov, S. 1973. pH buffering of pore water of recent anoxic marine sediments. *Limnol. Oceanogr.* **18,** 86–94.

Berg, J. 1979. Discussion of methods of investigating the food of fishes with reference to a preliminary study of the prey of *Gobiusculus flavescens* (Gobiidae). *Mar. Biol.* **50,** 263–273.

Bernard, F. R. 1974. Annual biodeposition and gross energy budget of mature Pacific oysters, *Crassostrea gigas. J. Fish. Res. Bd, Canada,* **31,** 185–190.

Berner, R. A. 1963. Electrode studies of hydrogen sulfide in marine sediments. *Geochim. Cosmochin. Acta,* **27,** 563–575.

Berner, R. A. 1976. The benthic boundary layer from the viewpoint of a geochemist. In *The Benthic Boundary Layer,* Ed. I. N. McCave. Plenum Press, New York, pp. 33–35.

Berner, R. A. 1980. *Early Diagenesis: A theoretical Approach.* Princeton Univ. Press, 241 pp.

Bernstein, B. B. and J. P. Meader. 1979. Temporal persistence of biological patch structure in an abyssal benthic community. *Mar. Biol.* **51,** 179–183.

Bernstein, B. B., B. E. Williams and K. H. Mann. 1981. The role of behavioral responses to predators in modifying urchins' *Strongylocentrotus droebachiensis* destructive grazing and seasonal foraging patterns. *Mar. Biol.* **63,** 39–49.

Bertalanffy, L. Von. 1938. A quantitative theory of organic growth. *Hum. Biol.* **10,** 181–213.

Bertru, G. 1971. La microzone oxydée et la consommation d'oxygene dans les sediments des etanges oligotrophes. *Arch. Hydrobiol.* **68,** 277–287.

Beverton, R. J. H. and S. J. Holt. 1957. On the dynamics of exploited fish populations. *Fish. Inv., Lond.,* Ser. II, **19,** 1–533.

Beyers, R. J. 1963. The metabolism of twelve laboratory microecosystems. *Ecol. Monogr.* **33,** 281–306.

Bezrukov, P. L., E. M. Emel'yanov, A. P. Lisitsyn and E. A. Romankevich. 1977. Organic carbon in the upper sediment layer of the worlds oceans. *Okeanologiya,* **17,** 850–854.

Bienfang, P. K. 1980. Herbivore diet affects fecal pellet settling. *Can. J. Fish. Aquat. Sci.* **37,** 1352–1357.

Bienfang, P. K. 1981. Sinking rates of heterogeneous, temperate phytoplankton populations. *J. Plankton Res.* **3,** 235–252.

Blackburn, M., R. M. Laurs, R. W. Owen and B. Zeitschel. 1970. Seasonal and areal changes in standing stocks of phytoplankton, zooplankton and micronekton in the eastern tropical Pacific. *Mar. Biol.* **7,** 14–31.

Blackburn, T. H., P. Klieber and T. Fenchel. 1975. Photosynthetic sulfide oxidation in marine sediments. *Oikos,* **26,** 103–108.

Blackman, F. F. 1905. Optima and limiting factors. *Ann. Bot.* **19,** 281–295.

Blaxter, J. H. S. 1963. The feeding of herring larvae and their ecology in relation to feeding. *Calif. Coop. Oceanic Fish Investig. Rep.* **10,** 79–88.

Bloom, S. A., J. L. Simon and V. D. Hunter. 1972. Animal–sediment relations and community analysis of a Florida estuary. *Mar. Biol.* **13,** 43–56.

Blueweiss, L., H. Fox, V. Kudzwa, B. Nakashima, R. Peters and S. Sams. 1978. Relationships between body size and some life history parameters. *Oecologia,* **37,** 257–252.

Blumer, M., M. M. Mullin and R. R. L. Guillard. 1970. A polyunsaturated hydrocarbon (3, 6, 9, 12, 15, 18-heneicosahexaene) in the marine food web. *Mar. Biol* **6,** 226–235.

Blumer, M., R. R. L. Guillard and T. Chase. 1971. Hydrocarbons of marine phytoplankton. *Mar. Biol.* **8,** 183–189.

Bodiou, J.-Y. and P. Chardy. 1973. Analyse en composantes principales du cycle annuel d'un peuplement de copépods harpacticoides des sables fins infralittoraux de Banyuls-sur-Mer. *Mar. Biol.* **20,** 27–34.

Bodungen, B. von K. non Bröckel, V. Smetacek and B. Zeitzschel. 1975. Ecological studies on the plankton in Kiel Bight. 1. Phytoplankton. *Merentutkimuslait. Julk/Havsforskningsinst. Skr.* **239.** 179–186.

Bonner, J. T. 1965. *Size and Cycle.* Princeton University Press, Princeton, New Jersey. 219 pp.

Bordovsky, O. Y. 1965. Accumulation and transformation of organic substance in marine sediments, Parts 1-3. *Mar. Geol.* **3,** 3–144.

Boswell, P. G. H. 1961. *Muddy Sediments.* Heffer Co., Cambridge, Mass. 140 pp.

Bouldin, D. R. 1968. Models for describing the diffusion of oxygen and other mobile constituents across the mud-water interface. *J. Ecol.* **56,** 77–87.

Bourget, E. and G. Lacroix. 1973. Aspects saisonniers de la fixation de l'epifaune benthique de l'étage infralittoral de l'estuaire du Saint-Laurent. *J. Fish. Res. Bd. Canada,* **30,** 867–880.

Boyd, C. M. 1973. Small-scale spatial patterns of marine zooplankton examined by an electronic *in situ* zooplankton detecting device. *Netherlands J. Sea Res.* **7,** 103–111.

Boyd, C. M. 1976. Selection of particle sizes by filter-feeding copepods: a plea for reason. *Limnol. Oceanogr.* **21,** 175–180.

Boyd, C. M. 1981. Experimental plankton food chains. In *Analysis of Marine Ecosystems,* Ed. A. R. Longhurst. Academic Press (London), pp. 627–649.

Boynton, W. R., W. M. Kemp, C. G. Osborne, K. R. Kaumeyer and M. C. Jenkins. 1981. Influence of water circulation rate on *in situ* measurements of benthic community respiration. *Mar. Biol.* **65,** 185–190.

Boysen Jensen, P. 1919. Valuation of the Limfjord 1. *Rep. Dan. biol. Stn.* **26,** 1–24.

Boysen Jensen, P. 1932. Die Stoffproduktion der Pflanzen. *Publ. Gustav Fischer (Jena).* 108 pp.

Braarud, T. 1963. Reproduction in the marine coccolithophorid *Coccolithus huxleyi* in culture. *Pubbl. staz. zool. Napoli.* **33,** 110–116.

Braarud, T. and B. Foyn. 1931. Beiträge zur Kenntnis des Stoffwechsels in Meer. *Avh. norske Vidensk Akad. Oslo,* **14,** 24 pp.

Breen, P. A. and K. H. Mann. 1976. Destructive grazing by sea urchins in Eastern Canada. *J. Fish. Res. Bd. Canada,* **33,** 1278–1283.

Brett, J. R. 1970. Fish—The energy cost of living. In *Marine Aquaculture,* Ed. W. J. McNiel. Oregon State University Press, pp. 37–52.

Brinkhuis, B. H., N. R. Temple and R. F. Jones. 1976. Photosynthesis and respiration of exposed salt-marsh fucoids. *Mar. Biol.* **34,** 349–359.

Brock, M. L. and T. D. Brock. 1968. The application of micro-autoradiographic techniques to ecological studies. *Mitt. Internat. Vevein. Limnol.* **15,** 1–29.

Brocksen, R. W., G. E. Davis and C. E. Warren. 1970. Analysis of trophic processes on the basis of density-dependent functions. In *Marine Food Chains,* Ed. J. H. Steele. Oliver & Boyd, Edinburgh, pp. 468–498.

Broda, E. 1978. *The Evolution of the Bioenergetic Processes.* Pergamon Press (Oxford). pp. 229.

Brooks, J. L. and S. I. Dodson. 1965. Predation, body size, and composition of plankton. *Science,* **150,** 28–35.

Brown, D. A. and T. R. Parsons. 1978. Relationship between cytoplasmic distribution of mercury and toxic effects to zooplankton and chum salmon (*Oncorhynchus keta*) exposed to mercury in a controlled experiment. *J. Fish. Res. Bd. Canada,* **35,** 880–884.

Brown, D. H., C. E. Gibby and M. Hickman, 1972. Photosynthetic rhythms in epipelic algal populations. *Br. Phycol. J.* **7,** 37–44.

Brown, D. L. and E. B. Tregunna. 1967. Inhibition of respiration during photosynthesis by some algae. *Can. J. Bot.* **45,** 1135–1143.

Brown, H. R. 1969. *Biochemical Microcalorimetry.* Academic Press, New York. 338 pp.

Brown, L. M., B. T. Hargrave and M. D. MacKinnon. 1981. Analysis of chlorophyll *a* in sediments by high-pressure liquid chromatography. *Can. J. Fish. and Aquatic Sci.* **38,** 205–214.

Brown, R. A. and H. L. Huffman, Jr. 1976. Hydrocarbons in open waters. *Science,* **191,** 847–849.

Brown, T. E. and F. L. Richardson. 1968. The effect of growth environment of the physiology of algae: light intensity. *J. Phycol.* **4,** 38–54.

Bruland, K. W. and M. W. Silver. 1981. Sinking rates of fecal pellets from gelatinous zooplankton (salps, pteropods, doliolids). *Mar. Biol.* **63,** 295–300.

Bryan, G. W. 1980. Recent trends in research on heavy-metal contamination in the sea. *Helgolander Meeresunters,* **33,** 6–25.

Bryan, J. R., J. P. Riley and P. J. LeB. Williams. 1976. A Winkler procedure for making precise measurements of oxygen concentration for productivity and related studies. *J. Exp. Mar. Biol. Ecol.* **21,** 191–197.

Buchanan, J. B. 1963. The bottom fauna communities and their sediment relationships off the Northumberland coast. *Oikos,* **14,** 154–175.

Buchanan, J. B., P. F. Kingston and M. Sheader. 1974. Long-term population trends of the benthic macrofauna in the offshore mud of the Northumberland coast. *J. Mar. Biol. Ass. U.K.* **54,** 785–795.

Buchanan, J. B. and R. M. Warwick. 1974. An estimate of benthic macrofaunal production in the offshore mud of the Northumberland coast. *J. Mar. Biol. Ass. U.K.* **54,** 197–222.

Buchanan, J. B., M. Sheader and P. F. Kingston. 1978. Sources of variability in the benthic macrofauna off the south Northumberland coast, 1971–1976. *J. Mar. Biol. Ass. U.K.* **58,** 191–209.

Buggeln, R. G. and J. S. Craigie. 1973. The physiology and biochemistry of *Chondrus crispus* Stackhouse, pp. 81–102. In *Chondrus crispus. Proceedings of the Nova Scotia Institute of Science*, Suppl. 27, Ed. M. J. Harey and McLachlan.

Bunker, H. J. 1936. A review of the physiology and biochemistry of the sulfur bacteria. *Dept. Sci. Ind. Res. Chem. Res. Spec. Rpt.* 3.

Bunt, J. S. 1964a. Primary productivity under sea ice in Antarctic waters. 1. Concentrations and photosynthetic activities of microalgae in the waters of McMurdo Sound, Antarctica. *Antarctic Res. Ser.* **1,** 13–26.

Bunt, J. S. 1964b. Primary productivity under sea ice in Antarctic waters. 2. Influence of light and other factors on photosynthetic activities of Antarctic marine microalgae. *Antarctic Res. Ser.* **1,** 27–31.

Bunt, J. S. and C. C. Lee. 1970. Seasonal primary production in Antarctic sea ice at McMurdo Sound in 1967. *J. Mar. Res.* **28,** 304–320.

Burke, M. V. and K. H. Mann. 1974. Productivity and production:biomass ratios of bivalve and gastropod populations in an eastern Canadian estuary. *J. Fish. Res. Bd. Canada,* **31,** 167–177.

Burke, V. and L. A. Baird. 1931. Fate of fresh water bacteria in the sea. *J. Bacteriol.* **21,** 287–298.

Burkholder, P. R. and L. M. Burkholder. 1956. Vitamin B_{12} in suspended solids and marsh muds collected along the coast of Georgia. *Limnol. Oceanogr.* **1,** 202–208.

Burkholder, P. R., L. M. Burkholder and L. P. Almodóvar. 1960. Antibiotic activity of some marine algae of Puerto Rico. *Botanica Mar.* **2,** 149–156.

Burkholder, P. R. and E. F. Mandelli. 1965. Productivity of microalgae in Antarctic sea ice. *Science,* **149,** 872–874.

Burkholder, P. R., A. Repark and J. Siebert. 1965. Studies on some Long Island Sound littoral communities of microorganisms and their primary productivity. *Bull. Torrey Bot. Club,* **92,** 378–402.

Burnett, B. R. 1981. Quantitative sampling of nannobiota (microbiota) of the deep-sea benthos. III. The bathyl San Diego Trough. *Deep-Sea Res.* **28,** 649–663.

Burns, N. M. and F. Rosa. 1980. *In situ* measurement of the settling velocity of organic carbon particles and 10 species of phytoplankton. *Limnol. Oceanogr.* **25,** 855–864.

Butler, E. I., E. D. S. Corner and S. M. Marshall. 1970. On the nutrition and metabolism of zooplankton. VII. Seasonal survey on nitrogen and phosphorus excretion by *Calanus* in the Clyde sea area. *J. Mar. Biol. Ass. U.K.* **50,** 525–560.

Cadée, G. C. 1979. Sediment reworking by the polychaete *Heteromastus filiformis* on a tidal flat in the Dutch Wadden Sea. *Neth. J. Sea Res.* **13,** 441–456.

Cadée, G. C. 1980. Reappraisal of the production and import of organic carbon in the western Wadden Sea. *Neth. J. Sea Res.* **14,** 305–322.

Caldwell, D. R. and T. M. Chriss. 1979. The viscous sublayer at the sea floor. *Science,* **205,** 1131–1132.

Calow, P. and C. R. Fletcher. 1972. A new radiotracer technique involving ^{14}C and ^{51}Cr for estimating the assimilation efficiences of aquatic, primary consumers. *Oecologia,* **9,** 155–170.

Calvin, M. and J. A. Baasham. 1962. *The Photosynthesis of Carbon Compounds.* Benjamin, New York. 127 pp.

Cammen, L. M. 1980a. The significance of microbial carbon in the nutrition of the deposit feeding polychaete *Nereis succinea. Mar. Biol.* **61,** 9–20.

Cammen, L. M. 1980b. Ingestion rate: an empirical model for aquatic deposit feeders and detritivores. *Oecologia,* **44,** 303–310.

Cammen, L. M. 1980c. A method for measuring ingestion rate of deposit feeders and its use with the polychaete *Nereis succinea. Estuaries,* **3,** 55–60.

Caperon, J. 1967. Population growth in micro-organisms limited by food supply. *Ecology,* **48,** 715–722.

Caperon, J. 1968. Population growth response of *Isochrysis galbana* to nitrate variation at limiting concentrations. *Ecology,* **49,** 866–872.

Caperon, J., S. A. Cattell and G. Krasnick. 1971. Phytoplankton kinetics in a subtropical estuary: eutrophication. *Limnol. Oceanogr.* **16,** 599–607.

Capriulo, G. M. and E. J. Carpenter. 1980. Grazing by 35 to 202 μm micro-zooplankton in Long Island Sound. *Mar. Biol.* **56,** 319–326.

Carefoot, T. H. 1967. Growth and nutrition of *Aplysia punctata* feeding on a variety of marine algae. *J. Mar. Biol. Ass. U.K.* **47,** 565–589.

Carey, A. G., Jr. 1981. A comparison of benthic infaunal abundance on two abyssal plains in the northeast Pacific Ocean. *Deep-Sea Res.* **28,** 467–469.

Carlucci, A. F. and P. M. Williams. 1965. Concentration of bacteria from seawater by bubble scavenging. *J. Cons. Int. Explor. Mer.* **30,** 28–33.

Carlucci, A. F. and P. M. McNally. 1969. Nitrification by marine bacteria in low concentrations of substrate and oxygen. *Limnol. Oceanogr.* **14,** 736–739.

Carlucci, A. F. and P. M. Bowes. 1970. Production of vitamin B_{12}, thiamine, and biotin by phytoplankton. *J. Phycol.* **6,** 351–357.

Carpenter, E. J. and R. R. L. Guillard. 1971. Intraspecific differences in nitrate half-saturation constants for three species of marine phytoplankton. *Ecology,* **52,** 183–185.

Cassie, R. M. 1959. Micro-distribution of plankton. *New Zealand J. Sci.* **2,** 398–409.

Cassie, R. M. 1962a. Frequency distribution models in the ecology of plankton and other organisms. *J. Anim. Ecol.* **31,** 65–92.

Cassie, R. M. 1962b. Microdistribution and other error components of C^{14} primary production estimates. *Limnol. Oceanogr.* **7,** 121–130.

Cassie, R. M. 1963. Multivariate analysis in the interpretation of numerical plankton data. *New Zealand J. Sci.* **6,** 36–59.

Cassie, R. M. 1968. Sample design. Zooplankton sampling. *Mon. Oceanogr. Method. Unesco (Paris),* **2,** 105–121.

Castenholz, R. W. 1961. The effect of grazing on marine littoral diatom populations. *Ecology,* **42,** 783–794.

Chambers, M. R. and H. Milne. 1975. The production of *Macoma balthica* (L.) in the Ythan estuary. *Estuar. Coast. Mar. Sci.* **3,** 133–144.

Chang, B. D. and T. R. Parsons. 1975. Metabolic studies on the amphipod *Anisogammarus pugettensis* in relation to its trophic position in the food web of young salmonids. *J. Fish. Res. Bd. Canada,* **32,** 243–247.

Chardy, P., M. Glemarec and A. Laurec. 1976. Application of

inertia methods to benthic marine ecology: practical implications of the basic options. *Estuar. Coast. Mar. Sci.* **4**, 179–205.

Chau, Y. K., L. Chuecas and J. P. Riley. 1967. The component combined amino acids of some marine phytoplankton species. *J. Mar. Biol. Ass. U.K.* **47**, 543–554.

Chave, K. E. 1965. Carbonates: association with organic matter in surface seawater. *Science,* **148,** 1723–1724.

Chave, K. E. 1970. Carbonate-organic interactions in sea water. *Symp. Organic Matter in Natural Waters,* Ed. D. W. Hood, University of Alaska, pp. 373–385.

Checkley, D. M. Jr. 1978. The egg production of a marine, planktonic copepod in relation to its food supply. Ph.D. Thesis, University of California, San Diego.

Chindonova, Y. G. 1959. The nutrition of certain groups of abyssal macroplankton in the north-western area of the Pacific Ocean. *Trudy Institute Okeanologii,* **30,** 166–189. (Nat. Inst. Oceanogr. Translation No. 131, Wormley, U.K.)

Christensen, J. P. and A. H. Devol. 1980. Adenosine triphosphate and adenylate energy change in marine sediments. *Mar. Biol.* **56,** 175–182.

Christian, R. R., K. Bancroft and W. J. Wiebe. 1975. Distribution of microbial adenosine triphosphate in salt marsh sediments at Sapelo Island, Georgia. *Soil Sci.* **119,** 89–96.

Christiansen, F. E. 1968. Nitrogen Excretion by some Planktonic Copepods of Bras-d'Or Lake, Nova Scotia, M.S. Thesis, Dalhousie University, Halifax. 107 pp.

Chuecas, L. and J. P. Riley. 1969. Component fatty acids of the total lipids of some marine phytoplankton. *J. Mar. Biol. Ass. U.K.* **49,** 97–116.

Clark, M. E., G. A. Jackson and W. J. North. 1972. Dissolved free amino acids in southern California coastal waters. *Limnol. Oceanogr.* **17,** 749–758.

Clark, R. B. and A. Milne. 1955. The sublittoral fauna of two sandy bays on the isle of Cumbrae, Firth of Clyde. *J. Mar. Biol. Ass. U.K.* **34,** 161–180.

Clarke, A. 1979. On living in cold water: K-strategies in Antarctic benthos. *Mar. Biol.* **55,** 111–119.

Clarke, G. L. 1965. *Elements of Ecology,* John Wiley & Sons, New York. 560 pp.

Clarke, G. L. and H. R. James. 1939. Laboratory analysis of the selective absorption of light by sea water. *J. Opt. Soc. Amer.* **29,** 43–55.

Clarke, G. L. and E. J. Denton. 1962. Light and animal life. In *The Sea, Ideas and Observations on Progress in the Study of the Seas,* Ed. M. M. Hill. Interscience (New York), **1,** 456–468.

Clarke, M. R. and N. Merrett. 1972. The significance of squid, whale and other remains from the stomachs of bottom-living deep sea fish. *J. Mar. Biol. Ass. U.K.* **52,** 599–603.

Clendenning, K. A. 1971. Organic productivity in kelp areas. In *The Biology of Giant Kelp Beds (Macrocystis) in California.* Ed. W. J. North. *Nova Hedwigia,* **32** (Suppl.), 259–263.

Clutter, R. I. and G. H. Theilacker. 1971. Ecological efficiency of a pelagic mysid shrimp. Estimates from growth, energy budget, and mortality studies. *Fish. Bull.* **69,** 93–115.

Cohen, E. B., M. D. Grosslein, M. P. Sissenwine, F. Steimle and W. R. Wright. 1982. Energy budget of Georges Bank. *Can. Spec. Publ. Fish. Aquat. Sci.* **59,** 95–107.

Cole, L. C. 1954. The population consequences of life history phenomena. *Quart. Rev. Biol.* **29,** 103–137.

Colebrook, J. M., R. S. Glover and G. A. Robinson. 1961. Continuous plankton records: contributions towards a plankton atlas of the north-eastern Atlantic and the North Sea. *Bull. Mar. Ecol.* **5,** 67–80.

Coles, S. L. 1969. Quantitative estimates of feeding and respiration for three scleractinian corals. *Limnol. Oceanogr.* **14,** 949–953.

Comita, G. W. 1968. Oxygen consumption in *Diaptomus. Limnol. Oceanogr.* **13,** 51–57.

Conan, G., M. Roux and M. Sibuet. 1981. A photographic survey of a population of the stalked crinoid *Diplocrinus (Annacrinus) wyvillethomsoni* (Echinodermata) from the baythal slope of the Bay of Biscay. *Deep-Sea Res.* **28,** 441–453.

Conover, R. J. 1966a. Assimilation of organic matter by zooplankton. *Limnol. Oceanogr.* **11,** 338–345.

Conover, R. J. 1966b. Factors affecting the assimilation of organic matter by zooplankton and the question of superfluous feeding. *Limnol. Oceanogr.* **11,** 346–354.

Conover, R. J. 1966c. Feeding on large particles by *Calanus hyperboreus* (Kröyer). In *Some Contemporary Studies in Marine Science,* Ed. H. Barnes, Allen & Unwin, London, pp. 187–194.

Conover, R. J. 1968. Zooplankton—life in a nutritionally dilute environment. *Amer Zoologist,* **8,** 107–118.

Conover, R. J. 1978. Transformation of organic matter. In *Marine Ecology,* vol. IV. *Dynamics,* Ed. O. Kinne, Wiley, New York, pp. 221–499.

Conover, R. J. and V. Francis. 1973. The use of radioactive isotopes to measure the transfer of materials in aquatic food chains. *Mar. Biol.* **18,** 272–283.

Conover, R. J. and C. M. Lalli. 1974. Feeding and growth in *Clione limacina* (Phipps), a pteropod mollusc. II. Assimilation, metabolism and growth efficiency. *J. Exp. Mar. Biol. Ecol.* **16,** 134–154.

Conover, R. J. and M. E. Huntley. 1980. General rules of grazing in pelagic ecosystems. In *Primary Productivity in the Sea,* Ed. P. G. Falkowski. Plenum Press (N.Y.), pp. 461–485.

Conway, H. L. 1977. Interactions of inorganic nitrogen in the uptake and assimilation by marine phytoplankton. *Mar. Biol.* **39,** 221–232.

Coombs, J., W. M. Darley, O. Holm-Hansen and B. E. Volcani. 1967a. Studies on the biochemistry and fine structure of silica shell formation in diatoms. Chemical composition of *Navicula pelliculsa* in silicon-starvation synchrony. *Plant Physiol.* **42,** 1601–1606.

Coombs, J., C. Spanis and B. E. Volcani. 1967b. Studies on the biochemistry and fine structure of silica shell formation in diatoms. Photosynthesis and respiration in silicon-starvation synchrony of *Navicula pelliculosa. Plant Physiol.* **42,** 1607–1611.

Coombs, J., P. J. Halicki, O. Holm-Hansen and B. E. Volcani. 1967c. Studies on the biochemistry and fine structure of silica shell formation in diatoms. II. Changes in the concentration of nucleoside triphosphates in silicon starvation synchrony of *Navicula pelliculosa* (Brèb). *Hilse. Expl. Cell Res.* **47,** 315–328.

Copeland, B. J. 1965. Evidence for regulation of community metabolism in a marine ecosystem. *Ecology,* **46,** 563–564.

Copping, A. E. and C. J. Lorenzen. 1980. Carbon budget of a marine phytoplankton-herbivore system with carbon-14 as a tracer. *Limnol. Oceanogr.* **25,** 873–882.

Corkett, C. J. and I. A. McLaren. 1978. The biology of *Pseudocalanus. Adv. Mar. Biol.* **15,** 1–231.

Corner, E. D. S. 1978. Pollution studies with marine plankton.

Part 1: Petroleum hydrocarbons and related compounds. *Adv. Mar. Biol.* **15,** 289–380.

Corner, E. D. S., C. B. Cowey and S. M. Marshall. 1965. Nitrogen excretion by *Calanus. J. Mar. Biol. Ass. U.K.* **45,** 429–442.

Corner, E. D. S. and B. S. Newell. 1967. On the nutrition and metabolism of zooplankton. IV. The forms of nitrogen excreted by *Calanus. J. Mar. Biol. Ass. U.K.* **47,** 113–120.

Corner, E. D. S., C. B. Cowey and S. M. Marshall. 1967. On the nutrition and metabolism of zooplankton. V. Feeding efficiency of *Calanus finmarchicus. J. Mar. Biol. Ass. U.K.* **47,** 259–270.

Cosper, T. C. and M. R. Reeve. 1975. Digestive efficiency of the chaetognath *Sagitta hispida* Conant. *J. Exp. Mar. Biol. Ecol.* **17,** 33–38.

Coull, B. C. 1972. Species diversity and fauna affinities of meiobenthic Copepoda in the deep sea. *Mar. Biol.* **14,** 48–51.

Coull, B. C., R. L. Ellison, J. W. Fleeger, R. P. Higgins, W. D. Hope, W. D. Hummon, R. M. Reiger, W. E. Sterrer, H. Thiel and J. H. Tietjen. 1977. Quantitative estimates of the meiofauna from the deep sea off North Carolina, U.S.A. *Mar. Biol.* **39,** 233–240.

Cowey, C. B. and E. D. S. Corner. 1963. On the nutrition and metabolism of zooplankton. II. The relationship between the marine copepod *Calanus helgolandicus* and particulate material in Plymouth sea water, in terms of amino acid composition. *J. Mar. Biol. Ass. U.K.* **43,** 495–511.

Cowey, C. B. and E. D. S. Corner. 1966. The amino-acid composition of certain unicellular algae, and of the faecal pellets produced by *Calanus finmarchicus* when feeding on them. In *Some Contemporary Studies in Marine Science*, Ed. H. Barnes, George Allen & Unwin, London, pp. 225–231.

Cretney, W. J., C. S. Wong, D. R. Green and C. A. Bawden. 1978. Long-term fate of heavy crude oil in a spill contaminated B.C. coastal bay. *J. Fish. Res. Bd.* **35,** 521–527.

Cretney, W. J., R. W. Macdonald, C. S. Wong, D. R. Green, B. Whitehouse and G. G. Geesey. 1981. Biodegradation of a chemically dispersed crude oil. In *Proceedings, Oil Spill Conference (Prevention, Behaviour, Control, Cleanup) Atlanta, Georgia.* Amer. Pet. Inst. Publ. No. 4334, pp. 37–43.

Crisp, D. J. 1971. Energy flow measurements. In *Methods for the Study of Marine Benthos,* Eds. N. A. Holme and A. D. McIntyre, I.B.P. Handbook, **16.** Oxford and Edinburgh: Blackwell Scientific Publications, pp. 197–279.

Cruz, A. A. de la and W. E. Poe. 1975. Amino acid content of marsh plants. *Estuar. Coast. Mar. Sci.* **3,** 243–246.

Culkin, F. and R. J. Morris. 1970. The fatty acid composition of two marine filter-feeders in relation to a phytoplankton diet. *Deep-Sea Res.* **17,** 861–865.

Cupp, E. E. 1943. *Marine Plankton Diatoms of the West Coast of North America*, University of California Press, Berkeley. 237 pp.

Curl, H. 1962. Standing crops of carbon, nitrogen, and phosphorus and transfer between trophic levels, in continental shelf waters south of New York. *Rapp. Proc.-Verb. Cons. int. Explor. Mer.* **153,** 183–189.

Currie, R. I. 1962. Pigments in zooplankton faeces. *Nature,* **193,** 956–957.

Curtis, M. A. and G. H. Petersen. 1978. Size-class heterogeneity within the spatial distributions of subarctic marine benthic populations. *Astarte,* **10,** 103–105.

Cushing, D. H. 1955. Production of a pelagic fishery in the sea. *Fish. Invest. London*, Ser. 2, **18,** 103.

Cushing, D. H. 1962. Patchiness. *Rapp. Proc.-Verb. Cons. int. Explor. Mer,* **153,** 152–164.

Cushing, D. H. 1964. The work of grazing in the sea. In *Grazing in Terrestrial and Marine Environments,* Ed. D. J. Crisp, Blackwell, London, pp. 207–225.

Cushing, D. H. 1971. Upwelling and fish production. *Adv. Mar. Biol.* **9,** 255–334.

Cushing, D. H. 1973. Production in the Indian Ocean and transfer from primary to secondary level. In *The Biology of the Indian Ocean,* Ed. B. Zeitzschel. Springer-Verlag (Berlin), pp. 475–486.

Cushing, D. H. 1975. *Marine Ecology and Fisheries,* Cambridge Univ. Press (Cambridge), pp. 278.

Cushing, D. H. and T. Vucetic. 1963. Studies on a *Calanus* patch. III. The quantity of food eaten by *Calanus finmarchicus. J. Mar. Biol. Ass. U.K.* **43,** 349–371.

Cushing, D. H. and H. F. Nicholson. 1966. Method of estimating algal production rates at sea. *Nature,* **212,** 310–311.

Cushman, J. A. 1931. The Foraminifera of the Atlantic Ocean. *Bull. U.S. Nat. Mus.* **104,** pp. 55.

Dagg, M. 1974. Loss of prey body contents during feeding by an aquatic predator. *Ecology,* **55,** 903–906.

Dagg, M. J. 1976. Complete carbon and nitrogen budgets for the carnivorous amphipod, *Calliopius laevivsculus* (Kroyer). *Int. Rev. ges. Hydrobiol,* **61,** 297–357.

Dahl, E. 1979. Deep-sea carrion-feeding amphipods: evolutionary patterns in niche adaptation. *Oikos,* **33,** 167–175.

Dal Pont, G. and B. Newell. 1963. Suspended organic matter in the Tasman Sea. *Aust. J. Mar. Freshw. Res.* **14,** 155–165.

Dale, N. G. 1974. Bacteria in intertidal sediments: factors related to their distribution. *Limnol. Oceanogr.* **19,** 509–518.

Dale, T. 1979. Total chemical and biological oxygen consumption of the sediments in Lindåspollene, Western Norway. *Mar. Biol.* **49,** 333–341.

Dales, R. P. 1960. On the pigments of the Chrysophyceae. *J. Mar. Biol. Ass. U.K.* **39,** 693–699.

Darley, W. M., C. T. Ohlman and B. B. Wimpee. 1979. Utilization of dissolved organic carbon by natural populations of epibenthic salt marsh diatoms. *J. Phycol.* **15,** 1–15.

Davies, J. M. 1975. Energy flow through the benthos in a Scottish sea loch. *Mar. Biol.* **31,** 353–362.

Davis, C. C. 1977. *Sagitta* as food for *Acartia. Astarte,* **10,** 1–3.

Davis, C. O., N. F. Breitner and P. J. Harrison. 1978. Continuous culture of marine diatoms under silicon limitation. 3. A model of Si-limited growth. *Limnol. Oceanogr.* **23,** 41–52.

Davis, N. and G. R. Van Blaricom. 1978. Spatial and temporal heterogeneity in a sand bottom epifaunal community of invertebrates in shallow water. *Limnol. Oceanogr.* **23,** 417–427.

Dawes, C. J., R. E. Moon and M. A. Davis. 1978. The photosynthetic and respiratory rates and tolerances of benthic algae from a mangrove and salt marsh estuary: a comparative study. *Est. Coastal Mar. Sci.* **6,** 175–185.

Day, J. H., J. G. Field and M. Potts Montgomery. 1971. The use of numerical methods to determine the distribution of the benthic fauna across the continental shelf at North Carolina. *J. Anim. Ecol.* **40,** 93–125.

Dayton, P. K. 1975. Experimental evaluation of ecological dominance in a rocky intertidal algal community. *Ecol. Monogr.* **45,** 137–159.

Dayton, P. K. and R. R. Hessler. 1972. Role of biological disturbance in maintaining diversity in the deep sea. *Deep-Sea Res.* **19,** 199–208.

Degens, E. T. 1970. Molecular nature of nitrogenous compounds in sea water and recent marine sediments. In *Organic Matter in Natural Waters,* Ed. D. W. Hood. University of Alaska, pp. 77–106.

Deibel, D. 1982. Laboratory-measured grazing and ingestion rates of the salp, *Thalia democratica* Forskal, and the doliolid, *Dolioletta gegenbauni* Uljanin, Tunicata, Thaliacea. *J. Plank. Res.* **4,** 189–201.

Demers, S., P. E. Lafleur, L. Legendre and C. L. Trump. 1979. Short-term covariability of chlorophyll and temperature in the St. Lawrence estuary. *J. Fish. Res. Bd. Canada,* **36,** 568–573.

Dempsey, M. J. 1981. Marine bacterial fouling: a scanning electron microscopy study. *Mar. Biol.* **61,** 305–315.

Desbruyères, D., J. Y. Bervas and A. Khripounoff. 1980. Un cas de colonization rapide d'un sediment profond. *Oceanol. Acta,* **3,** 285–281.

Deuser, W. G. 1971. Organic carbon budget of the Black Sea. *Deep-Sea Res.* **18,** 995–1004.

Deuser, W. G., E. H. Ross and R. F. Anderson. 1981. Seasonality in the supply of sediment to the deep Sargasso Sea and implications for the rapid transfer of matter to the deep ocean. *Deep-Sea Res.* **28,** 495–505.

Devol, A. H. and S. I. Ahmed. 1981. Are high rates of sulfate reduction associated with anaerobic oxidation of methane. *Nature,* **291,** 407–408.

Dickie, L. M. 1972. Food chains and fish production. *ICNAF Spec. Publ.* **8,** 201–221.

Dickie, L. M. and K. H. Mann. 1972. Aquatic ecological systems: the formal approach to holistic models (unpublished).

Dickson, R. and A. Lee. 1972. Recent hydro-meteorological trends in the North Atlantic fishing grounds. *Fish. Indust. Rev.* **2,** 1–8.

Dortch, Q. and S. I. Ahmed. 1979. Nitrate reductase and glutamate dehydrogenase activities in *Skeletonema costatum* as measures of nitrogen assimilation rates. *J. Plankton Res.* **1,** 169–186.

Doty, M. S. 1959. Phytoplankton photosynthetic periodicity as a function of latitude. *J. Mar. Biol. Ass. India,* **1,** 66–68.

Doty, M. S. and M. Oguri. 1957. Evidence for a photosynthetic daily periodicity. *Limnol Oceanogr.* **2,** 37–40.

Downing, J. A. 1979. Agggregation, transformation and the design of benthos sampling programs. *J. Fish. Res. Bd. Can.* **36,** 1454–1463.

Doyle, R. W. 1972. Genetic variation of *Ophiomusium lymani* (Echinodermata) populations in the deep sea. *Deep-Sea Res.* **19,** 661–664.

Doyle, R. W. 1979. Ingestion rate of a selective deposit feeder in a complex mixture of particles: testing the energy-optimization hypothesis. *Limnol. Oceanogr.* **24,** 867–874.

Drake, D. E. 1971. Suspended sediment and thermal stratification in Santa Barbara Channel, California. *Deep-Sea Res.* **18,** 763–769.

Driscoll, E. G. and D. E. Brandon. 1973. Mollusc–sediment relationships in northwestern Buzzards Bay, Massachusetts, U.S.A. *Malacologia,* **12,** 13–46.

Droop, M. R. 1968. Vitamin B_{12} and marine ecology. IV. The kinetics of uptake, growth and inhibition in *Monochrysis lutheri. J. Mar. Biol. Ass. U.K.* **48,** 689–733.

Droop, M. R. 1970. Vitamin B_{12} and marine ecology. V. Continuous culture as an approach to nutritional kinetics. *Helgoländer wiss. Meeresunters,* **20,** 629–636.

Droop, M. R., J. J. A. McLaughlin, I. J. Pinter and L. Provasoli. 1959. Specificity of some protophytes toward vitamin B-like compounds. *Preprints Int. Oceanogr. Congr. (New York),* pp. 916–918.

Dubinsky, Z. 1980. Light utilization efficiency in natural phytoplankton communities. In *Primary Productivity in the Sea,* Ed. P. G. Falkowsky. Plenum Publ., New York, pp. 83–97.

Dubinsky, Z. and T. Berman. 1976. Light utilization efficiencies of phytoplankton in Lake Kinneret (Sea of Galilee). *Limnol. Oceanogr.* **21,** 226–230.

Dubinsky, Z. and T. Berman. 1979. Seasonal changes in the spectral composition of downwelling irradiance in Lake Kinneret (Israel). *Limnol. Oceanogr.* **24,** 652–663.

Duff, D. C. B., D. L. Bruce and N. J. Antia. 1966. The antibacterial activity of marine planktonic algae. *Can. J. Microbiol.* **12,** 877–884.

Duff, S. and J. M. Teal. 1965. Temperature change and gas exchange in Nova Scotia and Georgia salt marsh muds. *Limnol. Oceanogr.* **10,** 67–73.

Dugdale, R. C. 1967. Nutrient limitation in the sea: dynamics, identification and significance. *Limnol. Oceanogr.* **12,** 685–695.

Dugdale, R. C., J. J. Goering and J. H. Ryther. 1964. High-nitrogen fixation rates in the Sargasso Sea and the Arabian Sea. *Limnol. Oceanogr.* **9,** 507–510.

Dugdale, R. C. and J. J. Goering. 1967. Uptake of new and regenerated forms of nitrogen in primary productivity. *Limnol. Oceanogr.* **12,** 196–206.

Dunbar, M. J. 1960. The evolution of stability in marine environments. Natural selection at the level of the ecosystem. *Am. Nat.* **94,** 129–136.

Dunbar, M. J. 1962. The life cycle of *Sagitta elegans* in arctic and subarctic seas, and the modifying effects of hydrographic differences in the environment. *J. Mar. Res.* **20,** 76–91.

Dunstan, W. M., L. P. Atkinson and J. Matoli. 1975. Simulation and inhibition of phytoplankton growth by low molecular weight hydrocarbons. *Mar. Biol.* **31,** 305–310.

Duursma, E. K. 1960. Dissolved Organic Carbon, Nitrogen and Phosphorus in the Sea, Ph.D. Theses, J. B. Wolters, Groningen. 147 pp.

Duursma, E. K. 1965. The dissolved organic constituents of sea water. In *Chemical Oceanography,* Ed. J. P. Riley and G. Skirrow, Academic Press, New York, pp. 433–475.

Duval, W. S. and G. H. Geen. 1976. Diel feeding and respiration rhythms in zooplankton. *Limnol. Oceanogr.* **21,** 823–829.

Duval W. S., L. C. Martin and R. P. Fink. 1981. A prospectus on the biological effects of oil spills. Unpubl. rep. by ESL Environmental Sciences Ltd. for Dome Petroleum Ltd., Calgary. pp. 92.

Eagle, R. A. 1975. Natural fluctuations in a soft bottom benthic community. *J. Mar. Biol. Ass. U.K.* **55,** 865–878.

Eckman, J. E. 1979. Small-scale patterns and processes in a soft-substratum, intertidal community. *J. Mar. Res.* **37,** 437–457.

Eckman, J. E., A. R. M. Nowell and P. A. Jumars. 1981.

Sediment destabilization by animal tubes. *J. Mar. Res.* **39,** 361–374.

Edwards, R. R. C. 1967. Estimates of the respiratory rate of young plaice (*Pleuronectes platessa* L.) in natural conditions using zinc-65. *Nature,* **216,** 1335–1337.

Edwards, R. R. C., D. M. Finlayson and J. H. Steele. 1969. The ecology of O-group plaice and common dabs in Loch Ewe. II. Experimental studies of metabolism. *J. Exp. Mar. Biol. Ecol.* **8,** 299–309.

Edwards, R. W. and H. L. J. Rolley. 1965. Oxygen consumption of river muds. *J. Ecol.* **53,** 1–19.

Edwards, S. F. and B. L. Welsh. 1982. Trophic dynamics of a mud snail (*Ilyanassa obsoleta* (Say)) population on an intertidal mudflat. *Est. Coastal and Shelf Sci.* **14,** 663–686.

Eglinton, G. and M. T. J. Murphy. 1969. *Organic Geochemistry: Methods and Results,* Springer-Verlag, New York. 828 pp.

Elliot, J. M. 1977. *Some Methods for the Statistical Analysis of Samples of Benthic Invertebrates.* Sci. Publ. 25 (2nd ed.), Biol. Ass. 144 pp.

Elmgren, R. 1975. Benthic meiofauna as indicator of oxygen conditions in the northern Baltic proper. *Merentutkimslait. Julk. Havsforskningsinst. Skr.* **239,** 265–271.

Elmgren, R. 1978. Structure and dynamics of Baltic benthos communities, with particular reference to the relationship between macro and meiofauna. *Kieler Meeresforsch.* **4,** 1–22.

Eltringham, S. K. 1971. *Life in Mud and Sand.* The English Universities Press Ltd., London. 218 pp.

Emerson, R. and C. M. Lewis. 1942. The photosynthetic efficiency of phycocyanin *Chroococcus,* and the problem of carotenoid participation in photosynthesis. *J. Gen. Physiol.* **25,** 579–595.

Emlen, J. M. 1973. *Ecology: an Evolutionary Approach,* Addison-Wesley, Massachusetts. 493 pp.

Eppley, R. W. 1972. Temperature and phytoplankton growth in the sea. *Fish. Bull.* **70,** 1063–1085.

Eppley, R. W. and P. R. Sloan. 1965. Carbon balance experiments with marine phytoplankton. *J. Fish. Res. Bd. Canada,* **22,** 1083–1097.

Eppley, R. W., R. W. Holmes and J. D. H. Strickland. 1967. Sinking rates of marine phytoplankton measured with a fluorometer. *J. Exp. Mar. Biol. Ecol.* **1,** 191–208.

Eppley, R. W., O. Holm-Hansen and J. D. H. Strickland. 1968. Some observations on the vertical migration of dinoflagellates. *J. Phycol.* **4,** 333–340.

Eppley, R. W. and J. D. H. Strickland. 1968. Kinetics of marine phytoplankton growth. In *Advances in Microbiology of the Sea,* Eds. M. R. Droop and E. J. F. Wood, Academic Press, London, Vol. 1. pp. 23–62.

Eppley, R. W., J. L. Coatsworth and L. Solorzano. 1969a. Studies of nitrate reductase in marine phytoplankton. *Limnol. Oceanogr.* **14,** 194–205.

Eppley, R. W., J. N. Rogers and J. J. McCarthy. 1969b. Half-saturation constants for uptake of nitrate and ammonium by marine phytoplankton. *Limnol. Oceanogr.* **14,** 912– 920.

Eppley, R. W. and W. H. Thomas. 1969. Comparison of half-saturation constants for growth and nitrate uptake of marine phytoplankton. *J. Phycol.* **5,** 375–379.

Eppley, R. W., P. Koeller and G. T. Wallace, Jr. 1978. Stirring influences on the phytoplankton species composition within enclosed columns of coastal sea water. *J. Exp. Mar. Biol. Ecol.* **32,** 219–239.

Eppley, R. W. and B. J. Peterson. 1979. Particulate organic matter flux and planktonic new production in the deep ocean. *Nature,* **282,** 677–680.

Esaias, W. E. and H. C. Curl, Jr. 1972. Effect of dinoflagellate bioluminescence on copepod ingestion rates. *Limnol. Oceanogr.* **17,** 901–906.

Estrada, M., I. Valiela and J. M. Teal. 1974. Concentration and distribution of chlorophyll in fertilized plots in a Massachusetts salt marsh. *J. Exp. Mar. Biol. Ecol.* **14,** 47–56.

Ewing, J. A. 1973. Wave-induced bottom currents on the outer shelf. *Mar. Geol.* **15,** M31–M35.

Fager, E. W. 1957. Determination and analysis of recurrent groups. *Ecology,* **38**(4), 586–595.

Fager, E. W. 1964. Marine sediments: effects of a tube-building polychaete. *Science,* **143,** 356–359.

Fager, E. W. and J. A. McGowan. 1963. Zooplankton species groups in the North Pacific. *Science,* **140,** 453–460.

Fager, E. W. and A. R. Longhurst. 1968. Recurrent group analysis of species assemblage of demersal fish in the Gulf of Guinea. *J. Fish. Res. Bd. Canada,* **25**(7), 1405–1421.

Fagerström, T. and A. Jernelov. 1972. Some aspects of the quantitative ecology of mercury. *Water Res.* **6,** 1193–1202.

Falkowsky, P. G. 1975. Nitrate uptake in marine Phytoplankton: (Nitrate, Chloride)-activated adenosine triphosphatase from *Skeletonema costatum* (Bacillariophyceae). *J. Phycol.* **11,** 323–326.

Falkowsky, P. G. and R. B. Rivkin. 1976. The role of glutamine synthetase in the incorporation of ammonium in *Skeletonema costatum* (Bacillariophyceae). *J. Phycol.* **12,** 448–450.

Falkowsky, P. G. and D. P. Stone. 1975. Nitrate uptake in marine phytoplankton: Energy sources and interaction with carbon fixation. *Mar. Biol.* **32,** 77–84.

Faller, A. J. 1971. Oceanic turbulence and the Langmuir circulation. *Ann. Rev. Ecol. Syst.* **2,** 201–236.

Farrington, J. W. and P. A. Meyer. 1975. Hydrocarbons in the marine environment. In *Environmental Chemistry: Air and Water Pollution,* Ed. H. S. Stoker and S. L. Seagar. Scott, Forsman, Glenview, Illinois, pp. 103–136.

Fasham, M. J. R. 1977. The application of some stochastic processes to the study of plankton patchiness. In *Spatial Pattern in Plankton Communities,* Ed. J. H. Steele. Plenum Press (New York), pp. 131–156.

Fasham, M. J. R., M. V. Angel and H. S. J. Roe. 1974. An investigation of the spatial pattern of zooplankton using the Longhurst–Hardy plankton recorder. *J. Exp. Mar. Biol. Ecol.* **16,** 93–112.

Fauchald, K. and P. A. Jumars. 1979. The diet of worms: a study of polychaete feeding guilds. *Oceanogr. Mar. Biol. Ann. Rev.* **17,** 193–284.

Favorite, F. and W. J. Ingraham, Jr. 1976. Sunspot activity and oceanic conditions in the northern North Pacific Ocean. *J. Oceanogr. Soc. Japan.* **32,** 107–115.

Fee, E. J. 1969. A numerical model for the estimation of photosynthetic production, integrated over time and depth, in natural waters. *Limnol. Oceanogr.* **14,** 906–911.

Fenchel, T. 1969. The ecology of marine microbenthos. IV. Structure and function of the benthic ecosystem, its chemical and physical factors and the microfauna communities with special reference to the ciliated protozoa. *Ophelia,* **6,** 1–182.

Fenchel, T. 1974. Intrinsic rate of natural increase: the relationship with body size. *Oecologia,* **14,** 317–326.

Fenchel, T. 1975. Factors determining the distribution patterns of mud snails (Hydrobiidae). *Oecologia,* **20,** 1–17.

Fenchel, T. 1982. Ecology of heterotrophic microflagellates. IV. Quantitative occurrence and importance as bacterial consumers. *Mar. Ecol. Prog. Ser.* **9,** 35–42.

Fenchel, T. and R. H. Riedl. 1970. The sulphide system: a new biotic community underneath the oxidized layer of marine sand bottoms. *Mar. Biol.* **7,** 255–268.

Fenchel, T. and B. J. Straarup. 1971. Vertical distribution of photosynthetic pigments and the penetration of light in marine sediments. *Oikos,* **22,** 172–182.

Fenchel, T., L. H. Kofoed and A. Lappalainen. 1975. Particle-size selection of two deposit feeders: the amphipod *Corophium volutator* and the prosobranch *Hydrobia ulvae. Mar. Biol.* **30,** 119–128.

Fenchel, T. and B. Barker-Jørgensen. 1977. Detritus food chains of aquatic ecosystems: the role of bacteria. In *Advances in Microbial Ecology,* Vol. 1, Ed. M. Alexander. Plenum Press, N.Y., pp. 1–58.

Ferguson, C. F. and J. K. B. Raymont. 1974. Biochemical studies on marine zooplankton. XII. Further investigations on *Euphausia superba* (Dana). *J. Mar. Biol. Ass. U.K.* **54,** 719–725.

Field, J. G. 1971. A numerical analysis of changes in the soft-bottom fauna along a transect across False Bay, South Africa. *J. Exp. Mar. Biol. Ecol.* **7,** 215–253.

Findlay, S. E. G. 1981. Small-scale spatial distribution of meiofauna on a mud and sandflat. *Est. Coastal and Shelf Sci.* **12,** 471–484.

Fitzgerald, G. P. 1968. Detection of limiting or surplus nitrogen in algae and aquatic weeds. *J. Phycol.* **4,** 121–126.

Foerster, R. E. 1968. The sockeye salmon, *Oncorhynchus nerka. Bull. Fish. Res. Bd. Canada, Bull.* **162,** pp. 422.

Foerster, R. E. and W. E. Ricker. 1941. The effect of reduction of predaceous fish on survival of young sockeye salmon at Cultus Lake. *J. Fish. Res. Bd. Canada,* **5,** 315–336.

Fogg, G. E. 1952. The production of extracellular nitrogenous substances by a blue-green alga. *Proc. Roy. Soc. (London).* B. **139,** 372–397.

Fogg, G. E. 1966. The extracellular products of algae. *Oceanogr. Mar. Biol. Ann. Rev.* **4,** 195–212.

Fogg, G. E., C. Nalewajko and W. D. Watt. 1965. Extracellular products of phytoplankton photosynthesis. *Proc. Roy. Soc. (London),* B, **162,** 517–534.

Fong, V. and K. H. Mann. 1980. Role of gut flora in the transfer of amino acids through a marine food chain. *Can. J. Fish. Aquat. Sci.* **37,** 88–96.

Ford, C. W. and E. Percival. 1965. The carbohydrates of *Phaedactylum tricornutum.* Part 1. Preliminary examination of the organism and characterization of low molecular weight material and of a glucan. *J. Chem. Soc.* (5), 7035–7042.

Foulds, J. B. and K. H. Mann. 1978. Cellulose digestion in the mysid shrimp *Mysis stenolepis* and its ecological implications. *Limnol. Oceanogr.* **23,** 760–766.

Fox, D. L., D. M. Updegraff and G. D. Novelli. 1944. Carotenoid pigments in the ocean floor. *Arch. Biochem.* **5,** 1–23.

Fox, D. L., S. C. Crane and B. M. McConnaughey. 1948. A biochemical study of the marine annelid worm, *Thoracophelia mucronata,* its food, biochromes and carotenoid metabolism. *J. Mar. Res.* **7,** 567–585.

Frankenberg, D. and R. J. Menzies. 1968. Some quantitative analyses of deep-sea benthos off Peru. *Deep-Sea Res.* **15,** 623–626.

Frazer, H. J. 1935. Experimental study of porosity and permeability of classic sediments. *J. Geol.* **43,** 910–1010.

Friedman, M. M. and J. R. Strickler. 1975. Chemoreceptors and feeding in calanoid copepods (Arthropoda: Crustacea). *Proc. Nat. Acad. Sci.* **72,** 4185–4188.

Fritsch, F. E. 1956. *The Structure and Reproduction of the Algae,* Vol. 1. Cambridge University Press, London. 791 pp.

Fritz, S. 1957. Solar energy on clear and cloudy days. *Sci. Month.* **84,** 55–65.

Frost, B. W. 1972. Effects of size and concentration of food particles on the feeding behaviour of the marine planktonic copepod *Calanus pacificus. Limnol. Oceanogr.* **17,** 805–815.

Frost, B. W. 1974. Feeding processes at lower trophic levels in pelagic communities. In *The Biology of the Oceanic Pacific,* Oregon State Univ. Press, pp. 59–77.

Frost, B. W. 1975. A threshold feeding behaviour in *Calanus pacificus. Limnol. Oceanogr.* **20,** 263–266.

Fujita, Y. 1970. Photosynthesis and plant pigments. *Bull. Plankton Soc. Japan,* **17,** 20–31.

Fukazawa, N., Ishimaru, T., M. Takahashi and Y. Fujita. 1980. A mechanism of 'red tide' formation. I. Growth rate estimate by DCMU-induced fluorescence increase. *Mar. Ecol. Prog. Ser.* **3,** 217–222.

Fulton, J. 1973. Some aspects of the life history of *Calanus plumchrus* in the Strait of Georgia. *J. Fish. Res. Bd. Canada,* **30,** 811–815.

Gage, J. 1972. Community structure of the benthos in Scottish sea-lochs. I. Introduction and species diversity. *Mar. Biol.* **14,** 281–297.

Gage, J. and A. D. Geekie. 1973. Community structure of the benthos in Scottish sea-lochs. II. Spatial pattern. *Mar. Biol.* **19,** 41–53.

Gage, J. and P. B. Tett. 1973. The use of log normal statistics to describe the benthos of Lochs Etine and Creran. *J. Anim. Ecol.* **42,** 373–382.

Gage, J., R. H. Lightfoot, M. Pearson and P. A. Tyler. 1980. An introduction to a sample time series of abyssal macrobenthos: methods and principle sources of variability. *Oceanol. Acta,* **3,** 169–176.

Gallagher, J. L. and F. C. Daiber. 1973. Diel rhythms in edaphic community metabolism in a Delaware salt marsh. *Ecology.* **54,** 1160–1163.

Gallardo, V. A. 1977. Large benthic microbial communities in sulfide biota under Peru–Chile subsurface countercurrent. *Nature,* **268,** 331–332.

Gärdefors, D. and L. Orrhage. 1968. Patchiness of some marine bottom animals. A methodological study. *Oikos,* **19,** 311–322.

Gardner, D. 1975. Observations on the distribution of dissolved mercury in the ocean. *Mar. Poll. Bull.* **6,** 43–46.

Gardner, D. and J. P. Riley. 1973. Distribution of dissolved mercury in the Irish Sea. *Nature,* **241,** 526–527.

Gardner, W. D. 1980a. Sediment trap dynamics and calibration: a laboratory evaluation. *J. Mar. Res.* **38,** 17–39.

Gardner, W. D. 1980b. Field assessment of sediment traps. *J. Mar. Res.* **38,** 41–52.

Garfield, P. C., T. T. Packard and L. A. Codispoti. 1979. Particulate protein in the Peru upwelling system. *Deep-Sea Res.* **26,** 623–639.

Gargas, E. 1970. Measurements of primary productions, dark

fixation and vertical distribution of the microbenthic algae in the Øresund. *Ophelia,* **8,** 251–253.

Gargas, E. 1971. 'Sun-shade' adaptation in microbenthic algae from the Øresund. *Ophelia,* **9,** 107–112.

Garrett, W. D. 1964. *The Organic Chemical Composition of the Ocean Surface.* NRL Report, 6201, 1–12.

Garrett, W. D. 1965. Collection of slick-forming materials from the sea surface. *Limnol. Oceanogr.* **10,** 602–605.

Garrett, W. D. 1967. The organic chemical composition of the ocean surface. *Deep-Sea Res.* **14,** 221–227.

Gartner-Kepkay, K. E., L. M. Dickie, K. R. Freeman and E. Zouros. 1980. Genetic differences and environments of mussel populations in the Maritime provinces. *Can. J. Fish. Aquat. Sci.* **37,** 775–782.

Gatten, R. R. and J. R. Sargent. 1973. Wax ester biosynthesis in calanoid copepods in relation to vertical migration. *Neth. J. Sea Res.* **7,** 150–158.

Gaudy, R. 1974. Feeding four species of pelagic copepods under experimental conditions. *Mar. Biol.* **25,** 125–141.

Gauld, D. T. 1951. The grazing rate of planktonic copepods. *J. Mar. Biol. Ass. U.K.* **29,** 695–706.

Gauld, D. T. 1966. The swimming and feeding of planktonic copepods. In *Some Contemporary Studies in Marine Science,* Ed. H. Barnes, Allen & Unwin, London, pp. 313–334.

George, J. L. and D. E. H. Frear. 1966. Pesticides in the Antarctic. *J. Appl. Ecol.* 3 (suppl.) 155–167.

Gerlach, S. A. 1971. On the importance of marine meiofauna for benthos communities. *Oecologia,* **6,** 176–190.

Gerlach, S. A. 1978. Food-chain relationships in subtidal silty sand marine sediments and the role of meiofauna in stimulating bacterial productivity. *Oecologia,* **33,** 55–69.

Gerlach, S. A. 1981. *Marine Pollution.* Springer-Verlag (New York), pp. 104–119.

Gibbs, M. and J. A. Schiff. 1960. Chemosynthesis: the energy relations of chemoautotrophic organisms. In *Plant Physiology,* vol. 1B. *Photosynthesis and Chemosynthesis,* Ed. F. C. Steward, Academic Press, London, pp. 279–319.

Gieskes, J. M. 1982. The practical salinity scale. 1978. A reply to comments by T. R. Parsons. *Limnol. Oceanogr.* **27,** 387–389.

Gifford, D. J., R. N. Bohrer and C. M. Boyd. 1981. Spines on diatoms. Do copepods care? *Limnol. Oceanogr.* **26,** 1057–1061.

Gilmartin, M. and N. Revelante. 1974. The 'Island Mass' effect on the phytoplankton and primary production of the Hawaiian Islands. *J. Exp. Mar. Biol. Ecol.* **16,** 181–204.

Gilmer, R. W. 1972. Free-floating mucus webs; a novel feeding adaptation for the open ocean. *Science,* **176,** 1239–1240.

Gilmer, R. W. 1974. Some aspects of feeding in thecosomatous pteropod molluscs. *J. Exp. Mar. Biol. Ecol.* **15,** 127–144.

Glasser, J. W. 1979. The role of predation in shaping and maintaining the structure of communities. *Am. Nat.* **113,** 631–640.

Glemarec, M. 1979. Les fluctuations temporelles des peupelements benthiques liess aux fluctuations climatiques. *Oceanol. Acta,* **2,** 365–371.

Glover, R. S., G. A. Robinson and J. M. Colebrook. 1974. Marine biological surveillance. *Environ. Change,* **2,** 395–402.

Gnaiger, E., G. Gluth and W. Wieser. 1978. pH fluctuations in an intertidal beach in Bermuda. *Limnol. Oceanogr.* **23,** 851–857.

Goering, J. J., R. C. Dugdale and D. W. Menzel. 1966. Estimates of *in situ* rates of nitrogen uptake by *Trichodesmium* sp. in the tropical Atlantic Ocean. *Limnol. Oceanogr.* **11,** 614–620.

Goering, J. J., D. D. Wallen and R. M. Nauman. 1970. Nitrogen uptake by phytoplankton in the discontinuity layer of the eastern subtropical Pacific Ocean. *Limnol. Oceanogr.* **15,** 789–796.

Goering, J. J., D. M. Nelson and J. A. Carter. 1973. Silicic acid uptake by natural populations of marine phytoplankton. *Deep-Sea Res.* **20,** 777–789.

Goldberg, E. D. 1971. River-ocean interactions. In *Fertility of the Sea,* Ed. J. D. Costlow, Gordon & Breach, New York, Vol. 1, pp. 143–156.

Goldberg, E. D. 1975. The mussel watch—a first step in global marine monitoring. *Mar. Poll. Bull.* **6,** 111.

Goldman, J. C. and E. J. Carpenter. 1974. A kinetic approach to the effect of temperature on algal growth. *Limnol. Oceanogr.* **19,** 756–766.

Goldman, J. C., J. J. McCarthy and D. G. Peavey. 1979. Growth rate influence on the chemical composition of phytoplankton in oceanic waters. *Nature,* **279,** 210–215.

Goodwin, T. W. 1955. Carotenoids. In *Modern Methods of Plant Analysis,* Eds. K. Paech and M. V. Traceay, Springer-Verlag (Berlin), Vol. 3, pp. 272–311.

Goodwin, T. W. 1957. The nature and distribution of carotenoids in some blue-green algae. *J. Gen. Microbiol.* **17,** 467–473.

Gophen, M. and R. P. Harris. 1981. Visual predation by a marine cyclopod copepod, *Corycaeus anglicus. J. Mar. Biol. Ass. U.K.* **61,** 391–399.

Gordon, D. C. 1970a. A microscopic study of organic particles in the North Atlantic Ocean. *Deep-Sea Res.* **17,** 175–185.

Gordon, D. C. 1970b. Some studies on the distribution and composition of particulate organic carbon in the North Atlantic Ocean. *Deep-Sea Res.* **17,** 233–243.

Gordon, D. C., Jr. 1971. Distribution of particulate organic carbon and nitrogen at an oceanic station in the central Pacific. *Deep-Sea Res.* **8,** 1127–1134.

Gordon, D. C., Jr. 1977. Variability of particulate organic carbon and nitrogen along the Halifax-Bermuda section. *Deep-Sea Res.* **24,** 257–270.

Gordon, D. C., J. Dale and P. D. Keizer. 1978. Importance of sediment working by the deposit-feeding polychaete *Arenicola marina* on the weathering rate of sediment-bound oil. *J. Fish. Res. Bd. Canada,* **35,** 591–603.

Gordon, J. D. M. 1979. Seasonal reproduction in deep-sea fish. In *Cyclic Phenomena in Marine Plants and Animals,* Ed. E. Naylor and R. G. Hartnoll. Pergamon Press, Oxford, pp. 223–228.

Gordon, L. I., P. K. Park, S. W. Hager and T. R. Parsons. 1971. Carbon dioxide partial pressures in north Pacific surface waters—time variations. *J. Oceanogr. Soc. Japan,* **27,** 81–90.

Gordon, W. G. and E. O. Whittier. 1966. In *Fundamentals of Dairy Chemistry,* Eds. B. H. Webb and A. H. Johnson, Avi Publishing Co., Westport, Conn., pp. 60.

Gran, H. H. and T. Braarud. 1935. A quantitative study of the phytoplankton in the Bay of Fundy and the Gulf of Maine including observations on hydrography, chemistry and turbidity. *J. Biol. Bd. Canada,* **1,** 219–467.

Grasshoff, K. 1976. *Methods of Seawater Analysis.* Verlag Chemie. pp. 317.

Grassle, J. F. 1977. Slow recolonization of deep sea sediments. *Nature,* **265,** 618–619.

Grassle, J. F., H. L. Sanders, R. R. Hessler, G. T. Rowe and T. McLellan. 1975. Pattern and zonation: a study of the bathyal megafauna using the research submersible *Alvin. Deep-Sea Res.* **22,** 475–481.

Grassle, J. F., H. L. Sanders and W. K. Smith. 1979. Faunal changes with depth in the deep sea benthos. *Ambio,* **6,** 47–50.

Gray, J. S. 1966. The attractive factor of intertidal sands to *Protodrilus symbiaticus. J. Mar. Biol. Ass. U.K.* **46,** 627–645.

Gray, J. S. 1967. Substrate selection by the archiannclid *Protodrilus hypoleucus* Armenante. *J. Exp. Mar. Biol. Ecol.* **1,** 47–54.

Gray, J. S. 1974. Animal-sediment relationships. *Oceanogr. Mar. Biol. Ann. Rev.* **12,** 223–261.

Gray, J. S. and F. B. Mirza. 1979. A possible method for detection of pollution-induced disturbance on marine benthic communities. *Mar. Poll. Bull.* **10,** 142–146.

Gray, J. S. and T. H. Pearson. 1982. Objective selection of sensitive species indicative of pollution-induced change in benthic communities. I. Comparative methodology. *Mar. Ecol. Prog. Ser.* **9,** 111–119.

Green, R. H. 1979. *Sampling Design and Statistical Methods for Environmental Biologists.* Wiley, New York. 257 pp.

Greenslate, J. 1974. Micro-organisms participate in the construction of manganese nodules. *Nature,* **294,** 181–183.

Greenwood, D. J. 1968. Measurement of microbial metabolism in soil, pp. 138–157. In *Ecology of Soil Bacteria,* Eds. T. R. G. Gray and D. Parkinson, Univ. Toronto Press.

Greenwood, P. J. 1980. Growth, respiration and tentative energy budgets for two populations of the sea urchin *Parenchinus angulosus* (Leske). *Est. Coastal Mar. Sci.* **10,** 347–367.

Greve, W. 1972. Ökologische Untersuchungen an *Pleurobrachia pileus* 2. Laboratoriums-untersuchungen. *Helgoländer Wiss. Meeresunters.* **23,** 141–164.

Greve, W. and T. R. Parsons. 1977. Photosynthesis and fish production: Hypothetical effects of climatic change and pollution. *Helgoländer wiss. Meeresunters.* **30,** 666–672.

Greze, V. N. and E. P. Baldina. 1964. *Trudy Sevastopol biol. Sta. 18,* Fish. Res. Bd. Canada, Translation 893 (Population dynamics and the annual production of *Acartia clausii* Giesbr. and *Centropages kroyeri* in the neritic zone of the Black Sea).

Grice, G. D. and V. R. Gibson. 1975. Occurrence, viability and significance of resting eggs of the calanoid copepod *Labidocera aestiva. Mar. Biol.* **31,** 335–337.

Grice, D. G. and M. R. Reeve (Eds.). 1982. *Marine Mesocosms.* Springer-Verlag (New York). pp. 430.

Griffin, J. J., H. Windom and E. D. Goldberg. 1968. The distribution of clay minerals in the world ocean. *Deep-Sea Res.* **15,** 433–459.

Grøntved, J. 1960. On the productivity of microbenthos and phytoplankton in some Danish fjords. *Meddr. Danm. Fisk. – og Havunders., N.S.* **3,** 55–92.

Gross, F. and E. Zeuthen. 1948. The buoyancy of plankton diatoms: a problem of cell physiology. *Proc. Roy. Soc. (London),* **135,** 382–389.

Guerinot, M. L., W. Fong and D. G. Patrequin. 1977. Nitrogen fixation (acetylene reduction) associated with sea urchins *(Strongylocentrotus droebachiensis)* feeding on seaweeds and eelgrass. *J. Fish. Res. Bd. Canada.* **34,** 416–420.

Guillard, R. R. L. 1963. Organic sources of nitrogen for marine diatoms. In *Symposium on Marine Microbiology,* Ed. C. H. Oppenheimer, C. C. Thomas, Springfield, Illinois, pp. 93–104.

Gulland, J. A. 1965. Survival of the youngest stages of fish, and its relation to year-class strength. *ICNAF Spec. Publ.* **6,** 363–371.

Gulland, J. A. 1970. The fish resources of the oceans. *FAO Fish. Tech. Pap.* No. 97, p. 423.

Gundersen, K. 1968. The formation and utilization of reducing power in aerobic chemo-autotrophic bacteria. *Zeitschrift. Allg. Mikrobiol.* **8,** 445–457.

Gunkel, W. and G. Gassman. 1980. Oil, oil dispersants and related substances in the marine environment. *Helg. Meeres.* **33,** 164–181.

Haas, L. W. and K. L. Webb. 1979. Nutritional mode of several non-pigmented and microflagellates from the York River Estuary, Virginia. *J. Exp. Mar. Biol. Ecol.* **39,** 125–134.

Haedrich, R. L. and G. T. Rowe. 1977. Megafaunal biomass in the deep sea. *Nature,* **269,** 141–142.

Haedrich, R. L., G. T. Rowe and P. T. Polloni. 1980. The megabenthic fauna in the deep sea south of New England, U.S.A. *Mar. Biol.* **57,** 165–179.

Haertel, L., C. Osterberg, H. Curl, Jr. and P. K. Park. 1969. Nutrient and plankton ecology of the Columbia River estuary. *Ecology,* **50,** 962–978.

Haines, E. B. 1977. The origins of detritus in Georgia salt marsh estuaries. *Oikos,* **29,** 254–260.

Haines, E. B. and C. L. Montague. 1979. Food sources of estuarine invertebrates analyzed using $^{13}C/^{12}C$ ratios. *Ecology,* **60,** 48–56.

Halim, Y. 1960. Observations on the Nile bloom of phytoplankton in the Mediterranean. *J. Cons. Int. Explor. Mer,* **26,** 57–67.

Hall, K. J., W. C. Weimer and G. F. Lee. 1970. Amino acids in an estuarine environment. *Limnol. Oceanogr.* **15,** 162–164.

Hall, K. J., P. M. Kleiber and I. Yesaki. 1972. Heterotrophic uptake of organic solutes by microorganisms in the sediment. *Mem. Ist. Ital. Idiobiol.* **29** (Suppl). 441–447.

Hallberg, R. O. 1968. Some factors of significance in the formation of sedimentary metal sulfides. *Stockholm Contr. Geol.* **15,** 39–66.

Halldal, P. 1958. Pigment formation and growth in blue-green algae in crossed gradients of light intensity and temperature. *Physiol. Plant.* **11,** 401–420.

Halldal, P. 1974. Light and photosynthesis of different marine algal groups. In *Optical Aspects of Oceanography,* Eds. N. G. Jerlov and E. Steemann Nielsen. Academic Press, London, pp. 345–360.

Halldal, P. 1981. Solar energy capturing by marine plants. In *Biological and Chemical Utilization of Solar Energy,* Eds. K. Shibata, A. Imamura and A. Ikegami. Japan Scientific Societies Press, Tokyo, pp. 48–58 (Japanese).

Haltiner, G. J. and F. L. Martin. 1957. *Dynamical and Physical Meteorology,* McGraw Hill, New York. 470 pp.

Hama, T., T. Miyazaki, Y. Ogawa, T. Iwakuma, M. Takahashi, A. Otsuki and S. Ichimura. 1983. Measurement of photosynthetic production of marine phytoplankton population by using a stable 13C isotope. *Mar. Biol.* **73,** 31–37.

Hamilton, R. D. and O. Holm-Hansen. 1967. Adenosine triphosphate content of marine bacteria. *Limnol. Oceanogr.* **12,** 319–324.

Hamilton, R. D. and J. E. Preslan. 1970. Observations on heterotrophic activity in the eastern tropical Pacific. *Limnol. Oceanogr.* **15,** 395–401.

Handa, N. 1969. Carbohydrate metabolism in the marine diatom *Skeletonema costatum. Mar. Biol.* **4,** 208–214.

Handa, N. 1970. Dissolved and particulate carbohydrates. *Symp. Organic Matter in Natural Waters,* Ed. D. W. Hood, University of Alaska, pp. 129–152.

Handa, N. and H. Tominaga. 1969. A detailed analysis of carbohydrates in marine particulate matter. *Mar. Biol.* **2,** 228–235.

Handa, N. and K. Yanagi. 1969. Studies on water extractable carbohydrates of the particulate matter from the northwest Pacific Ocean. *Mar. Biol.* **4,** 197–207.

Handa, N. and E. Tanoue. 1980. Carbon cycle in the ocean. *La Mer,* **18,** 190–199.

Harbison, G. R. and R. W. Gilmer. 1976. The feeding rates of the pelagic tunicate *Pegea confederata* and two other salps. *Limnol. Oceanogr.* **21,** 517–528.

Harbison, G. R., L. B. Madin and N. R. Swanberg. 1978. On the natural history and distribution of oceanic ctenophores. *Deep-Sea Res.* **25,** 233–256.

Harder, W. 1968. Reaction of plankton organisms to water stratification. *Limnol. Oceanogr.* **13,** 156–168.

Harding, G. C. H. 1974. The food of deep-sea copepods. *J. Mar. Biol. Ass. U.K.* **54,** 141–155.

Harding, G. A. 1977. Surface area of the euphausiid *Thysanoessa raschi* and its relation to body length, weight and respiration. *J. Fish. Res. Bd. Canada,* **34,** 225–231.

Hardy, A. C. 1936. The continuous plankton recorder. *Discovery Rep.* **11,** 457–510.

Hargrave, B. T. 1969. Similarity of oxygen uptake by benthic communities. *Limnol. Oceanogr.* **14,** 801–805.

Hargrave, B. T. 1970. The effect of a deposit-feeding amphipod on the metabolism of benthic microflora. *Limnol. Oceanogr.* **15,** 21–30.

Hargrave, B. T. 1971. An energy budget for a deposit-feeding amphipod. *Limnol. Oceanogr.* **16,** 99–103.

Hargrave, B. T. 1972a. Aerobic decomposition of sediment and detritus as a function of particle surface area and organic content. *Limnol. Oceanogr.* **17,** 583–596.

Hargrave, B. T. 1972b. Prediction of egestion by the deposit-feeding amphipod *Hyalella azteca. Oikos,* **23,** 116–124.

Hargrave, B. T. 1973. Coupling carbon flow through some pelagic and benthic communities. *J. Fish. Res. Bd. Canada,* **30,** 1317–1326.

Hargrave, B. T. 1976. The central role of invertebrate feces in sediment decomposition. In *The Role of Terrestrial and Aquatic Organisms in Decomposition Processes,* Eds. A. MacFayden and J. M. Anderson. Blackwell Scientific Publ., Oxford, pp. 301–321.

Hargrave, B. T. 1980. Factors affecting the flux of organic matter to sediments in a marine bay. In *Marine Benthic Dynamics,* Eds. K. R. Tenore and B. C. Coull. Univ. S. Carolina Press, pp. 243–263.

Hargrave, B. T. and G. H. Green. 1970. Effects of copepods grazing on two natural phytoplankton populations *(A. tonsa). J. Fish. Res. Bd. Canada,* **27,** 1395–1403.

Hargrave, B. T., G. A. Phillips and S. Taguchi. 1976. Sedimentation measurements in Bedford Basin, 1973–1974. *Fish. Mar. Ser. Rept.* **608,** 129 pp.

Hargrave, B. T. and G. A. Phillips. 1977. Oxygen uptake by microbial communities on solid surfaces. In *Aquatic Microbial Communities,* Ed. J. Cairns, Jr., Garland Publ. Inc., New York

Hargrave, B. T. and G. A. Phillips. 1981. Annual *in situ* carbon dioxide and oxygen flux across a subtidal marine sediment. *Est. Coastal Mar. Sci.* **12,** 725–737.

Harris, E. 1959. The nitrogen cycle in Long Island Sound. *Bull. Bingham Oceanogr.* **17**(1), 31–65.

Harris, G. P. 1980. The measurement of photosynthesis in natural populations of phytoplankton. In *The Physiological Ecology of Phytoplankton,* Ed. I. Morris, Blackwell Scientific Publ., Oxford, pp. 129–187.

Harris, R. P. 1973. Feeding, growth, reproduction and nitrogen utilization by the harpacticoid copepod, *Trigriopus brevicornis. J. Mar. Biol. Ass. U.K.* **53,** 785–800.

Harrison, P. G. and K. H. Mann. 1975. Detritus formation from eelgrass (*Zostera marina* L.): the relative effects of fragmentation, leaching and decay. *Limnol. Oceanogr.* **20,** 924–934.

Harrison, P. J. and C. D. Davis. 1979. The use of outdoor phytoplankton continuous cultures to analyse factors influencing species selection. *J. Exp. Mar. Biol. Ecol.* **41,** 9–23.

Hartwig, E. O. 1976. Nutrient cycling between the water column and a marine sediment. I. Organic carbon. *Mar. Biol.* **34,** 285–295.

Hartwig, E. O. 1978. Factors affecting respiration and photosynthesis by the benthic community of a subtidal siliceous sediment. *Mar. Biol.* **46,** 283–293.

Harvey, G. R., H. P. Miklas, U. T. Bowen and W. G. Steinhauer. 1973. Observations on the distribution of chlorinated hydrocarbons in Atlantic Ocean organisms. *J. Mar. Res.* **32,** 103–118.

Harvey, G. W. 1966. Microlayer collection from the sea surface: A new method and initial results. *Limnol. Oceanogr.* **11,** 608–613.

Harvey, H. W. 1957. *The Chemistry and Fertility of Sea Water,* 2nd edn. Cambridge University Press. 234 pp.

Harvey, P. H., J. S. Ryland and P. J. Hayward. 1976. Pattern analysis in bryozoan and spirorbid communities. II. Distance sampling methods. *J. Exp. Mar. Biol. Ecol.* **21,** 99–108.

Hatch, M. D. and C. R. Slack. 1970. Photosynthetic CO_2-fixation pathways. *Ann. Rev. Plant Physiol.* **21,** 141–162.

Hattori, A. 1962a. Light-induced reduction of nitrate, nitrite and hydroxylamine in a blue-green alga, *Anabaena cylindrica. Plant and Cell Physiol.* **3,** 355–369.

Hattori, A. 1962b. Adaptive formation of nitrate reducing system in *Anabaena cylindrica. Plant and Cell Physiol.* **3,** 371–377.

Hattori, A. and E. Wada. 1971. Nitrite distribution and its regulating processes in the equatorial Pacific Ocean. *Deep-Sea Res.* **18,** 557–568.

Haug, A., S. Myklestad and E. Sakshaug. 1973. Studies on the phytoplankton ecology of Trondheimsfjord. I. The chemical composition of phytoplankton populations. *J. Exp. Mar. Biol. Ecol.* **11,** 15–26.

Haxo, F. T. 1960. The wavelength dependence of photosynthesis and the role of accessory pigments. In *Comparative Biochemistry of Photoreactive Systems,* Ed. M. B. Allen, Academic Press, New York, pp. 339–360.

Haxo, F. T. and D. C. Fork. 1959. Photosynthetically active accessory pigments of cryptomonads. *Nature,* **184,** 1051–1052.

Healey, F. P. 1980. Slope of monad equation as an indicator of advantage in nutrient competition. *Microbial. Ecol.* **5,** 281–286.

Healey, M. C. 1982. Multispecies, multistock aspects of Pacific salmon management. *Canadian Spec. Publ. Fish. Aqu. Sci.* **59**, 119–126.

Hecky, R. E. and P. Kilham. 1974. Environmental control of phytoplankton cell size. *Limnol. Oceanogr.* **19**, 361–366.

Hedges, J. I. and D. C. Mann. 1979. The lignin geochemistry of marine sediments from the southern Washington coast. *Geochim. Cosmochim. Acta.* **43**, 1809–1818.

Hedgpeth, J. W. 1957. Classification of marine environments. *Mem. Geol. Soc. Amer.* **67**, 17–28.

Heezen, B. C., M. Tarp and M. Ewing. 1959. The floors of the ocean. I. North Atlantic. *Geol. Soc. Amer. Spec. Paper* **65**, 122 pp.

Heinrich, A. K. 1962. The life histories of plankton animals and seasonal cycles of plankton communities in the oceans. *J. Cons. Int. Explor. Mer,* **27**, 15–24.

Heip, C. 1976. The spatial pattern of *Cyprideis torosa* (Jones 1850) (Crustacea: Ostracoda). *J. Mar. Biol. Ass. U.K.* 179–189.

Heip, C. and P. Engels. 1974. Comparing species diversity and evenness indices. *J. Mar. Biol. Ass. U.K.* **54**, 559–563.

Hellebust, J. A. 1965. Excretion of some organic compounds by marine phytoplankton. *Limnol. Oceanogr.* **10**, 192–206.

Hellebust, J. A. 1970. The uptake and utilisation of organic substances by marine phytoplankters. In *Organic Matter in Natural Waters,* Ed. D. W. Hood, University of Alaska, pp. 225–256.

Hempel, G. 1978. North Sea fisheries and fish stocks—a review of recent changes. *Rapp. P. Réun. Cons. int. explor. Mer.* **173**, 145–167.

Hempel, G. and H. Weikert. 1972. The neuston of the subtropical and boreal North-eastern Atlantic Ocean. A review. *Mar. Biol.* **13**, 70–88.

Henriksen, K., J. I. Hansen and T. H. Blackburn. 1981. Rates of nitrification, distribution of nitrifying bacteria and nitrate fluxes in different types of sediment from Danish waters. *Mar. Biol.* **61**, 299–304.

Herbert, R. A. 1975. Heterotrophic nitrogen fixation in shallow estuarine sediments. *J. Exp. Mar. Biol. Ecol.* **18**, 215–225.

Heron, A. C. 1972. Population ecology of a colonizing species: the pelagic tunicate *Thalia democratica. Oecologia,* **10**, 269–293.

Heron, A. C. 1973. A specialized predator–prey relationship between the copepod *Sapphirina angusta* and the pelagic tunicate *Thalia democratica. J. Mar. Biol. Ass. U.K.* **53**, 429–435.

Hersey, J. B. and R. H. Backus. 1962. Sound scattering by marine organisms. In *The Sea,* Ed. M. N. Hill, Interscience New York, Vol. 1, pp. 498–539.

Hewitt, E. J. 1957. Some aspects of micronutrient element metabolism in plants. *Nature,* **180**, 1020–1022.

Heyraud, M. 1979. Food ingestion and digestive transit time in the euphasuiid *Meganyctiphanes norvegica* as a function of animal size. *J. Plankton Res.* **1**, 301–311.

Hickey, J. J. and D. W. Anderson. 1968. Chlorinated hydrocarbons and egshell changes in raptorial and fish-eating birds. *Science,* **162**, 271–273.

Hickling, C. F. 1970. Estuarine fish farming. *Adv. Mar. Biol.* **8**, 119–213.

Hill, R. and C. P. Whittingham. 1955. *Photosynthesis,* Methuen's Monographs. John Wiley, New York. 165 pp.

Himmelman, J. H. and T. H. Carefoot. 1975. Seasonal changes in calorific value of three Pacific coast seaweeds, and their significance to some marine invertebrate herbivores. *J. Exp. Mar. Biol. Ecol.* **18**, 139–151.

Hinchcliffe, P. R. and J. P. Riley. 1972. The effect of diet and the component fatty acid composition of *Artemia salina. J. Mar. Biol. Ass. U.K.* **52**, 203–211.

Hinga, K. R. 1979. The food requirements of whales in the southern hemisphere. *Deep-Sea Res.* **26**, 569–577.

Hinga, K. R., J. McN. Sieburth and G. R. Heath. 1979. The supply and use of organic material at the deep-sea floor. *J. Mar. Res.* **37**, 557–579.

Hjort, J. 1914. Fluctuations in the great fisheries of northern Europe viewed in the light of biological research. *Rapp. Proc. Verbaux,* **20**, 1–228.

Hobson, L. A. 1966. Some influences of the Columbia River effluent on marine phytoplankton during January 1961. *Limnol. Oceanogr.* **11**, 223–234.

Hobson, L. A. 1971. Relationships between particulate organic carbon and micro-organisms in upwelling areas off Southwest Africa. *Invest. Pesq.* **35**, 195–208.

Hock, C. W. 1940. Decomposition of chitin by marine bacteria. *Biol. Bull* **79**, 199–206.

Hodson, R. E., F. Azam and R. F. Lee. 1977. Effects of four oils on marine bacterial populations: Controlled experimental ecosystem pollution experiment. *Bull. Mar. Sci.* **27**, 119–126.

Hofman, E. E., J. M. Klinck and G. A. Päffenhofer. 1981. Concentrations and vertical fluxes of zooplankton fecal pellets on a continental shelf. *Mar. Biol.* **61**, 327–335.

Hofmann, T. and H. Lees. 1952. The biochemistry of the nitrifying organisms. 2. The free-energy efficiency of *Nitrosomonas. Biochem. J.* **52**, 140–142.

Hogetsu, K., M. Sakamoto and H. Sumikawa. 1959. On the high photosynthetic activity of *Skeletonema costatum* under the strong light intensity. *Bot. Mag., Tokyo,* **72**, 421–422.

Holland, A. F., W. K. Mountford, M. H. Hiegel, K. R. Kaumeyer and J. A. Mihursky. 1980. Influence of predation on infaunal abundance in Upper Chesapeake Bay, U.S.A. *Mar. Biol.* **57**, 221–235.

Holme, N. A. 1950. Population dispersion in *Tellina tennis* da Costa. *J. Mar. Biol. Ass. U.K.* **29**, 267–280.

Holme, N. A. and A. D. McIntyre. 1971. (Eds.) *Methods for the Study of Marine Benthos. I.B.P. Handbook,* **16.** Blackwell Scientific Publications, Oxford. 334 pp.

Holmes, R. W. and T. M. Widrig. 1956. The enumeration and collection of marine phytoplankton. *J. Cons. Int. Explor. Mer,* **22**, 21–32.

Holm-Hansen, O. 1969a. Determination of microbial biomass in ocean profiles. *Limnol Oceanogr.* **14**, 740–747.

Holm-Hansen, O. 1969b. Algae: amounts of DNA and organic carbon in single cells. *Science,* **163**, 87–88.

Holm-Hansen, O. 1970. ATP levels in algal cells as influenced by environmental conditions. *Plant. Cell Physiol.* **11**, 689–700.

Holm-Hansen, O. and C. R. Booth. 1966. The measurement of adenosine triphosphate in the ocean and its ecological significance. *Limnol. Oceanogr.* **11**, 510–519.

Holm-Hansen, O., J. D. H. Strickland and P. M. Williams. 1966. A detailed analysis of biologically important substances in a profile off southern California. *Limnol. Oceanogr.* **11**, 548–561.

Holm-Hansen, O., W. H. Sutcliffe, Jr. and J. Sharp. 1968. Measurement of deoxyribonucleic acid in the ocean and its ecological significance. *Limnol. Oceanogr.* **13**, 507–514.

Honeyman, D. 1874. The ocean (account of a public lecture). *Acadian Recorder,* Sat., Feb. 14.

Honjo, S. 1980. Material fluxes and modes of sedimentation in the mesopelagic and bathypelagic zones. *J. Mar. Res.* **38,** 53–97.

Honjo, S., S. J. Manganini and J. J. Cole. 1982. Sedimentation of biogenic matter in the deep ocean. *Deep-Sea Res.* **29,** 609–625.

Howarth, R. W. and J. M. Teal. 1979. Sulfate reduction in a New England salt marsh. *Limnol. Oceanogr.* **24,** 999–1013.

Hughes, R. N. 1969. A study of feeding in *Scrobicularia plana. J. Mar. Biol. Ass. U.K.* **49,** 805–823.

Hughes, R. N. 1970. An energy budget for a tidal flat population of the bivalve *Scrobicularia plaua* Da Costa. *J. Anim. Ecol.* **39,** 357–381.

Hughes, R. N. 1971a. Ecological energetics of the keyhole limpet *Fissurella barbadensis* Gmelin. *J. Exp. Mar. Biol. Ecol.* **6,** 167–178.

Hughes, R. N. 1971b. Ecological energetics of *Nerita* (Archaeogastropoda, Neritacea) populations on Barbados, West Indies. *Mar. Biol.* **11,** 12–22.

Hughes, R. N., D. L. Peer and K. H. Mann. 1972. Use of multivariate analysis to identify functional components of the benthos in St. Margaret's Bay, Nova Scotia. *Limnol. Oceanogr.* **17**(1), 111–121.

Hughes, T. G. 1975. The sorting of food particles by *Abra* sp. (Bivalvia: Tellinacea). *J. Exp. Mar. Biol. Ecol.* **20,** 137–156.

Hulberg, L. W. and J. S. Oliver. 1980. Caging manipulations in marine soft-bottom communities: importance of animal interactions on sedimentary habitat modification. *Can. J. Fish. Aquat. Sci.* **37,** 1130–1139.

Hulbert, E. M. 1970. Competition for nutrients by marine phytoplankton in oceanic, coastal, and estuarine regions. *Ecology,* **51,** 475–484.

Hulbert, E. M., J. H. Ryther and R. R. L. Guillard. 1960. The phytoplankton of the Sargasso Sea off Bermuda. *J. Cons. int. explor. Mer,* **25,** 115–128.

Humphrey, G. F. 1975. The photosynthesis: Respiration ratio of some unicellular marine algae. *J. Exp. Mar. Biol. Ecol.* **18,** 111–119.

Humphreys, W. F. 1979. Production and respiration in animal populations. *J. Anim. Ecol.* **48,** 427–453.

Hunding, C. 1973. Diel variation in oxygen production and uptake in a microbenthic littoral community of a nutrient-poor lake. *Oikos,* **24,** 352–360.

Hunding, C. and B. T. Hargrave. 1973. A comparison of benthic microalgal production measured by C^{14} and oxygen methods. *J. Fish. Res. Bd. Canada,* **30,** 309–312.

Hurd, L. E., V. M. Mellinger, L. L. Wolfe and S. J. McNaughton. 1971. Stability and diversity of three trophic levels in terrestrial successional ecosystems. *Science,* **173,** 1134–1136.

Hurlbert, S. H. 1971. The nonconcept of species diversity: a critique and alternative parameters. *Ecology,* **52**(4), 577–586.

Hutchinson, G. E. 1938. On the relation between the oxygen deficit and the productivity and typology of lakes. *Int. Rev. ges. Hydrobiol.* **36,** 336–355.

Hutchinson, G. E. 1957. *A Treatise on Limnology*, Vol. I, John Wiley & Sons, Inc., New York. 1015 pp.

Hutchinson, G. E. 1961. The paradox of the plankton. *Amer. Nat.* **95,** 137–145.

Hutchinson, G. E. 1969. Eutrophication, past and present. In *Eutrophication: Causes, Consequences, Correctives,* Nat. Acad. Sci., Washington, D.C., pp. 17–26.

Hylleberg, J. 1972. Carbohydrases of some marine invertebrates with notes on their food and on the natural occurrence of the carbohydrates studied. *Mar. Biol.* **14,** 130–142.

Hylleberg, J. 1975. Selective feeding by *Abarenicola pacifica* with notes on *Abarenicola vagabunda* and a concept of gardening in lugworms, *Ophelia,* **14,** 113–137.

Hylleberg, J. 1976. Resource partitioning on the basis of hydrolytic enzymes in deposit-feeding mud snails (Hydrobiidae). II. Studies on niche overlap. *Oecologia,* **23,** 115–125.

Hylleberg, J. and V. Gallucci. 1975. Selectivity in feeding by the deposit-feeding bivalve *Macoma nasuta. Mar. Biol.* **32,** 167–178.

Hyman, L. H. 1959a. *The Invertebrates: Smaller Coelomate Groups.* McGraw-Hill, New York. 783 pp.

Hyman, L. H. 1959b. *The Invertebrates: Protozoa through Ctenophora.* McGraw-Hill, New York. 726 pp.

Ichimura, S. 1956a. On the ecological meaning of transparency for the production of matter in phytoplankton community of lake. *Bot. Mag., Tokyo,* **69,** 219–226.

Ichimura, S. 1956b. On the standing crop and productive structure of phytoplankton community in some lakes of central Japan. *Bot. Mag., Tokyo,* **69,** 7–16.

Ichimura, S. 1960. Diurnal fluctuation of chlorophyll content in lake water. *Bot. Mag., Tokyo,* **73,** 217–224.

Ichimura, S. 1967. Environmental gradient and its relation to primary productivity in Tokyo Bay. *Records Oceanogr. Works. Japan,* **9,** 115–128.

Ichimura, S. and Y. Saijo. 1958. On the application of ^{14}C-method to measuring organic matter production in the lake. *Bot. Mag., Tokyo,* **17,** 174–180.

Ichimura, S., Y. Saijo and Y. Aruga. 1962. Photosynthetic characteristics of marine phytoplankton and their ecological meaning in the chlorophyll method. *Bot. Mag., Tokyo,* **75,** 212–220.

Ichimura, S. and Y. Aruga. 1964. Photosynthetic natures of natural algal communities in Japanese waters. In *Recent Researches in the Fields of Hydrosphere, Atmosphere and Nuclear Geochemistry.* Eds. Y. Miyake and T. Koyoma, Maruzen, Tokyo, pp. 13–37.

Ichimura, S., S. Nagasawa and T. Tanaka. 1968. On the oxygen and chlorophyll maxima found in the metalimnion of a mesotrophic lake. *Bot. Mag., Tokyo,* **81,** 1–10.

Idso, S. B. and R. G. Gilbert. 1974. On the universality of the Poole and Atkins Secchi disk-light extinction equation. *J. Appl. Ecol.* **11,** 399–401.

Ikeda, T. 1970. Relationship between respiration rate and body size in marine plankton animals as a function of the temperature of habitat. *Bull. Fac. Fish. Hokkaido Univ.* **21,** 91–112.

Ikeda, T. 1977. Feeding rates of planktonic copepods from a tropical sea. *J. Exp. Mar. Biol. Ecol.* **29,** 263–277.

Ikusima, I. 1967. Ecological studies on the productivity of aquatic plant communities. III. Effect of depth on daily photosynthesis in submerged macrophytes. *Bot. Mag., Tokyo,* **80,** 57–67.

Iles, T. D. and M. Sinclair. 1982. Atlantic Herring: stock discreteness and abundance. *Science,* **215,** 627–633.

Isaacs, J. D. 1973. Potential trophic biomasses and trace-substance concentrations in unstructured marine food webs. *Mar. Biol.* **22,** 97–104.

Isaacs, J. D. and R. A. Schwartzlose. 1975. Active animals of the deep-sea floor. *Sci. Am.* **233,** 84–91.

Itoh, K. 1970. A consideration on feeding habits of planktonic copepods in relation to the structure of their oral parts. *Bull. Plankton Soc. Japan,* **17,** 1–10.

Iturriaga, R. 1979. Bacterial activity related to sedimentary particulate matter. *Mar. Biol.* **55,** 157–169.

Iturriaga, R. and A. Zsolnay. 1981. Transformation of some dissolved organic compounds by a natural heterotrophic population. *Mar. Biol.* **62,** 125–129.

Iverson, R. L., L. K. Coachman, R. T. Cooney, T. S. English, J. J. Goering, G. L. Hunt Jr., M. C. Macauley, C. P. McRoy, W. S. Reeburg and T. E. Whitledge. 1979. Ecological significance of fronts in the southeastern Bering Sea. In *Ecological Processes in Coastal and Marine Systems,* R. J. Livingston. Plenum Press., N.Y., pp. 437–468.

Ivlev, V. S. 1944. The time of hunting and the path followed by the predater in relation to the density of the prey population. *Zool. Zh.* **23**(4), 139–145 (translation by L. Birkett, Lowestoft).

Ivlev, V. S. 1945. The biological productivity of waters. *Usp. Sovrem. Biol.* **19,** 98–120.

Ivlev, V. S. 1961. *Experimental Ecology of the Feeding of Fishes.* translated by D. Scott. Yale Univ. Press, New Haven, 302 pp.

Ivleva, I. V. 1970. The influence of temperature on the transformation of matter in marine invertebrates. In *Marine Food Chains,* Ed. J. H. Steele. Oliver & Boyd (Edinburgh), pp. 96–112.

Jacobs, R. P. W. M. 1980. Effects of the 'Amoco Cadiz' oil spill on the seagrass community at Roscoff with special reference to the benthic infauna. *Mar. Ecol. Prog. Ser.* **2,** 207–212.

Jannasch, H. W. 1958. Studies on planktonic bacteria by means of a direct membrane filter method. *J. Gen. Microbiol.* **18,** 609–620.

Jannasch, H. W. 1960. Versuche über denitrifikation und die Verfügbarkeit des Sauerstoffes in Wassen und Schlamm. *Arch. F. Hydrobiol.* **56,** 355–369.

Jannasch, H. W. 1970. Threshold concentrations of carbon sources limiting bacterial growth in sea water. In *Organic Matter in Natural Waters,* Ed. D. W. Hood, University of Alaska, pp. 321–328.

Jannasch, H. W. 1972. New approaches to assessment of microbial activity in polluted waters. In *Water Pollution Microbiology,* Ed. R. Mitchell, J. Wiley, New York, pp. 291–303.

Jannasch, H. W. and P. H. Pritchard. 1972. The role of inert particulate matter in the activity of aquatic microorganisms. *Mem. Ist. Ital. Idiobiol.* **29**(Suppl.), 289–308.

Jannasch, H. W. and C. O. Wirsen. 1973. Deep-sea microorganisms: *in situ* response to nutrient enrichment. *Science,* **180,** 641–643.

Jannasch, H. W. and C. O. Wirsen. 1979. Chemosynthetic primary production at East Pacific sea floor spreading centers. *Bioscience,* **29,** 592–598.

Jansson, B. O. 1968. Quantitative and experimental studies of the interstitial fauna in four Swedish sandy beaches. *Ophelia,* **5,** 1–71.

Jassby, A. D. and T. Platt. 1976. Mathematical formulation of the relationship between photosynthesis and light for phytoplankton. *Limnol. Oceanogr.* **21,** 540–547.

Jeffrey, S. W. 1961. Paper chromatographic separation of chlorophylls and carotenoids from marine algae. *Biochem. J.* **80,** 336–342.

Jeffrey, S. W. 1969. Properties of two spectrally different components in cholorophyll *c* preparations. *Biochim. Biophys. Acta.* **177,** 456–467.

Jeffrey, S. W. 1974. Profiles of photosynthetic pigments in the ocean using thin-layer chromatography. *Mar. Biol.* **26,** 101–110.

Jeffrey, S. W. 1980. Algal pigment systems. In *Primary Productivity in the Sea,* Ed. P. G. Falkowski. Plenum Pub. Corp. (New York), pp. 33–58.

Jeffrey, S. W. and M. B. Allen. 1964. Pigments, growth and photosynthesis in cultures of two chrysomonads. *Coccolithus huxleyi* and a *Hymenomonas* sp. *J. Gen. Microbiol.* **36,** 277–288.

Jeffrey, S. W. and G. M. Hallegraeff. 1980. Studies of phytoplankton species and photosynthetic pigments in a warm core eddy of the East Australian Current. I. Summer populations. *Mar. Ecol. Prog. Ser.* **3,** 285–294.

Jeffries, H. P. 1962. Environmental characteristics of Raritan Bay, a polluted estuary. *Limnol. Oceanogr.* **7,** 21–31.

Jeffries, H. P. 1969. Seasonal composition of temperate plankton communities: free amino acids. *Limnol. Oceanogr.* **14,** 41–52.

Jeffries, H. P. 1970. Seasonal composition of temperate plankton communities: fatty acids. *Limnol. Oceanogr.* **15,** 419–426.

Jeffries, H. P. 1972. Fatty-acid ecology of a tidal marsh. *Limnol. Oceanogr.* **17,** 433–440.

Jeffries, H. P. 1975. Diets of juvenile Atlantic menhaden *(Brevoortia tyrannus)* in three estuarine habitats as determined from fatty acid composition of gut contents. *J. Fish. Res. Bd. Canada,* **32,** 587–592.

Jensen, S. and A. Jernelov. 1967. Biosyntes av metylkvicksilver. *Biocid information.* **10,** 3–4.

Jerlov, N. G. 1968. *Optical Oceanography.* Elsevier, New York. 194 pp.

Jerlov, N. G. 1976. *Marine Optics,* Elsevier Scientific Publ. pp. 231.

Jitts, H. R. 1963. The simulated *in situ* measurement of oceanic primary production. *Aust. J. Mar. Freshw. Res.* **14,** 139–147.

Johannes, R. E. 1964a. Phosphorus excretion as related to body size in marine animals: microzooplankton and nutrient regeneration. *Science,* **146,** 923–924.

Johannes, R. E. 1964b. Uptake and release of phosphorus by a benthic marine amphipod. *Limnol. Oceanogr.* **9,** 235–242.

Johannes, R. E. 1965. Influence of marine protozoa on nutrient regeneration. *Limnol. Oceanogr.* **10,** 434–442.

Johannes, R. E. 1968. Nutrient regeneration in lakes and oceans. In *Advances in Microbiology of the Sea,* Eds. M. R. Droop and E. J. Ferguson Wood, Academic Press, New York, pp. 203–213.

Johannes, R. E. and K. L. Webb. 1965. Release of dissolved amino acids by marine zooplankton. *Science,* **150,** 76–77.

Johannes, R. E. and M. Satomi. 1967. Measuring organic matter retained by aquatic invertebrates. *J. Fish. Res. Bd. Canada,* **24,** 2467–2471.

Johannes, R. E., S. L. Coles and N. T. Kuenzel. 1970. The role of zooplankton in the nutrition of some scleractinian corals. *Limnol. Oceanogr.* **15,** 579–586.

Johannes, R. E. and K. L. Webb. 1970. Release of dissolved organic compounds by marine and fresh water invertebrates. In *Organic Matter in Natural Waters,* Ed. D. W. Hood, University of Alaska, pp. 257–273.

Johnson, M. G. 1974. Production and productivity. In *The Benthos of Lakes,* Ed. R. O. Brinkhurst, Macmillan Press, London, pp. 46–64.

Johnson, M. G. and R. O. Brinkhurst. 1971a. Production of benthic macroinvertebrates of Bay of Quinte and Lake Ontario. *J. Fish. Res. Bd. Canada,* **28,** 1699–1714.

Johnson, M. G. and R. O. Brinkhurst. 1971b. Benthic community metabolism in Bay of Quinte and Lake Ontario. *J. Fish. Res. Bd. Canada,* **28,** 1715–1725.

Johnson, P. W. and J. McN. Sieburth. 1979. Chroococoid cyanobacteria in the sea: A ubiquitous and diverse phototrophic biomass. *Limnol. Oceanogr.* **24,** 928–935.

Johnson, R. G. 1972. Conceptual models of benthic communities. In *Models in Paleobiology.* Ed. T. J. M. Schopf, Freeman & Cooper, San Francisco, pp. 149–150.

Johnson, R. G. 1974. Particulate matter at the sediment-water interface in coastal environments. *J. Mar. Res.* **32,** 313–330.

Johnston, R. 1962. An equation for the depth distribution of deep-sea zooplankton and fishes. *Rapp. Proc.-Verb. Cons. int. Explor. Mer,* **153,** 217–219.

Johnston, R. 1963. Seawater, the natural medium of phytoplankton. I. General features. *J. Mar. Biol. Ass. U.K.* **43,** 427–456.

Joint, I. R. 1978. Microbial production of an estuarine mudflat. *Est. and Coastal Mar. Sci.* **7,** 185–195.

Jones, G. E. 1963. Suppression of bacterial growth by sea water. In *Symposium on Marine Microbiology,* Ed. C. H. Oppenheimer, C. C. Thomas, Springfield, Illinois, pp. 572–579.

Jones, G. E. 1964. Effect of chelating agents on the growth of *Escherichia coli* in sea water. *J. Bacteriol.* **87,** 483–499.

Jones, G. E. 1967. Precipitates from autoclaved sea water. *Limnol. Oceanogr.* **13,** 165–167.

Jones, G. E. 1971. The fate of freshwater bacteria in the sea. *Developments in Indust. Microbiol.* **12,** 141–151.

Jones, H. P., J. Y. Monnat, C. J. Cadbury and T. J. Stowe. 1978. Birds oiled during the *Amoco Cadiz* incident—an interim report. *Mar. Pollut. Bull.* **9,** 307–310.

Jones, J. G. and B. M. Simon. 1975. Some observations on the fluorometric determination of glucose in freshwater. *Limnol. Oceanogr.* **20,** 882–887.

Jones, K. 1974. Nitrogen fixation in a salt marsh. *J. Ecol.* **62,** 553–565.

Jones, N. S. 1950. Bottom fauna communities. *Biol. Rev.* **25,** 281–313.

Jones, N. S. 1956. The fauna and biomass of a muddy sand deposit off Port Erin, I.O.M. *J. Anim. Ecol.* **25,** 217–252.

Jones, R. 1976. Growth of fishes. In *Ecology of the Seas,* Ed. D. H. Cushing and J. J. Walsh. Blackwell Scientific (Oxford), pp. 251–279.

Jørgensen, B. B. 1977. The sulfur cycle of a coastal marine sediment (Limfjorden, Denmark). *Limnol. Oceanogr.* **22,** 814–832.

Jørgensen, B. B. and T. Fenchel. 1974. The sulfur cycle of a marine sediment model system. *Mar. Biol.* **24,** 189–201.

Jørgensen, C. B. 1966. *Biology of Suspension Feeding,* Pergamon Press, London. 357 pp.

Jørgensen, E. G. 1964. Adaptation to different light intensities in the diatom *Cyclostella memeghiniana* Küts. *Physiol. Plant.* **17,** 136–145.

Josefson, A. B. 1981. Persistence and structure of two deep macrobenthic communities in the Skagerrak (West coast of Sweden). *J. Exp. Mar. Biol. Ecol.* **50,** 63–97.

Jumars, P. A. 1975a. Environmental grain and polychaete species' diversity in a bathyal benthic community. *Mar. Biol.* **30,** 253–266.

Jumars, P. A. 1975b. Methods for measurement of community structure in deep-sea macrobenthos. *Mar. Biol.* **30,** 245–252.

Jumars, P. A. 1978. Spatial autocorrelation with R.U.M. (Remote Underwater Manipulation): Vertical and horizontal structure of a bathyal benthic community. *Deep-Sea Res.* **25,** 589–604.

Jumars, P. A. 1980. Rank correlation and concordance tests in community analysis: an inappropriate null hypothesis. *Ecology,* **61,** 1553–1554.

Jumars, P. A. and R. R. Hessler. 1976. Hadal community structure: implications from the Aleutian Trench. *J. Mar. Res.* **34,** 547–560.

Kajihara, M. 1971. Settling velocity and porosity of large suspended particles. *J. Oceanogr. Soc. Japan,* **27,** 158–162.

Kamykovski, D. 1974. Possible interactions between phytoplankton and semidiurnal internal tides. *J. Mar. Res.* **32,** 67–89.

Kane, J. E. 1967. Organic aggregates in the surface waters of the Ligurian Sea. *Limnol. Oceanogr.* **12,** 287–294.

Kanneworff, E. 1965. Life cycle, food, and growth of the amphipod *Ampelisca macrocephala* Liljeborg from the Øresund. *Ophelia,* **2,** 305–318.

Kanwisher, J. W. 1966. Photosynthesis and respiration in some seaweeds. In *Some Contemporary Studies in Marine Science,* Ed. H. Barnes, Allen & Unwin Ltd., London, pp. 407–420.

Karl, D. M., P. A. LaRock, J. M. Morse and W. Sturges. 1976. Adenosine triphosphate in the North Atlantic ocean and its relationships to the oxygen minimum. *Deep-Sea Res.* **23,** 81–88.

Kato, K. 1956. Chemical investigations on marine humus in bottom sediments. *Mem. Fac. Fisheries, Hokkaido Univ.* **4,** 91–209.

Kay, D. C. and A. E. Brafield. 1973. The energy relations of the polychaete *Neanthes* (=*Nereis*) *virens* Sars. *J. Anim. Ecol.* **42,** 673–692.

Kayama, M. 1964. Fatty acid metabolism of fishes. *Bull. Japanese Soc. Sci. Fish.* **30,** 647–659 (in Japanese).

Kepkay, P. E. and J. A. Novitsky. 1980. Microbial control of organic carbon in marine sediments: coupled chemoautotrophy and heterotrophy. *Mar. Biol.* **55,** 261–266.

Kerfoot, W. B. 1970. Bioenergetics of vertical migration. *Amer. Nat.* **104,** 529–546.

Kerfoot, W. C. 1977. Implications of copepod predation. *Limnol. Oceanogr.* **22,** 316–325.

Kerr, S. R. 1974. Theory of size distribution in ecological communities. *J. Fish. Res. Bd. Canada,* **31,** 1859–1862.

Kerr, S. R. and N. V. Martin. 1970. Trophic-dynamics of lake trout production systems. In *Marine Food Chains,* Ed. J. H. Steele, Oliver & Boyd, Edinburgh, pp. 365–376.

Kerr, S. R. and W. P. Vass. 1973. Pesticide residues in aquatic invertebrates. In *Environmental Pollution by Pesticides,* Ed. C. A. Edwards. Plenum Press (London), pp. 134–180.

Ketchum, B. H. 1939. The absorption of phosphate and nitrate by illuminated cultures of *Nitzschia closterium. Am. J. Bot.* **26,** 399–407.

Ketchum, B. H. 1967. Phytoplankton nutrients in estuaries. In *Estuaries,* Ed. G. H. Lauff, AAAS, Washington, pp. 329–335.

Kevern, N. R. 1966. Feeding rate of carp estimated by a radioisotopic method. *Trans. Amer. Fish. Soc.* **95,** 363–371.

Khmeleva, N. N. 1972. Intensity of generative growth in crustaceans. *Dokl. Akad. Nauk SSSR,* **207,** 707–710.

Khripounoff, A. and M. Sibuet. 1980. La nutrition d'echinoderms abyssaux I. Alimentation des holothuries. *Mar. Biol.* **60,** 17–26.

Kibby, H. V. 1971. Effect of temperature on the feeding behaviour of *Daphnia rosea. Limnol. Oceanogr.* **16,** 580–581.

Kiefer, D. and J. D. H. Strickland. 1970. A comparative study of photosynthesis in seawater samples incubated under two types of light attenuator. *Limnol. Oceanogr.* **15,** 408–412.

Kierstead, H. and L. B. Slobodkin. 1953. The size of water masses containing plankton blooms. *J. Mar. Res.* **12,** 141–147.

Kils, U. 1982. (unpublished). The unique position of krill in the Antarctic system. (Symposium presentation) Joint Oceanographic Assembly, Halifax, Nova Scotia, August 1982.

Kimata, M., H. Kadota, Y. Hata and T. Tajima. 1955. Studies on the marine sulfite-reducing bacteria. I. Distribution of marine sulfate-reducing bacteria in the coastal waters receiving a considerable amount of pulp-mill drainage (Japanese, English summary). *Bull. Japan Soc. Sci. Fisheries,* **21,** 102–108.

King, K. 1979. The life history and vertical distribution of the chaetognath, *Sagitta elegans,* in Dabob Bay, Washington. *J. Plank. Res.* **1,** 153–167.

Kishino, M. 1981. Energy balance of underwater radiation related to marine photosynthesis. In *Biological and Chemical Utilization of Solar Energy,* Ed. K. Shibata, A. Iwamura and A. Ikegami. Japan Scientific Societies Press, Tokyo, pp. 59–69 (in Japanese).

Kleiber, M. 1961. *The Fire of Life, an Introduction to Animal Energetics,* John Wiley & Sons, Inc., New York, N.Y. 454 pp.

Klein, G., E. Rachor and S. A. Gerlach. 1975. Dynamics and productivity of two populations of the benthic tube-dwelling amphipod *Ampelisca brevicornis* Costa in Helgoland Bight. *Ophelia,* **14,** 139–159.

Knight-Jones, E. W. and J. Moyse. 1961. Intraspecific competition in sedentary marine animals. *Symp. Soc. Exp. Biol.* **15,** 72–95.

Koblentz-Mishke, O. J., V. V. Volkovisnky and J. G. Kabanova. 1970. Plankton primary production of the world ocean. In *Scientific Exploration of the South Pacific,* Standard Book No. 309-01755-6. Nat. Acad. Sci., Wash., pp. 183–193.

Koehl, M. A. R. and J. R. Strickler. 1981. Copepod feeding currents: Food capture at low Reynolds number. *Limnol. Oceanogr.* **26,** 1062–1073.

Koeller, P. and T. R. Parsons. 1977. The growth of young salmonids *(Onchorhynchus keta)*: Controlled ecosystem pollution experiment. *Bull. Mar. Sci.* **27,** 114–118.

Kofoed, L. H. 1975. The feeding biology of *Hydrobia rentrosa* Montagu. II. Allocation of the components of the carbon-budget and the significance of the secretion of dissolved organic material. *J. Exp. Mar. Biol. Ecol.* **19,** 243–256.

Kok, B. 1960. Efficiency of photosynthesis. In *Handbuch der Pflanzenphysiologie,* Ed. W. Ruhland Springer Verlag, New York, Vol. 5, Part 1, pp. 563–633.

Kok, B. and G. Hoch. 1961. Spectral changes in photosynthesis. In *A Symposium on Light and Life,* Eds. W. D. McElroy and B. Glass, The Johns Hopkins Press, Baltimore, pp. 397–423.

Komar, P. D., A. P. Morse, L. F. Small and S. W. Fowler. 1981. An analysis of sinking rates of natural copepod and euphausiid pellets. *Limnol. Oceanogr.* **26,** 172–180.

Korinek, J. 1927. Ein Beitrag zur Mikrobiologie des Meeres *Zentralbl. Bakteriol.* **71,** 73–79.

Kranck, K. 1975. Sediment deposition from flocculated suspensions. *Sedimentology,* **22,** 111–123.

Kranck, K. 1980. Variability of particulate matter in a small coastal inlet. *Can. J. Fish. Aquat. Sci.* **37,** 1209–1215.

Krause, H. R. 1961. Einige Bemerkungen über den postmortalen Abbau von Süsswasser—Zooplankton unter laboratoriums —und Freiland bedingungen. *Arch. Hydrobiol.* **57,** 539–543.

Krause, H. R. 1962. Investigation of the decomposition of organic matter in natural waters. *FAO Fish. Biol. Rep.* No. **34,** 19 pp.

Krause, H. R., L. Mochel and M. Stegmann. 1961. Organische Sauren als geloste Intermediarprodukte des postmortalen Abbaues von Süsswasser. *Zooplankton Naturwissenschaften,* **48,** 434–435.

Kriss, A. E., M. N. Lebedeva and I. N. Mitzkevich. 1960. Microorganisms as indicators of hydrological phenomena in seas and oceans. II. Investigation of the deep circulation of the Indian Ocean using microbiological methods. *Deep-Sea Res.* **6,** 173–183.

Kroopnick P. 1974. The dissolved O_2–CO_2–^{13}C system in the eastern equatorial Pacific. *Deep-Sea Res.* **21,** 211–227.

Kuenzler, E. J. 1961a. Structure and energy flow of a mussel population in a Georgia salt marsh. *Limnol. Oceanogr.* **6,** 191–204.

Kuenzler, E. J. 1961b. Phosphorus budget of a mussel population. *Limnol. Oceanogr.* **6,** 400–415.

Kuenzler, E. J. 1965. Glucose-6-phosphate utilization by marine algae. *J. Phycol.* **1,** 156–164.

Kuenzler, E. J. 1970. Dissolved organic phosphorus excretion by marine phytoplankton. *J. Phycol.* **6,** 7–13.

Kuenzler, E. J. and B. H. Ketchum. 1962. Rate of phosphorus uptake by *Phaeodactylum tricornutum. Biol. Bull.* **123,** 134–145.

Kuenzler, E. J. and J. P. Perras. 1965. Phosphatase of marine algae. *Biol. Bull.* **128,** 271–284.

Kullenberg, G. (Ed.). 1982. *Pollution Transfer and Transport in the Sea,* Vol. II. CRC Press Inc. (Boca Raton, Florida), pp. 1–65.

Kuznetsov, S. I. 1959. *Die Rolle der Mikroorganismen im Stoffkreislauf der Seen*, VEB Deutscher Verlag der Wissenschafften, Berlin. 301 pp.

Kuznetsov, S. I. 1968. Recent studies on the role of microorganisms in the cycling of substances in lakes. *Limnol. Oceanogr.* **13,** 211–224.

Lacaze, J. C. 1974. Ecotoxicology of crude oils and the use of experimental marine ecosystems. *Mar. Poll. Bull.* **5,** 153–156.

Lackey, B. 1961. Bottom sampling and environmental niches. *Limnol. Oceanogr.* **6,** 271–279.

Laevastu, T. and F. Favorite. 1981. Holistic simulation models of shelf-seas ecosystems. In *Analysis of Marine Ecosystems,* Ed. A. R. Longhurst. Academic Press (London), pp. 701–727.

LaFond, E. C. and K. G. LaFond. 1971. Oceanography and its relation to marine organic production. In *Fertility of the Sea,* Ed. J. D. Costlow, Gordon & Breach (New York), Vol. 1, pp. 241–265.

Lalli, C. M. 1970. Structure and function of the buccal apparatus of *Clione limacina* (Phipps) with a review of feeding in gymnosomatous pteropods. *J. Exp. Mar. Biol. Ecol.* **4,** 101–118.

Lalli, C. M. 1972. Food and feeding of *Paedocliona doliiformis* Danforth, a neotonous gymnosomatous pteropod. *Biol. Bull.* **143,** 392–402.

Lalli, C. M. and R. J. Conover. 1973. Reproduction and development of *Paedoclione doliiformis*, and a comparison with *Clione limacina* (Opisthobranchia: Gymnosomata). *Mar. Biol.* **19,** 13–22.

Lam, R. K. and B. W. Frost. 1976. Model of copepod filtering response to changes in size and concentration of food. *Limnol Oceanogr.* **21,** 490–500.

Lamport, W. 1974. A method for determining food selection by zooplankton. *Limnol. Oceanogr.* **19,** 995–997.

Lance, J. 1962. Effects of water of reduced salinity on the vertical migration of zooplankton. *J. Mar. Biol. Ass. U.K.* **42,** 131–154.

Larkin, P. A. 1977. An epitaph for the concept of maximum sustained yield. *Trans. Amer. Fish. Soc.* **106,** 1–11.

Larman, V. N. and P. A. Gabbott. 1975. Settlement of cyprid larvae of *Balanus balanoides* and *Elminius modestus* induced by extracts of adult barnacles and other marine animals. *J. Mar. Biol. Ass. U.K.* **55,** 183–190.

Larsen, H. 1962. Halophilism. In *The Bacteria. A Treatise on Structure and Function. IV. The Physiology of Growth,* Eds. I. C. Gunsalus and R. Y. Stanier, Academic Press, New York, pp. 297–342.

Lasker, R. 1966. Feeding, growth, respiration and carbon utilization of a euphausiid crustacean. *J. Fish. Res. Bd. Canada,* **23,** 1291–1317.

Lasker, R. 1975. Field criteria for survival of anchovy larvae: the relation between inshore chlorophyll maximum layers and successful first feeding. *Fish. Bull.* **73,** 453–462.

Lasker, R., J. B. J. Wells and A. D. McIntyre. 1970. Growth reproduction, respiration and carbon utilization of the sand-dwelling harpacticoid copepod, *Asellopsis intermedia. J. Mar. Biol. Ass. U.K.* **50,** 147–160.

Laws, E. A. 1975. The importance of respiration losses in controlling the size distribution of marine phytoplankton. *Ecology,* **56,** 419–426.

Leach, J. H. 1970. Epibenthic algal production in an intertidal mudflat. *Limnol. Oceanogr.* **15,** 514–521.

Leatherland, T. M., J. D. Burton, F. Culkin, M. J. McCartney and R. J. Morris. 1973. Concentrations of some trace metals in pelagic organisms and of mercury in Northeast Atlantic Ocean water. *Deep-Sea Res.* **20,** 679–685.

LeBrasseur, R. J. and J. Fulton. 1967. A guide to zooplankton of the north western Pacific Ocean. *Fish. Res. Bd. Canada. Circular No. 84,* 34 pp.

LeBrasseur, R. J. and O. D. Kennedy. 1972. The fertilization of Great Central Lake. II. Zooplankton standing stock. *Fish. Bull.* **70,** 25–36.

LeBrasseur, R. J., C. D. McAllister and T. R. Parsons. 1979. Addition of nutrients to a lake leads to greatly increased catch of salmon. *Environ. Conserv.* **6,** 187–190.

Lee, H. and R. C. Swartz. 1980. Biological processes affecting the distribution of pollutants in marine sediments, Part II. Biodeposition and bioturbation, pp. 555–606. In *Contaminants and Sediments*, Vol. 2, Ed. R. A. Baker, Ann Arbor Science Publ., Ann Arbor, Michigan.

Lee, R. F., J. C. Nevenzel and G. A. Paffenhöfer. 1970. Wax esters in marine copepods. *Science,* **167,** 1510–1511.

Lee, R. F., J. Hirota and A. M. Barnett. 1971. Distribution and importance of wax esters in marine copepods and other zooplankton. *Deep-Sea Res.* **18,** 1147–1165.

Lee, R. F., J. C. Nevenzel and G. A. Paffenhöfer. 1971. Importance of wax esters and other lipids in the marine food chain: Phytoplankton and copepods. *Mar. Biol.* **9,** 99–108.

Legendre, L. 1981. Hydrodynamic control of marine phytoplankton productions: The paradox of stability. In *Ecohydrodynamics,* Ed. J. Nihoul. Elsevier, pp. 191–207.

Lehman, J. T. 1976. The filter-feeder as an optimal forager, and the predicted shapes of feeding curves. *Limnol. Oceanogr.* **21,** 501–516.

Lekan, J. F. and R. E. Wilson. 1978. Spatial variability of phytoplankton biomass in the surface waters off Long Island. *Est. Coastal Mar. Sci.* **6,** 239–251.

Leong, R. J. H. and C. P. O'Connell. 1969. A laboratory study of particulate and filter feeding of northern anchovy *(Engraulis mordas). J. Fish. Res. Bd. Canada,* **26,** 557–582.

Leppäkoski, E. 1973. Benthic recolonization of the Bornholm Basin (Southern Baltic) in 1969–71. *Thalassia Jugosl.* **7**(1), 171–179.

Lerman, A., K. L. Carder and P. R. Betzer. 1977. Elimination of five suspensoides in the oceanic water column. *Earth and Planet Sci. Letters,* **37,** 6170.

Levi, D. and T. Wyatt. 1971. On the dependence of phaeopigment abundance on grazing by herbivores. *Thal. Jugoslavica,* **7,** 181–184.

Levings, C. D. 1973. Dominance and movements of winter flounder *(Pseudopleuronectes americanus)* at St. Margaret's Bay, Nova Scotia. *Naturaliste Can.* **100,** 337–345.

Levings, C. D. 1975. Analyses of temporal variations in the structure of a shallow-water benthic community in Nova Scotia. *Int. Rev. ges. Hydrobiol.* **60,** 449–470.

Levinton, J. 1972. Spatial distribution of *Necula proxima* Say (Protobranchia): an experimental approach. *Biol. Bull.* **143,** 175–183.

Levinton, J. S. 1975. Levels of genetic polymorphism at two enzyme encoding loci in eight species of the genus *Macoma* (Mollusca: Bivalvia). *Mar. Biol.* **33,** 41–47.

Levinton, J. S. 1979. Deposit-feeders, their resources and the study of resource limitation. In *Ecological Processes in Coastal and Marine Systems,* Ed. R. J. Livingstone, Plenum Press, N.Y., pp. 117–141.

Levy, E. M. 1980. Oil pollution and seabirds: Atlantic Canada 1976–77 and some implications for northern environments. *Mar. Pollut. Bull.* **11,** 51–56.

Lewin, J. C. 1957. Silicon metabolism in diatoms. IV. Growth and frustule formation in *Navicula pelliculosa. Can. J. Microbiol.* **3,** 427–433.

Lewin, J. 1966. Silicon metabolism in diatoms. V. Germanium dioxide, a specific inhibitor of diatom growth. *Phycologia,* **6,** 1–12.

Lewin, J. C., R. A. Lewin and D. E. Philpott. 1958. Observations on *Phaeodactylum tricornutum. J. Gen. Microbiol.* **18,** 418–426.

Lewin, J. C. and R. A. Lewin. 1960. Auxotrophy and heterotrophy in marine littoral diatoms. *Can. J. Microbiol.* **6,** 127–134.

Lewis, A. G. 1967. An enrichment solution for culturing the early development stages of the planktonic marine copepod *Euchaeta japonica.* Marukawa. *Limnol. Oceanogr.* **12,** 147–148.

Lewis, A. G, R. Ramnarine and M. S. Evans. 1971. Natural chelators—an indication of activity with the calanoid copepod *Euchaeta japonica. Mar. Biol.* **11,** 1–4.

Lewis, R. W. 1969. The fatty acid composition of Arctic

marine phytoplankton and zooplankton with special reference to minor acids. *Limnol. Oceanogr.* **14**, 35–40.

Lie, U. 1969. Standing crop of benthic infauna in Puget Sound and off the coast of Washington. *J. Fish. Res. Bd. Canada,* **26,** 55–62.

Lie, U. and R. A. Evans. 1973. Long-term variability in the structure of subtidal benthic communities in Puget Sound, Washington, U.S.A. *Mar. Biol.* **21,** 122–126.

Leibig, J. 1840. *Chemistry in its Application to Agriculture and Physiology,* Taylor & Walton, London (4th ed., 1847), pp. 352.

Lightfoot, R. H., P. A. Tyler and J. D. Gage. 1979. Seasonal reproduction in deep-sea bivalves and brittle stars. *Deep-Sea Res.* **26A,** 967–973.

Lisivnenko, L. N. 1961. Plankton and the food of larval Baltic herring, in the Gulf of Riga. *Trudy N--I. Instituta Rybnogo Khoziaistva Soveta Narodnogo Khoziaistva Latviiskori SSR,* **3,** 105–138 (Fish. Res. Bd. Canada, Trans. No. 444, pp. 36, 1963).

List, R. J. 1958. *Smithsonian Meteorological Tables,* 6th revised edition, Vol. 114, Publication 4014, Smithsonian Institution, Washington, D.C. 527 pp.

Litchfield, C. D. and G. D. Floodgate. 1975. Biochemistry and microbiology of some Irish Sea sediments: II. Bacteriological analyses. *Mar. Biol.* **30,** 97–103.

Llyod, M. and R. J. Ghelardi. 1964. A table for calculating the 'equitability' component of species diversity. *J. Anim. Ecol.* **33,** 217–225.

Lloyd, M., J. H. Zar and J. R. Karr. 1968. On the calculation of information—theoretical measures of diversity. *Amer. Midland Nat.* **79,** 257–272.

Loeblich, A. R., III. 1966. Aspects of the physiology and biochemistry of the Pyrrhophyta. *Phykos. Prof. Iyengar Memorial Volume.* **5,** 216–255.

Longbottom, M. R. 1970. The distribution of *Arenicola marina* (L) with particular reference to the effects of particle size and organic matter of the sediments. *J. Exp. Mar. Biol. Ecol.* **5,** 138–157.

Longhurst, A. R., A. D. Reith, R. E. Bower and D. L. R. Seibert. 1966. A new system for the collection of multiple serial plankton samples. *Deep-Sea Res.* **13,** 213–222.

Lonsdale, P. 1977. Clustering of suspension-feeding macrobenthos near abyssal hydro-thermal vents at oceanic spreading centers. *Deep-Sea Res.* **24,** 857–863.

Lopez, G. R. and L. H. Kofoed. 1980. Epipsammic browsing and deposit-feeding in mud snails (Hydrobiidae). *J. Mar. Res.* **38,** 585–599.

Lord, J. M., G. A. Codd and M. J. Merritt. 1970. The effect of light quality on glycolate formation and excretion in algae. *Plant Physiol.* **46,** 885–856.

Lorenzen, C. J. 1963. Diurnal variation in photosynthetic activity of natural phytoplankton populations. *Limnol. Oceanogr.* **8,** 56–62.

Lorenzen, C. J. 1965. A note on the chlorophyll and phaeophytin content of the chlorophyll maximum. *Limnol. Oceanogr.* **10,** 482–483.

Lorenzen, C. J. 1966. A method for the continuous measurement of *in vivo* chlorophyll concentrations. *Deep-Sea Res.* **13,** 223–227.

Lorenzen, C. J. 1972. Extinction of light in the ocean by phytoplankton. *J. Cons. Int. Explor. Mer.* **34,** 262–267.

Lorenzen, C. J. 1976. Primary production in the sea. In *Ecology of the Seas,* Eds. D. H Cushing and J. J. Walsh, Blackwell Scientific Publications, Oxford. 467 pp.

Lui, N. S. T. and O. A. Roels. 1972. Nitrogen metabolism of aquatic organisms. II. The assimilation of nitrate, nitrite and ammonia by *Biddulphia aurita. J. Phycol.* **8,** 259–264.

MacArthur, R. H. 1955. Fluctuations of animal populations and a measure of community stability. *Ecology,* **36,** 533–536.

MacArthur, R. H. and E. O. Wilson. 1967. *The Theory of Island Biogeography.* Princeton Univ. Press, Princeton, N.J. 203 pp.

MacIsaac, J. J. and R. C. Dugdale. 1969. The kinetics of nitrate and ammonia uptake by natural populations of marine phytoplankton. *Deep-Sea Res.* **16,** 45–57.

MacIsaac, J. J. and R. C. Dugdale. 1972. Interactions of light and inorganic nitrogen in controlling nitrogen uptake in the sea. *Deep-Sea Res.* **19,** 209–232.

Mackas, D. L. and C. M. Boyd. 1979. Spectral analysis of zooplankton spatial heterogeneity. *Science,* **204,** 62–64.

MacKay, N. J., M. N. Kazacos, R. H. Williams and M. I. Leedow. 1975. Selenium and heavy metals in black marlin. *Mar. Poll. Bull.* **6,** 57–61.

MacKinnon, J. C. 1973. Analysis of energy flow and production in an unexploited marine flatfish population. *J. Fish. Res. Bd. Canada,* **30,** 1717–1728.

MacLeod, R. A. 1965. The question of the existence of specific marine bacteria. *Bact. Rev.* **29,** 9–23.

Makarova, N. P. and V. Ye Zaika. 1971. Relationship between animal growth and quantity of assimilated food. *Hydrobiol. J.* **7,** 1–8.

Malone, T. C. 1971. Diurnal rhythms in netplankton and nannoplankton assimilation ratios. *Mar. Biol.* **10,** 285–289.

Malone, T. C. 1971a. The relative importance of nannoplankton and netplankton as primary producers in the california current system. *Mar. Biol.* **10,** 285–289.

Malone, T. C. 1971b. The relative importance of nannoplankton and netplankton as primary producers in tropical oceanic and neritic phytoplankton communities. *Limnol. Oceanogr.* **16,** 633–639.

Malone, T. C. 1975. Environmental control of phytoplankton cell size. *Limnol. Oceanogr.* **20,** 490.

Maly, E. J. 1969. A laboratory study of the interaction between the predatory rotifer *Aspalanchina* and *Paramecium. Ecology,* **50,** 59–73.

Maly, E. J. 1970. The influence of predation on the adult sex ratios of two copepod species. *Limnol. Oceanogr.* **15,** 566–573.

Mann, K. H. 1965. Energy transformations by a population of fish in the River Thames, *J. Anim. Ecol.* **34,** 253–275.

Mann, K. H. 1969. The dynamics of aquatic ecosystems. *Adv. Ecol. Res.* **6,** 1–81.

Mann, K. H. 1972a. Macrophyte production and detritus food chains in coastal waters. *Mem. Ist. Ital. Idrobiol.* **29,** Suppl., 353–383.

Mann, K. H. 1972b. Ecological energetics of the sea-weed zone in a marine bay on the Atlantic coast of Canada. II. Productivity of the seaweeds. *Mar. Biol.* **14,** 199–209.

Mann, K. H. 1973. Seaweeds: their productivity and strategy for growth. *Science,* **182,** 975–981.

Mann, K. H. 1977. Destruction of kelp-beds by sea urchins: a cyclical phenomenon or irreversible degradation. *Helgolander wiss. Meeres.* **30,** 455–467.

Mann, K. H. and R. B. Clark. 1978. Long-term effects of oil spills on marine intertidal communities. *J. Fish. Res. Bd. Canada,* **35,** 791–795.

Marcotte, B. M. 1974. Two new harpacticoid copepods from the

North Adriatic and a revision of the genus *Paramphiacella. Zool. J. Lin. Soc.* **55**(1), 65–82.
Mare, M. F. 1942. A study of a marine benthic community with special references to the micro-organisms. *J. Mar. Biol. Ass. U.K.* **25**, 517–554.
Margalef, D. R. 1951. Diversidad de especies en les communidades naturales. *Publ. Inst. Biol. apl., Barcelona,* **9**, 5–27.
Margalef, D. R. 1957. La teoria de la informacion en ecologia. *Mem. Real. Acad. Ciencias y Artes de Barcelona,* **23**, 373–449. (Translation by Wendell Hall, General Systems, Yearbook, **3**, 36–71, 1965.)
Margalef, D. R. 1958. Temporal succession and spatial heterogeneity in phytoplankton. *Perceptives in Marine Biology,* Ed. A. A. Buzzati-Traverso, University California Press, Berkeley, pp. 323–349.
Margalef, D. R. 1961. Correlations entre certains caractères synthétiques des populations de phytoplankton. *Hydrobiologia,* **18**, 155–164.
Margalef, D. R. 1965. Ecological correlations and the relationship between primary productivity and community structure. In *Primary Productivity in Aquatic Environments,* Ed. C. H. Goldman, University of California Press, Berkeley. *Mem. Ist Ital. Idrobiol,* **18** Suppl., 355–364.
Margalef, R. 1969. Diversity and stability: A practical proposal and a model of interdependence. In *Diversity and Stability in Ecological Systems,* Eds. G. M. Woodwell and H. H. Smith, *Brookhaven Symp. Biol.* **22**, 25–37.
Margalef, R. 1978. Life-forms of phytoplankton as survival alternatives in an unstable environment. *Oceanologica Acta,* **1**, 493–509.
Marr, J. W. S. 1962. The natural history and geography of the Antarctic krill (*Euphausia superba* Dana). *Discovery Rep.* **32**, 33–464.
Marshall, N. 1970. Food transfer through the lower trophic levels of the benthic environment. In *Marine Food Chains,* Ed. J. H. Steele, Oliver & Boyd, Edinburgh, pp. 52–66.
Marshall, P. T. 1958. Primary production in the Arctic. *J. Cons. Int. Explor. Mer.* **23**, 173–177.
Marshall, S. M. and A. P. Orr. 1948. Further experiments on the fertilization of a sea loch (Loch Criglin). The effect of different plant nutrients on the phytoplankton. *J. Mar. Biol. Ass. U.K.* **27**, 360–379.
Marshall, S. M. and A. P. Orr. 1955. *The Biology of a Marine Copepod, Calanus finmarchicus* (Gunnerus), Oliver & Boyd, Edinburgh. 188 pp.
Martens, C. S. 1974. A method for measuring dissolved gases in pore water. *Limnol. Oceanogr.* **19**, 525–530.
Martens, C. S. and R. A. Berner. 1974. Methane production in the interstitial waters of sulfate-depleted marine sediments. *Science,* **185**, 1167–1169.
Martens, C. S. and M. B. Goldhaber. 1978. Early diagenesis in transitional sedimentary environments of the White Oak Estuary, North Carolina. *Limnol. Oceanogr.* **23**, 428–441.
Martens, C. S., G. W. Kipphut and J. V. Klump. 1980. Sediment-water chemical exchange in the coastal zone traced by *in situ* Radon-222 flux measurements. *Science,* **208**, 285–288.
Martin, J. H. 1968. Phytoplankton–zooplankton relationships in Narragansett Bay. III. Seasonal changes in zooplankton excretion rates in relation to phytoplankton abundance. *Limnol. Oceanogr.* **13**, 63–71.
Marumo, R. and O. Asaoka. 1974. Trichodesmium in the East China Sea. I. Distribution of *Trichodesmium thiebautii* Gomont during 1961–1967. *J. Oceanogr. Soc. Japan,* **30**, 298–303.
Massé, H. 1972. Quantitative investigations of sand bottom macrofauna along the Mediterranean north-west coast. *Mar. Biol.* **15**, 209–220.
Mathieson, A. C. and J. S. Prince. 1973. Ecology of *Chondrus crispus* Stackhouse. In *Chondrus crispus,* Ed. M. J. Harvey and J. McLachlan, Proc. N.S. Inst. of Sci. **27**, pp. 53–80.
Mauchline, J. 1972. The biology of bathypelagic organisms, especially Crustacea. *Deep-Sea Res.* **19**, 753–780.
Mauchline, J. 1973. The broods of British Mysidacea (Crustacea). *J. Mar. Biol. Ass. U.K.* **53**, 801–817.
Mayzaud, P. 1973. Respiration and nitrogen excretion of zooplankton. II. Studies of the metabolic characteristics of starved animals. *Mar. Biol.* **21**, 19–28.
Mayzaud, P. and J. L. M. Martin. 1975. Some aspects of the biochemical and mineral composition of marine plankton. *J. Exp. Mar. Biol. Ecol.* **17**, 297–310.
McAllister, C. D. 1970. Zooplankton rations, phytoplankton mortality and the estimation of marine production. In *Marine Food Chains,* Ed. J. H. Steele, Oliver & Boyd, Edinburgh, pp. 419–457.
McAllister, C. D., N. Shah and J. D. H. Stickland. 1974. Marine phytoplankton photosynthesis as a function of light intensity: a comparison of methods. *J. Fish. Res. Bd. Canada,* **21**, 159–181.
McAllister, C. D., R. J. LeBrasseur and T. R. Parsons. 1972. Stability of enriched aquatic ecosystems. *Science,* **175**, 562–564.
McCaffrey, R. J., A. C. Myers, E. Darey, G. Morrison, M. Bender, N. Luedtke, D. Cullen, P. Froelich and G. Klinkhammer. 1980. The relation between pore water chemistry and benthic fluxes of nutrients and manganese in Narrgansett Bay, Rhode Island. *Limnol. Oceanogr.* **25**, 31–44.
McCarthy, J. J. 1970. A urease method for urea in seawater. *Limnol. Oceanogr.* **15**, 309–313.
McCarthy, J. J. and R. W. Eppley. 1972. A comparison of chemical, isotopic, and enzymatic methods for measuring nitrogen assimilation of marine phytoplankton. *Limnol. Oceanogr.* **17**, 371–382.
McCave, I. N. 1975. Vertical flux of particles in the ocean. *Deep-Sea Res.* **22**, 491–502.
McEachran, J. D., D. F. Boesch and J. A. Musick. 1976. Food division within two sympatric species—pairs of skates (Pisces: Rajidae). *Mar. Biol.* **35**, 301–317.
McGowan, J. A. 1971. Oceanic biogeography of the Pacific. In *The Micropaleontology of Oceans,* Eds. B. M. Funnell and W. R. Riedel, Cambridge University Press, pp. 3–74.
McIntyre, A. and A. W. H. Bé. 1967. Modern Coccolithophoridae of the Atlantic Ocean. 1. Placoliths and cyrtoliths. *Deep-Sea Res.* **14**, 561–597.
McIntyre, A. D. 1961. Quantitative differences in the fauna of boreal mud associations. *J. Mar. Biol. Ass. U.K.* **41**, 599–616.
McIntyre, A. D. 1969. Ecology of marine meiobenthos. *Biol. Rev.* **44**, 245–290.
McIntyre, A. D., A. L. S. Munro and J. H. Steele. 1970. Energy flow in a sand ecosystem. In *Marine Food Chains,* Ed. J. H. Steele, Oliver & Boyd, Edinburgh, pp. 19–31.
McLaren, I. A. 1963. Effects of temperature on growth of

zooplankton and the adaptive value of vertical migration. *J. Fish. Res. Bd. Canada,* **20,** 685–727.

McLaren, I. A. 1965. Some relationships between temperature and egg size, body size, development rate and fecundity of the copepod *Pseudocalanus. Limnol. Oceanogr.* **10,** 528–538.

McLaren, I. A. 1966a. Predicting development rate of copepod eggs. *Biol. Bull.* **131,** 457–469.

McLaren, I. A. 1966b. Adaptive significance of large size and long life of the chaetognath, *Sagitta elegans* in the Arctic. *Ecology,* **47,** 852–855.

McLaren, I. A. 1974. Demographic strategy of vertical migration by a marine copepod. *Amer. Nat.* **108,** 91–102.

McLaughlin, J. J. A. and P. A. Zahl. 1966. Endozoic algae. In *Symbiosis,* Ed. S. M. Henry, Academic Press, New York, pp. 257–297.

McNaught, D. C. and A. D. Hasler. 1964. Rate of movement of populations of *Daphnia* in relation to changes in light intensity. *J. Fish. Res. Bd. Canada,* **21,** 291–318.

Meadows, P. S. and J. I. Campbell. 1972. Habitat selection by aquatic invertebrates. In *Advances in Marine Biology,* Eds. F. S. Russell and M. Yonge, Academic Press, London and New York, 271–382.

Meland, S. M. 1962. Marine alginate-decomposing bacteria from north Norway. *Nytt Mag. Bot.* **10,** 53–80.

Menge, B. A. 1972. Foraging strategy of a starfish in relation to actual prey availability and environmental predictability. *Ecol. Monogr.* **42,** 25–50.

Menshutkin, V. V., M. E. Vinogradov and E. A. Shushkina. 1974. Mathematical model of the pelagic ecosystem in the Sea of Japan. *Oceanology,* **14,** 717–723 (English).

Menzel, D. W. 1964. The distribution of dissolved organic carbon in the Western Indian Ocean. *Deep-Sea Res.* **11,** 757–765.

Menzel, D. W. 1966. Bubbling of sea water and the production of organic particles: a re-evaluation. *Deep-Sea Res.* **13,** 963–966.

Menzel, D. W. 1967. Particulate organic carbon in the deep sea. *Deep-Sea Res.* **14,** 229–238.

Menzel, D. W. 1974. Primary productivity, dissolved and particulate organic matter and the sites of oxidation of organic matter. In *The Sea,* Ed. E. Goldberg, Interscience, New York, Vol. 5, pp. 659–678.

Menzel, D. W. and J. H. Ryther, 1960. The annual cycle of primary production in the Sargasso Sea off Bermuda. *Deep-Sea Res.* **6,** 351–367.

Menzel, D. W. and J. H. Ryther. 1961. Nutrients limiting the production of phytoplankton in the Sargasso Sea, with special reference to iron. *Deep-Sea Res.* **7,** 276–281.

Menzel, D. W., E. M. Hulbert and J. H. Ryther. 1963. The effects of enriching Sargasso Sea water on the production and species composition of the phytoplankton. *Deep-Sea Res.* **10,** 209–219.

Menzel, D. W. and J. H. Ryther. 1964. The composition of particulate organic matter in the western North Atlantic, *Limnol. Oceanogr.* **9,** 179–186.

Menzel, D. W. and J. J. Goering. 1966. The distribution of organic detritus in the ocean. *Limnol. Oceanogr.* **11,** 333–337.

Menzel, D. W., J. Anderson and A. Radtke. 1970. Marine phytoplankton vary in their response to chlorinated hydrocarbons. *Science,* **167,** 1724–1726.

Menzie, C. A. 1980. A note on the Hynes method of estimating secondary production. *Limnol. Oceanogr.* **25,** 770–773.

Menzies, R. J., R. Y. George and G. T. Rowe. 1973. *Abyssal Environment and Ecology of the World Oceans,* Wiley–Interscience, New York. 488 pp.

Merriman, D. 1965. Edward Forbes-Manxman. *Prog. Oceanogr.* **3,** 191–206.

Miller, D., C. M. Brown, T. H. Parson and S. O. Stanley. 1979. Some biologically important low molecular weight organic acids in the sediments of Loch Eil. *Mar. Biol.* **50,** 375–382.

Miller, R. J. and K. H. Mann. 1973. Ecological energetics of the seaweed zone in a marine bay on the Atlantic coast of Canada. III., Energy transformations by sea urchins. *Mar. Biol.* **18,** 99–114.

Mills, E. L. 1967. The biology of an ampeliscid amphipod crustacean sibling species pair. *J. Fish. Res. Bd. Canada,* **24,** 305–355.

Mills, E. L. 1969. The community concept in marine zoology, with comments on continua and instability in some marine communities: a review. *J. Fish. Res. Bd. Canada,* **26,** 1415–1428.

Mills, E. L. 1973. H.M.S. *Challenger, Halifax,* and the Reverend Dr. Honeyman. *Dal. Rev.* **53,** 529–545.

Mills, E. L. 1975. Benthic organisms and the structure of marine ecosystems. *J. Fish. Res. Bd. Canada,* **32,** 1657–1663.

Mills, E. L. and R. O. Fournier. 1979. Fish production and the marine ecosystem of the Scotian Shelf, Eastern Canada. *Mar. Biol.* **54,** 101–108.

Mills, E. L., K. Pittman and F. C. Tan. 1982. Food web structure on the Scotian Shelf, Eastern Canada, a study using ^{14}C as a food-chain tracer. *I.C.E.S. Symp. on Biological Productivity of Continental Shelves in the Temperate Zone of the North Atlantic.* Kiel, 1981.

Minderman, G. 1968. Addition, decomposition and accumulation of organic matter in forests. *J. Ecol.* **56,** 335–362.

Mishima, J. and E. P. Odum. 1963. Excretion rate of Zn^{65} by *Littorina irrorata* in relation to temperature and body size. *Limnol. Oceanogr.* **8,** 39–44.

Mitchell, C. L. 1979. Bioeconomics of commercial fish management. *J. Fish Res. Bd. Canada,* **36,** 699–704.

Møhlenberg, F. and H. U. Riisgård. 1978. Efficiency of particle retention in 13 species of suspension feeding bivalves. *Ophelia,* **17,** 239–246.

Møhlenberg, F. and H. U. Riisgård. 1979. Filtration rate, using a new indirect technique, in thirteen species of suspension-feeding bi-valves. *Mar. Biol.* **54,** 143–147.

Moore, H. B. 1938. Algal production and the food requirements of a limpet. *Proc. Malacol. Soc.* **23,** 117–118.

Moore, J. W. 1976. The proximate and fatty acid composition of some estuarine crustaceans. *Estuar. Coast. Mar. Sci.* **4,** 215–224.

Moore, L. R. 1969. Geomicrobiology and geomicrobiological attack on sedimented organic matter. In *Organic Geochemistry,* Eds. G. Eglinton and M. T. J. Murphy, Springer-Verlag, New York, pp. 265–303.

Moore, M. M., D. R. Livingstone, P. Donkin, B. L. Bayne, J. Widdows and D. M. Lowe. 1980. Mixed function oxygenases and xenobiotic detoxication/toxication systems in bivalve molluscs. *Helgoländer Meeresunters.* **33,** 278–291.

Mootz, C. A. and C. E. Epifanio. 1974. An energy budget for *Menippe mercenaria* larvae fed *Artemia* nauplii. *Biol. Bull.* **146,** 44–55.

Morcos, A. S. 1973. A table for the ionic composition of sea water based on 1967 atomic weights. *J. Cons. Int. Explor. Mer,* **35**, 94–95.

Morel, A. 1978. Available, usable, and stored radiant energy in relation to marine photosynthesis. *Deep-Sea Res.* **25**, 673–688.

Morel, A. and R. C. Smith. 1974. Relation between total quanta and total energy for aquatic photosynthesis. *Limnol. Oceanogr.* **19**, 591–600.

Morgan, E. 1970. The effect of environmental factors on the distribution of the amphipod *Pectenogammarus planicrurus*, with particular reference to grain size. *J. Mar. Biol. Ass. U.K.* **50**, 769–785.

Morris, R. J. 1972. The occurrence of wax esters in crustaceans from the Northeast Atlantic Ocean. *Mar. Biol.* **16**, 102–107.

Morse, J. W. 1974. Calculation of diffusive fluxes across the sediment-water interface. *J. Geophys. Res.* **79**, 5045–5048.

Moshkina, L. V. 1961. Photosynthesis by dinoflagellatae of the Black Sea. *Fiziologiya Rastenii,* **8**, 172–177 (in Russian). *Plant Physiol.* **8**, 129–132. (English translation.)

Moss, B. 1968. The chlorophyll *a* content of some benthic algal communities. *Arch. Hydrobiol.* **65**, 51–62.

Mullin, M. M. 1963. Some factors affecting the feeding of marine copepods of the genus *Calanus. Limnol. Oceanogr.* **8**, 239–250.

Mullin, M. M. 1965. Size fractionation of particulate organic carbon in the surface waters of the western Indian Ocean. *Limnol. Oceanogr.* **10**, 459–462.

Mullin, M. M. and E. R. Brooks. 1967. Laboratory culture, growth rate, and feeding behavior of a planktonic marine copepod. *Limnol. Oceanogr.* **12**, 657–666.

Mullin, M. M. and E. R. Brooks. 1970. Growth and metabolism of two planktonic marine copepods as influenced by temperature and type of food. In *Marine Food Chains,* Ed. J. H. Steele, Oliver & Boyd, Edinburgh, pp. 74–95.

Mullin, M. M. and P. M. Evans. 1974. The use of a deep tank in plankton ecology. II. Efficiency of a plankton food chain. *Limnol. Oceanogr.* **19**, 902–911.

Mullin, M. M., E. F. Stewart and F. J. Fuglister. 1975. Ingestion by planktonic grazers as a function of concentration of food. *Limnol. Oceanogr.* **20**, 259–262.

Murray, J. 1895. General observations on the distribution of marine organisms. *Rept. Sci. Res. Voy. H.M.S. 'Challenger', Summary Sci. Res.* **2**, 1431–1462.

Murray, J. W. and V. Grundmanis. 1980. Oxygen consumption in pelagic marine sediments. *Science,* **209**, 1527–1530.

Murray, S. P. 1970. Settling velocities and vertical diffusion of particles in turbulent water. *J. Geophys. Res.* **75**, 1647–1654.

Muus, B. 1967. The fauna of Danish estuaries and lagoons. *Meddr. Danm. Fisk. -og. Havunders., N.S.,* **5**, 1–316.

Myers, J. 1953. In *The Metabolism of Algae,* by G. E. Fogg, Methuen, London. 149 pp.

Myers, A. A., T. Southgate and T. F. Cross. 1980. Distinguishing the effects of oil pollution from natural cyclical phenomena on the biota of Bantry Bay, Ireland. *Mar. Poll. Bull.* **11**, 204–207.

Myklestad, S. 1974. Production of carbohydrates by marine planktonic diatoms. I. Comparison of nine different species in culture. *J. Exp. Mar. Biol. Ecol.* **15**, 261–274.

Myklestad, S., A. Haug and B. Larsen. 1972. Production of carbohydrates by the marine diatom *Chaetoceros affins* var. *willei* (Gran) Hustedt. II. Preliminary investigation of the extracellular polysaccharide. *J. Exp. Mar. Biol. Ecol.* **9**, 137–144.

Mysak, L. A., W. H. Hsieh and T. R. Parsons. 1982. On the relationship between interannual baroclinic waves and fish populations in the Northeast Pacific. *Biol. Ocean.* **3**, 63–103.

Nakajima, K. and S. Nishizawa. 1968. Seasonal cycles of chlorophyll and seston in the surface water of the Tsugaru Strait area. *Records Oceanogr. Works, Japan,* **9**, 219–246.

Nakajima, K. and S. Nishizawa. 1972. Exponential decrease in particulate carbon concentration in a limited depth interval in the surface layer of the Bering Sea. In *Biological Oceanography of the Northern North Pacific Ocean,* Ed. A. Y. Takenouti *et al.,* Idemitsu Shoten, Tokyo, Japan, pp. 495–505.

Nakanishi, M. and M. Monsi. 1965. Effect of variation in salinity on photosynthesis of phytoplankton growing in estuaries. *J. Fac. Sci., Univ. Tokyo, Sec. III. Botany,* **9**, 19–42.

Nassogne, A. 1970. Influence of food organisms on the development and culture of pelagic copepods. *Helgoländer wiss. Meeresunters.* **20**, 333–345.

Nees, J. C. 1949. A Contribution to Aquatic Population Dynamics, Ph.D. Thesis, University of Wisconsin, Madison.

Nemoto, T. 1967. Feeding pattern of euphausiids and differentiations in their body characterisitcs. *Inf. Bull. Plankton. Japan* (Comm. No. Dr. Y. Matsue), pp. 157–171.

Nemoto, T. and K. Ishikawa. 1969. Organic particulate and aggregate matters stained by histological reagents in the East China Sea. *J. Oceanogr. Soc. Japan,* **25**, 281–290.

Nemoto, T., K. Kamada and K. Hara. 1972. Fecundity of a euphausiid crustacean, *Nematoscelis difficilis,* in the North Pacific Ocean. *Mar. Biol.* **14**, 41–47.

Nesis, K. N. 1965. Biocoenoses and biomass of benthos of the Newfoundland—Labrador region. *Trudy VNIRO,* **57**, 453–489 (*Fish. Res. Bd. Canada Trans.* No. 2951).

Newell, B. S. 1967. The determination of ammonia in seawater. *J. Mar. Biol. Ass. U.K.* **47**, 271–280.

Newell, G. E. and R. C. Newell. 1963. *Marine Plankton,* Hutchinson Educational Ltd., London (Revised edition). 1966. pp. 221.

Newell, R. 1965. The role of detritus in the nutrition of two marine deposit feeders, the prosobranch *Hydrobia ulvae* and the bivalve *Macoma balthica. Proc. Zool. Soc. London,* **144**, 25–45.

Newell, R. C. 1970. *Biology of Intertidal Animals,* Logos Press, London. 555 pp.

Newell, R. C. and A. Roy. 1973. A statistical model relating the oxygen consumption of a mollusk (*Littorina littorea*) to activity, body size, and environmental conditions. *Physiol. Zool.* **46**, 253–275.

Newhouse, J., M. S. Doty and R. T. Tsuda. 1967. Some diurnal features of a neritic surface plankton population. *Limnol. Oceanogr.* **12**, 207–212.

Newmann, G. G. and D. E. Pollock. 1974. Growth of the rock lobster *Jasus lalandii* and its relationship to benthos. *Mar. Biol.* **24**, 339–346.

Nichols, F. H. 1975. Dynamics and energetics of three deposit-feeding benthic invertebrate populations in Puget Sound, Washington, *Ecol. Monogr.* **45**, 57–82.

Nickles, J. S., R. J. Bobbie, R. F. Martz, G. A. Smith, D. C. White and N. L. Richards. 1981. Effect of silicate grain shape, structure and location on the biomass and community structure of colonizing marine microbiota. *Appl. Environ. Microbiol.* **43**, 1262–1268.

Nicolaisen, W. and E. Kanneworff. 1969. On the burrowing and feeding habits of the amphipods *Bathyporeia pilosa* Lindström and *Bathyporeia sarsi* Watkin. *Ophelia,* **6**, 231–250.

Nishizawa, S. 1969. Suspended material in the sea. II. Re-evaluation of the hypotheses. *Bull. Plankton Soc. Japan,* **16,** 1–42.

Nishizawa, S. 1971. Concentration of organic and inorganic material in the surface skin at the equator, 155°W. *Bull. Plankton Soc. Japan,* **18,** 42–44.

Nissenbaum, A., M. J. Baedecker and I. R. Kaplan. 1972. Studies on dissolved organic matter from interstitial water of a reducing marine fjord. In *Advancers in Organic Geochemistry,* 1971, Eds. H. R. v. Gaertner and H. Wehner, Pergamon Press, Oxford, pp. 427–440.

Nixon, S. W., C. A. Oviatt, J. Garber and V. Lee. 1976. Diel metabolism and nutrient dynamics in a salt marsh embayment. *Ecology,* **57,** 740–750.

Nixon, S. W., J. R. Kelly, B. N. Furnas, C. A. Oviatt and S. S. Hale. 1980. Phosphorous regeneration and the metabolism of coastal marine bottom communities. In *Marine Benthic Dynamics,* Ed. K. K. Tenore and B. Ċ. Coull. Univ. S. Carolina Press, pp. 219–242.

North, B. B. 1975. Primary amines in California coastal waters: Utilization by phytoplankton. *Limnol. Oceanogr.* **20,** 20–27.

North, B. B. and G. C. Stephens. 1971. Uptake and assimilation of amino acids by *Platymonas.* II. Increased uptake in nitrogen-deficient cells. *Biol. Bull.* **140,** 242–254.

North, B. B. and G. C. Stephens. 1972. Amino acid transport in *Nitzschia ovalis* Arnott. *J. Phycol.* **8,** 64–68.

Novitsky, J. A. and P. E. Kepkay. 1981. Patterns of microbial heterotrophy through changing environments in a marine sediment. *Mar. Ecol. Prog. Ser.* **4,** 1–7.

Nowell, A. R. M., P. A. Jumars and J. E. Eckman. 1981. Effect of biological activity on the entrainment of marine sediments. *Mar. Geol.* **42,** 133–153.

O'Connell, C. P. 1972. The interrelation of biting and filtering in the feeding activity of the northern anchovy *(Engraulis mordax). J. Fish. Res. Bd. Canada,* **29,** 285–293.

O'Connell, C. P. and J. P. Zweifel. 1972. A laboratory study of particulate and filter feeding of the Pacific mackerel *Scomber japonicus. Fish. Bull.* **70,** 973–981.

Odum, E. P. 1971. *Fundamentals of Ecology,* W. B. Saunders, Philadelphia. 202 pp.

Odum, E. P. and A. A. De La Cruz. 1963. Detritus as a major component of ecostystems. *A.I.B.S. Bull.* **13,** 39–40.

Odum, E. P. and A. A. De La Cruz. 1967. Particulate organic detritus in a Georgia salt marsh-estuarine ecosystem. In *Estuaries,* Ed. G. H. Lauff, Amer. Ass. Adv. Sci. Publ. **83,** Washington, pp. 383–388.

Odum, H. T. 1967. Biological circuits and the marine systems of Texas. In *Pollution and Marine Ecology,* Ed. T. A. Olsen and F. J. Burgess. John Wiley & Sons, New York, pp. 99–157.

Odum, W. E. 1970. Utilization of the direct grazing and plant detritus food chains by the striped mullet *Mugil cephalus.* In *Marine Food Chains,* Ed. J. H. Steele, Oliver & Boyd, Edinburgh, pp. 222–240.

Ogura, N. 1972a. Decomposition of dissolved organic matter derived from dead phytoplankton. In *Biological Oceanography of the Northern Pacific Ocean,* Ed. A. Y. Takenouti, Idemitsu Shoten (Tokyo), pp. 507–515.

Ogura, N. 1972b. Rate and extent of decomposition of dissolved organic matter in surface seawater. *Mar. Biol.* **13,** 89–93.

OhEocha, C. and M. Raftery, 1959. Phycoerythrins and phycocyanins of cryptomonads. *Nature,* **184,** 1049–1051.

Ohle, W. 1956. Bioactivity, production, and energy utilization of lakes. *Limnol. Oceanogr.* **1,** 139–149.

Okubo, A. 1977. Horizontal dispersion and critical scales for plankton patches. In *Spatial Pattern in Plankton Communities,* Ed. J. H. Steele. Plenum Press (New York), pp. 21–42.

Oliver, J. D. 1982. Taxonomic scheme for the identification of marine bacteria. *Deep-Sea Res.* **29,** 795–798.

Olson, J. S. 1963. Energy storage and the balance of producers and decomposers in ecological systems. *Ecology,* **44,** 322–331.

Omori, M. 1965. The distribution of zooplankton in the Bering Sea and northern North Pacific, as observed by high-speed sampling of the surface waters, with special reference to the copepods. *J. Oceanogr. Soc. Japan,* **21,** 18–27.

Omori, M. 1969. Weight and chemical composition of some important oceanic zooplankton in the North Pacific Ocean, *Mar. Biol.* **3,** 4–10.

Omori, M. and W. M. Hamner. Patchy distribution of zooplankters: Behaviour, population assessment and sampling problems. *Limnol. Oceanogr.* (in press).

Orcutt, D. M. and G. W. Patterson. 1975. Sterol fatty acid and elemental composition of diatoms grown in chemically defined media. *Comp. Biochem. Physiol.* **50B,** 579–583.

Orr, W. L., Emery, K. O. and J. R. Grady. 1958. Preservation of chlorophyll derivatives in sediments off Southern California. *Bull. Amer. Assoc. Petrol. Geol.* **42,** 925–962.

Osman, R. W. 1977. The establishment and development of a marine epifaunal community. *Ecol. Monogr.* **47,** 37–63.

Otsuki, A. and T. Hanya. 1968. On the production of dissolved nitrogen-rich organic matter. *Limnol. Oceanogr.* **13,** 183–185.

Owen, R. W. and B. Zeizschel. 1970. Phytoplankton production: seasonal changes in the oceanic eastern tropical Pacific. *Mar. Biol.* **7,** 32–36.

Paasche, E. 1962. Coccolith formation. *Nature,* **193,** 1094–1095.

Paasche, E. 1966. Action spectrum of coccolith formation. *Physiol. Plant,* **19,** 770–779.

Paasche, E. 1973. Silicon and the ecology of marine plankton diatoms. II. Silicate-uptake kinetics in five diatom species. *Mar. Biol.* **19,** 262–269.

Pace, M. L., S. Shimmel and W. M. Darley. 1979. The effect of grazing by a gastropod, *Nassarius obsoletus,* on the benthic microbial community of a salt marsh mudflat. *Est. Coastal. Mar. Sci.* **9,** 121–124.

Packard, T. T. 1971. The measurement of respiratory electron-transport activity in marine phytpolankton. *J. Mar. Res.* **29,** 235–244.

Packard, T. 1979. Half-saturation constants for nitrate reductase and nitrate translocation in marine phytoplankton. *Deep-Sea Res.* **26A,** 321–326.

Paffenhöfer, G. A. 1970. Cultivation of *Calanus helgolandicus* under controlled conditions. *Helgoländer wiss. Meeresunters,* **20,** 346–359.

Paffenhöfer, G. A. 1971. Grazing and ingestion rates of nauplii, copepodids and adults of the marine planktonic copepod *Calanus helgolandicus. Mar. Biol.* **11,** 286–298.

Paffenhöfer, G. A. and J. D. H. Strickland. 1970. A note on the feeding of *Calanus helgolandicus* on detritus. *Mar. Biol.* **5,** 97–99.

Paffenhöfer, G. A. and S. C. Knowls. 1979. Ecological implications of fecal pellet size, production and consumption by copepods. *J. Mar. Res.* **37,** 35–49.

Paine, R. T. 1965. Natural history, limiting factors and energetics of the opisthobranch *Navanax inermis. Ecology,* **46,** 603–619.

Paine, R. T. 1966. Food web complexity and species diversity. *Amer. Nat.* **100**(910), 65–75.

Paine, R. T. 1969. A note on trophic complexity and community stability. *Amer. Nat.* **103**(929). 91–93.

Paine, R. T. 1971a. A short term experimental investigation of resource partitioning in a New Zealand rocky intertidal habitat. *Ecology,* **52**(6), 1096–1106.

Paine, R. T. 1971b. Energy flow in a natural population of the herbivorous gastropod *Tegula funebralis. Limnol. Oceanogr.* **16,** 86–98.

Paine, R. T. and R. L. Vadas. 1969. The effects of grazing by sea urchins, *Strongylocentrotus spp.,* on benthic algal populations. *Limnol. Oceanogr.* **14,** 710–719.

Palmer, M. A. and B. C. Coull. 1980. The prediction of development rate and the effect of temperature for the meiobenthic copepod *Micro-arthridion littorale* (Poppe). *J. Exp. Mar. Biol. Ecol.* **48,** 7383.

Paloheimo, J. E. and L. M. Dickie. 1965. Food and growth of fishes. I. A growth curve derived from experimental data. *J. Fish. Res. Bd. Canada,* **22,** 521–542.

Paloheimo, J. E. and L. M. Dickie. 1966a. Food and growth of fishes. II. Effects of food and temperature on the relation between metabolism and body weight. *J. Fish. Res. Bd. Canada,* **23,** 869–908.

Paloheimo, J. E. and L. M. Dickie. 1966b. Food and growth of fishes. III. Relation among food, body size and growth efficiency. *J. Fish. Res. Bd. Canada,* **23,** 1209–1248.

Pamatmat, M. M. 1968. Ecology and metabolism of a benthic community on an intertidal sand flat. *Int. Rev. ges. Hydrobiol,* **53,** 211–298.

Pamatmat, M. M. 1971. Oxygen consumption by the sea bed. VI. Seasonal cycle of chemical oxidation and respiration in Puget Sound. *Int. Rev. ges. Hydrobiol.* **56,** 769–793.

Pamatmat, M. M. 1975, *In situ* metabolism of benthic communities. *Cah. Biol. Mar.* **16,** 613–633.

Pamatmat, M. M. 1979. Anaerobic heat production of bivalves (*Polymesoda caroliniana* and *Modiolus demissus*) in relation to temperature, body size and duration of anoxia. *Mar. Biol.* **53,** 223–229.

Pamatmat, M. M. and K. Banse. 1969. Oxygen consumption by the seabed. II. *In situ* measurements to a depth of 180 m. *Limnol. Oceanogr.* **14,** 250–259.

Pamatmat, M. M. and A. Bhagwat. 1973. Anaerobic metabolism in Lake Washington sediments. *Limnol. Oceanogr.* **18,** 611–627.

Pamatmat, M. M. and H. R. Skjoldal. 1974. Dehydrogenase activity and adenosine triphosphate concentration of marine sediments in Lindaspollene, Norway. *Sarsia,* **56,** 1–12.

Pamatmat, M. M., G. Graf, W. Bengtsron and C. S. Novak. 1981. Heat production, A. T. P. concentration and electron transport activity of marine sediments. *Mar. Ecol. Prog. Ser,* **4,** 135–148.

Pandian, T. J. 1975. Mechanisms of heterotrophy. In *Marine Ecology,* II, Ed. O. Kinne, Wiley-Interscience, London, New York pp. 61–249.

Park, K. 1967. Nutrient regeneration of preformed nutrients off Oregon. *Limnol. Oceanogr.* **12,** 353–357.

Parker, P. L., C. Van Baalen and L. Maurer. 1967. Fatty acids in eleven species of blue-green algae: geochemical significance. *Science,* **155,** 707–708.

Parker, R. A. 1974. Empirical functions relating metabolic processes in aquatic systems to environmental variables. *J. Fish. Res. Bd. Canada,* **31,** 1550–1552.

Parker, R. H. 1975. *The Study of Benthic Communities: a Model and a Review,* Elsevier Scientif. Publ. Com., Amsterdam. 279 pp.

Parker, R. R. 1971. Size selective predation among juvenile salmonid fishes in a British Columbia inlet. *J. Fish. Res. Bd. Canada,* **28,** 1503–1510.

Parsons, T. R. 1961. On the pigment composition of eleven species of marine phytoplankters. *J. Fish. Res. Bd. Canada,* **18,** 1017–1025.

Parsons, T. R. 1969. The use of particle size spectra in determining the structure of a plankton community. *J. Oceanogr. Soc. Japan,* **25,** 172–181.

Parsons, T. R. 1976. The structure of life in the sea. In *The Ecology of the Seas,* Eds. D. H. Cushing and J. J. Walsh, Blackwell Scientific Publications, Oxford, pp. 81–97.

Parsons, T. R. 1979. Some ecological, experimental and evolutionary aspects of the upwelling ecosystem. *South African J. Sci.* **75,** 536–540.

Parsons, T. R., K. Stephens and J. D. H. Strickland. 1961. On the chemical composition of eleven species of marine phytoplankters. *J. Fish. Res. Bd. Canada,* **18,** 1001–1016.

Parsons, T. R. and J. D. H. Strickland. 1962a. Oceanic detritus. *Science,* **136,** 313–314.

Parsons, T. R. and J. D. H. Strickland. 1962b. On the production of particulate organic carbon by heterotrophic processes in sea water. *Deep-Sea Res.* **8,** 211–222.

Parsons, T. R., R. J. LeBrasseur and J. D. Fulton. 1967. Some observations on the dependence of zooplankton grazing on the cell size and concentration of phytoplankton blooms. *J. Oceanogr. Soc. Japan,* **23,** 10–17.

Parsons, T. R., R. J. LeBrasseur, J. D. Fulton and O. D. Kennedy. 1969. Production studies in the Strait of Georgia. Part II. Secondary production under the Fraser River plume, February to May, 1967. *J. Exp. Mar. Biol. Ecol.* **3,** 39–50.

Parsons, T. R. and R. J. LeBrasseur. 1970. The availability of food to different trophic levels in the marine food chain. In *Marine Food Chains,* Ed. J. H. Steele, Oliver & Boyd, Edinburgh, pp. 325–343.

Parsons, T. R. and H. Seki. 1970. Importance and general implications of organic matter in aquatic environments. In *Organic Matter in Natural Waters,* Ed. D. W. Hood, University of Alaska, pp. 1–27.

Parsons, T. R., K. Stephens and M. Takahashi. 1972. The fertilization of Great Central Lake. I. Effect of primary production. *Fish. Bull.* **70,** 13–23.

Parsons, T. R. and M. Takahashi. 1973. Environmental control of phytoplankton cell size. *Limnol. Oceanogr.* **18,** 511–515.

Parsons, T. R., B. O. Jasson, A. R. Longhurst and G. Saetersdal (Eds.). 1978. Marine ecosystems and fisheries oceanography. *Rapp. Proc.-Verbaux Réunion,* **173,** 5–58.

Parsons, T. R., J. Stronach, G. A. Borstad, G. Louttit and R. I. Perry. 1981. Biological fronts in the Strait of Georgia, British Columbia, and their relation to recent measurements of primary productivity. *Mar. Ecol.-Prog. Ser.* **6,** 237–242.

Parsons, T. R. and P. J. Harrison. 1983. Nutrient cycling in aquatic ecosystems, marine environment. *Encyclopedia of Plant Physiology* **12D,** 77–105.

Patten, B. C. 1959. An introduction to the cybernetics of the ecosystem: the trophic-dynamic aspect. *Ecology,* **40,** 221–231.

Patten, B. C. 1961. Preliminary method for estimating stability in plankton. *Science,* **134,** 1010–1011.

Patten, B. C. 1962a. Species diversity in net phytoplankton of Raritan Bay. *J. Mar. Res.* **20,** 57–75.

Patten, B. C. 1962b. Improved method for estimating stability in plankton. *Limnol. Oceanogr.* **7,** 266–268.

Patten, B. C. 1968. Mathematical models of plankton production. *Int. Rev. ges. Hydrobiol.* **53,** 357–408.

Patton, S., P. T. Chandler, E. B. Kalan, A. R. Loeblich III, G. Fuller and A. A. Benson, 1967. Food value of red tide (*Gonyaulax polyedra*). *Science,* **158,** 789–790.

Payne, J. F. 1976. Field evaluation of benzopyrene hydroxyllase induction as a monitor for marine pollution. *Science,* **191,** 945–946.

Payne, J. F. 1977. Mixed function oxidases in marine organisms in relation to petroleum hydrocarbon metabolism and detection. *Mar. Poll. Bull.* **8,** 112–116.

Pearre, S. Jr. 1980. Feeding by Chaetognatha: The relation of prey size to predator size in several species. *Mar. Ecol.* **3,** 125–134.

Pearson, T. H. 1980. Marine pollution effects of pulp and paper industry wastes. Helgoländer Meeresunters. **33,** 340–365.

Pease, A. K. 1976. Studies of the relationship of RNA/DNA ratios and the rate of protein synthesis to growth in the oyster, *Crassostrea virginica. Fish. Mar. Ser. Tech. Rpt.* **622,** 78 pp.

Peer, D. L. 1970. Relation between biomass, productivity and loss to predators in a population of a marine benthic polychaete *Pectinaria hypoborea. J. Fish. Res. Bd. Canada,* **27,** 2143–2153.

Pentreath, R. J. 1976. Some further studies on the accumulation and retention of ^{65}Zn and ^{54}Mn by the plaice, *Pleuronectes platessa* L. *J. Exp. Mar. Biol. Ecol.* **21,** 179–189.

Percy, J. A. 1979. Seasonal changes in organic composition and caloric content of an Arctic marine amphipod, *Onisimus (=Boeckosimus) affinis* H. J. Hansen. *J. Exp. Mar. Biol. Ecol.* **40,** 183–192.

Perkin, R. G. and E. L. Lewis. 1980. The practical salinity scale 1978: Fitting the data IEEE (Inst. Electr. Electron. Eng.). *J. Oceanic. Eng.* **5,** 9–16.

Perkins, E. J. 1963. Penetration of light into littoral soils. *J. Ecol.* **51,** 687–692.

Perry, M. J. 1976. Phosphate utilization by an oceanic diatom in phosphorus limited chemostat culture and in the oligotrophic waters of the central North Pacific. *Limnol. Oceanogr.* **21,** 88–107.

Peters, R. H., and F. H. Rigler. 1973. Phosphorus release by *Daphnia. Limnol. Oceanogr.* **18,** 821–839.

Petersen, C. G. J. 1913. Valuation of the sea. II. The animal communities of the sea bottom and their importance for marine zoogeography. *Rep. Dan. Biol. Stn.* **21,** 1–44.

Peterson, B. J., R. W. Howarth, F. Lipschultz and D. Ashendorf. 1980. Salt marsh detritus: an alternative interpretation of stable carbon isotope ratios and the fate of *Spartina alterniflora. Oikos,* **34,** 173–177.

Peterson, C. H. 1975. Stability of species and of community for the benthos of two lagoons. *Ecology,* **56,** 958–965.

Peterson, C. H. 1979. Predation, competitive exclusion and diversity in the soft-sediment benthic communities of estuaries and lagoons. In *Ecological Processes in Coastal and Marine Systems,* Ed. R. J. Livingston. Plenum Press, pp. 233–264.

Peterson, R. 1975. The paradox of the plankton: An equilibrium hypotheses. *Amer. Nat.* **109,** 35–48.

Petipa, T. S. 1966. Relationship between growth, energy metabolism and ration in *Acartia clausi* Giesbr. Physiology of marine animals. Akademiya Nauk SSR. *Oceanogr. Comm.* 82–91 (translated by M. A. Paranjape, Univ. Wash.).

Petipa, T. S. and N. P. Makarova. 1969. Dependence of phytoplankton production on rhythm and rate of elimination. *Mar. Biol.* **3,** 191–195.

Petipa, T. S., E. V. Pavlova and G. N. Midonov. 1970. The food web structure, utilization and transport of energy by trophic levels in the planktonic communities. In *Marine Food Chains,* Ed. J. H. Steele, Oliver & Boyd, Edinburgh, pp. 142–167.

Phillips, D. J. H. 1977. The use of biological indicator organisms to monitor trace metal pollution in marine and estuarine environments—a review. *Environ. Pollut.* **13,** 281–317.

Pickard, G. L. 1964. *Descriptive Physical Oceanography.* Pergamon Press, Oxford. 199 pp.

Pielou, E. C. 1966. The measurement of diversity in different types of biological collections. *J. Theoret. Biol.* **13,** 131–144.

Pingree, R. D. 1978. Mixing and stabilization of phytoplankton distributions on the Northwest European Continental shelf. In *Spatial Pattern in Plankton Communities,* Ed. J. H. Steele. Plenum Press (New York), pp. 181–220.

Pingree, R. D., P. M. Holligan and G. T. Mardell. 1978. The effect of vertical stability on phytoplankton distributions in the summer on the Northwest European Shelf. *Deep-Sea Res.* **25,** 1011–1028.

Platt, T. 1969. The concept of energy efficiency in primary production. *Limnol. Oceanogr.* **14,** 653–659.

Platt, T. 1971. The annual production by phytoplankton in St. Margaret's Bay. Nova Scotia. *J. Cons. Int. Explor. Mer.* **33,** 324–334.

Platt, T. 1972. Local phytoplankton abundance and turbulence. *Deep-Sea Res.* **19,** 183–187.

Platt, T., V. M. Brawn and B. Irwin. 1969. Caloric and carbon equivalents of zooplankton biomass. *J. Fish. Res. Bd. Canada,* **26,** 2345–2349.

Platt, T., L. M. Dickie and R. W. Trites. 1970. Spatial heterogeneity of phytoplankton in a near-shore environment. *J. Fish. Res. Bd. Canada,* **27,** 1453–1473.

Platt, T., A. Prakash and B. Irwin. 1972. Phytoplankton nutrients and flushing of inlets on the coast of Nova Scotia. *Naturaliste Can.* **99,** 253–261.

Platt, T. and B. Irwin. 1973. Caloric content of phytoplanton. *Limnol. Oceanogr.* **18,** 306–310.

Platt, T., K. L. Denman and A. D. Jassby. 1975. *The Mathematical Representation and Prediction of Phytoplankton Productivity,* Tech. Rep. 523, Fish Mar. Sev., Environ. Canada. 110 pp.

Platt, T. and D. V. Subba Rao. 1975. Primary production of marine microphytes. In *Photosynthesis and Productivity in Different Environments.* Cambridge Univ. Press (U.K.), pp. 249–280.

Platt, T. and A. D. Jassby. 1976. The relationship between photosynthesis and light for natural assemblages of coastal marine phytoplankton. *J. Phycol.* **12,** 421–430.

Platt, T. and K. Denman. 1978. The structure of pelagic marine ecosystems. *Rapp. P. Réun. Comm. int. explor. Mer* **173,** 60–65.

Platt, T., K. H. Mann and R. E. Ulanowicz. 1981. *Mathematical Models in Biological Oceanography. Unesco Mono. Ocean. Method,* **7,** 156 pp.

Pomeroy, L. R. 1959. Algal productivity in salt marshes of Georgia. *Limnol. Oceanogr.* **4,** 386–397.

Pomeroy, L. R. 1970. The strategy of mineral cycling. *Ann. Rev. Ecol. Syst.* **1,** 171–190.

Pomeroy, L. R. 1974, The ocean's food web, a changing paradigm. *Bioscience,* **24,** 499–504.

Pomeroy, L. R. 1979. Secondary production mechanisms of continental shelf communities. In *Ecological Processes in Coastal and Marine Systems,* Ed. R. J. Livingston. Plenum Press, N.Y., pp. 163–186.

Pomeroy, L. R., H. M. Matthews and H. S. Min. 1963. Excretion of phosphate and soluble organic phosphorus compounds by zooplankton. *Limnol. Oceanogr.* **8,** 50–55.

Pomeroy, L. R. and R. E. Johannes. 1968. Occurrence and respiration of ultraplankton in the upper 500 metres of the ocean. *Deep-Sea Res.* **15,** 381–391.

Pomeroy, L. R. and D. Deibel. 1980. Aggregation of organic matter by pelagic tunicates. *Limnol. Oceanogr.* **25,** 643–652.

Porter, J. R. 1946. *Bacterial Chemistry and Physiology,* Chapman & Hall, London, pp. 1073.

Porter, K. G. 1973. Selective grazing and differential digestion of algae by zooplankton. *Nature,* **244,** 179–180.

Postma, H. 1967. Sediment transport and sedimentation in the estuarine environment. In *Estuaries.* Ed. G. H. Lauff, *Amer. Ass. Adv. Sci. Publ.* **83,** 158–179.

Poulet, S. A. and G. Ouellet. 1982. The role of amino acids in the chemosensory swarming and feeding of marine copepods. *J. Plank. Res.* **4,** 341–361.

Powell, E. N., M. A. Crenshaw and R. M. Rieger. 1980. Adaptations to sulfide in sulfide-system meiofauna. End products of sulfide detoxification in three turbrllarians and a gastrotrich. *Mar. Ecol. Prog. Ser.* **2,** 169–177.

Prakash, A. 1971. Terrigenous organic matter and coastal phytoplankton fertility. In *Fertility of the Sea,* Ed. J. D. Costlow, Gordon & Breach, New York, Vol. 2, pp. 351–368.

Prakash, A. and M. A. Rashid. 1968. Influence of humic substances on the growth of marine phytoplankton: dinoflagellates. *Limnol. Oceanogr.* **13,** 598–606.

Pratt, D. W. 1966. Competition between *Skeletonema costatum* and *Olithodiscus luteus* in Narragansett Bay and in culture. *Limnol. Oceanogr.* **11,** 447–455.

Price, R. and R. M. Warwick. 1980. The effect of temperature on the respiration rate of meiofauna. *Oecologia,* **44,** 145–148.

Provasoli, L. 1958. Nutrition and ecology of protozoa and algae. *Ann. Rev. Microbiol.* **12,** 279–308.

Provasoli, L., J. J. A. McLaughlin and M. R. Droop. 1957. The development of artificial media for marine algae. *Archiv. Mikrobiol.* **25,** 392–428.

Provasoli, L., K. Shiraishi and J. R. Lance 1959. Nutritional idiosyncrasies of *Artemia* and *Tigriopus* in monoxenic culture. *Ann. New York Acad. Sci.* **77,** 250–261.

Provasoli, L. and J. J. A. McLaughlin. 1963. Limited heterotrophy of some photosynthetic dinoflagellates. In *Symposium on Marine Microbiology,* Ed. C. H. Oppenheimer, C. C. Thomas, Springfield, Illinois, pp. 105–113.

Pshenin, L. N. 1963. Distribution and ecology of *Azotobacter* in the Black Sea. In *Symposium on Marine Microbiology,* Ed. C. H. Oppenheimer, C. C. Thomas, Springfield, Illinois, pp. 383–391.

Purcell, J. E. 1981a. Feeding ecology of *Rhyzophysa eysenhardti,* a siphonphore predator of fish larvae. *Limnol. Oceanogr.* **26,** 424–432.

Purcell, J. E. 1981b. Dietary composition and diel feeding patterns of epipelagic siphonophores. *Mar. Biol.* **65,** 83–90.

Pütter, A. 1909. *Die Ernahrung der Wassertiere und der Stoffhaushalt der Gewasser.* Fischer, Jena. 168 pp.

Rabinowitch, E. I. 1951. *Photosynthesis and Related Processes,* Vol. II, Part I, Interscience, New York, pp. 1211–2088.

Rabinowitch, E. and Godvindjee. 1969. *Photosynthesis.* John Wiley, 273 pp.

Rachor, E. and S. A. Gerlach. 1975. Variations in macrobenthos in the German Bight. *Symp. on the Changes in the North Sea Fish Stocks and their Causes.* I.C.E.S., No. 11, 16 pp.

Radford, P. J., I. R. Joint and A. R. Hiby. 1981. Simulation models of individual production processes. In *Analysis of Marine Ecosystems,* Ed. A. R. Longhurst. Academic Press (London), pp. 677–700.

Rassoulzadegan, F. and M. Etienne. 1981. Grazing rate of the tintinnid *Stenosemella ventricosa* (Clap. and Lachm.) Jorg on the spectrum of the naturally occurring particulate matter from a Mediterranean neritic area. *Limnol. Oceanogr.* **26,** 258–270.

Rau, G. H. and J. I. Hedges. 1979. Carbon-13 depletion in a hydro-thermal vent mussel: suggestion of a chemosynthetic food source. *Science,* **203,** 648–649.

Raymont, J. E. G. and R. J. Conover. 1961. Further investigations on the carbohydrate content of marine zooplankton. *Limnol. Oceanogr.* **6,** 154–164.

Raymont, J. E. G., R. T. Srinivasagam and J. K. B. Raymont. 1969a. Biochemical studies on marine zooplankton. VII Observations on certain deep sea zooplankton. *Int. Rev. ges. Hydrobiol.* **54,** 357–365.

Raymont, J. E. G., R. T. Srinivasagam and J. K. B. Raymont. 1969b. Biochemical studies on marine zooplankton. VI. Investigations on *Meganyctiphanes norvegica* (M. Sars). *Deep-Sea Res.* **16,** 141–156.

Raymont, J. E. G., C. F. Ferguson and J. K. B. Raymont. 1973. Biochemical studies on marine zooplankton. XI. The amino acid composition of some local species. *Suppl. Publ. Mar. Biol. Ass. India,* pp. 91–99.

Redfield, A. C. 1934. On the proportions of organic derivatives in sea water and their relation to the composition of plankton. *James Johnstone Memorial Volume (Liverpool),* pp. 176.

Redfield, A. C. 1942. The processes determining the concentration of oxygen, phosphate and other organic derivatives within the depths of the Atlantic Ocean. *Pap. Phys. Oceanogr. Met.* **9,** 1–22.

Redfield, A. C. 1955. The hydrography of the Gulf of Venezuela. Papers Marine Biol. Oceanogry. *Deep-Dea Res. Suppl.* **3,** 115–133.

Redfield, A. C. 1958. The biological control of chemical factors in the environment. *Amer. Sci.* **46,** 205–221.

Reeve, M. R. 1963. The filter-feeding of *Artemia.* I. In pure cultures of plant cells. *J. Exptl. Biol.* **40,** 195–205.

Reeve, M. R. 1964. Feeding of zooplankton, with special reference to some experiments with *Sagitta. Nature,* **201,** 211–213.

Reeve, M. R. and T. C. Cosper. 1975. Chaetognatha. In *Reproduction of Marine Invertebrates,* Ed. A. C. Giese and J. S. Pearse, Academic Press (New York), pp. 157–184.

Reeve, M. R. and M. A. Walter. 1978. Nutritional ecology of ctenophores — a review of recent research. *Adv. Mar. Biol.* **15,** 249–287.

Reeve, M. R., M. A. Walter and T. Ikeda. 1978. Laboratory studies of ingestion and food utilization in lobate and tentaculate ctenophores. *Limnol. Oceanogr.* **23,** 740–751.

Reimers, C. E. 1982. Organic matter in anoxic sediments off central Peru: relations of porosity, microbial decomposition and deformation properties. *Mar. Geol.* **46,** 175–197.

Reineck, H. E. 1968. Die Sturmflutlagen. In *Sedimentologie, Faunenzonierung und Faziesabfolge vor der Ostküste der inneren Deutschen* Bucht. Ed. H. E. Reineck, Senckenberg. leth. **49,** 270–272.

Remsen, C. C. 1971. The distribution of urea in coastal and oceanic waters. *Limnol. Oceanogr.* **16,** 732–740.

Reshkin, S. J. and G. A. Knauer. 1979. Light stimulation of phosphate uptake in natural assemblages of phytoplankton. *Limnol. Oceanogr.* **24,** 1121–1124.

Revsbech, N. P., J. Sørensen, T. H. Blackburn and J. P. Lomholt. 1980. Distribution of oxygen in marine sediments measured with microelectrodes. *Limnol. Oceanogr.* **25,** 403–411.

Rex, M. A. 1981. Community structure in the deep-sea benthos. *Ann. Rev. Ecol. Syst.* **12,** 331–353.

Rhee, G.-Y. and I. J. Gotham. 1981a. The effect of environmental factors on phytoplankton growth: Temperature and the interactions of temperature with nutrient limitation. *Limnol. Oceanogr.* **26,** 635–648.

Rhee, G.-Y. and I. J. Gotham. 1981b. The effect of environmental factors on phytoplankton growth: Light and the interactions of light with nitrate limitation. *Limnol. Oceanogr.* **26,** 649–659.

Rhoads, D. C. 1973. The influence of deposit-feeding benthos on water turbidity and nutrient recycling. *Amer. J. Sci.* **273,** 1–22.

Rhoads, D. C. 1974. Organism-sediment relations on the muddy sea floor. *Oceanogr. Mar. Biol. Ann. Rev.* **12,** 263–300.

Rhoads, D. C. and D. K. Young. 1970. The influence of deposit-feeding organisms on sediment stability and community trophic stucture. *J. Mar. Res.* **28**(2), 150–178.

Rhoads, D. C. and D. K. Young. 1971. Animal-sediment relations in Cape Cod Bay, Massachusetts. II. Reworking by *Malpedia oolitica* (Holothuroidea). *Mar. Biol.* **11,** 255–261.

Rhoads, D. C., P. L. McCall and J. Y. Yingst. 1978. Disturbance and production on the estuarine sea floor. *Amer. Sci.* **66,** 577–586.

Rice, A. L., R. G. Aldred, E. Darlington and R. A. Wild. 1982. The quantitative estimation of the deep-sea megabenthos: a new approach to an old problem. *Oceanologica Acta* **5,** 63–72.

Rice, M. A., K. Wallis and G. C. Stephens. 1980. Influx and net flux of amino acids into larval and juvenile European flat oysters, *Ostrea edulis* (L.). *J. Exp. Mar. Biol. Ecol.* **48,** 51–59.

Richards, S. W. and G. A. Riley. 1967. The benthic epifauna of Long Island Sound. *Bull. Bingham Oceanogr. Coll.* **19,** 89–135.

Richardson, P. 1976. Gulf Stream rings. *Oceanus,* **19,** 65–68.

Richardson, P., R. Armstrong and C. R. Goldman. 1970. Contomporaneous disequilibrium, a new hypothesis to explain the "paradox of the plankton", *Proc. Nat. Acad. Sci.* **67,** 1710–1714.

Richman, S. 1958. The transformation of energy by *Daphnia pulex. Ecol. Monogr.* **28,** 273–291.

Richman, S. and J. N. Rogers. 1969. The feeding of *Calanus helgolandicus* on synchronously growing populations of the marine diatom *Ditylium brightwelli. Limnol. Oceanogr.* **14,** 701–709.

Ricker, W. E. 1937. Statistical treatment of sampling processes useful in the enumeration of plankton organisms. *Arch. Hydrobiol (Plankt.),* **31,** 68–84.

Ricker, W. E. 1958. Handbook of computations for biological statistics of fish populations. *Fish. Res. Bd. Bull.* **119,** pp. 300.

Ricker, W. E. 1968. Food from the sea. In *Resources and Man,* Nat. Acad. Sci.-Nat. Res. Council, W. H. Freeman Co., San Francisco, pp. 87–108.

Ricker, W. E. 1971. Ed. *Methods for Assessment of Fish Production in Fresh Waters,* I.B.P. Handbook, **3,** Oxford and Edinburgh: Blackwell Scientific Publications. 326 pp.

Ricketts, T. R. 1966a. On the chemical composition of some unicellular algae. *Phytochem.* **5,** 67–76.

Ricketts, T. R. 1966b. The carotenoids of the phytoflagellate, *Micromonas pusilla. Phytochem.* **5,** 571–580.

Ricketts, T. R. 1967a. The pigment composition of some flagellates possessing scaly flagella. *Phytochem.* **6,** 669–676.

Ricketts, T. R. 1967b. Further investigations into the pigment composition of green flagellates possessing scaly flagella. *Phytochem.* **6,** 1375–1386.

Ricketts, T. R. 1970. The pigments of the Prasinophyceae and related organisms. *Phytochem.* **9,** 1835–1842.

Riedl, R. J., N. Huang and R. Machan. 1972. The subtidal pump, a mechanism of intertidal water exchange by wave action. *Mar. Biol.* **13**(3), 210–221.

Riedl, R. J. and R. Machan. 1972. Hydrodynamic patterns in lotic intertidal sands and their bioclimatological implications. *Mar. Biol.* **13**(3), 179–209.

Rigler, F. H. 1971. Feeding rates. Zooplankton. In *A Manual on Methods for the Assessment of Secondary Productivity in Fresh Waters,* Eds. W. T. Edmondson and G. G. Winberg, Blackwell Scientific, Edinburgh, pp. 228–256.

Riley, G. A. 1946. Factors controlling phytoplankton populations on Georges Bank. *J. Mar. Res.* **6,** 54–73.

Riley, G. A. 1947b. A theoretical analysis of the zooplankton population of Georges Bank. *J. Mar. Res.* **6,** 104–113.

Riley, G. A. 1951. Oxygen, phosphate and nitrate in the Atlantic Ocean. *Bull. Bingham Oceanogr. Coll.* **13,** 1–126.

Riley, G. A. 1956a. Oceanography of Long Island Sound, 1952–54. II. Physical Oceanography. *Bull. Bingham Oceanogr. Coll.* **15,** 15–46.

Riley, G. A. 1956b. Oceanography of Long Island Sound, 1952–54. IX. Production and utilization of organic matter. *Bull. Bingham Oceanogr. Coll.* **15,** 324–343.

Riley, G. A. 1963. Organic aggregates in sea water and the dynamics of their formation and utilization. *Limnol. Oceanogr.* **8,** 372–381.

Riley, G. A. 1970. Particulate and organic matter in sea water. *Adv. Mar. Biol.* **8,** 1–118.

Riley, G. A. 1975. Transparency-chlorophyll relations. *Limnol. Oceanogr.* **20,** 150–152.

Riley, G. A., H. Stommel and D. A. Bumpus. 1949. Quantitative ecology of the plankton of the western North Atlantic. *Bull. Bingham Oceanogr. Coll.* **12,** 1–169.

Riley, J. P. and T. R. S. Wilson. 1967. The pigments of some marine phytoplankton species. *J. Mar. Biol. Ass. U.K.* **47,** 351–362.

Riley, J. P. and D. A. Segar. 1969. The pigments of some further marine phytoplankton species. *J. Mar. Biol. Ass. U.K.* **49,** 1047–1056.

Riley, J. P. and D. A. Segar, 1970. The seasonal variation of the free and combined dissolved amino acids in the Irish Sea. *J. Mar. Biol. Ass. U.K.* **50,** 713–720.

Riley, J. P. and I. Roth. 1971. The distribution of trace elements in some species of phytoplankton grown in culture. *J. Mar. Biol. Ass. U.K.* **51,** 63–72.

Riley, J. P. and R. Chester, 1971. *Introduction to Marine Chemistry,* Academic Press, London. 465 pp.

Riley, J. P. and G. Skirrow. 1975. *Chemical Oceanography,* Vol. 1–4, Academic Press, New York.

Risebrough, R. W., D. W. Menzel, D. J. Martin and H. S. Olcott. 1967. DDT residues in Pacific sea birds: a persistent insecticide in marine food chains. *Nature,* **216,** 589–591.

Rittenberg, S. C., K. O. Emery and W. L. Orr. 1955. Regeneration of nutrients in sediments of marine basins. *Deep-Sea Res.* **3,** 23–45.

Riznyk, R. Z. and H. K. Phinney. 1972. Manometric assessment of intertidal microalgal production in two estuarine sediments. *Oecologia,* **10,** 193–203.

Robertson, A. I. 1979. The relationship between annual production: biomass ratios and lifespans for marine macrobenthos. *Oecologia,* **38,** 193–202.

Romanenko, W. I. 1964a. Potential capacity of the microflora of sludge sediments for heterotrophic assimilation of carbon dioxide and for chemosynthesis. *Mikrobiologiya.* **33,** 134–139 (in Russian).

Romanenko, W. I. 1964b. Heterotrophic assimilation of CO_2 by the aquatic microflora. *Mikrobiologiya,* **33,** 679–683 (in Russian).

Rosenberg, R. 1973. Succession in benthic macrofauna in a Swedish fjord subsequent to the closure of a sulphite pulp mill. *Oikos,* **24,** 1–16.

Rosenberg, R. 1974. Spatial dispersion of an estuarine benthic faunal community. *J. Exp. Mar. Biol. Ecol.* **15,** 69–80.

Rosenthal, H. and G. Hempel. 1970. Experimental studies in feeding and food requirements of herring larvae (*Clupea harengus* L.). In *Marine Food Chains,* Ed. J. H. Steele, Oliver & Boyd, Edinburgh, pp. 344–364.

Rosenzweig, M. L. and R. H. MacArthur. 1963. Graphical representation and stability conditions of predator-prey interactions. *Amer. Nat.* **97,** 209–223.

Rowe, G. T. 1971a. Observations on bottom currents and epibenthic populations in Hatteras submarine canyon. *Deep-Sea Res.* **18,** 569–581.

Rowe, G. T. 1971b. Benthic biomass and surface productivity. In *Fertility of the Sea,* Ed. J. D. Costlow, Gordon & Breach, New York, pp. 441–454.

Rowe, G. T., G. Keller, H. Edgerton, N. Staresinic and J. MacIlvaine. 1974a. Time-lapse photography of the biological reworking of sediments in Hudson submarine canyon. *J. Sed. Petrology,* **44**(2), 549–552.

Rowe, G. T., P. T. Palloni and S. G. Horner. 1974b. Benthic biomass estimates from the northwestern Atlantic Ocean and the Northern Gulf of Mexico. *Deep-Sea Res.* **21,** 641–650.

Rowe, G. T., C. H. Clifford, K. L. Smith, Jr. and P. C. Hamilton. 1975. Benthic nutrient regeneration and its coupling to primary productivity in coastal waters. *Nature,* **255,** 215–217.

Runge, J. A. 1980. Effects of hunger on the feeding behaviour of *Calanus pacificus. Limnol. Oceanogr.* **25,** 134–145.

Russel, F. S., A. J. Southward, G. T. Boalch and E. I. Butler. 1971. Changes in biological conditions in the English Channel off Plymouth during the last half century. *Nature,* **234,** 468–470.

Ryther, J. H. 1954. The ratio of photosynthesis to respiration in marine plankton algae and its effect upon the measurement of productivity. *Deep-Sea Res.* **21,** 134–139.

Ryther, J. H. 1956. Photosynthesis in the ocean as a function of light intensity. *Limnol. Oceanogr.* **1,** 61–70.

Ryther, J. H. 1963. IV. Biological Oceanography. 17. Geographic variations in productivity. In *The Sea,* Ed. M. N. Hill, Interscience Publishers, New York, Vol. 2, pp. 347–380.

Ryther, J. H. 1965. The measurement of primary production. *Limnol. Oceanogr.* **1,** 72–84.

Ryther, J. H. 1969. Photosynthesis and fish production in the sea. The production of organic matter and its conversion to higher forms of life vary throughout the world ocean. *Science,* **166,** 72–76.

Ryther, J. H. and D. D. Kramer. 1961. Relative iron requirement of some coastal and off-shore plankton algae. *Ecology,* **42,** 444–446.

Ryther, J. H. and R. R. L. Guillard. 1962. Studies of marine planktonic diatoms. III. Some effects of temperature on respiration of five species. *Can. J. Microbiol.* **8,** 447–453.

Ryther, J. H. and D. W. Menzel. 1965. On the production, composition, and distribution of organic matter in the West Arabian Sea. *Deep-Sea.* **12,** 199–209.

Saijo, Y. 1969. Chlorophyll pigments in the deep sea. *Bull. Japanese Soc. Fish. Oceanogr.* Special No. (Prof. Uda's Commemorative Papers), pp. 179–182.

Saijo, Y. and S. Ichimura. 1962. Some considerations on photosynthesis of phytoplankton from the point of view of productivity measurement. *J. Oceanogr. Soc. Japan, 20th Anniv. Vol.,* pp. 687–693.

Saila, S. B., R. A. Pibanowski and D. S. Vaughan. 1976. Optimum allocation strategies for sampling benthos in the New York Bight. *Estuar. Coast. Mar. Sci.* **4,** 119–128.

Sakamoto, M. 1966. The chlorophyll amount in the euphotic zone in some Japanese lakes and its significance in the photosynthetic production of phytoplankton community. *Bot. Mag., Tokyo,* **79,** 77–88.

Saks, N. M. and E. G. Kahn, 1979. Substrate competition between a salt marsh diatom and a bacterial population. *J. Phycol.* **15,** 17–21.

Sakshang, E. and O. Holm-Hansen. 1977. Chemical composition of *Skeletonema costatum* and *Pavlora (Monochrysis) lutheri* as a function of nitrate-, phosphate- and iron-limited growth. *J. Exp. Mar. Biol. Ecol.* **29,** 1–34.

Samuelsson, G. and G. Öquist. 1977. A method for studying photosynthetic capacities of unicellular algae based on *in vivo* chlorophyll fluorescence. *Physiol. Plant.* **40,** 315–319.

Sanders, H. L. 1956. Oceanography of Long Island Sound, 1952–1954. X. The biology of marine bottom communities. *Bull. Bingham Ocean. Coll.* **15,** 345–414.

Sanders, H. L. 1958. Benthic studies in Buzzards Bay. I. Animal-sediment relationships. *Limnol. Oceanogr.* **3**(3), 245–258.

Sanders, H. L. 1960. Benthic studies in Buzzards Bay. III. The structure of the soft-bottom community. *Limnol. Oceanogr.* **5,** 138–153.

Sanders, H. L. 1968. Marine benthic diversity: a comparative study. *Amer. Nat.* **102,** 243–282.

Sanders, H. L. and R. R. Hessler. 1969. Ecology of the deep-sea benthos. *Science,* **163,** 1419–1424.

Sasaki, H. and S. Nishizawa. 1981. Vertical flux profiles of particulate material in the sea off Sanriku. *Mar. Ecol. Prog. Ser.* **6,** 191–201.

Satomi, M. and L. R. Pomeroy. 1965. Respiration and phosphorus excretion in some marine populations. *Ecology,* **46,** 877–881.

Schaefer, M. B. 1965. The potential harvest of the sea. *Trans. Amer. Fish. Soc.* **94,** 123–128.

Schafer, C. T. 1971. Sampling and spatial distribution of benthonic foraminifera. *Limnol. Oceanogr.* **16,** 944–951.

Schiemer, F., A. Duncan and R. Z. Klekowski. 1980. A bioenergetic study of a benthic nematode, *Plectus palustris* de Man 1880, throughout its life cycle. II. Growth, fecundity and energy budgets at different densities of bacterial food and general ecological considerations. *Oecologia,* **44,** 205–212.

Schnack, S. B. 1979. Feeding of *Calanus helgolandicus* on phytoplankton mixtures. *Mar. Ecol. Prog. Ser.* **1,** 41–47.

Schnute, J. and D. Fournier. 1980. A new approach to length-frequency analysis: growth structure. *Can. J. Fish. Aquat. Sci.* **37,** 1337–1351.

Schwinghamer, P. 1981. Characteristic size distributions of integral benthic communities. *Can. J. Fish. Aquat. Sci.* **38,** 1255–1263.

Schwinghamer, P., F. Tan and D. C. Gordon, Jr. 1983. Stable carbon isotope studies of the Pecks Cove mudflat ecosystem in the Cumberland Basin, Bay of Fundy. *Can. J. Fish. Aquat. Sci.* **40,** (Suppl. 1), 262–272.

Seki, H. 1965a. Microbial studies on the decomposition of chitin in the marine environment. IX. Rough estimation of chitin decomposition in the ocean. *J. Oceanogr. Soc. Japan.* **21,** 253–260.

Seki, H. 1965b. Decomposition of chitin in marine sediments. *J. Oceanogr. Soc. Japan,* **21,** 261–268.

Seki, H. 1968. Relation between production and mineralization of organic matter in Aburatsubo Inlet, Japan. *J. Fish. Res. Bd. Canada,* **25,** 625–637.

Seki, H. 1972. Formation of Anoxic Zones in Seawater. In *Biological Oceanography of the Northern North Pacific Ocean,* Eds. Y. Takenouchi *et al.,* Idemitsu Shoten, Tokyo, pp. 487–493.

Seki, H., J. Skelding and T. R. Parsons. 1968. Observations on the decomposition of a marine sediment. *Limnol. Oceanogr.* **13,** 440–447.

Seki, H., K. V. Stephens and T. R. Parsons. 1969. The contribution of allochthonous bacteria and organic materials from a small river into a semi-enclosed sea. *Arch. Hydrobiol,* **66,** 37–47.

Seki, H., N. Toshisuke and H. Otobe. 1974. Petroleumlytic bacteria in different water masses of the Pacific Ocean in January, 1973. *La Mer,* **12,** 16–19.

Self, R. F. L. and P. A. Jumars. 1978. New resource axes for deposit feeders? *Mar. Res.* **36,** 627–641.

Semina, J. J. 1972. The size of phytoplankton cells in the Pacific Ocean. *Int. Revue ges. Hydrobiol.* **57,** 177–205.

Shanks, A. L. and J. D. Trent. 1980. Marine snow: sinking rates and potential role in vertical flux. *Deep-Sea Res.* **27,** 137–143.

Shannon, C. E. and W. Weaver. 1963. *The Mathematical Theory of Communication,* University of Illinois Press, Urbana, 125 pp.

Sharp, J. H. 1973. Size classes of organic carbon in seawater. *Limnol. Oceanogr.* **18,** 441–447.

Shelbourne, J. E. 1957. The feeding and condition of plaice larvae in good and bad plankton patches. *J. Mar. Biol. Ass. U.K.* **36,** 539–552.

Sheldon, R. W. and T. R. Parsons. 1967a. *A Practical Manual on the use of the Coulter Counter in Marine Science,* Coulter Electronics Sales Co., Toronto. 66 pp.

Sheldon, R. W. and T. R. Parsons. 1967b. A continuous size spectrum for particulate matter in the sea. *J. Fish. Res. Bd. Canada,* **24,** 909–915.

Sheldon, R. W., T. P. T. Evelyn and T. R. Parsons. 1967. On the occurrence and formation of small particles in sea water. *Limnol. Oceanogr.* **12,** 367–375.

Sheldon, R. W. and W. H. Sutcliffe, Jr. 1969. Retention of marine particles by screens and filters. *Limnol. Oceanogr.* **14,** 441–444.

Sheldon, R. W., A Prakash and W. H. Sutcliffe, Jr. 1972. The size distribution of particles in the ocean. *Limnol. Oceanogr.* **17**(3), 327–340.

Sheldon, R. W., W. H. Sutcliffe, Jr. and A. Prakash. 1973. The production of particles in the surface waters of the ocean with particular reference to the Sargasso Sea. *Limnol. Oceanogr.* **18,** 719–733.

Sheldon, R. W., W. H. Sutcliffe Jr. and M. A. Paranjape. 1977. Structure of pelagic food chain and relationship between plankton and fish production. *J. Fish. Res. Bd. Canada,* **34,** 2344–2353.

Sheldon, R. W., W. H. Sutcliffe Jr. and K. Drinkwater. 1982. Fish production in multispecies fisheries. *Can. Spec. Publ. Fish. Aquat. Sci.* **59,** 28–38.

Shepard, F. P. 1954. Nomeclature based on sand-silt-clay ratios. *J. Sed. Petrol.* **24,** 151–158.

Sherman, K., C. Jones, L. Sullivan, W. Smith, P. Berrien and L. Ejsymont. 1981. Congruent shifts in sand eel abundance in western and eastern North Atlantic ecosystems. *Nature,* **291,** 486–489.

Shimada, B. M. 1958. Diurnal fluctuations in photosynthetic rate and chlorophyll *a* content of phytoplankton from eastern Pacific waters. *Limnol. Oceanogr.* **3,** 336–339.

Sholkovitz, E. and A. Soutar. 1975. Changes in the composition of the bottom water of the Santa Barbara Basin: effect of turbidity currents. *Deep-Sea Res.* **22,** 13–21.

Sieburth, J. McN. 1960. Acrylic acid, an 'antibiotic' principle in *Phaeocystis* blooms in Antarctic waters. *Science,* **132,** 676–677.

Sieburth, J. McN. 1961. Antibiotic properties of acrylic acid, a factor in the gastrointestinal antibiosis of polar marine animals. *J. Bacteriol.* **82,** 72–79.

Sieburth, J. McN. 1964. Antibacterial substances produced by marine algae. In *Developments in Industrial Microbiology,* Soc. Industr. Microbiol., Washington, D.C., pp. 124–134.

Sieburth, J. McN. 1968. Observations on bacteria planktonic in Narragansett Bay, Rhode Island; A résumé. Proceedings of the U.S.-Japan seminar on marine microbiology. *Bull. Misaki Marine Biol. Inst. Kyoto Univ.* **12,** 49–64.

Sieburth, J. McN. 1969. Studies on algal substances in the sea. III. The production of extracellular organic matter by littoral marine algae. *J. Exp. Mar. Biol. Ecol.* **3,** 290–309.

Sieburth, J. McN. 1971. Distribution and activity of oceanic bacteria. *Deep-Sea Res.* **18,** 1111–1121.

Sieburth, J. M. 1979. *Sea Microbes.* Oxford University Press, New York. 491 pp.

Sieburth, J. McN. and A. Jensen. 1969. Studies on algal substances in the sea. II. The formation of Gelbstoff (humic material) by exudates of Phaeophyta. *J. Exp. Mar. Biol. Ecol.* **3,** 275–289.

Sieburth, J. McN., P. J. Willis, K. M. Johnson, C. M. Burney, D. M. Lavoie, K. R. Hinga, D. A. Caron, F. W. French, P. W. Johnson and P. Davis. 1976. Dissolved organic matter and heterotrophic microneuston in surface microlayers of the North Atlantic. *Science,* **194,** 1415–1418.

Sieburth, J. McN., V. Smetacek and J. Lenz. 1978. Pelagic ecosystem structure: Heterotrophic compartments of plankton and their relationship to plankton size fractions. *Limnol. Oceanogr.* **23,** 1256–1263.

Siegel, B. Z., S. M. Siegel and F. Thorarinsson. 1973. Icelandic Geothermal activity and the mercury of the Greenland Icecap. *Nature,* **241,** 526.

Silvert, W. L. 1981. Principles of ecosystem modelling. In *Analysis of Marine Ecosystems.* Ed. A. R. Longhurst. Academic Press, London, pp. 651–676.

Sinclair, M. 1978. Summer phytoplankton variability in the lower St. Lawrence estuary. *J. Fish. Res. Bd. Canada,* **35,** 1171–1185.

Skopintsev, B. A. 1966. Some aspects of the distribution and composition of organic matter in the waters of the ocean. *Oceanology,* **6,** 441–450. (Fish. Res. Bd. Canada, Transl. No. 930.)

Skopintsev, B. A. 1971. Recent advances in the study of organic matter in oceans. *Oceanology,* **11,** 775–789.

Slawyk, G., Y. Collos and J. C. Auclair. 1977. The use of the 13C and 15N isotopes for the simultaneous measurement of carbon and nitrogen turnover rates in marine phytoplankton. *Limnol. Oceanogr.* **22,** 925–932.

Sloan, P. R. and J. D. H. Strickland. 1966. Heterotrophy of four marine phytoplankters at low substrate concentrations. *J. Phycol.* **2,** 29–32.

Slobodkin, L. B. 1961. *Growth and Regulation of Animal Populations* (Ch. 12), Holt, Rinehart, & Winston, New York. 184 pp.

Slovacek, R. E. and P. J. Hannan. 1977. *In vivo* fluorescence determinations of phytoplankton chlorophyll *a. Limnol. Oceanogr.* **22,** 919–925.

Smayda, T. J. 1958. Biogeographical studies of marine phytoplankton. *Oikos,* **9,** 158–191.

Smayda, T. J. 1963. Succession of phytoplankton and the ocean as an holocoenotic environment. In *Symposium on Marine Microbiology,* Ed. C. H. Oppenheimer, C. C. Thomas, Springfield, Illinois, pp. 260–274.

Smayda, T. J. 1964. Enrichment experiments using the marine centric diatom *Cyclotella nana* (Clone 13-1) as an assay organism. In *Proceedings of Symposium on Experimental Marine Ecology,* Occasional Publication No. 2, Graduate School of Oceanography, University of Rhode Island, pp. 25–32.

Smayda, T. J. 1970a. The suspension and sinking of phytoplankton in the sea. *Oceanogr. Mar. Biol. Ann. Rev.* **8,** 353–414.

Smayda, T. J. 1970b. Growth potential bioassay of water masses using diatom cultures: Phosphorescent Bay (Puerto Rico) and Caribbean Waters. *Helgoländer wiss. Meeresunters.* **20,** 172–194.

Smayda, T. J. and B. J. Boleyn. 1966. Experimental observations on the flotation of marine diatoms. II. *Skeletonema costatum* and *Rhizosolenia setigera. Limnol. Oceanogr.* **11,** 18–34.

Smetacek, V., B. Von Bodungen, K. Von Brockel and B. Zeitschel. 1976. The plankton tower. II. Release of nutrients from sediments due to changes in the density of bottom water. *Mar. Biol.* **34,** 373–378.

Smetacek, V., K. von Brockel, B. Zeitzschel and W. Zenk. 1978. Sedimentation of particulate matter during a phytoplankton spring bloom in relation to the hydrographic regime. *Mar. Biol.* **47,** 211–226.

Smith, E. L. 1936. Photosynthesis in relation to light and carbon dioxide. *Proc. Nat. Acad. Science, Wash.* **22,** 504–511.

Smith, F. E. 1954. Quantitative aspects of population growth. In *Dynamics of Growth Processes,* Ed. E. Boell, Princeton University Press, Princeton, New Jersey, pp. 274–294.

Smith, F. E. and E. R. Baylor. 1953. Color responses in the Cladocera and their ecological significance. *Amer. Nat.* **87,** 49–55.

Smith, K. L., Jr. 1973. Respiration of a sublittoral community. *Ecology,* **54,** 1065–1075.

Smith, K. L., Jr. 1974. Oxygen demands of San Diego trough sediments: an *in situ* study. *Limnol. Oceanogr.* **19,** 939–944.

Smith, K. L., Jr. 1978. Benthic community respiration in the N.W. Atlantic Ocean: *in situ* measurements from 40 to 5200 meters. *Mar. Biol.* **47,** 337–348.

Smith, K. L., Jr., G. A. White, M. B. Laver and J. A. Haugsness. 1978. Nutrient exchange and oxygen consumption by deep-sea benthic communities: preliminary *in situ* measurements. *Limnol. Oceanogr.* **23,** 997–1005.

Smith, K. L., Jr., G. A. White and M. B. Laver. 1979. Oxygen uptake and nutrient exchange of sediments measured *in situ* using a free vehicle grab respirometer. *Deep-Sea Res.* **26A,** 337–346.

Smith, K. L., Jr. and M. B. Laver. 1981. Respiration of the bathypelagic fish, *Cyclothone acclimidens. Mar. Biol.* **61,** 261–266.

Smith, R. I. 1964. *Keys to Marine Invertebrates of the Woods Hole Region,* Contrib. No. 11, Systematics-Ecology Program. Mar. Biol. Lab., Woods Hole, Mass. 208 pp.

Sokolova, G. A. and G. I. Karavaiko. 1964. *Physiology and geochemical activity of thiobacilli.* Akademiya Nauk SSSR Institut Mikrobiologii, Moskva [translated from Russian, Israel Program for Scientific Translation (Jerusalem), 1968]. 283 pp.

Sokolova, M. N. 1972. Trophic structure of deep-sea macrobenthos. *Mar. Biol.* **16,** 1–12.

Solórzano, L. and J. D. H. Strickland. 1968. Polyphosphate in sea water. *Limnol. Oceanogr.* **13,** 515–518.

Sorokin, Yu. I. 1961. Heterotrophic carbon dioxide assimilation by micro-organisms. *Zhurnal Obshchei Biologii,* **22,** 265–272 (in Russian).

Sorokin, Yu. I. 1964a. On the trophic role of chemosynthesis in water bodies. *Int. Rev. ges. Hydrobiol.* **49,** 307–324.

Sorokin, Yu. I. 1964b. A quantitative study of the microflora in the central Pacific Ocean. *J. Cons. Int. Explor. Mer,* **29,** 25–40.

Sorokin, Yu. I. 1964c. On the primary production and bacterial activities in the Black Sea. *J. Cons. Int. Explor. Mer,* **29,** 41–60.

Sorokin, Yu. I. 1965. On the trophic role of chemosynthesis and bacterial biosynthesis in water bodies. *Mem. Ist. Ital. Idiobiol.* **18**(Suppl.), 187–205.

Sorokin, Yu. I. 1966. On the carbon dioxide uptake during the cell synthesis by microorganisms. *Zeitschrift Allg. Mikrobiol.* **6,** 69–73.

Sorokin, Yu. I. 1969. On the trophic role of chemosynthesis and bacterial biosynthesis in water bodies. In *Primary Productivity in Aquatic Environments.* Ed. C. R. Goldman, Mem. Ist. Ital. Idrobiol., 18 Suppl., University of California Press, Berkeley, pp. 187–205.

Sorokin. Yu. I. 1970. Determination of the activity of hetero-

trophic microflora in the ocean using C^{14} containing organic matter. *Mikrobiologiya,* **39,** 149–156 (in Russian).
Sorokin, Y. I. 1981. Microheterotrophic organisms in marine ecosystems. In *Analysis of Marine Ecosystems,* Ed. A. R. Longhurst. Academic Press (London), pp. 293–342.
Sorokin, Yu. I., T. S. Petipa and Ye. V. Pavlova. 1970. Quantitative estimate of marine bacterioplankton as a source of food. *Oceanology,* **10,** 253–260. (Trans. Scripta Technia inc. for Amer. Geophys. Union.)
Sournia, M. A. Rythme nychéméral du rapport 'intensité photosynthétique chlorophylle' dans le plancton marin. *C. R. Acad. Sci. Paris,* **265,** 1000–1003.
Sournia, M. A. 1969. Cycle annuel du phytoplancton et de la production primaire dans les mers tropicales. *Mar. Biol.* **3,** 287–303.
Sournia, A. 1974. Circadian periodicities in natural populations of marine phytoplankton. *Adv. Mar. Biol.* **12,** 325–386.
Sournia, A. 1982. Form and function in marine phytoplankton. *Biol. Rev.* **57,** 347–394.
Southward, A. J. 1974. Changes in the plankton community of the Western English Channel. *Nature,* **249,** 180–181.
Southward, A. J. and E. C. Southward. 1972. Observations on the role of dissolved organic compounds in the nutrition of benthic invertebrates II. Uptake by other animals living in the same habitat as Pogonophores, and by some littoral Polychaeta. *Sarsia,* **48,** 61–70.
Southward, A. J. and E. C. Southward. 1974. Observations on the role of dissolved organic compounds in the nutrition of benthic invertebrates. III. Uptake in relation to organic content of the habitat. *Sarsia,* **50,** 29–46.
Southward, A. J. and E. C. Southward. 1978. Recolonization of rocky shores in Cornwall after use of toxic dispersants to clean up the *Torrey Canyon* spill. *J. Fish. Res. Bd. Canada,* **35,** 682–706.
Spencer, C. P. 1954. Studies on the culture of a marine diatom. *J. Mar. Biol. Ass. U.K.* **33,** 265–290.
Spoehr, H. A. and H. W. Milner. 1949. The chemical composition of *Chlorella,* effect of environmental conditions. *Plant Physiol.* **24,** 120–149.
Spratt, T. A. B. and E. Forbes. 1847. *Travels in Lycia, Milyas, the Cibyratis,* London, 2 vols.
Starr, T. J. 1956. Relative amounts of vitamin B_{12} in detritus from oceanic and estuarine environments near Sapelo Island, Georgia. *Ecology,* **37,** 658–664.
Stavn, R. H. 1971. The horizontal-vertical distribution hypothesis: Langmuir circulation and *Daphnia* distributions. *Limnol. Oceanogr.* **16,** 453–466.
Steele, J. H. 1962. Environmental control of photosynthesis in sea. *Limnol. Oceanogr.* **7,** 137–150.
Steele, J. H. 1974. *The Structure of Marine Ecosystems,* Harvard Univ. Press, Cambridge, Mass. 128 pp.
Steele, J. H. 1976. Patchiness. In *The Ecology of the Seas,* Ed. D. H. Cushing and J. J. Walsh. Blackwell Scientific (Oxford), pp. 98–115.
Steele, J. H. 1979. Some problems in the management of marine resources. In *Applied Biology,* IV Ed. T. H. Coaker. Academic Press (New York), pp. 103–140.
Steele, J. H. and C. S. Yentsch 1960. The vertical distribution of chlorophyll. *J. Mar. Biol. Ass. U.K.* **39,** 217–226.
Steele, J. H. and I. E. Baird. 1968. Production ecology of a sandy beach. *Limnol. Oceanogr.* **13,** 14–25.
Steele, J. H., A. L. S. Munro and G. S. Giese. 1970. Environmental factors controlling the epipsammic flora on beach and sublittoral sands. *J. Mar. Biol. Ass. U.K.* **50,** 907–918.
Steele, J. H. and M. M. Mullin. 1977. Zooplankton dynamics. In *The Sea: Ideas and observations on progress in the study of the seas.* Eds. E. D. Goldberg, I. N. McCave, J. J. O'Brien and J. H. Steele. John Wiley & Sons (New York), vol. 6, pp. 857–890.
Steemann Nielsen, E. 1952. The use of radioactive carbon (C^{14}) for measuring organic production in the sea. *J. Cons. Int. Explor. Mer.* **18,** 117–140.
Steemann Nielsen, E. 1958. Experimental methods for measuring organic production in the sea. *Rapp. Proc.-Verb. Cons. Int. Explor. Mer.* **144,** 38–46.
Steemann Nielsen, E. 1961. Chlorophyll concentration and rate of photosynthesis in *Chlorella vulgaris. Physiol. Plant.* **14,** 868–876.
Steemann Nielsen, E. 1962. On the maximum quantity of plankton chlorophyll per surface unit of a lake or the sea. *Int. Rev. ges. Hydrobiol.* **47,** 333–338.
Steemann Nielsen, E. 1965. On the determination of the activity in ^{14}C ampoules for measuring primary production. *Limnol. Oceanogr.,* Suppl. **10:** R247–252.
Steemann Nielsen, E. and V. Kr. Hansen. 1959a. Light adaptation in marine phytoplankton populations and its interrelation with temperature. *Physiol. Plant.* **12,** 353–370.
Steemann Nielsen, E. and V. Kr. Hansen, 1959b. Measurements with the carbon-14 technique of the respiration rates in natural populations of phytoplankton. *Deep-Sea Res.* **5,** 222–233.
Steemann Nielsen, E. and V. Kr. Hansen. 1961. Influence of surface illumination on plankton photosynthesis in Danish waters (56°N) throughout the year. *Physiol. Plant.* **14,** 595–613.
Steemann Nielsen, E. and E. G. Jorgensen. 1962. The physiological background for using chlorophyll measurements in hydrobiology and a theory explaining daily variations in chlorophyll concentration. *Arch. Hydrobiol.* **58,** 349–357.
Steemann Nielsen, E., V. Kr. Hansen and E. G. Jorgensen. 1962. The adaptation to different light intensities in *Chlorella vulgaris* and the time dependence on transfer to a new light intensity. *Physiol. Plant.* **15,** 505–517.
Steemann Nielsen, E. and T. S. Park. 1964. On the time course in adapting to low light intensities in marine phytoplankton. *J. Cons. Int. Explor. Mer,* **29,** 19–24.
Steemann Nielsen, E. and E. G. Jørgensen. 1962. The of algae. I. General part. *Physiol. Plant.* **21,** 401–413.
Steemann Nielsen, E. and M. Willemoës. 1971. How to measure the illumination rate when investigating the rate of photosynthesis of unicellular algae under various light conditions. *Int. Rev. ges. Hydrobiol.* **56,** 541–556.
Stefánsson, U. and F. A. Richards. 1964. Distributions of dissolved oxygen, density and nutrients off the Washington and Oregon coasts. *Deep-Sea Res.* **11,** 355–380.
Stephens, G. C. 1967. Dissolved organic material as a nutritional source for marine and estuarine invertebrates. In *Estuaries,* Ed. G. H. Lauff, Am. Ass. Adv. Sci. **83,** 367–373.
Stephens, K. 1970. Automated measurement of dissolved nutrients. *Deep-Sea Res.* **17,** 393–396.
Stephens, K., R. H. Sheldon and T. R. Parsons. 1967. Seasonal variations in the availability of food for benthos in a coastal environment. *Ecology,* **48**(5), 852–855.
Steven, D. M. and R. Glombitza. 1972. Oscillatory variation of

a phytoplankton population in a tropical ocean. *Nature,* **237,** 105–107.

Stevenson, L. H. 1978. A case for bacterial dormancy in aquatic systems. *Microb. Ecol.* **4,** 127–133.

Stevenson, L. H., C. E. Millwood and B. H. Hebeler. 1974. Aerobic, heterotrophic bacterial populations in estuarine water and sediments. In *Effect of the Ocean Environment on Microbial Activities,* Eds. R. R. Cowell and R. Y. Morita, Univ. Park Press, Baltimore, pp. 268–285.

Stewart, J. E. and L. J. Marks. 1978. Distribution and abundance of hydrocarbon-utilizing bacteria in sediments of Chedabucto Bay, Nova Scotia, in 1976. *J. Fish. Res. Bd. Canada,* **35,** 581–584.

Stommel, H. 1949. Trajectories of small bodies sinking slowly through convection cells. *J. Mar. Res.* **8,** 24–29.

Strain, H. H. 1951. The pigments of algae. In *Manual of Phycology,* Chronica Botanica, Waltham, Mass., pp. 243–262.

Strain, H. H. 1958. *Chloroplast Pigments and Chromatographic Analysis,* 32nd Priestley Lecture, Pennsylvania State University Press, 180 pp.

Strain, H. H. 1966. Fat-soluble chloroplast pigments: their identification and distribution in various Australian plants. In *Biochemistry of Chloroplasts,* Ed. T. W. Goodwin, Academic Press, New York, Vol. 1, pp. 387–406.

Strathmann, R. R. 1967. Estimating the organic carbon content of phytoplankton from cell volume or plasma volume. *Limnol. Oceanogr.* **12,** 411–418.

Strickland, J. D. H. 1958. Solar radiation penetrating the ocean. A review of requirements, data and methods of measurement, with particular reference to photosynthetic productivity, *J. Fish. Res. Bd. Canada,* **15,** 453–493.

Strickland, J. D. H. 1960. Measuring the production of marine phytoplankton. *Fish. Res. Bd. Canada Bull.* **122,** 172.

Strickland, J. D. H. 1965. Production of organic matter in the primary stages of the marine food chain. In *Chemical Oceanography.* Eds. J. P. Riley and G. Skirrow, Academic Press, London, Vol. 1, pp. 477–610.

Strickland, J. D. H. 1968. A comparison of profiles of nutrient and chlorophyll concentrations taken from discrete depths and by continuous recording. *Limnol. Oceanogr.* **13,** 388–391.

Strickland, J. D. H. and K. H. Austin. 1960. On the forms, balance and cycle of phosphorous observed in the coastal and oceanic waters of the Northeastern Pacific. *J. Fish. Res. Bd. Canada,* **17,** 337–345.

Strickland, J. D. H., R. W. Eppley and B. Rojas de Mendiola. 1969. Phytoplankton populations, nutrients and photosynthesis in Peruvian coastal waters. *Bol. Inst. del Mar del Peru,* **2,** 1–45.

Strickland, J. D. H. and T. R. Parsons. 1972. A practical handbook of seawater analysis. *Fish. Res. Bd. Canada, Bull.* **167,** 311.

Strickler, J. R. and A. K. Bal. 1973. Setae of the first antennae of the copepod *Cyclops scutifer* (Sars): Their structure and importance. *Proc. Nat. Acad. Sci.* **70,** 2656–2659.

Stross, R. G., S. W. Chisholm and T. A. Downing. 1973. Causes of daily rhythms in photosynthetic rates of phytoplankton. *Biol. Bull.* **145,** 200–209.

Stull, E. A., E. de Amezaga and C. R. Goldman. 1973. The contribution of individual species of algae to primary productivity of Castle Lake, California. *Verh. Int. Verein. Limnol.* **18,** 1776–1783.

Suess, E. 1976. Nutrients near the depositional interface. In *The Benthic Boundary Layer,* Ed. I. N. McCave, Plenum Press, New York, pp. 57–79.

Suess, E. 1980. Particulate organic carbon flux in the oceans: surface productivity and oxygen utilization. *Nature,* **288,** 260–263.

Sugiura. Y. 1965. On the reserved nutrients matters. *Bull. Soc. Franco-japonaise d'Oceanographie,* **2,** 7–11.

Sullivan, B. K. 1980. *In situ* feeding behaviour of *Sagitta elegans* and *Eukrohnia hamtor* (Chaetognatha) in relation to the vertical distribution and abundance of prey at Ocean Station 'P'. *Limnol. Oceanogr.* **25,** 317–326.

Sushchenya, L. M. 1970. Food rations, metabolism and growth of crustaceans. In *Marine Food Chains,* Ed. J. H. Steele, Oliver & Boyd, Edinburgh, pp. 127–141.

Sutcliffe, W. H. Jr. 1969. Relationship between growth rate and ribonucleic acid concentration in some invertebrates. *J. Fish. Res. Bd. Canada,* **27,** 606–609.

Sutcliffe, W. H. Jr. 1973. Correlations between seasonal river discharge and local landings of American lobster (*Homarus americanus*) and Atlantic halibut (*Hippoglossus hippoglossus*) in the Gulf of St. Lawrence. *J. Fish. Res. Bd. Canada,* **30,** 856–859.

Sutcliffe, W. H. Jr., E. R. Baylor and D. W. Menzel. 1963. Sea surface chemistry and Langmuir circulation. *Deep-Sea Res.* **10,** 233–243.

Sutherland, J. P. 1977. Effect of *Schizoporella* (Ectoprocta) removal on the fouling community at Beaufort, North Carolina, U.S.A. In *Ecology of Marine Benthos,* B. C. Coull. Univ. S. Carolina Press, pp. 155–176.

Suyama, M., K. Nakajima and J. Nonaka. 1965. Studies on the protein and non-protein nitrogenous constituents of *Euphausia. Bull. Japanese Soc. Sci. Fish.* **31,** 302–306 (Japanese with English summary).

Suzuki, N. and K. Kato. 1953. Studies on suspended 'marine snow' in the sea. Part I. Sources of marine snow. *Bull. Fac. Fish., Hokkaido University,* **4,** 132–137.

Sverdrup, H. U. 1953. On conditions for the vernal blooming of phytoplankton. *J. Cons. Explor. Mer,* **18,** 287–295.

Sverdrup, H. U., M. W. Johnson and R. H. Fleming. 1946. *The Oceans, their Physics, Chemistry and General Biology,* Prentice-Hall, New York. 1087 pp.

Sysoeva, T. K. and A. A. Degtereva. 1965. The relation between the feeding of cod larvae and pelagic fry and the distribution and abundance of their principal food organisms. *ICNAF Spec. Publ.* **6,** 411–416.

Taghon, G. L. 1982. Optimal foraging by deposit-feeding invertebrates: roles of particle size and organic coating. *Oecologia,* **52,** 295–304.

Taghon, G. L., R. F. L. Self and P. A. Jumars. 1978. Predicting particle selection by deposit feeders: a model and its implications. *Limnol. Oceanogr.* **23,** 752–759.

Taghon, G. L., A. R. M. Nowell and P. A. Jumars. 1981. Induction of suspension feeding in spionid polychaetes by high particulate fluxes. *Science,* **210,** 562–564.

Taguchi, S. and K. Nakajima. 1971. Plankton and seston in the sea surface of three inlets of Japan. *Bull. Plankton Soc. Japan,* **18,** 20–36.

Taguchi, S. and B. T. Hargrave. 1978. Loss rates of suspended material sedimented in a marine bay. *J. Fish. Res. Board Can.* **35,** 1614–1620.

Takahashi, M., S. Shimura, Y. Yamaguchi and Y. Fujita. 1971. Photoinhibition of phytoplankton photosynthesis as a function of exposure time. *J. Oceanogr. Soc. Japan,* **27,** 43–50.

Takahashi, M., K. Satake and N. Nakamoto. 1972. Chlorophyll profile and photosynthetic activity in the north and equatorial Pacific Ocean. *J. Oceanogr. Soc. Japan.* **28,** 27–36.

Takahashi, M. and T. R. Parsons. 1972. The maximization of the standing stock and primary productivity of marine phytoplankton under natural conditions. *India J. Mar. Sci.* **1.**

Takahashi, M., K. Fujii and T. R. Parsons. 1973. Simulation study of phytoplankton photosynthesis and growth in the Fraser River estuary. *Mar. Biol.* **19,** 102–116.

Takahashi, M. and F. Nash. 1973. The effect of nutrient enrichment on algal photosynthesis in Great Central Lake, British Columbia, Canada. *Arch. Hydrobiol.* **71,** 166–182.

Takahashi, M. and T. Ikeda. 1975. Excretion of ammonia and inorganic phosphorus by *Euphausia pacifica* and *Metridia pacifica* at different concentrations of phytoplankton. *J. Fish. Res. Bd. Canada,* **32,** 2189–2195.

Takahashi, M. and N. Fukazawa. 1982. A mechanism of "red tide" formation. II. Effect of selective nutrient stimulation on the growth of different phytoplankton species in natural water. *Mar. Biol.* **70,** 267–273.

Talling, J. F. 1957a. The phytoplankton population as a compound photosynthetic system. *New Phytol.* **56,** 133–149.

Talling, J. F. 1957b. Photosynthetic characteristics of some freshwater plankton diatoms in relation to underwater radiation. *New Phytol.* **56,** 29–50.

Talling, J. F. 1960. Comparative laboratory and field studies of photosynthesis by a marine planktonic diatom. *Limnol. Oceanogr.* **5,** 62–77.

Talling, J. F. 1973. The application of some electrochemical methods to the measurement of photosynthesis and respiration in freshwaters. *Freshw. Biol.* **3,** 335–362.

Tamiya, H., E. Hase, K. Shibata, A Mitsuya, T. Iwamura, T. Nihei and T. Sasa. 1953. Kinetics of growth of *Chorella* with special reference to its dependence on quantity of available light and on temperature. In *Algal Culture: from Laboratory to Pilot Plant,* Ed. J. S. Burlow, Carnegie Inst. Publ. No. 600, pp. 204–232.

Tanabe, S., R. Tatsukawa, M. Kawano and H. Hidaka. 1982. Global distribution and atmospheric transport of chlorinated hydrocarbons: HCH (BHC) isomers and DDT compounds in Western Pacific, Eastern Indian and Antarctic Oceans. *J. Ocean. Soc. Japan,* **38,** 137–148.

Tanada, T. 1951. The photosynthetic efficiency of carotenoid pigments in *Navicula minima. Amer. J. Bot.* **38,** 276–283.

Taniguchi, A. 1973. Phytoplankton-zooplankton relationships in the Western Pacific Ocean and adjacent seas. *Mar. Biol.* **21,** 115–121.

Tanoue, E. and N. Handa. 1980. Vertical transport of organic materials in the Northern North Pacific as determined by sediment trap experiment. Part I. Fatty acid composition. *J. Oceanogr. Soc. Japan,* **36,** 231–245.

Taylor, L. R. 1961. Aggregation, variance and the mean. *Nature,* **189,** 732–735.

Taylor, L. R. 1980. New light on the variance/mean view of aggregation and transformation: comment. *Can. J. Fish. Aquat. Sci.* **37,** 1330–1332.

Taylor, W. R. 1964. Light and photosynthesis in intertidal diatoms. *Helgol. Wiss. Meeresunters.* **10,** 29–37.

Teal, J. M. 1962. Energy flow in the salt marsh ecosystem of Georgia. *Ecology,* **43,** 614–624.

Teal, J. M. and J. Kanwisher. 1961. Gas exchange in a Georgia salt marsh. *Limnol. Oceanogr.* **6,** 388–399.

Tenore, K. R. 1981. Organic nitrogen and caloric content of detritus I. Utilization by the deposit-feeding polychaete, *Capitella capitata. Est. Coastal and Shelf Sci.* **12,** 39–47.

Tenore, K. R. and D. L. Rice. 1980. A review of trophic factors affecting secondary production of deposit-feeders. In *Marine Benthic Dynamics,* K. R. Tenore and B. C. Coull, Univ. S. Carolina Press, pp. 325–340.

Tett, P. B. 1973. The use of log-normal statistics to describe phytoplankton population from the Firth of Lorne area. *J. Exp. Mar. Biol. Ecol.* **11,** 121–136.

The Ring Group. 1981. Gulf Stream cold-core rings: Their physics, chemistry and biology. *Science,* **212,** 1091–1100.

Theede, H., A. Ponat, K. Hiroki and C. Schlieper. 1969. Studies on the resistance of marine bottom invertebrates to oxygen deficiency and hydrogen sulphide. *Mar. Biol.* **2,** 325–337.

Thiel, H. 1975. The size structure of the deep-sea benthos. *Int. Rev. ges. Hydrobiol,* **60,** 575–606.

Thiel, H. 1981. Meiobenthos and nanobenthos of the deep sea. In *The Sea,* Vol. 8, Ed. G. T. Rowe.

Thistle, D. 1978. Harpacticoid dispersion patterns: implications for deep-sea diversity maintenance. *J. Mar. Res.* **36,** 377–397.

Thomas, W. H. 1968. Nutrient requirements and utilization: Algae. In *Metabolism,* Eds. P. L. Altman and D. S. Dittmer, Fed. Am. Soc. Exptl. Biol., Bethesda, Md., pp. 210–228.

Thomas, W. H. 1970a. On nitrogen deficiency in tropical Pacific oceanic phytoplankton: photosynthetic parameters in poor and rich water. *Limnol. Oceanogr.* **15,** 380–385.

Thomas, W. H. 1970b. Effect of ammonium and nitrate concentration on chlorophyll increases in natural tropical Pacific phytoplankton populations. *Limnol. Oceanogr.* **15,** 386–394.

Thomas, W. H. and R. W. Owen, Jr. 1971. Estimating phytoplankton production from ammonium and chlorophyll concentrations in nutrient-poor water of the eastern tropical Pacific Ocean. *Fish. Bull.* **69,** 87–92.

Thomas, W. H. and A. N. Dodson. 1974a. Effect of interactions between temperature and nitrate supply on the cell-division rates of two marine phytoflagellates. *Mar. Biol.* **24,** 213–217.

Thomas, W. H. and A. N. Dodson. 1974b. Inhibition of diatom photosynthesis by germanic acid: separation of diatom productivity from total marine primary productivity. *Mar. Biol.* **27,** 11–19.

Thomas, W. H. and D. L. R. Seibert. 1977. Effects of copper on phytoplankton standing crop and productivity: controlled ecosystem pollution experiment. *Bull. Mar. Sci.* **27,** 23–33.

Thorestenson, D. C. and F. T. MacKenzie. 1974. Time variability of pore water chemistry in recent carbonate sediments, Devil's Hole, Harrington Sound, Bermuda. *Geochim. Cosmochim. Acta,* **38,** 1–19.

Thorson, G. 1950.Reproductive and larval ecology of marine bottom invertebrates. *Biol. Rev.* **25,** 1–45.

Thorson, G. 1957. Bottom communities. *Mem. Geol. Soc. Amer.* **67**(1), 461–534.

Thorson, G. 1966. Some factors influencing the recruitment and establishment of marine benthic communities. *Netherlands J. Sea. Res.* **3,** 267–293.

Throndsen, J. 1973. Motility in some marine nanoplankton flagellates. *Norw. J. Zool.* **21,** 193–200.

Thurston, M. H. 1979. Scavenging abyssal amphipods from the north-east Atlantic Ocean. *Mar Biol.* **51,** 55–68.

Tilman, D. 1977. Resource competition between planktonic algae: An experimental and theoretical approach. *Ecology,* **58,** 338–348.

Tilman, D., M. Mattson and S. Langer. 1981. Competition and nutrient kinetics along a temperature gradient: An experimental test of a mechanistic approach to niche theory. *Limnol. Oceanogr.* **26,** 1020–1033.

Tilton, R. C. 1968. The distribution and characterization of marine sulfur bacteria. *Rev. Int. Oceanogr. Med.* **9,** 237–253.

Tilton, R. C., A. B. Cober and G. E. Jones. 1967. Marine thiobacilli. I. Isolation and distribution. *Can. J. Microbiol.* **13,** 1521–1528.

Titman, D. 1976. Ecological competition between algae: Experimental confirmation of resource-based competition theory. *Science,* **192,** 463–465.

Tolbert, N. E. 1974. Photorespiration. In *Algal Physiology and Biochemistry.* Ed. W. D. Stewart, Univ. Calif., pp. 474–504.

Tominaga, H. and S. Ichimura. 1966. Ecological studies on the organic matter production in a mountain river ecosystem. *Bot. Mag., Tokyo,* **79,** 815–829.

Tooming, H. 1970. Mathematical description of net photosynthesis and adaptation processes in the photosynthetic apparatus of plant communities. In *Prediction and Measurement of Photosynthetic Productivity, Ed. I. Malek et al.,* Pudoc, Wageningen, pp. 103–113.

Torres, J. J., B. W. Belman and J. J. Childress. 1979. Oxygen consumption rates of midwater fishes as a function of depth of occurrence. *Deep-Sea Res.* **26,** 185–197.

Tracey, M. L., K. Nelson, D. Hedgecock, R. A. Shleser and M. L. Pressick. 1975. Biochemical genetics of lobsters: genetic variation and the structure of american lobster (*Homarus americanus*) populations. *J. Fish. Res. Bd. Canada,* **32,** 2091–2101.

Tranter, D. J. and B. S. Newell, 1963. Enrichment experiments in the Indian Ocean. *Deep-Sea Res.* **10,** 1–9.

Travers, M. 1971. Diversité du microplancton du Golfe de Marseille en 1964. *Mar. Biol.* **8,** 308–343.

Trevallion, A. 1971. Studies on *Tellina tennis* da Costa. III. Aspects of general biology and energy flow. *J. Exp. Mar. Biol. Ecol.* **7,** 95–122.

Trevallion, A., R. R. C. Edwards and J. H. Steele. 1970. Dynamics of a benthic bivalve. In *Marine Food Chains,* Ed. J. H. Steele, Oliver & Boyd, Edinburgh, pp. 285–295.

Tsikhon-Lukanina, Ye. A. and T. A. Lukasheva. 1970. Conversion of food energy in the young of some marine isopods. *Oceanology,* **10,** 553–556.

Tsyban, A. V. 1971. Marine bacterioneuston. *J. Oceanogr. Soc. Japan,* **27,** 56–66.

Turner, J. T. 1977. Sinking rates of fecal pellets from marine copepod *Pontella meadii. Mar. Biol.* **40,** 249–259.

Turner, R. 1973. Wood boring bivalves, opportunistic species in the deep sea. *Science,* **180,** 1377–1379.

Turpin, D. H. 1980. Processes in nutrient based phytoplankton ecology. Ph.D. thesis, Univ. British Columbia, Vancouver, B.C., Canada, pp. 131.

Turpin, D. H. and P. J. Harrison. 1978. Fluctuations in free amino acid pools of *Gymnodinium simplex* (Dinophyceae) in response to ammonia perturbation: Evidence for glutamine syntethase pathway. *J. Phycol.* **14,** 461–464.

Turpin, D. H. and P. J. Harrison. 1979. Limiting nutrient patchiness and its role in phytoplankton ecology. *J. Exp. Mar. Biol. Ecol.* **39,** 151–166.

Tyler, J. E. 1975. The *in situ* quantum efficiency of natural phytoplankton populations. *Limnol. Oceanogr.* **20,** 976–980.

UNEP/UNESCO. 1980. *River Inputs to Ocean Systems,* Ed. J. M. Martin, J. D. Burton and D. Eisma. Unesco. pp. 384.

UNESCO. 1973. A guide to the measurement of marine primary production under some special conditions. In *Monographs on Oceanographic Methodology,* **3,** Paris, 73 pp.

V-Stosh, H. A., G. Theil and K. V. Kowallik. 1973. Entwicklungs-geschichtliche Untersuchungen an zentrischen Diatomeen V Bau und Lebenszyklus von *Chaetoceros didymum* mit Beobachtungen uber einige audre Arten der Gattung. *Helgoländer wiss. Meeresunters.* **25,** 384–445.

Vaccaro, R. F. and H. W. Jannasch. 1967. Variations in uptake kinetics for glucose by natural populations in seawater. *Limnol. Oceanogr.* **12,** 540–542.

Vaccaro, R. F., S. E. Hicks, H. W. Jannasch and F. G. Carey. 1968. The occurrence and role of glucose in seawater. *Limnol. Oceanogr.* **13,** 356–360.

Vahl, O. 1981. Energy transformations by the Iceland scallop, *Chlamys islandica* (O. F. Muller), from 70°N. II. The population energy budget. *J. Exp. Mar. Biol. Ecol.* **53,** 297–303.

Valiela, I. and J. M. Teal, 1974. Nutrient limitation in salt marsh vegetation. In *Ecology of Halophytes,* Eds. R. J. Reimold and W. H. Queen, Academic Press, New York and London, pp. 547–563.

Valiela, I., J. M. Teal and N. Y. Persson. 1976. Production dynamics of experimentally enriched salt marsh vegetation: below-ground biomass. *Limnol. Oceanogr.* **21,** 245–252.

Vance, R. R. 1973. On reproductive strategies in marine benthic invertebrates. *Am. Nat.* **107,** 339–352.

Vanderploeg, H. A. and R. L. Ondricek. 1982. Intersetule distances are a poor predictor of particle-retention efficiency in *Diaphomus sicilis. J. Plankton Res.* **4,** 237–244.

Van Raalte, C. D., I. Valiela and J. M. Teal. 1976. Production of epibenthic salt marsh algae: light and nutrient limitation. *Limnol. Oceanogr.* **21,** 862–872.

Venrick, E. L. 1972. Small-scale distributions of oceanic diatoms. *Fish. Bull.* **70,** 363–372.

Venrick, E. L., J. A. McGowan and A. W. Mantyla. 1973. Deep maxima of photosynthetic chlorophyll in the Pacific Ocean. *Fish Bull.* **71,** 41–52.

Vidal, J. 1980a. Physioecology of zooplankton. I. Effects of phytoplankton concentration, temperature, and body size on the growth rate of *Calanus pacificus* and *Pseudocalanus* sp. *Mar. Biol.* **56,** 111–134.

Vidal, J. 1980b. Physioecology of zooplankton. III. Effects of phytoplankton concentration, temperature, and body size on metabolic rate of *Calanus pacificus. Mar. Biol.* **56,** 195–202.

Vidal, J. 1980c. Physioecology of zooplankton. IV. Effects of phytoplankton concentration, temperature, and body size on the net production efficiency of *Calanus pacificus. Mar. Biol.* **56,** 203–211.

Vilks, G., E. H. Anthony and W. T. Williams. 1970. Application of association-analysis to distribution studies of recent Foraminifera. *Can. J. Earth. Sci.* **7**(6), 1462–1469.

Vinogradov, A. P. 1953. The elementary chemical composition of marine organisms. Translation by Efron and Selton. *Memoir. Sears Found. Mar. Res.* **2,** 130–146.

Vinogradov, M. E. Vertical migrations of zooplankton and their importance for the nutrition of abyssal pelagic fauna. *Trudy Inst. Okean.* **13,** 71–76.

Vinogradov, M. E. 1962. Feeding of deep-sea zooplankton. *Rapp. Proc. Verb. Cons. Perm. Int. Explor. Mer,* **153,** 114–120.

Vinogradov, M. E. 1968. *Vertical Distribution of the Oceanic Zooplankton,* Nauka Publ. House, Moscow, (Transl. Israel. Prog. for Sci. Transl. Jerusalem, 1970.) 320 pp.

Vinogradov, M. E., I. I. Gitelzon and Yu. I. Sorokin. 1970. The vertical structure of a pelagic community in the tropical ocean. *Mar. Biol.* **6,** 187–194.

Vinogradov, M. E., V. V. Menshutkin and E. A. Shushkina. 1972. On mathematical simulation of a pelagic ecosystem in tropical waters of the ocean. *Mar. Biol.* **16,** 261–268.

Vinogradov, M. E., V. F. Krapivin, V. V. Menshutkin, B. S. Fleyshman and E. A. Shuskina. 1973. Mathematical model of the functions of the pelagical ecosystem in tropical regions (from the 50th voyage of the R/V *Vityaz*). *Oceanology,* **13,** 704–717 (English translation).

Vinogradova, N. G. 1962. Some problems of the study of deep-sea bottom fauna. *J. Oceanogr. Soc. Jap., 20th Ann.,* pp. 724–741.

Vinogradova, Z. A. and V. V. Koval'skiy. 1962. Elemental composition of the Black Sea plankton. *Dokl. Akad. Nauk, SSSR,* **147,** 1458–1460.

Vlymen, W. J. 1970. Energy expenditure of swimming copepods. *Limnol. Oceanogr.* **15,** 348–356.

Vogel, K. and B. J. D. Meeuse. 1968. Characterization of the reserve granules from the dinoflagellate *Thecadinium inclinatum* Balech. *J. Phycol.* **4,** 317–318.

Volkman, J. K., D. J. Smith, G. Eglinton, T. E. V. Forsberg and E. D. S. Corner. 1981. Sterol and fatty acid composition of four marine Haptophycean algal. *J. Mar. Biol. Ass. U.K.* **61,** 509–527.

Volkmann, C. M. and C. H. Oppenheimer. 1962. The microbial decomposition of organic carbon in surface sediments of marine bays of the Central Texas Gulf Coast. *Publ. Inst. Mar. Sci.* **8,** 80–96.

Vollenweider, R. A. 1965. Calculation models of photosynthesis—depth curves and some implications regarding day rate estimates in primary production measurements. In *Primary Productivity in Aquatic Environments,* Ed. C. R. Goldman, Mem. Ist. Ital., Idrobiol., *18,* Suppl.: The University of California Press, Berkeley, pp. 425–457.

Vollenweider, R. A. (Ed.). 1969. A manual on methods for measuring primary production in aquatic environments including a chapter on bacteria. *IBP Handbook No. 12.* F. A. Davis Co., Philadelphia. 213 pp.

Vollenweider, R. A. and A. Nauwerck. 1961. Some observations on the ^{14}C-method for measuring primary production. *Verh. int. Limnol.* **14,** 134–139.

Von Arx, W. S. 1962. *An Introduction to Physical Oceanography,* Addison-Wesley, London, 135 pp.

Wada, E. and A. Hattori. 1971. Nitrite metabolism in the euphotic layer of the central North Pacific Ocean. *Limnol. Oceanogr.* **16,** 766–722.

Wailes, G. H. 1937. *Canadian Pacific Fauna.* 1. Protozoa. 1a. Lobosa, 1b. Reticulosa, 1c. Heliozoa, 1d. Radiolaria, Biological Board of Canada, Toronto. 14 pp.

Wailes, G. H. 1939. *Canadian Pacific Fauna.* 1. Protozoa. 1e. Magsigophora, Biological Board of Canada, Toronto, 45 pp.

Wailes, G. H. 1943. *Canadian Pacific Fauna.* 1. Protozoa. 1f. Ciliata, 1g. Suctoria, Biological Board of Canada, Toronto, pp. 46.

Waksman, S. A. 1933. On the distribution of organic matter in the sea bottom and the chemical nature and origin of marine humus. *Soil. Sci.* **36,** 125–147.

Waksman, S. A., C. L. Carey and H. W. Reuszer. 1933. Marine bacteria and their role in the cycle of life in the sea. I. Decomposition of marine plant and animal residues by bacteria. *Biol. Bull.* **65,** 57–79.

Waksman, S. A. and M. Hotchkiss. 1938. On the oxidation of organic matter in marine sediments by bacteria. *J. Mar. Res.* **15,** 101–118.

Wallen, D. G. and G. H. Geen. 1971a. Light quality in relation to growth, photosynthetic rates and carbon metabolism in two species of marine plankton algae. *Mar. Biol.* **10,** 34–43.

Wallen, D. G. and G. H. Geen. 1971b. Light quality and concentration of proteins, RNA, DNA and photosynthetic pigments in two species of marine plankton algae. *Mar. Biol.* **10,** 44–51.

Wallen, D. G. and G. H. Geen. 1971c. The nature of the photosynthate in natural phytoplankton populations in relation to light quality. *Mar. Biol.* **10,** 157–168.

Walne, P. R. 1972. The influence of current speed, body size and water temperature on the filtration rate of five species of bivalve. *J. Mar. Biol. Ass. U.K.* **32,** 345–374.

Walsh, J. J. 1971. Relative importance of habitat variable in predicting the distribution of phytoplankton at the ecotone of the antarctic upwelling ecosystem. *Ecol. Monographs,* **41,** 291–309.

Walsh, J. J. 1972. Implications of a systems approach to oceanography. *Science,* **176,** 969–975.

Walsh, J. J. 1976. Models of the sea. In *The Ecology of the Seas,* Ed. D. H. Cushing and J. J. Walsh. Blackwell Scientific, pp. 388–407.

Walsh, J. J. 1980. Shelf sea ecosystems. In *The Analysis of Marine Ecosystems,* Ed. A. R. Longhurst, Academic Press, London, pp. 159–196.

Walsh, J. J. and R. C. Dugdale. 1971. A simulation model of the nitrogen flow in the Peruvian upwelling system. *Inv. Resq.* **35,** 309–330.

Walsh, J. J. and C. P. McRoy. 1979. Component 8: ecosystem analysis and synthesis in the southeast Bering Sea. PROBES (Processes and Resources of the Bering Sea Shelf). *Prog. Rept. 1979,* vol. 1, pp. 281–317.

Walsh, J. J., G. T. Rowe, R. L. Iverson and C. P. McRoy. 1981. Biological export of shelf carbon as a sink of the global CO_2 cycle. *Nature,* **291,** 196–201.

Walter, M. D. 1973. Fressverhalten und Darminhaltsuntersuchungen bei Sipunculiden. *Helgoländer wiss. Meeresunters,* **25,** 486–494.

Wangersky, P. J. 1965. The organic chemistry of sea water. *Amer. Scient.* **53,** 358–374.

Wangersky, P. J. 1974. Particulate organic carbon: sampling variability. *Limnol. Oceanogr.* **19,** 980–984.

Wangersky, P. J. 1976. Particulate organic carbon in the Atlantic and Pacific oceans. *Deep-Sea Res.* **23,** 457–465.

Wangersky, P. J. and D. C. Gordon, 1965. Particulate carbonate, organic carbon, and Mn^{++} in the open ocean. *Limnol. Oceanogr.* **10,** 544–550.

Warwick, R. M. and J. B. Buchanan. 1970. The meiofauna off the coast of Northumberland. I. The structure of the nematode population. *J. Mar. Biol. Ass. U.K.* **50,** 129–146.

Warwick, R. M. and R. Price. 1975. Macrofauna production in an estuarine mudflat. *J. Mar. Biol. Ass. U.K.* **55,** 1–18.

Warwick, R. M. and R. J. Uncles. 1980. Distribution of benthic macrofauna associations in the Bristol Channel in relation to tidal stress. *Mar. Ecol. Prog. Ser.* **3,** 97–103.

Watanabe, T. and R. G. Ackman. 1974. Lipids and fatty acids of the American *Crassostrea virginica* and European flat *Ostrea edulis* oysters from a common habitat, before and after feeding with *Dicrateria inoruata* or *Isochrysis galbana. J. Fish. Res. Bd. Canada,* **31,** 403–409.

Waters, T. F. 1969. The turnover ratio in production ecology of freshwater invertebrates. *Amer. Nat.* **103,** 173–185.

Waters, T. F. 1979. Influence of benthos life history upon estimation of secondary production. *J. Fish. Res. Bd. Canada,* **36,** 1425–1430.

Watson, S. W. 1965. Characteristics of a marine nitrifying bacterium, *Nitrosocystis oceanus* sp. nov. *Limnol. Oceanogr. Suppl.* **10** (Redfield 75th Anniv. Vol.), R274–R289.

Watson, S. W. 1971. Taxonomic considerations of the family Nitrobacteracea Buchanan. *Int. J. System. Bacteriol.* **21,** 254–270.

Watt, W. D. 1966. Release of dissolved organic material from the cells of phytoplankton populations. *Proc. Roy. Soc.* (*London*), B, **164,** 521–551.

Webb, J. 1969. Biologically significant properties of submerged marine sands. *Proc. Roy. Soc. (London),* B, **174,** 355–402.

Webb, K. L. and R. E. Johannes. 1967. Studies of the release of dissolved free amino acids by marine zooplankton. *Limnol. Oceanogr.* **12,** 376–382.

Webb, K. L. and R. E. Johannes. 1969. Do marine crustaceans release dissolved amino acids? *Comp. Biochem. Physiol.* **29,** 875–878.

Webb, K. L. and C. F. D'Elia. 1980. Nutrient and oxygen redistribution during a spring neap tidal cycle in a temperate estuary. *Science,* **207,** 983–985.

Webster, T. J. M., M. Paranjape and K. H. Mann. 1975. Sedimentation of organic matter in St. Margaret's Bay, Nova Soctia. *J. Fish. Res. Bd. Canada,* **32,** 1399–1407.

Weiss, P. 1969. The living system: determinism stratified. In *Beyond Reductionism,* Eds. A. Koestler and J. Smythies. Hutchinson, London, pp. 3–55.

Welch, E. B. and G. W. Isaac. 1967. Chlorophyll variation with the tide and with plankton productivity in an estuary. *J. Water Poll. Cont. Fed.* **39,** 360–366.

Welch, H. E. 1968. Relationships between assimilation efficiencies and growth efficiencies for aquatic consumers. *Ecology,* **49,** 755–759.

Wells, P. G. and J. B. Sprague. 1976. Effects of crude oil on American lobster (*Homarus americanus*) larvae in the laboratory. *J. Fish. Res. Bd. Canada,* **33,** 1604–1614.

Westlake, D. F. 1963. Comparisons of plant productivity. *Biol. Rev.* **38,** 385–425.

Westlake, D. F. 1965. Some problems in the measurement of radiation under water: A review. *Photochemistry and Photobiology,* **4,** 849–868.

Wheeler, E. H. 1967. Copepod detritus in the deep sea. *Limnol. Oceanogr.* **12,** 697–701.

Whitfield, M. 1969. *Eh* as an operational parameter in estuarine studies. *Limnol. Oceanogr.* **14,** 547–558.

Whitlatch, R. B. 1974. Food-resource partioning in the deposit feeding polychaeta *Pectinaria gouldii. Biol. Bull.* **147,** 227–235.

Whitlatch, R. B. 1980 Patterns of resource utilization and coexistance in marine intertidal deposit-feeding communities. *J. Mar. Res.* **38,** 743–765.

Whitlatch, R. B. 1981. Animal–sediment relationships in intertidal marine benthic habitats: some determinants of deposit-feeding species diversity. *J. Exp. Mar. Biol. Ecol.* **53,** 31–45.

Whitlatch, R. B. and R. G. Johnson. 1974. Methods for staining organic matter in marine sediments. *J. Sed. Petrol.* **44,** 1310–1312.

Wickett, W. P. 1967. Ekman transport and zooplankton concentration in the North Pacific Ocean. *J. Fish. Res. Bd.* **24,** 581–594.

Wickstead, J. H. Food and feeding in pelagic copepods. *Proc. Zool. Soc. London,* **139,** 545–555.

Wiebe, P. H. 1970. Small-scale spatial distribution in oceanic zooplankton. *Limnol. Oceanogr.* **15,** 205–217.

Wiebe, P. H. 1971. A field investigation of the relationship between length of tow, size of net and sampling error. *J. Cons. Int. Explor. Mer,* **34,** 110–117.

Wiebe, P. H. and W. R. Holland. 1968. Plankton patchiness: effects on repeated net tows. *Limnol. Oceanogr.* **13,** 315–321.

Wiebe, W. J. 1979. Anaerobic benthic microbial processes: changes from the estuary to the continental shelf. p. 469–488. In *Ecological Processes in Coastal and Marine Ecosystems,* Ed. R. J. Livingston. Plenum Press.

Wieser, W. 1960. Benthic studies in Buzzards Bay. II. The meiofauna. *Limnol. Oceanogr.* **5,** 121–137.

Wieser, W. and M. Zech. 1976. Dehydrogenases as tools in the study of marine sediments. *Mar. Biol.* **36,** 113–122.

Wigley, R. L. and A. D. McIntyre. 1964. Some quantitative comparisons of offshore meiobenthos and macrobenthos south of Martha's Vineyard. *Limnol. Oceanogr.* **9,** 485–493.

Wildish, D. J. 1977. Factors controlling marine and estuarine sublittoral macrofauna. *Helgo. Wiss. Meers.* **30,** 445–454.

Wildish, D. J. and D. D. Kristmanson. 1979. Tidal energy and sublittoral macrobenthic animals in estuaries. *J. Fish. Res. Canada,* **36,** 1197–1206.

Wildish, D. J. and D. Peer. 1981. Methods for estimating secondary production in marine Amphipoda. *Can. J. Fish. Aquat. Sci.* **38,** 1019–1026.

Wildish, D. S. and N. J. Poole. 1970. Cellulase activity in *Orchestia gammarella* (Pallas). *Comp. Biochem. Physiol.* **33,** 713–716.

Wilhm, J. L. 1968. Use of biomass units in Shannon's Formula. *Ecology,* **49,** 153–156.

Williams, P. J. LeB. 1970. Heterotrophic utilization of dissolved organic compounds in the sea. I. Size distribution of population and relationship between respiration and incorporation of growth substrates. *J. Mar. Biol. Ass. U.K.* **50,** 859–870.

Williams, P. J. LeB. 1975. Biological and chemical aspects of dissolved organic materials in sea water. In *Chemical Oceanography,* Ed. J. P. Riley and G. Skirrow, Academic Press, pp. 301–363.

Williams, P. J. LeB. and R. W. Gray. 1970. Heterotrophic utilization of dissolved organic compounds in the sea. II. Observations on the responses of heterotrophic marine populations to abrupt increases in amino acid concentration. *J. Mar. Biol. Ass. U.K.* **50,** 871–881.

Williams, P. M. 1965. Fatty acids derived from lipids of marine origin. *J. Fish. Res. Bd. Canada,* **22,** 1107–1122.

Williams, P. M. 1968. Organic and inorganic constituents of the Amazon River. *Nature,* **218,** 937–938.

Williams, P. M. and K. S. Chan. 1966. Distribution and speciation of iron in natural waters: Transition from river water to a marine environment, British Columbia, Canada. *J. Fish. Res. Bd. Canada,* **23,** 575–593.

Williams, P. M., H. Oeschger and P. Kinney. 1969. Natural radiocarbon activity of the dissolved organic carbon in the North-East Pacific Ocean. *Nature,* **224,** 256–258.

Williams, P. M. and L. I. Gordon. 1970. Carbon-13: carbon-12 ratios in dissolved and particulate organic matter in the sea. *Deep-Sea Res.* **17,** 19–27.

Williams, P. M., J. A. McGowan and M. Stuiver. 1970. Bomb carbon-14 in deep-sea organisms. *Nature,* **227,** 375–376.

Williams, R. 1972. The abundance and biomass of the interstitial fauna of a graded series of shell-gravels in relation to the available space. *J. Anim. Ecol.* **41,** 623–646.

Williamson, M. H. 1961. A method for studying the relation of plankton variations to hydrography. *Bull. Mar. Ecol.* **5,** 224–229.

Wilson, D. P. 1951. A biological difference between natural sea waters. *J. Mar. Biol. Ass. U.K.* **30,** 1–20.

Winberg, G. C. 1968. *Methods for the Estimation of Production of Aquatic Animals* (trans. by A. Duncan, 1971). Adv. Ecol. Res. Academic Press, London—New York. 175 pp.

Winsor, C. P. and G. L. Clarke. 1940. A statistical study of variation in the catch of plankton nets. *J. Mar. Res.* **3,** 1–34.

Winter, J. E. 1969. Über den Einfluss der Nahrungskonzentration und anderer Faktoren auf Filtrierleistung und Nahrungsausnutzung der Muscheln *Arctica islandica* and *Modiolus modiolus. Mar. Biol.* **4,** 87–135.

Winter, J. E. 1978. A review of the knowledge of suspension-feeding in lamellibranchiate bivalves, with special reference to artificial aquaculture systems. *Aquaculture,* **13,** 1–33.

Wishner, K. F. 1980. The biomass of the deep-sea benthopelagic plankton. *Deep-Sea Res.* **27,** 203–216.

Withers, P. C. 1978. Models of diffusion-mediated gas exchange in animal burrows. *Am. Nat.* **112,** 1101–1112.

Wolff, T. 1979. Macrofaunal utilization of plant remains in the deep sea. *Sarsia* **64,** 117–136.

Wong, C. S., D. R. Green and W. J. Cretney. 1976. Distribution and source of tar on the Pacific Ocean. *Mar. Pollut. Bull.* **7,** 102–105.

Wood, E. J. F. 1953. Heterotrophic bacteria in marine environments of eastern Australia. *Aust. J. Mar. Freshw. Res.* **4,** 160–200.

Wood, E. J. F. 1958. The significance of marine microbiology. *Bact. Rev.* **22,** 1–19.

Wood, E. J. F. 1965. *Marine Microbial Ecology,* Reinhold, New York, pp. 243.

Wood, H. G. and C. H. Werkmann. 1935. The utilization of CO_2 by the propionic acid bacteria in the dissimulation of glycerol. *J. Bacteriol.* **30,** 332.

Wood, H. G. and C. H. Werkman. 1936. The utilization of CO_2 in the disimulation of glycerol by the propionic acid bacteria. *Biochem. J.* **30,** 48–53.

Wood, H. G. and C. H. Werkman. 1940. The relationship of the bacterial utilization of CO_2 to succinic acid formation. *Biochem. J.* **34,** 129–138.

Wood, J. M., F. Scott Kennedy and C. G. Rosen. 1968. Synthesis of methyl-mercury compounds by extracts of a methnogenic bacterium. *Nature,* **220,** 173–174.

Wood, L. H. and K. E. Chua. 1973. Glucose flux at the sediment-water interface of Toronto Harbour, Lake Ontario, with reference to pollution stress. *Can. J. Microbiol.* **19,** 413–420.

Woodhead, D. S. 1980. Marine disposal of radioactive wastes. *Helgoländer Meeresunters.* **33,** 122–137.

Woodin, S. A. 1974. Polychaete abundance patterns in a marine soft-sediment environment: the importance of biological interactions. *Ecol. Monogr.* **44,** 171–181.

Woodin, S. A. 1976. Adult-larval interactions in dense infaunal assemblages: patterns of abundance. *J. Mar. Res.* **34,** 25–41.

Woodin, S. A. 1978. Refuges, disturbance and community structure: a marine soft-bottom example. *Ecology,* **59,** 274–284.

Woodwell, G. M., C. F. Wurster and P. A. Isaacson. 1967. DDT residues in an east coast estuary: A case of biological concentration of a persistent insecticide. *Science* **156,** 821–823.

Wooster, W. S. and J. L. Reid, Jr. 1963. Eastern boundary currents. In *The Seas,* Ed. M. N. Hill, Interscience Publishers, New York, Vol. 2, pp. 253–280.

Wright, R. T. 1964. Dynamics of a phytoplankton community in an ice-covered lake. *Limnol. Oceanogr.* **9,** 163–178.

Wright, R. T. and J. E. Hobbie. 1965. The uptake of organic solutes in lake water. *Limnol. Oceanogr.* **10,** 22–28.

Wright, R. T. and J. E. Hobbie. 1966. Use of glucose and acetate by bacteria and algae in aquatic ecosystems. *Ecology,* **47,** 447–464.

Wurster, C. F. 1968. DDT reduces photosynthesis by marine phytoplankton. *Science,* **159,** 1474–1475.

Wurster, C. F. and D. B. Wingate. 1968. DDT residues and declining reproduction in the Bermuda petrel. *Science,* **159,** 979–981.

Wyatt, T. 1980. The growth season in the sea. *J. Plank. Res.* **2,** 81–97.

Yamada, M. and T. Ota. 1970. Studies on the lipid of plankton. IV. Unsaponifiable matter of lipid of *Calanus plumchrus. J. Japan Oil Chem. Soc.* **19,** 377–382. (Fish. Res. Bd. Canada Trans. No. 1590.)

Yentsch, C. S. 1965. Distribution of chlorophyll and phaeophytin in the open ocean. *Deep-Sea Res.* **12,** 653–666.

Yentsch, C. S. and J. H. Ryther. 1957. Short-term variations in phytoplankton chlorophyll and their significance. *Limnol. Oceanogr.* **2,** 140–142.

Yentsch, C. S. and J. H. Ryther. 1959. Absorption curves of acetone extracts of deep water particulate matter. *Deep-Sea Res.* **6,** 72–74.

Yentsch, C. S. and R. W. Lee. 1966. A study of photosynthetic light reactions, and a new interpretation of sun and shade phytoplankton. *J. Mar. Res.* **24,** 319–337.

Yetka, J. E. and W. J. Wiebe, 1974. Ecological application of antibiotics as respiratory inhibitors of bacterial populations. *Appl. Microbiol.* **28,** 1033–1039.

Yingst, J. Y. 1976. The utilization of organic matter in shallow marine sediments by an epibenthic deposit-feeding holothurian. *J. Exp. Mar. Biol. Ecol.* **23,** 55–69.

Yingst, J. Y. 1978. Patterns of micro- and meiofaunal abundance in marine sediments, measured with the adenosine triphosphate assay. *Mar. Biol.* **47,** 41–54.

Yingst, J. Y. and D. C. Rhoads. 1980. The role of bioturbation in the enhancement of bacterial growth rates in marine sediments, In *Marine Benthic Dynamics,* Ed. K. R. Tenore and B. C. Coull. Univ. of S. Carolina Press, pp. 407–421.

Young, D. R., A. J. Mearns, T. K. Jan and R. P. Eganhouse. 1981. The cesium-potassium ratio and trace metal biomagnification in two contaminated marine food webs. In

Oceans 81 Conference Record, Vol. 1. Publ. Inst. Electrical and Electronic Engin. (Wash., D.C.). pp. 570–574.

Youngblood, W. W., M. Blumer, R. L. Guillard and F. Fiore. 1971. Saturated and unsaturated hydrocarbons in marine benthic algae. *Mar. Biol.* **8,** 190–201.

Zaika, V. E. 1973. *Specific Production of Aquatic Invertebrates.* Halsted Press, John Wiley & Sons, Inc., New York, 154 pp.

Zaitzev, Yu. P. 1961. Surface pelagic biocoenose of the Black Sea. *Zool. Zh.* **40,** 818–825.

Zedler, J. B. 1980. Algal mat productivity: comparisons in a salt marsh. *Estuaries,* **3,** 122–131.

Zeitzschel, B. 1970. The quantity, composition and distribution of suspended particulate matter in the Gulf of California. *Mar. Biol.* **7,** 305–318.

Zeitzschel. B. 1980. Sediment-water interactions in nutrient dynamics, In *Marine Benthic Dynamics,* Ed. K. R. Tenore and B. C. Coull, Univ. S. Carolina Press, pp. 195–218.

Zenkevitch, L. A., A. Filatova, G. M. Belyaev, T. S. Lukyanova and I. A. Suetova. 1971. Quantitative distribution of zoobenthos in the world ocean. *Bull. der Moskauer Gen. der Naturforscher, Abt. Biol.* **76,** 27–33.

Zeuthen, E. 1970. Rate of living as related to body size in organisms. *Polskie Archiwum Hydrobiologii,* **17,** 21–30.

ZoBell, C. E. 1946. *Marine Microbiology,* Chronica Botanica, Waltham, Mass. 240 pp.

ZoBell, C. E. 1962. Geochemical aspects of the microbial modification of carbon compounds. In *Advances in Organic Geochemistry,* Pergamon Press, London, pp. 1–18.

ZoBell, C. E. 1968. Bacterial life in the deep sea. In *Proceedings of the U.S.-Japan Seminar on Marine Microbiology,* August 1966 in Tokyo. *Bull. Misaki Marine Biol. Kyoto Inst. Univ.,* No. 12, 77–96.

Zsolnay, A. 1977. Hydrocarbon content and chlorophyll correlation in the waters between Nova Scotia and the Gulf Stream. *Deep-Sea Res.* **24,** 199–207.

INDEX